VOLUME EIGHTY FOUR

ADVANCES IN BOTANICAL RESEARCH

Secondary Endosymbioses

ADVANCES IN BOTANICAL RESEARCH

Series Editors

Jean-Pierre Jacquot
Professor, Membre de L'Institut Universitaire de France, Unité Mixte de Recherche INRA, UHP 1136 "Interaction Arbres Microorganismes", Université de Lorraine, Faculté des Sciences, Vandoeuvre, France

Pierre Gadal
Honorary Professor, Université Paris-Sud XI, Institut Biologie des Plantes, Orsay, France

VOLUME EIGHTY FOUR

ADVANCES IN BOTANICAL RESEARCH

Secondary Endosymbioses

Volume Editor

YOSHIHISA HIRAKAWA
Faculty of Life and Environmental Sciences,
University of Tsukuba, Tsukuba, Japan

ELSEVIER

ACADEMIC PRESS
An imprint of Elsevier

Academic Press is an imprint of Elsevier
125 London Wall, London, EC2Y 5AS, United Kingdom
The Boulevard, Langford Lane, Kidlington, Oxford OX5 1GB, United Kingdom
50 Hampshire Street, 5th Floor, Cambridge, MA 02139, United States
525 B Street, Suite 1800, San Diego, CA 92101-4495, United States

First edition 2017

ISBN: 978-0-12-802651-9
ISSN: 0065-2296

For information on all Academic Press publications
visit our website at https://www.elsevier.com/books-and-journals

Publisher: Zoe Kruze
Acquisition Editor: Sam Mahfoudh
Editorial Project Manager: Joanna Collett
Production Project Manager: James Selvam
Senior Cover Designer: Miles Hitchen

Typeset by SPi Global, India

CONTENTS

CONTRIBUTORS

Chris Bowler
IBENS, Ecole Normale Supérieure, CNRS, Inserm, PSL Research University, Paris, France

Fabien Burki
Science for Life Laboratory, Program in Systematic Biology, Uppsala University, Uppsala, Sweden

Richard G. Dorrell
IBENS, Ecole Normale Supérieure, CNRS, Inserm, PSL Research University, Paris, France

Zoltán Füssy
Institute of Parasitology, Biology Centre of the Czech Academy of Sciences, České Budějovice, Czech Republic

Lucia Hadariová
Faculty of Science, Charles University in Prague, Prague, Czech Republic

Vladimír Hampl
Faculty of Science, Charles University in Prague, Prague, Czech Republic

Sanduni V. Hapuarachchi
Research School of Biology, Australian National University, Acton, ACT, Australia

Yoshihisa Hirakawa
Faculty of Life and Environmental Sciences, University of Tsukuba, Tsukuba, Japan

Štěpánka Hrdá
Faculty of Science, Charles University in Prague, Prague, Czech Republic

Luděk Kořený
University of Cambridge, Cambridge, United Kingdom

Miroslav Oborník
Institute of Parasitology, Biology Centre of the Czech Academy of Sciences; Faculty of Science, University of South Bohemia, České Budějovice; Institute of Microbiology of the Czech Academy of Sciences, Centre Algatech, Třeboň, Czech Republic

Naoko T. Onodera
Dalhousie University, Halifax, NS, Canada

David R. Smith
University of Western Ontario, London, ON, Canada

Goro Tanifuji
National Museum of Nature and Science, Tsukuba, Ibaraki, Japan

Yoshinori Tsuji
Kwansei Gakuin University, Sanda, Hyogo, Japan

Giel G. van Dooren
Research School of Biology, Australian National University, Acton, ACT, Australia

Anna M.G. Vanclová
Faculty of Science, Charles University in Prague, Prague, Czech Republic

Ross F. Waller
University of Cambridge, Cambridge, United Kingdom

Masaki Yoshida
University of Tsukuba, Tsukuba, Ibaraki, Japan

PREFACE

It is widely known that two major organelles, plastid and mitochondrion, have evolved by integration of bacterial endosymbionts. Half a century since Lynn Margulis proposed the endosymbiotic theory, recent advance of molecular studies has revealed the complex evolution history of plastids. The common ancestor of Archaeplastida including land plants and several algae obtained a plastid by the uptake of a photosynthetic prokaryote related to extent cyanobacteria, and this event is termed "primary" endosymbiosis. After that, certain phagotrophic protists have engulfed plastid-bearing algae, which are named "secondary" endosymbioses. Those eukaryote-to-eukaryote endosymbioses have occurred multiple times in different lineages, which have led to the spread of complex plastids, the so-called secondary plastids, in diverse eukaryotes. Thus, it can be said that secondary endosymbioses have been a driving force for the diversification of complex life on Earth. To date, eight major groups (stramenopiles, dinoflagellates, apicomplexans, chromerids, haptophytes, cryptophytes, euglenophytes, and chlorarachniophytes) are known to possess secondary plastids. This volume entitled "Secondary Endosymbioses" consists of 10 chapters written by expert scientists for respective groups with secondary plastids.

The first chapter, written by Fabien Burki, provides the overview of secondary plastid evolution. He summarizes previous studies of molecular phylogeny and comparative genomics and explains several inferred models of plastid evolution. The second chapter (by David Smith) describes the primary endosymbiosis as an introduction for secondary endosymbioses. He illustrates the origin of primary endosymbiosis, plastid genome evolution, and loss of photosynthesis in some lineages of Archaeplastida. Chapter 3 (by Richard Dorrell and Chris Bowler) focuses on stramenopiles that is a supergroup including the popular microalgae diatoms. This chapter describes the morphology, biochemistry, genomic diversity, and endosymbiotic origin of stramenopile plastids. Chapter 4 (by Ross Waller and Luděk Kořený) summarizes dinoflagellates that possess extraordinary complicated plastids. They mainly describe some species of this group that replaced their original plastids by higher order endosymbioses with certain primary and secondary plastid-bearing algae. Chapter 5 (by Giel van Dooren and Sanduni Hapuarachchi) deals with apicomplexan parasites, some of which are known as the causative agents of malaria, that contain nonphotosynthetic secondary plastid, termed apicoplasts. They illustrate the membrane biology, biogenesis, and metabolic functions of the apicoplasts. Chapter 6 (by Zoltán Füssy and Miroslav Oborník) focuses on chromerids that are photosynthetic relatives of apicomplexans. They describe the morphology, phylogeny, and cellular biochemistry of chromerids, as well as the inferred tertiary origin of chromerid plastids. Chapter 7 (by Yoshinori Tsuji and Masaki Yoshida) describes haptophytes that are a dominant marine primary producer. They illustrate the morphological diversity of calcified scales and the unique carbon metabolisms in haptophytes. Chapter 8 (by Goro Tanifuji and Naoko Onodera) deals with cryptophytes whose plastids possess a vestigial endosymbiont's nucleus, termed nucleomorph, which presents the smoking gun of eukaryote-to-eukaryote secondary endosymbiosis. This chapter mainly illustrates the endosymbiotic evolution of nucleomorph genomes. Secondary plastids can be divided into red- and green-algal lineages based on integrated algal endosymbionts. Chapters 3–8 illustrate organisms with red-algal secondary plastids, whereas the last two chapters explain green-algal secondary plastids. Chapter 9 written by Vladimír Hampl and colleagues summarizes euglenophyte plastids in terms of their origin, genomes, and metabolisms. The last chapter describes chlorarachniophytes that are another organism possessing a nucleomorph. I summarize the current understanding of the evolutionary history of chlorarachniophytes in terms of morphology, phylogeny, and genomics.

As mentioned earlier, this volume covers current aspects of plastid evolution. I believe that this volume is useful to graduate students and researchers in a broad range of fields including plant physiology, cell biology, genomics, and evolutionary biology. Finally, I would like to thank all contributors for joining to produce this volume. I would also like to thank Kazuya Takahashi for providing a cover picture (dinoflagellate), and Takuro Nakayama and Yuji Inagaki for constructive comments on specific chapters of this volume.

Yoshihisa Hirakawa

CHAPTER ONE

The Convoluted Evolution of Eukaryotes With Complex Plastids

Fabien Burki[1]
Science for Life Laboratory, Program in Systematic Biology, Uppsala University, Uppsala, Sweden
[1]Corresponding author: e-mail address: fabien.burki@ebc.uu.se

Contents

Abstract

The textbook version of how plastids were established by endosymbiosis and subsequently diversified is like a well-oiled machine: a cyanobacterial endosymbiont was taken up by a heterotrophic cell and transformed over time into a bona fide photosynthetic organelle (plastid), ultimately giving rise to all plants and algae. The reality, however, is much more complicated and this chapter attempts to describe recent advances in the field of plastid evolution brought to light by disciplines such as phylogenomics, comparative genomics, and cell biology. If (almost) all plastids may ultimately trace back to the same original endosymbiotic event, the very large diversity of plastids we observe today can only be explained by multiple layers of endosymbioses. That is, plastids were passed between distantly related eukaryotic lineages multiple times, essentially creating a phylogenetic imbroglio where plastids appear monophyletic but hosts are not. The burning question then is: how can we best fit plastid and host data into a comprehensive evolutionary framework? Focusing not only on the so-called complex plastids (the product of eukaryote-to-eukaryote endosymbioses) and the lineages that host them but

Advances in Botanical Research, Volume 84
ISSN 0065-2296
http://dx.doi.org/10.1016/bs.abr.2017.06.001

also on the many related plastid-lacking lineages and orphan taxa, I discuss the emergence of new models of plastid evolution. These models generalize the notion of serial endosymbioses to explain the scattered distribution of plastids in the eukaryotic tree of life. As such, they make new testable predictions as to how complex algae are connected through endosymbiotic gene transfer, but testing this will require first to determine the real magnitude of this process.

1. INTRODUCTION

The origin of photosynthetic organelles (i.e. plastids) was a landmark event in the history of life. It was the launching point of the astonishing diversification of photosynthetic eukaryotes, including macroscopic land plants and large algae (e.g. seaweed) but also countless forms of microscopic algae. Eukaryotic phototrophs are not only diverse, they thrive in most environments given a bit of light and can be so abundant that even tiny unicellular algae are seen from space during large blooming events. Despite the evolutionary and ecological importance of plants and algae—clearly, our planet would not be the same without photosynthetic eukaryotes—our understanding of how this diversity came to be is still limited. Central to the question of plastid origin is the endosymbiotic theory (see, e.g., Archibald, 2015a and references therein), which posits that plastids arose from the integration of a free-living cyanobacterium within a full-fledged eukaryotic host some 1–2 billion years ago (Falcón, Magallón, & Castillo, 2010; Parfrey, Lahr, Knoll, & Katz, 2011; Yoon, Hackett, Ciniglia, Pinto, & Bhattacharya, 2004). Over a century after its formulation, this influential theory that explains the birth of all endosymbiotic organelles (i.e. plastids but also mitochondria) is now beyond refute (Archibald, 2015a; Martin, Garg, & Zimorski, 2015). Much more controversial is how the process of endosymbiosis has driven the evolution and diversification of photosynthetic eukaryotes and the plastids they harbour (Archibald, 2015a, 2015b; Gould, Maier, & Martin, 2015; Keeling, 2009; Lane & Archibald, 2008; Zimorski, Ku, Martin, & Gould, 2014).

At the heart of the issue is the observation that although all plastids may be closely related, the algae that host them are not. Indeed, plastids have not only been vertically transmitted along the species tree, but they have also spread a number of times laterally between distantly related groups of eukaryotes (Archibald, 2009; Gould, Waller, & McFadden, 2008; Keeling, 2013; Reyes-Prieto, Weber, & Bhattacharya, 2007; Stiller, 2014). Today's plastids are found across a very large fraction of the eukaryotic tree of life (Burki, 2014), so much so that resolving the deep structure of the tree is inextricably linked to resolving the mysteries of plastid evolution. In this chapter, I will discuss plastid evolution not only from the perspectives of both plastid and host lineages but also from the many non-photosynthetic lineages that are interspaced in the evolutionary tree. The focus will be on higher order endosymbioses, i.e., endosymbioses that took place between eukaryotic cells and that resulted in the lateral spread of plastids across the tree. Despite many remaining uncertainties about how, and how often, plastids spread by eukaryote-to-eukaryote endosymbioses, these are exciting times for the field of plastid evolution. In recent years, comparative genomics and phylogenomics of a wide array of algae and heterotrophic microbial eukaryotes (protists) have led to new models of plastid evolution. These models represent a new conceptual framework for interpreting the evolution of plastid by endosymbioses.

2. DIVERSITY OF PLASTIDS

Mitochondria and plastids are the only definite cytosolic compartments of endosymbiotic origin (Martin et al., 2015 and references therein). Both were established as organelles from once free-living bacteria—in the case of plastids from an ancestor of modern-day cyanobacteria (Gould et al., 2008; Palmer, 2003; Reyes-Prieto et al., 2007). But unlike mitochondria, which originated very early in eukaryote evolution and evolved only by vertical inheritance, plastids originated later and were acquired by different groups of eukaryotes via multiple layers of endosymbioses (Archibald, 2015a). This resulted in different 'kinds' of plastids with respect to which cells were the recipient and donors in the respective endosymbiotic events. Accordingly, plastids have been classified as 'primary' or 'complex' (i.e. secondary, tertiary, and so on) to represent where in the layers of endosymbioses they stand (Fig. 1). As we will see, it is often not so straightforward assigning a plastid to one of these categories, as this requires one to trace back the endosymbiotic history of that plastid in the broad context of eukaryote evolution. For now, I will just make a distinction between 'primary' and 'complex' to describe the main differences between plastids and how this relates to their evolutionary history.

2.1 Primary Plastids

First off are primary plastids. Although they are not the focus of this chapter, primary plastids deserve more than a few words because they represent the

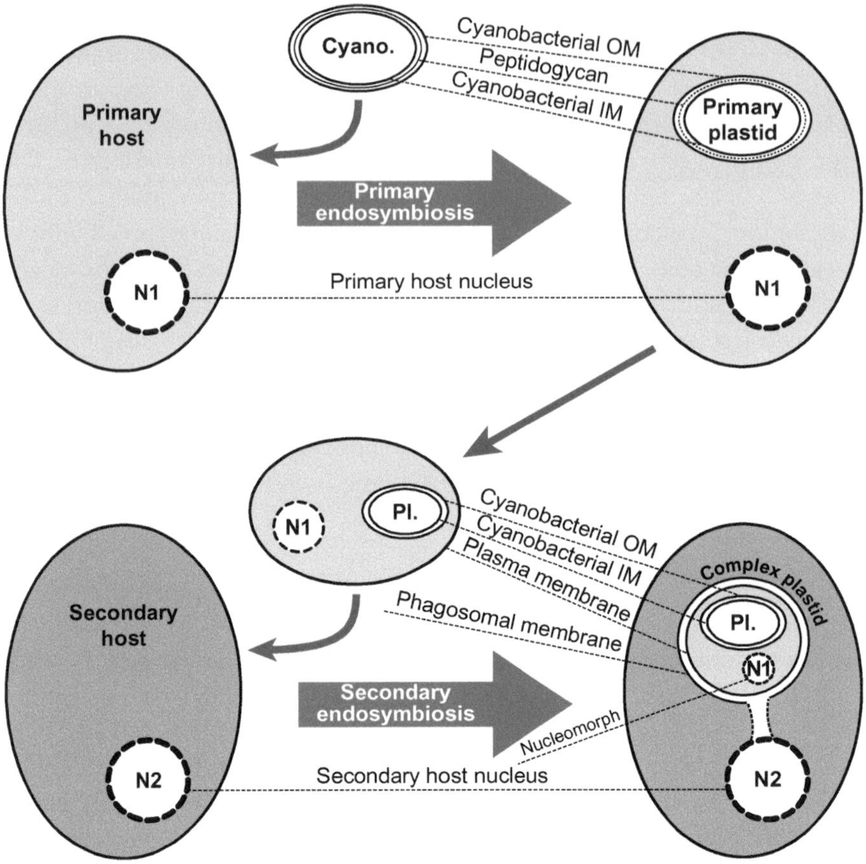

Fig. 1 Origin of primary and complex plastids by endosymbiosis. Primary endosymbiosis is the process by which a cyanobacterium is turned into a plastid in a heterotrophic eukaryote host. The primary plastids have retained both cyanobacterial membranes and, in the case of glaucophytes, a layer of peptidoglycan (this layer was presumably lost in red and green algae). Secondary and other higher order endosymbioses took place between two eukaryotes. As a result, complex plastids are bounded by additional membranes. In this diagram, a classical four-membrane-bound plastid is depicted, where the two outermost membranes are generally interpreted as the plasma membrane of the endosymbiont and the phagosomal membrane of the secondary host, respectively. A nucleomorph is also shown; this remnant of the primary host cell nucleus is present only in chlorarachniophytes and cryptophytes. In cryptophytes, haptophytes, and ochrophytes the outermost plastid membrane is contiguous with the nuclear envelope. Abbreviations: *Cyano.*, cyanobacteria; *IM*, inner membrane; *N*, nucleus; *OM*, outer membrane; *Pl.*, plastid.

point of entry of photosynthesis in eukaryotes (Archibald, 2009; Keeling, 2010). These plastids are the only ones that descend directly from a cyanobacterium, i.e., they were established by primary endosymbiosis following the uptake of the cyanobacterial progenitor of all plastids by a eukaryotic host (Fig. 1). Primary plastids reside in the host cytosol and are surrounded by two membranes corresponding to the two membranes of the cyanobacterial cell (Gould et al., 2008). Assuming that the endosymbiont initially entered the host cell by phagocytosis, for example, as a food source, this means that the phagosomal membrane was lost during the establishment of primary plastids, leaving only the prokaryotic membranes (Cavalier-Smith, 1982). Three lineages are known to harbour primary plastids (Table 1), namely red algae, glaucophytes, and green algae (from which the land plants emerged) (Delwiche & Timme, 2011). More often than not, the origin of primary plastids is viewed as a singular event (see, e.g., Palmer, 2000), leading to the establishment of the clade Archaeplastida (or Plantae), which includes the three lineages containing primary plastids (Adl et al., 2005, 2012; Cavalier-Smith, 1998). The underlying assumption of Archaeplastida is that a single cyanobacterial endosymbiosis took place in a common ancestor of this group (Palmer, 2003; Reyes-Prieto et al., 2007).

Table 1 Algal Lineages and Characteristics of Their Plastids

Lineage	Belong to[a]	Plastid Membranes[b]	Putative Plastid Origin[c]
Glaucophytes	Archaeplastida	2 (with peptidoglycan)	Cyanobacteria
Red algae	Archaeplastida	2	Cyanobacteria
Green algae	Archaeplastida	2	Cyanobacteria
Paulinella chromatophora	Rhizaria	2 (with peptidoglycan)	Cyanobacteria
Chlorarachniophytes	Rhizaria	4	Green algae
Euglenids	Excavates	3	Green algae
Cryptophytes	Cryptista	4 (with ER connection)[d]	Red algae
Haptophytes	Haptista	4 (with ER connection)[d]	Red algae
Ochrophytes	Stramenopiles	4 (with ER connection)[d]	Red algae
Myzozoans	Alveolates	2–5	Red algae

[a]Main assemblage (no specific taxonomic rank) containing the plastid lineage.
[b]Number of membranes dividing the host cytosol from the plastid stroma.
[c]All plastids ultimately trace back to cyanobacteria, so this column only refers to the direct source of plastids in the lineage under consideration.
[d]The plastid outermost membrane is continuous with the ER in these three lineages.

For the purpose of this chapter, I will briefly note that Archaeplastida is viewed as a paradigmatic clade by most researchers, although there is still no comprehensive support for this group (Howe, Barbrook, Nisbet, Lockhart, & Larkum, 2008; Larkum, Lockhart, & Howe, 2007; Nozaki, 2005; Stiller & Hall, 1997; Stiller, Reel, & Johnson, 2003). Certainly, evidence in favour of a common origin of primary plastids has accumulated steadily in the past 40 years; for example, most phylogenetic analyses based on plastid genes/genomes show undisputable monophyly (Chu, Qi, Yu, & Anh, 2004; Criscuolo & Gribaldo, 2011; Li, Lopes, Foster, Embley, & Cox, 2014; Ponce-Toledo et al., 2017; Rodríguez-Ezpeleta et al., 2005) and the same core set of translocon components (TIC and TOC complexes) is present in all three lineages of Archaeplastida to import nucleus-encoded proteins into the plastid (Bhattacharya, Archibald, Weber, & Reyes-Prieto, 2007; Gould et al., 2008; Gross & Bhattacharya, 2009; Shi & Theg, 2013). Importantly, however, the monophyly of plastids does not necessarily entail the monophyly of hosts (Mackiewicz & Gagat, 2014), and indeed even in the most recent phylogenomic analyses of hundreds of nuclear markers, archaeplastidan host lineages consistently fail to group together with significant support (Brown et al., 2013; Burki et al., 2016; Yabuki et al., 2014). Accordingly, other scenarios for the origin of plastids in Archaeplastida have been proposed, including a single primary endosymbiosis, followed by one or more higher order endosymbioses (Kim & Maruyama, 2014); or multiple primary endosymbioses involving closely related cyanobacteria, followed by convergent evolution of plastid features (Stiller, 2014). These are not mainstream scenarios but should not be brushed aside just yet. As we will see in greater details later on, the field of plastid evolution is constantly adjusting in light of new data and sometimes this means large adjustments. For primary plastids, one of the greatest potentials for new insights on their origin comes from an unassuming (at first sight) alga. *Paulinella chromatophora*, that is the name of this alga, is evolutionarily unrelated to Archaeplastida, yet it harbours what are now considered bona fide photosynthetic organelles that are also direct descendent of cyanobacterial endosymbionts (but different from the ancestors of Archaeplastida plastids) (Gagat, Bodył, & Mackiewicz, 2016; Marin, Nowack, & Melkonian, 2005). Known as chromatophores, these organelles are remarkable because they clearly represent an independent, much more recent acquisition of photosynthetic capabilities in eukaryotes. Despite their much younger age (~60–200 Ma), new genomic data showed that chromatophores have evolved along the same general lines as primary plastids, including some levels of genetic integration (i.e. endosymbiotic gene transfer (EGT) to the host nucleus, see below) and the establishment of a machinery for importing nucleus-encoded proteins into the chromatophores (Mackiewicz, Bodył, & Gagat, 2012; Nowack & Grossman, 2012; Nowack, Melkonian, & Glöckner, 2008; Nowack et al., 2011), demonstrating the potential for independent acquisitions of plastids from cyanobacteria (Bodył, Mackiewicz, & Gagat, 2012).

2.2 Complex Plastids

The other major type of plastid, and main focus of this chapter, is often referred to as complex plastids. They reside within the host's endomembrane system and are called complex because unlike primary plastids, they are surrounded by (generally) four membranes (Table 1; Archibald, 2009; Gentil, Hempel, Moog, Zauner, & Maier, 2017; Gould et al., 2015, 2008; Keeling, 2013). The significance of these additional membranes came to light 40 years ago, when the concept of secondary endosymbiosis was first proposed (Gibbs, 1978; Taylor, 1974). In secondary endosymbiosis, a eukaryotic host engaged in an endosymbiotic relationship not with a prokaryote but with another eukaryote, itself containing a primary plastid (Fig. 1). The resulting complex plastids are surrounded not only by the same two cyanobacterial membranes found in primary plastids but also by two outer eukaryotic membranes whose precise origin remains unclear. Often considered homologous to the plasma membrane of the endosymbiotic alga and the phagosomal membrane of the secondary host (Fig. 1; Bodył, 2005; Cavalier-Smith, 2000), other possibilities exist (Gould et al., 2015). In addition to plastids with four membranes, other plastids classically considered to be of secondary origin are surrounded by three membranes (Table 1), in which case either one of the outer membranes was lost. Remarkably, plastids with between two and five membranes can also result from yet an additional round of endosymbiosis. In what is known as tertiary endosymbiosis, the organelle is derived from the uptake of a secondary plastid-containing alga, a process that has taken place on at least three independent occasions (Hackett, Maranda, Yoon, & Bhattacharya, 2003; Inagaki, Dacks, Doolittle, Watanabe, & Ohama, 2000; Tengs et al., 2000). Finally, the diversity of complex plastids would not be complete without mentioning serial secondary plastids, where secondary plastids have been replaced by new plastids derived from primary plastid-bearing algae (Kamikawa et al., 2015).

Given the above morphological diversity of complex plastids, and adding to this the fact that their evolution has spanned several hundred million years (Parfrey et al., 2011; Yoon et al., 2004), it comes to no surprise that untangling the spread of complex plastids across the eukaryotic tree has been, and still is, so controversial. Complex plastids come in two flavours: green algal- and red algal-derived. On the green side, there is no longer any doubt that two independent secondary endosymbioses led to the complex plastids of green algal origin. It has not always been such a clear-cut call (Cavalier-Smith, 1999), but in this case the host lineages were repeatedly shown to be from two very different parts of the eukaryotic tree (Keeling, 2004; Keeling et al., 2005) and the origin of the plastids themselves could be pinpointed to distinct groups of green algae (Rogers, Gilson, Su, McFadden, & Keeling, 2007). The two lineages with green secondary plastids are the photosynthetic euglenids (e.g. the lab species *Euglena gracilis*), belonging to the major eukaryotic group excavates (Simpson, 2003), and the chlorarachniophytes (e.g. the amoeboflagellate algal model *Bigelowiella natans*), belonging to the major group Rhizaria (Nikolaev et al., 2004). On the red side, the origin and spread of plastids have been much more difficult to resolve. Red algal-derived plastids account for a very large diversity of the photosynthetic lineages on the planet, as well as groups that have secondarily lost photosynthetic capabilities but retained their plastids. As depicted in Fig. 2, no less than four main groups possess plastids derived from red algae, namely myzozoans (photosynthetic alveolates containing dinoflagellates, the parasitic apicomplexans, chromerids, and colpodellids), ochrophytes (photosynthetic stramenopiles, including the well-known kelps and diatoms), cryptophytes (e.g. the model alga *Guillardia theta*), and haptophytes (e.g. the bloom-forming *Emiliania huxleyi*). I will herein refer to these main photosynthetic lineages as MOCHa (for myzozoans, ochrophytes, cryptophytes, haptophytes).

3. PLASTID VS HOST DATA: DIVERGENT VIEW ON THE TREE OF EUKARYOTES

3.1 Plastid Evidence Support a Single Origin

The difficulties in determining the evolutionary history of red complex plastids stem mainly from inconsistencies between plastid and host data (Archibald, 2015b; Keeling, 2013; Stiller, 2014). Indeed, just as in the case of Archaeplastida, plastid data overwhelmingly support a single origin of red algal-derived plastids in an early ancestor of MOCHa. Plastid phylogenies,

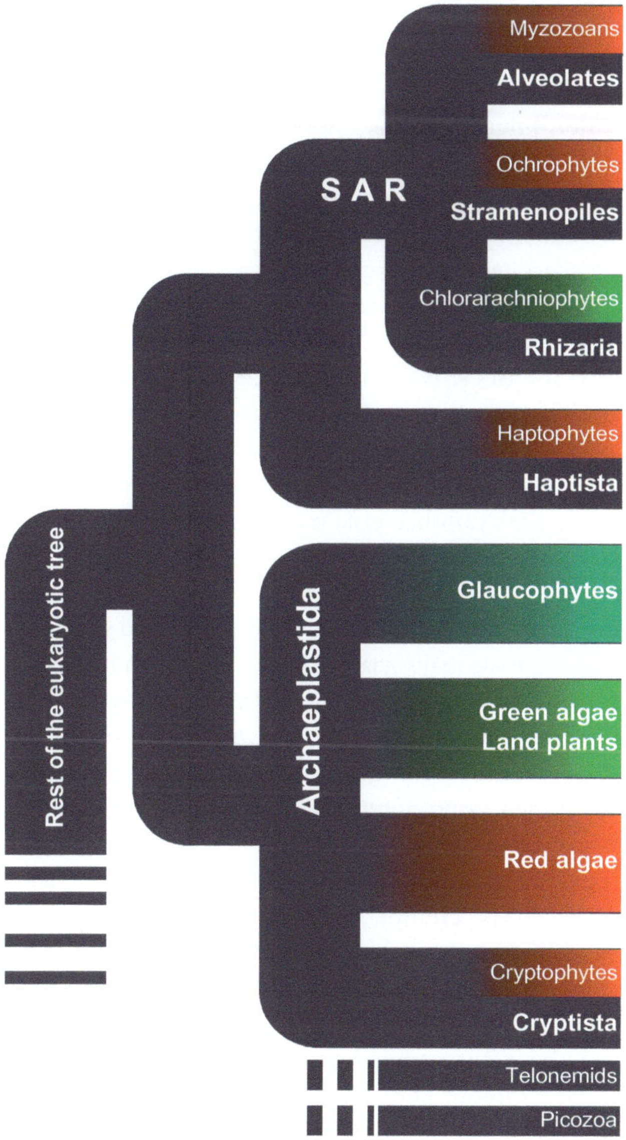

Fig. 2 Partial representation of the eukaryotic tree of life. This diagram shows a consensus tree for the relationships between the SAR group, Haptista, Cryptista, and Archaeplastida. The overall relationships are based on recent phylogenomic papers (see text for references). The multifurcation in SAR and Archaeplastida illustrates the uncertainties of the branching order in these groups. The colours represent lineages with plastids. When the colours fill the branches (in glaucophytes, green algae + land plants, and red algae), it shows that all members of these groups harbour a plastid. Otherwise the groups contain a mixture of plastid-containing and plastid-lacking taxa; this is the case for Rhizaria (which include the green algal-derived plastid-bearing chlorarachniophytes) and alveolates, stramenopiles, Haptista, and Cryptista (which together include the MOCHa lineages: myzozoans, ochrophytes, cryptophytes, haptophytes). Telonemids and Picozoa are also represented in this tree, because they are two of the most contentious plastid-lacking orphan lineages, but they potentially hold clues to better understand the evolution of plastids.

for example, are generally unambiguous in recovering the monophyly of MOCHa (Janouškovec, Horák, Oborník, Lukeš, & Keeling, 2010; Ševčíková et al., 2015). But even more compelling evidence supporting a single origin comes from comparisons of the protein import system across the outer plastid membranes, specifically across the second membrane from the outside. Plastid genomes encode only a small fraction of the proteins needed for plastid functions, and this works because the full organellar metabolism is sustained mainly by nucleus-encoded plastid-targeted proteins (Archibald, 2015b; Bock & Timmis, 2008; Timmis, Ayliffe, Huang, & Martin, 2004). Thus, these proteins must cross the three or four membranes of complex plastids to reach their action site, and how they do this has been intensely researched. Not surprisingly, the two innermost membranes corresponding to the cyanobacterial membranes of the primary plastid precursor are crossed using the same TIC–TOC translocons found in Archaeplastida (Gould et al., 2008; Stork, Lau, Moog, & Maier, 2013). The outermost membrane of four-membraned red and green complex plastids is crossed cotranslationally via the host's endomembrane system, which again is not surprising given the location of these plastids within the lumen of the endoplasmic reticulum (ER) (Bolte et al., 2009; Heiny, Pautz, Recker, & Przyborski, 2014; Liu et al., 2016). But strikingly, the last membrane corresponding to the second outermost membrane (the key membrane here) appears to be crossed using the same import machinery in all investigated red lineages with complex plastids (Felsner et al., 2011; Hempel, Bullmann, Lau, Zauner, & Maier, 2009; Hempel, Felsner, & Maier, 2010), and this import system seems different than that in green complex plastids (Hirakawa, Burki, & Keeling, 2012). Named SELMA (for symbiont-specific ERAD-like machinery), this unique multiprotein complex was demonstrably co-opted just once from the ER-associated protein degradation (ERAD) system of the red algal endosymbiont, thus constituting a robust character for the single origin of all MOCHa plastids (Gentil et al., 2017; Gould et al., 2015).

However, and this is a key argument, plastid data alone represent only partial evidence for the single origin of MOCHa plastids, because alternative biological scenarios can explain the monophyly of plastids without assuming the monophyly of hosts (the same went for Archaeplastida). Host phylogeny is therefore a critical piece in the puzzle, but it has not always been weighted accordingly in models of plastid evolution. Two aspects of host phylogeny deserve particular attention: (1) our (still) poor understanding of the host relationships and (2) the many plastid-lacking lineages that would interrupt the monophyly of MOCHa, should MOCHa share a common ancestor.

3.2 Host Lineages Are Not Monophyletic

Over the past 15 years, the combination of genomic-scale data sets (i.e. genomes and transcriptomes) with phylogenetic principles has become a routine approach to address questions where single genes have failed (Delsuc, Brinkmann, & Philippe, 2005). I will be quick to point out that this phylogenomic approach is not perfect, entailing its suite of artefacts, but when carefully applied phylogenomics can unleash the increased phylogenetic power contained in genomic data sets (Philippe et al., 2011). And this has led to some large-scale rearrangements of the eukaryotic tree of life few would have predicted (see, e.g., Burki, 2014 for a recent review). Most importantly from the perspective of complex plastid evolution, the latest phylogenomic analyses all recovered MOCHa host lineages scattered across large parts of the tree in a clearly polyphyletic configuration (Fig. 2; Baurain et al., 2010; Burki et al., 2009, 2016; Burki, Okamoto, Pombert, & Keeling, 2012). This is not to say that the phylogenetic origins of MOCHa are resolved, quite the contrary, but it is fair to say that it has become increasingly unlikely that they all form a monophyletic assemblage. So, what do we know? In the next few paragraphs, I will briefly review the most relevant advances and open questions with respect to the phylogeny of MOCHa host lineages.

Let us start with one of the largest groups of eukaryotes, the SAR group, which has resolved big chunks of the eukaryotic tree into a monophyletic clade (Fig. 2; Burki et al., 2007; Hackett et al., 2007; Rodríguez-Ezpeleta et al., 2007). SAR stands for stramenopiles, alveolates, and Rhizaria and thus includes two of the MOCHa lineages (myzozoans and ochrophytes) and one of the lineages with green complex plastid (chlorarachniophytes). In addition, each of the three constituting SAR groups also includes a myriad of related plastid-lacking lineages. Just to name a few: the oomycetes, labyrinthulids, and *Blastocystis* in stramenopiles (Derelle, López-García, Timpano, & Moreira, 2016); the ciliates and colponemids in alveolates (Janouškovec, Tikhonenkov, Mikhailov, et al., 2013); foraminifers, acanthareans, and ascetosporeans in Rhizaria (Sierra et al., 2016). Although SAR was proposed only 10 years ago, it has quickly become one of the best-supported major groups of eukaryotes (Burki, 2014). This support is important because it means that stramenopiles and alveolates are somehow related to the exclusion of the other two MOCHa members, cryptophytes and haptophytes, and this relationship is robust. Within SAR, the branching order is less clear, but a recent look at this issue favoured a close relationship between alveolates and Rhizaria, with stramenopiles being sister to that group (He, Sierra, Pawlowski, & Baldauf, 2016).

Moving on to cryptophytes and haptophytes, we are suddenly faced with two of the most contentious groups to place in the global tree of eukaryotes. Once specifically united under the banner of the Hacrobia hypothesis (Okamoto, Chantangsi, Horák, Leander, & Keeling, 2009), which mainly derived from plastid data but also early hints from host data, these two groups have since rarely been recovered together with better-sampled host lineages. Instead, haptophytes seem most closely related to centrohelids, a group of heliozoan plastid-lacking taxa that has long remained enigmatic (Burki et al., 2016; Cavalier-Smith, Chao, & Lewis, 2015); the grouping of haptophytes and centrohelids is known as Haptista (Fig. 2). Interestingly, the position of Haptista within the eukaryotic tree recently received decent support for a sister relationship to SAR, albeit this position has yet to stand the test of time (Burki et al., 2016). Unlike haptophytes, which origin appears on track to be resolved, cryptophytes remain one of the great mysteries in the tree of eukaryotes. As for the rest of MOCHa lineages, cryptophytes are also related to a bunch of seemingly plastid-lacking taxa, namely *Palpitomonas*, katablepharids, and goniomonads, which successively diverged in this order within a robust group called Cryptista (Burki, Okamoto, et al., 2012; Yabuki et al., 2014). But the phylogenetic position of Cryptista in the big eukaryotic picture remains uncertain, and no strong hints exist as to where it precisely belongs. The only long-term tendency is that it might be in some way related to Archaeplastida. This was already observed early on in rRNA phylogenies that showed weak relationships of cryptophytes and goniomonads (i.e. cryptomonads) to glaucophytes (e.g. Moreira et al., 2007), and ambiguous affinities to Archaeplastida continue to be reported from phylogenomic studies (Brown et al., 2013; Burki et al., 2016; Burki, Okamoto, et al., 2012). Regardless of its precise origin, what we can say today with some certainty is that (1) it is unlikely that Cryptista is related to any of the other MOCHa lineages, (2) Cryptista likely diverged early on in eukaryote evolution, and (3) Cryptista thus holds crucial clues in the quest to untangle the origin and spread of red algal-derived plastids across the eukaryotic tree.

This account of MOCHa host phylogeny would not be complete without mentioning at least two more orphan lineages. Orphans are defined here as lineages we know exist, but their evolutionary origin has remained mysterious. In that sense, Cryptista is an orphan lineage, but other orphans are even more enigmatic. This is the case of the telonemids and picobiliphytes-turned-Picozoa, two lineages that likely represent phylums of their own (Fig. 2; Not et al., 2007; Seenivasan, Sausen, Medlin, & Melkonian,

2013; Shalchian-Tabrizi et al., 2006). Although originally described as photosynthetic, picozoans were later reassessed to lack plastids based on the draft genomes of three single cells (Yoon et al., 2011). Telonemids are also non-photosynthetic organisms, and there is no reason to believe that they harbour cryptic plastids. The phylogenetic position of both lineages has been addressed by several single and multigene analyses, but no attempt was conclusive. Still, telonemids and picozoans have been repeatedly inferred in the vicinity of MOCHa lineages, or in one way or another branched off deeply in the tree, for example, near Archaeplastida (Burki et al., 2009, 2016; Burki, Okamoto, et al., 2012; Cuvelier et al., 2008; Shalchian-Tabrizi et al., 2006). Therefore, we will need to keep a close eye on these two orphan lineages in the future, but also on others that have yet to be discovered, because a better understanding of their position in the tree will undoubtedly provide new clues on plastid evolution.

4. HOW OFTEN HAVE RED ALGAL-DERIVED PLASTIDS SPREAD?

Let us take a step back and look at how the above puzzling inconsistencies between plastid and host data have been reconciled over the past 20 years to explain the observed distribution across the tree of complex plastids. Since the case of known green complex plastids is settled, this discussion will be about the origin and spread of red algal-derived plastids.

4.1 The Chromalveolate Hypothesis

As we have seen, MOCHa taxa are *all* more closely related to various plastid-lacking lineages than they are to each other (Fig. 2). To explain this distribution of organisms, the field of plastid evolution has been heavily influenced by a conceptual framework that favours plastid loss over gain. This framework is described under the chromalveolate hypothesis (Cavalier-Smith, 1999), which was an ambitious and provocative solution to deal with the wide diversity of red algal-containing organisms. The guiding principle of the chromalveolate hypothesis is that establishing a new plastid involves a series of genetic and cellular transformations deemed so complex that the occurrence of endosymbiosis should be minimized (Cavalier-Smith, 1999). More precisely, this hypothesis was built around the inferred difficulties in establishing a protein-targeting system in a nascent plastid (Keeling, 2009). Bluntly said, if establishing such a system is evolutionarily unlikely to happen more than once, then it is more parsimonious to infer a single origin in a common ancestor. Consequently, under this framework, similar plastids are viewed as vertically inherited from common ancestry and derived plastid-lacking taxa are inferred to have secondarily lost their plastids.

The chromalveolate hypothesis is an elegant way to explain the origin of red complex plastids, and plastid characters generally support it; the evolution of SELMA, just to name one strong character that we already discussed, is most easily interpreted under this framework (Gould et al., 2015). In recent years, however, the general improvement of the tree of eukaryotes coupled with the placement in it of several once orphan lineages has diminished the enthusiasm for this hypothesis. Indeed, the number of plastid losses required to fit plastid-lacking taxa within the chromalveolate hypothesis has increased in such a way that continuing to rationalize a single origin of red algal plastids in a hypothetical common ancestor has become very complicated (Archibald, 2015b; Lane & Archibald, 2008). To the best of our current knowledge, the most favourable scenario for this hypothesis would be that SAR, Haptista, and Cryptista form a mega-assemblage to the exclusion of all other extant eukaryotic groups (Fig. 3). Should that be the case and given the topology in Fig. 3, the minimum count of plastid losses required under the chromalveolate hypothesis is 10, that is, 3 in alveolates, 2 in stramenopiles, 1 in an ancestor of Rhizaria, 1 in Haptista, and 3 in Cryptista. Importantly, these are very conservative numbers derived from a quick count based on some of the most recent phylogenies (Burki et al., 2016; Derelle et al., 2016; Janouškovec, Tikhonenkov, Mikhailov, et al., 2013; Yabuki et al., 2014) and are based only on well-resolved relationships within these groups. In reality the number of losses would certainly be higher due to many lineages not accounted for here (for example, all MAST lineages in stramenopiles (Massana, del Campo, Sieracki, Audic, & Logares, 2014)).

4.2 Plastid Loss and Plastid Dependency

In these circumstances, and in light of new genomic data, the premise that plastid loss is more likely to happen than plastid gain is worthy of a fresh look. Although ultimately this is an issue that rests heavily on conceptual choices, there are in fact little empirical data to favour loss over gain. Rather, the opposite seems true (Stiller, 2014). There is no doubt that the capacity to photosynthesize was lost many times during eukaryote evolution (Figueroa-Martinez, Nedelcu, Smith, & Reyes-Prieto, 2015; Keeling, 2013). But does that mean that plastids were lost too in secondarily

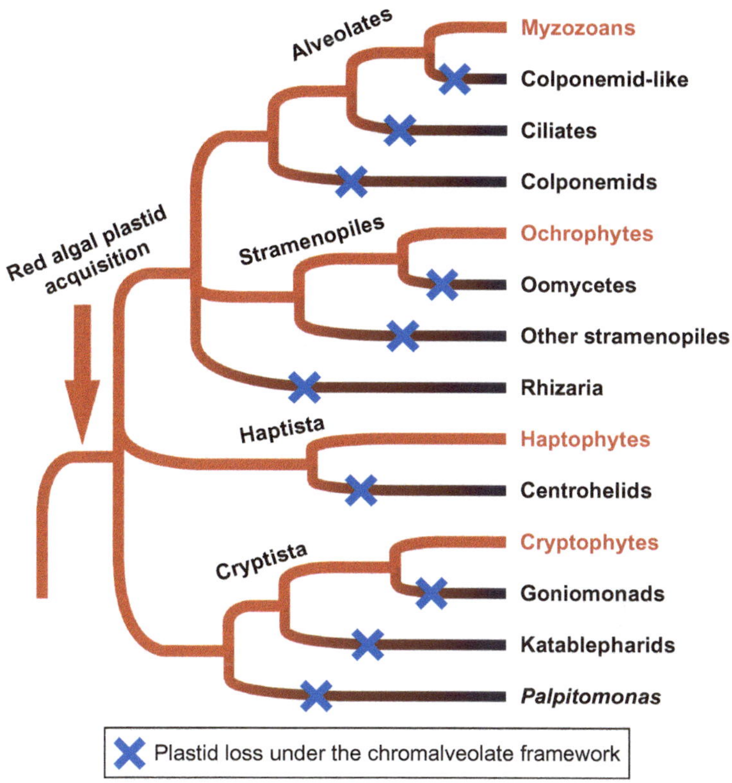

Fig. 3 Hypothetical tree best-fitting the chromalveolate hypothesis. The relationships within each group are based on recent publications (see main text for references). The relationships between each group are uncertain (except for the SAR group) and so are represented as multifurcating. Under the chromalveolate hypothesis, the endosymbiosis that led to the establishment of complex red plastids in MOCHa took place in the common ancestor of SAR, Haptista, and Cryptista. To the best of our current knowledge, at least 10 plastid losses must have occurred to explain the current distribution of complex red plastids.

nonphotosynthetic organisms? The answer to that question is a firm no. Myzozoa represents a great testing ground for addressing this question because more than anywhere else photosynthesis has been demonstrably lost on multiple occasions, i.e., in all apicomplexans, colpodellids, and about half of dinoflagellates (Janouškovec et al., 2015). In these groups, however, empirical evidence of cryptic plastid retention now exists in almost all nonphotosynthetic organisms where plastids have been searched for with some dedication. These include the well-characterized apicoplasts in apicomplexans (McFadden, Reith, Munholland, & Lang-Unnasch, 1996;

Waller & McFadden, 2005), but also as-yet unseen nonphotosynthetic plastids in colpodellids and dinoflagellates (Janouškovec et al., 2017, 2015).

Some might ask then, how can we be so sure of the presence of a plastid if it has never been observed? The following reasoning is based on the principles of metabolic dependency, which has been discussed in detail elsewhere (Janouškovec et al., 2017, 2015). In short, reconstructed cytosolic and plastidial pathways of several key plastid functions revealed unexpected metabolic dependency on plastids (other than photosynthesis) in all investigated free-living colpodellids and nonphotosynthetic dinoflagellates. Among all pathways investigated, the common retention of the isoprenoid plastidial pathway and absence of cytosolic counterparts were proposed as the strongest evidence for the absolute metabolic plastid dependency in these organisms, and thus for the cryptic presence of never-seen plastids. This is a critical new result that forces us to reevaluate the global importance of plastid loss. To this day, only two demonstrable cases of plastid loss are known across the entire eukaryotic diversity, in the apicomplexan *Cryptosporidium* and the basal dinoflagellate *Hematodinium* (Abrahamsen, 2004; Gornik et al., 2015). Importantly, these are both parasites, a lifestyle in which salvaging of metabolites from their host could alleviate the metabolic dependence on a plastid. For free-living organisms, the outright loss of a plastid has in fact never been confirmed, probably due to absolute dependency on plastid metabolites that cannot be recovered from food or the environment (Janouškovec et al., 2017).

It seems thus not that 'easy' to discard of a plastid once it was established, or at least it does not appear to be a rampant process in evolution. In contrast, endosymbioses are much more common, even if we just consider the ones that gave rise to plastids (Archibald, 2009; Gould et al., 2008). In comparison, just the two well-defined secondary endosymbioses with green algae in chlorarachniophytes and euglenids already equal the number of plastid losses that are backed up with some data (Abrahamsen, 2004; Gornik et al., 2015). So when we put into the mix the minimal number of losses required under the chromalveolate framework (10 as argued in Fig. 3, but in reality much more), it becomes obvious that an alternative, better-fitting model is needed to explain the spread of red complex plastids.

4.3 Serial Eukaryote-to-Eukaryote Endosymbioses

Not one but several models have been recently proposed to describe the evolution of red complex plastids (e.g. Bodył, Stiller, & Mackiewicz, 2009; Dorrell & Smith, 2011; Petersen et al., 2014; Sanchez Puerta &

Delwiche, 2008; Stiller et al., 2014). Although they differ in their specifics, they all attempt to take into account the phylogenetic incongruences between plastid and host data (i.e. monophyletic plastids in polyphyletic hosts; see also Baurain et al., 2010) and do not make the assumption that plastid loss is less of an evolutionary burden than plastid gain. Instead, the scattered distribution of red algal-derived plastids in the tree is explained by additional endosymbiotic events that effectively bypass plastid-lacking lineages. In other words the idea of higher order, eukaryote-to-eukaryote endosymbioses is generalized in order to allow more frequent transfers of plastids between distantly related eukaryotes (Bodył et al., 2009; Dorrell & Smith, 2011; Petersen et al., 2014; Sanchez Puerta & Delwiche, 2008; Stiller et al., 2014). Consequently, some of what were considered secondary plastids under the chromalveolate hypothesis are here viewed as tertiary or even quaternary plastids. For example, Fig. 4 depicts

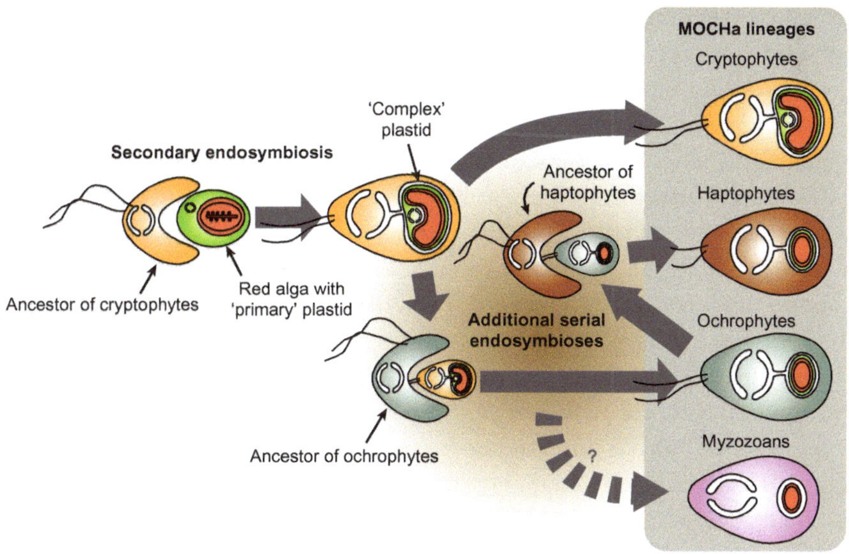

Fig. 4 Origin and evolution of complex red plastid by serial eukaryote-to-eukaryote endosymbioses. Several models have recently been proposed that all follow the same logic: red algal plastids were passed between distantly related lineages by secondary, tertiary, and even quaternary endosymbioses. This specific diagram follows the recent proposition by Stiller et al. (2014), where cryptophytes were the original recipient of complex red plastids (secondary endosymbiosis), which were then passed to ochrophytes and haptophytes by higher order endosymbioses. This model has the advantage to derive from phylogenetically independent empirical data (the overall similarity between genomes), yet it fits well with current plastid and nuclear phylogenies. Note that the plastids in myzozoans were not included in this model.

the explicit model of Stiller et al. (2014), where only an ancestor of cryptophytes was the original recipient of red algal plastids. From there plastids moved to ochrophytes by tertiary endosymbiosis and to haptophytes by quaternary endosymbiosis. This model stands out because it is based on direct data comparison (overall similarity between genomes) rather than more speculative proposals of serial plastid endosymbioses. It also reinforces the position of cryptophytes at the source of the dispersal of red complex plastids, a 'cryptophyte-first' scenario that is gaining in popularity (Bodył et al., 2009; Burki et al., 2016). More than any other groups of eukaryotes, cryptophytes thus emerge as key players in early plastid dispersal and if they are truly the only lineage that experienced a secondary endosymbiosis with red algae, it will be crucial to monitor how the phylogenetic position of Cryptista unfold, especially with respect to Archaeplastida.

All things considered, the serial endosymbiotic models of plastid evolution are currently our best bet to explain the data at hand. This is definitely true when regarding the scattered distribution of plastid-containing lineages in the tree, but it is also true when tracking the history of the plastids themselves. Indeed, one easily overlooked aspect of these models is that, from the perspective of plastid data, passing plastids of the same ancestry from donors to recipients does not fundamentally differ from a single plastid origin followed by vertical inheritance. SELMA, for example, which represents a very robust character against the multiple origins of red complex plastids (Gould et al., 2015), remains compatible under the serial endosymbiotic models. Evidence for this is indirect but a comparison with the TIC–TOC protein import machinery shows how this could be (Archibald, 2015b). In complex algae, the genes encoding the various components of these systems have clearly been relocated to the host nucleus more than once, in chlorarachniophytes and euglenids as well as in MOCHa lineages (Curtis et al., 2012; Gould et al., 2008; Hirakawa et al., 2012). There is thus a realistic possibility that SELMA genes followed a similar path to the different host nucleus from serially acquired endosymbionts, and that the SELMA machinery itself was targeted to the second outermost membrane in a convergent manner. Yet, in spite of all the benefits and explaining of the above serial endosymbiotic models, they also come with other complications that would be unwise to simply dismiss. Most obviously, each round of additional endosymbiosis necessarily involves accompanying losses of membranes, whose likelihood is hard to judge (see the discussion in Gould et al., 2015). For example, how can it be that the topology of plastid membranes in haptophytes, supposedly of quaternary origin in the model of Stiller et al. (2014), is identical to that of secondary plastids in cryptophytes in the same model. This is for now a mystery, as is the precise order of endosymbioses in serial models, and until new empirical data unlock the situation the evolution of red complex plastids will remain a lot about conceptual preferences.

5. COMPLEX ALGAE AND GENOME MOSAICISM

The problematic of the spread of plastids by endosymbiosis is intrinsically linked to another debate that has been raging in the field of plastid evolution and beyond, namely how much endosymbiosis has impacted the nuclear genomes of complex algae (Archibald, 2015b; Burki et al., 2014; Elias & Archibald, 2009; Moreira & Deschamps, 2014). A correlate of plastid establishment is EGT, a process that takes place during endosymbiotic integration and results in the flow of genes from endosymbiont to host (Curtis et al., 2012; Martin et al., 2002; Timmis et al., 2004). As for mitochondria, EGT partly explains the small size of plastid genomes compared to their free-living prokaryotic relatives, generally encoding less than 10% of the coding capacity of extant cyanobacteria (Janouškovec, Tikhonenkov, Burki, et al., 2013; Kim & Maruyama, 2014; Lee et al., 2016; Qiu, Lee, Yoon, & Bhattacharya, 2017). In the nuclear genomes of complex algae, two origins of EGT are expected (Fig. 5, blue and orange arrows, respectively): (1) cyanobacterial genes that can be traced back to the plastid, either directly or via the endosymbiont nucleus (i.e. the nucleomorph, which has generally vanished but is still present in chlorarachniophytes and cryptophytes), and (2) cyanobacterial-unrelated genes that were transferred between nuclei during and after endosymbiosis. The nuclear genomes of complex algae are thus expected to be a mix and match of genes with various origins, at least of cyanobacterial and algal origins in addition to the host DNA (Archibald, 2015b). On top of that, another well-known source of foreign genes in nuclear genomes is horizontal gene transfer (HGT), in which genes are passed from a donor that is not an endosymbiont (Keeling & Palmer, 2008; Nowack et al., 2016; Price et al., 2012; Schönknecht et al., 2013). Although the prevalence of HGT in algae was shown to be not nearly as high as that of EGT (Ku et al., 2015), it is still likely that HGT has contributed even moderately to the genome mosaicism in at least some algae.

The pressing question, then, is how many foreign genes (and more generally foreign DNA) reside in the nuclear genomes of complex algae?

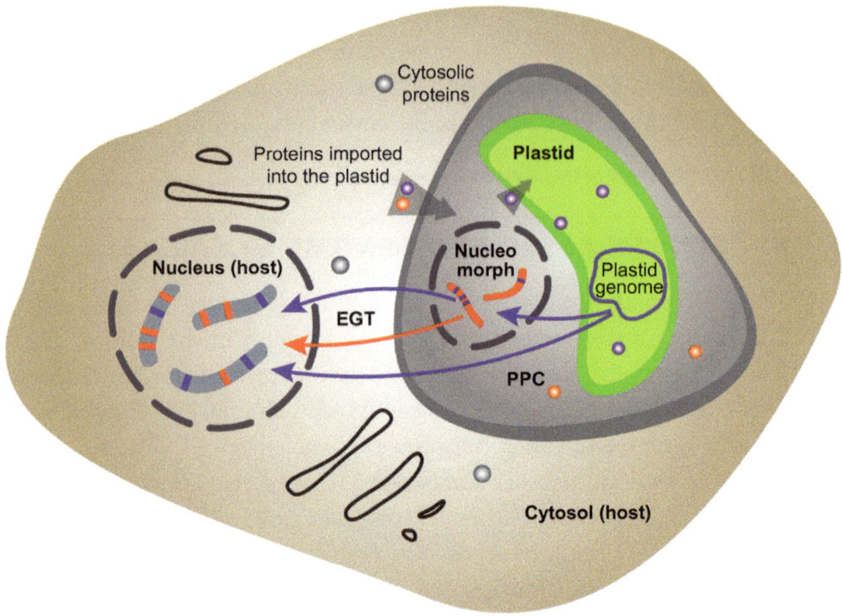

Fig. 5 Endosymbiotic gene transfer (EGT) results in genome mosaicism in algae. This example illustrates a generic cell with a four-membrane-bound complex plastid, resulting from the uptake of a primary alga or serial endosymbioses. Two 'kinds' of EGT are represented: the *blue arrows* show the flow of cyanobacterial genes from plastid to host, whereas the *orange arrow* only shows the transfer between nuclei of non-cyanobacterial genes. As a result of EGT, most plastid proteins are nucleus-encoded and imported into the plastid through specialized import machineries. A nucleomorph is depicted here, but in most algae this remnant of the endosymbiont nucleus has been lost. Abbreviations: *PPC*, periplastidal compartment (corresponds to the endosymbiont's cytosol).

Despite remarkable new insights in recent years, the honest answer to this question is that we still do not really know. As too often when protists are involved, the current limitations mainly come from the poor taxonomic diversity available in genomic databases, which lowers the ability of our analytical tools to derive clear signal (Sibbald & Archibald, 2017). This is made worse by the antiquity of the evolutionary events under scrutiny here; most plastid endosymbioses are so old that it becomes very tricky to extract genuine signal over background noise. In these circumstances, attempts to quantify the endosymbiotic signal in algae and determine its precise evolutionary origin have been rather inconsistent. This is best illustrated by the wide range of interpretations deriving from similar data but using different search criteria. For example, of the c.1700 genes deemed of green algal ancestry in diatoms (Moustafa et al., 2009), only 286 were still consistent with EGT in a later study, and mostly of red algal or unresolved origin (Deschamps & Moreira, 2012). Similarly, of the 513 putative red or green algal genes (~1:1 ratio) in the chromerid *Chromera velia* (Woehle, Dagan, Martin, & Gould, 2011), only 51 were still considered as EGT genes (with a majority of red algal-derived) in a reanalysis of the same data set (Burki, Flegontov, et al., 2012).

Other studies of various complex algae have shown the same trend, i.e., that the *detectable* endosymbiotic (algal) footprint in nuclear genomes is modest, ranging from none to about 7% of the analysable genes (Burki et al., 2014; Curtis et al., 2012; Hehenberger, Burki, Kolisko, & Keeling, 2016). These figures should be taken with a grain of salt, but for now, they constitute a reasonable working hypothesis that is backed up by a recent large-scale analysis of over 28,000 eukaryotic and prokaryotic gene families across the tree of life (Ku et al., 2015). By looking at the phyletic distribution within each clustered gene family, the authors were able to extract a signal that rose above the background for the introduction of about 1500 cyanobacterial genes into eukaryotes. Because these genes are absent in species lacking plastids, this signal was logically interpreted as the footprint of endosymbiosis corresponding to the origin of plastids in Archaeplastida (Ku et al., 2015). But, interestingly, out of these 1500 cyanobacterial genes, only 406 were found in complex algae, which corresponded to just 1.4% of the thousands of gene families analysed. True, this number only represents a partial view of the global EGT signal as it only includes genes that ultimately trace back to cyanobacteria, i.e., it does not include nucleus-to-nucleus EGT of noncyanobacterial origins. Still, it is in remarkable agreement with the algal signal recovered previously, for example, the 353 and 508 proteins of endosymbiotic origin in the genomes of the chlorarachniophyte *B. natans* and the cryptophyte *G. theta*, respectively (Curtis et al., 2012)

The main lesson from these studies is that although it is undeniable that the genomes of complex algae contain a certain proportion of genes contributing to an overall algal signal, best interpreted as the genomic footprint of endosymbiosis, the strength of this signal remains to be determined with more certainty. But regardless of the strength, the next logical question is, of course, to look at what this signal comprises. Most studies thus far have searched for a general *algal* pattern in nuclear genomes of complex algae;

in other words a combined green, red, and, to a lesser extent, glaucophyte phylogenetic signal that deviates from the expected vertical ancestry of a given lineage, and that may or may not have its roots in cyanobacteria (e.g. Burki, Flegontov, et al., 2012; Burki et al., 2014; Curtis et al., 2012; Deschamps & Moreira, 2012; Moustafa et al., 2009). In doing so, these studies acknowledge the fact that current databases are too limited for more targeted searches, but they also complicate the interpretation of this signal (Zimorski et al., 2014). Indeed, since no lineage is known to harbour green algal- and red algal-derived plastids in the same cell, how can we interpret the observation that genes of seemingly green and red algal origin are usually found together in complex algae? There are several reasons for this codetection of green and red algal genes, and the main one is likely methodological as we said earlier (see discussion in, e.g., Archibald, 2015b; Moreira & Deschamps, 2014). Other reasons might be linked to lifestyle, for example, mixotrophy that would favour HGT from food sources (Archibald, Rogers, Toop, Ishida, & Keeling, 2003). Or, as some have proposed, the large green signal in red algal plastid-containing organisms such as diatoms is evidence of cryptic endosymbiosis (Frommolt et al., 2008; Moustafa et al., 2009).

Regardless of the interpretation, it is clear that the 500 or less genes composing the currently detectable algal signal are insufficient to account for all the functions carried out in a plastid. Plastid proteomes are predicted to amount to about 700–1000 proteins in complex algae, even double that when the nucleomorphs and primary host cytosol have been retained (Curtis et al., 2012; Gruber, Rocap, Kroth, Armbrust, & Mock, 2015). Furthermore, EGT was repeatedly predicted to have also contributed to biochemical pathways in the host (Burki et al., 2014; Curtis et al., 2012; Martin et al., 2002), so some of the detected EGT signal does not actually service the plastid. One possibility that might compensate for some of the missing count comes from the observation that host-derived proteins, not algal-derived, function in the plastid (Curtis et al., 2012). However, as Archibald recently put it '... even accounting for proteins encoded in the plastid and nucleomorph genomes, there are nowhere near enough host nucleus-encoded algal proteins to service the plastid' in the current estimation of the endosymbiotic footprint (Archibald, 2015b). Consequently, if it is clear that algal genes have invaded the nuclear genomes of complex algae, mostly via EGT and probably also some levels of HGT, beyond that the betting is still on with respect to the magnitude of the endosymbiotic contribution to host genomes.

6. CONCLUDING REMARKS

The evolution of plastids is convoluted, and at the moment, there is no model that fits all evidence. The last 10 years have seen a great deal of work in phylogenomics, comparative genomics, and cell biology that radically changed our perception of how plastids may have evolved. It is true that in many cases new data have brought apparent complexity rather than reliable solutions, but this just shows that new empirical evidence is needed to verify a priori assumptions. It will take time to reassemble all evidence into a comprehensive framework. In doing so, it will be important to continue the often underappreciated job of species discovery. We have barely scratched the surface of the massive diversity of microbes, yet it is only by incorporating all players that we will be able to fill crucial gaps in our models of the convoluted evolution of plastids.

ACKNOWLEDGEMENTS

This work was supported by a grant from Science for Life Laboratory (SciLifeLab); F.B. is a SciLifeLab fellow. F.B. would like to thank Drs David R. Smith and Yoshihisa Hirakawa for providing critical comments on an earlier version of the chapter.

REFERENCES

Abrahamsen, M. S. (2004). Complete genome sequence of the apicomplexan, *Cryptosporidium parvum*. *Science, 304*(5669), 441–445.

Adl, S. M., Simpson, A. G. B., Farmer, M. A., Andersen, R. A., Anderson, O. R., Barta, J. R., et al. (2005). The new higher level classification of eukaryotes with emphasis on the taxonomy of protists. *The Journal of Eukaryotic Microbiology, 52*(5), 399–451.

Adl, S. M., Simpson, A. G. B., Lane, C. E., Lukeš, J., Bass, D., Bowser, S. S., et al. (2012). The revised classification of eukaryotes. *The Journal of Eukaryotic Microbiology, 59*(5), 429–493.

Archibald, J. M. (2009). The puzzle of plastid evolution. *Current Biology, 19*(2), R81–R88.

Archibald, J. M. (2015a). Endosymbiosis and eukaryotic cell evolution. *Current Biology, 25*(19), R911–R921.

Archibald, J. M. (2015b). Genomic perspectives on the birth and spread of plastids. *Proceedings of the National Academy of Sciences of the United States of America, 112*(33), 10147–10153.

Archibald, J. M., Rogers, M. B., Toop, M., Ishida, K.-I., & Keeling, P. J. (2003). Lateral gene transfer and the evolution of plastid-targeted proteins in the secondary plastid-containing alga Bigelowiella natans. *Proceedings of the National Academy of Sciences of the United States of America, 100*(13), 7678–7683.

Baurain, D., Brinkmann, H., Petersen, J., Rodríguez-Ezpeleta, N., Stechmann, A., Demoulin, V., et al. (2010). Phylogenomic evidence for separate acquisition of plastids in cryptophytes, haptophytes, and stramenopiles. *Molecular Biology and Evolution, 27*(7), 1698–1709.

Bhattacharya, D., Archibald, J. M., Weber, A. P. M., & Reyes-Prieto, A. (2007). How do endosymbionts become organelles? Understanding early events in plastid evolution. *BioEssays*, *29*(12), 1239–1246.
Bock, R., & Timmis, J. N. (2008). Reconstructing evolution: Gene transfer from plastids to the nucleus. *BioEssays*, *30*(6), 556–566.
Bodył, A. (2005). Do plastid-related characters support the chromalveolate hypothesis? *Journal of Phycology*, *41*(3), 712–719.
Bodył, A., Mackiewicz, P., & Gagat, P. (2012). Organelle evolution: Paulinella breaks a paradigm. *Current Biology*, *22*(9), R304–R306.
Bodył, A., Stiller, J. W., & Mackiewicz, P. (2009). Chromalveolate plastids: Direct descent or multiple endosymbioses? *Trends in Ecology & Evolution*, *24*(3), 119–121. author reply 121–122.
Bolte, K., Bullmann, L., Hempel, F., Bozarth, A., Zauner, S., & Maier, U. G. (2009). Protein targeting into secondary plastids. *The Journal of Eukaryotic Microbiology*, *56*(1), 9–15.
Brown, M. W., Sharpe, S. C., Silberman, J. D., Heiss, A. A., Lang, B. F., Simpson, A. G. B., et al. (2013). Phylogenomics demonstrates that breviate flagellates are related to opisthokonts and apusomonads. *Proceedings of the Royal Society B: Biological Sciences*, *280*(1769), 20131755.
Burki, F. (2014). The eukaryotic tree of life from a global phylogenomic perspective. *Cold Spring Harbor Perspectives in Biology*, *6*(5), a016147.
Burki, F., Flegontov, P., Oborník, M., Cihlar, J., Pain, A., Lukeš, J., et al. (2012). Reevaluating the green versus red signal in eukaryotes with secondary plastid of red algal origin. *Genome Biology and Evolution*, *4*(6), 626–635.
Burki, F., Imanian, B., Hehenberger, E., Hirakawa, Y., Maruyama, S., & Keeling, P. J. (2014). Endosymbiotic gene transfer in tertiary plastid-containing dinoflagellates. *Eukaryotic Cell*, *13*(2), 246–255.
Burki, F., Inagaki, Y., Bråte, J., Archibald, J. M., Keeling, P. J., Cavalier-Smith, T., et al. (2009). Large-scale phylogenomic analyses reveal that two enigmatic protist lineages, Telonemia and Centroheliozoa, are related to photosynthetic chromalveolates. *Genome Biology and Evolution*, *1*, 231–238.
Burki, F., Kaplan, M., Tikhonenkov, D. V., Zlatogursky, V., Minh, B. Q., Radaykina, L. V., et al. (2016). Untangling the early diversification of eukaryotes: A phylogenomic study of the evolutionary origins of Centrohelida, Haptophyta and Cryptista. *Proceedings of the Royal Society B: Biological Sciences*, *283*(1823), 20152802.
Burki, F., Okamoto, N., Pombert, J.-F., & Keeling, P. J. (2012). The evolutionary history of haptophytes and cryptophytes: Phylogenomic evidence for separate origins. *Proceedings of the Royal Society B: Biological Sciences*, *279*(1736), 2246–2254.
Burki, F., Shalchian-Tabrizi, K., Minge, M., Skjaeveland, A., Nikolaev, S. I., Jakobsen, K. S., et al. (2007). Phylogenomics reshuffles the eukaryotic supergroups. *PLoS One*, *2*(8), e790.
Cavalier-Smith, T. (1982). The origins of plastids. *Biological Journal of the Linnean Society*, *17*(3), 289–306.
Cavalier-Smith, T. (1998). A revised six-kingdom system of life. *Biological Reviews of the Cambridge Philosophical Society*, *73*(3), 203–266.
Cavalier-Smith, T. (1999). Principles of protein and lipid targeting in secondary symbiogenesis: Euglenoid, dinoflagellate, and sporozoan plastid origins and the eukaryote family tree. *The Journal of Eukaryotic Microbiology*, *46*(4), 347–366.
Cavalier-Smith, T. (2000). Membrane heredity and early chloroplast evolution. *Trends in Plant Science*, *5*(4), 174–182.
Cavalier-Smith, T., Chao, E. E., & Lewis, R. (2015). Multiple origins of Heliozoa from flagellate ancestors: New cryptist subphylum Corbihelia, superclass Corbistoma, and monophyly of Haptista, Cryptista, Hacrobia and Chromista. *Molecular Phylogenetics and Evolution*, *93*, 331–362.

Chu, K. H., Qi, J., Yu, Z., & Anh, V. (2004). Origin and phylogeny of chloroplasts revealed by a simple correlation analysis of complete genomes. *Molecular Biology and Evolution, 21*(1), 200–206.

Criscuolo, A., & Gribaldo, S. (2011). Large-scale phylogenomic analyses indicate a deep origin of primary plastids within cyanobacteria. *Molecular Biology and Evolution, 28*(11), 3019–3032.

Curtis, B. A., Tanifuji, G., Burki, F., Gruber, A., Irimia, M., Maruyama, S., et al. (2012). Algal genomes reveal evolutionary mosaicism and the fate of nucleomorphs. *Nature, 492*(7427), 59–65.

Cuvelier, M. L., Ortiz, A., Kim, E., Moehlig, H., Richardson, D. E., Heidelberg, J. F., et al. (2008). Widespread distribution of a unique marine protistan lineage. *Environmental Microbiology, 10*(6), 1621–1634.

Delsuc, F., Brinkmann, H., & Philippe, H. (2005). Phylogenomics and the reconstruction of the tree of life. *Nature Reviews Genetics, 6*(5), 361–375.

Delwiche, C. F., & Timme, R. E. (2011). Plants. *Current Biology, 21*(11), R417–R422.

Derelle, R., López-García, P., Timpano, H., & Moreira, D. (2016). A phylogenomic framework to study the diversity and evolution of stramenopiles (=heterokonts). *Molecular Biology and Evolution, 33*(11), 2890–2898.

Deschamps, P., & Moreira, D. (2012). Re-evaluating the green contribution to diatom genomes. *Genome Biology and Evolution, 4*(7), 795–800.

Dorrell, R. G., & Smith, A. G. (2011). Do red and green make brown?: Perspectives on plastid acquisitions within chromalveolates. *Eukaryotic Cell, 10*(7), 856–868.

Elias, M., & Archibald, J. M. (2009). Sizing up the genomic footprint of endosymbiosis. *BioEssays, 31*(12), 1273–1279.

Falcón, L. I., Magallón, S., & Castillo, A. (2010). Dating the cyanobacterial ancestor of the chloroplast. *The ISME Journal, 4*(6), 777–783.

Felsner, G., Sommer, M. S., Gruenheit, N., Hempel, F., Moog, D., Zauner, S., et al. (2011). ERAD components in organisms with complex red plastids suggest recruitment of a preexisting protein transport pathway for the periplastid membrane. *Genome Biology and Evolution, 3*, 140–150.

Figueroa-Martinez, F., Nedelcu, A. M., Smith, D. R., & Reyes-Prieto, A. (2015). When the lights go out: The evolutionary fate of free-living colorless green algae. *New Phytologist, 206*(3), 972–982.

Frommolt, R., Werner, S., Paulsen, H., Goss, R., Wilhelm, C., Zauner, S., et al. (2008). Ancient recruitment by chromists of green algal genes encoding enzymes for carotenoid biosynthesis. *Molecular Biology and Evolution, 25*(12), 2653–2667.

Gagat, P., Bodył, A., & Mackiewicz, P. (2016). Breaking the endosymbiont-organelle border: The case of *Paulinella chromatophora*. *Endocytobiosis and Cell Research, 27*(3), 45–51.

Gentil, J., Hempel, F., Moog, D., Zauner, S., & Maier, U. G. (2017). Review: Origin of complex algae by secondary endosymbiosis: A journey through time. *Protoplasma, 52*, 1–9.

Gibbs, S. P. (1978). The chloroplasts of *Euglena* may have evolved from symbiotic green algae. *Canadian Journal of Botany, 56*(22), 2883–2889.

Gornik, S. G., Febrimarsa, Cassin, A. M., MacRae, J. I., Ramaprasad, A., Rchiad, Z., et al. (2015). Endosymbiosis undone by stepwise elimination of the plastid in a parasitic dinoflagellate. *Proceedings of the National Academy of Sciences of the United States of America, 111*(18), 5767–5772.

Gould, S. B., Maier, U. G., & Martin, W. F. (2015). Protein import and the origin of red complex plastids. *Current Biology, 25*(12), R515–R521.

Gould, S. B., Waller, R. F., & McFadden, G. I. (2008). Plastid evolution. *Annual Review of Plant Biology, 59*(1), 491–517.

Gross, J., & Bhattacharya, D. (2009). Mitochondrial and plastid evolution in eukaryotes: An outsiders' perspective. *Nature Reviews Genetics, 10*(7), 495–505.

Gruber, A., Rocap, G., Kroth, P. G., Armbrust, E. V., & Mock, T. (2015). Plastid proteome prediction for diatoms and other algae with secondary plastids of the red lineage. *The Plant Journal, 81*(3), 519–528.

Hackett, J. D., Maranda, L., Yoon, H. S., & Bhattacharya, D. (2003). Phylogenetic evidence for the cryptophyte origin of the plastid of Dinophysis (Dinophysiales, Dinophyceae). *Journal of Phycology, 39*(2), 440–448.

Hackett, J. D., Yoon, H. S., Li, S., Reyes-Prieto, A., Rümmele, S. E., & Bhattacharya, D. (2007). Phylogenomic analysis supports the monophyly of cryptophytes and haptophytes and the association of Rhizaria with chromalveolates. *Molecular Biology and Evolution, 24*(8), 1702–1713.

He, D., Sierra, R., Pawlowski, J., & Baldauf, S. L. (2016). Reducing long-branch effects in multi-protein data uncovers a close relationship between Alveolata and Rhizaria. *Molecular Phylogenetics and Evolution, 101*, 1–7.

Hehenberger, E., Burki, F., Kolisko, M., & Keeling, P. J. (2016). Functional relationship between a dinoflagellate host and its diatom endosymbiont. *Molecular Biology and Evolution, 33*(9), 2376–2390.

Heiny, S. R., Pautz, S., Recker, M., & Przyborski, J. M. (2014). Protein traffic to the *Plasmodium falciparum* apicoplast: Evidence for a sorting branch point at the Golgi. *Traffic, 15*(12), 1290–1304.

Hempel, F., Bullmann, L., Lau, J., Zauner, S., & Maier, U. G. (2009). ERAD-derived preprotein transport across the second outermost plastid membrane of diatoms. *Molecular Biology and Evolution, 26*(8), 1781–1790.

Hempel, F., Felsner, G., & Maier, U. G. (2010). New mechanistic insights into pre-protein transport across the second outermost plastid membrane of diatoms. *Molecular Microbiology, 76*(3), 793–801.

Hirakawa, Y., Burki, F., & Keeling, P. J. (2012). Genome-based reconstruction of the protein import machinery in the secondary plastid of a chlorarachniophyte alga. *Eukaryotic Cell, 11*(3), 324–333.

Howe, C. J., Barbrook, A. C., Nisbet, R. E. R., Lockhart, P. J., & Larkum, A. W. D. (2008). The origin of plastids. *Philosophical Transactions of the Royal Society B: Biological Sciences, 363*(1504), 2675–2685.

Inagaki, Y., Dacks, J. B., Doolittle, W. F., Watanabe, K. I., & Ohama, T. (2000). Evolutionary relationship between dinoflagellates bearing obligate diatom endosymbionts: Insight into tertiary endosymbiosis. *International Journal of Systematic and Evolutionary Microbiology, 50*(6), 2075–2081.

Janouškovec, J., Gavelis, G. S., Burki, F., Dinh, D., Bachvaroff, T. R., Gornik, S. G., et al. (2017). Major transitions in dinoflagellate evolution unveiled by phylotranscriptomics. *Proceedings of the National Academy of Sciences of the United States of America, 114*(2), E171–E180.

Janouškovec, J., Horák, A., Oborník, M., Lukeš, J., & Keeling, P. J. (2010). A common red algal origin of the apicomplexan, dinoflagellate, and heterokont plastids. *Proceedings of the National Academy of Sciences of the United States of America, 107*(24), 10949–10954.

Janouškovec, J., Liu, S.-L., Martone, P. T., Carré, W., Leblanc, C., Collén, J., et al. (2013). Evolution of red algal plastid genomes: Ancient architectures, introns, horizontal gene transfer, and taxonomic utility of plastid markers. *PLoS One, 8*(3), e59001.

Janouškovec, J., Tikhonenkov, D. V., Burki, F., Howe, A. T., Kolisko, M., Mylnikov, A. P., et al. (2015). Factors mediating plastid dependency and the origins of parasitism in apicomplexans and their close relatives. *Proceedings of the National Academy of Sciences of the United States of America, 112*(33), 10200–10207.

Janouškovec, J., Tikhonenkov, D. V., Mikhailov, K. V., Simdyanov, T. G., Aleoshin, V. V., Mylnikov, A. P., et al. (2013). Colponemids represent multiple ancient alveolate lineages. *Current Biology, 23*(24), 2546–2552.

Kamikawa, R., Tanifuji, G., Kawachi, M., Miyashita, H., Hashimoto, T., & INAGAKI, Y. (2015). Plastid genome-based phylogeny pinpointed the origin of the green-colored plastid in the dinoflagellate Lepidodinium chlorophorum. *Genome Biology and Evolution*, 7(4), 1133–1140.

Keeling, P. J. (2004). Diversity and evolutionary history of plastids and their hosts. *American Journal of Botany*, *91*, 1481–1493.

Keeling, P. J. (2009). Chromalveolates and the evolution of plastids by secondary endosymbiosis. *The Journal of Eukaryotic Microbiology*, *56*(1), 1–8.

Keeling, P. J. (2010). The endosymbiotic origin, diversification and fate of plastids. *Philosophical Transactions of the Royal Society B: Biological Sciences*, *365*(1541), 729–748.

Keeling, P. J. (2013). The number, speed, and impact of plastid endosymbioses in eukaryotic evolution. *Annual Review of Plant Biology*, *64*(1), 27.1–27.25.

Keeling, P. J., Burger, G., Durnford, D. G., Lang, B. F., Lee, R. W., Pearlman, R. E., et al. (2005). The tree of eukaryotes. *Trends in Ecology & Evolution*, *20*(12), 670–676.

Keeling, P. J., & Palmer, J. D. (2008). Horizontal gene transfer in eukaryotic evolution. *Nature Reviews Genetics*, *9*, 605–618.

Kim, E., & Maruyama, S. (2014). A contemplation on the secondary origin of green algal and plant plastids. *Acta Societatis Botanicorum Poloniae*, *83*(4), 331–336.

Ku, C., Nelson-Sathi, S., Roettger, M., Sousa, F. L., Lockhart, P. J., Bryant, D., et al. (2015). Endosymbiotic origin and differential loss of eukaryotic genes. *Nature*, *524*(7566), 427–432.

Lane, C. E., & Archibald, J. M. (2008). The eukaryotic tree of life: Endosymbiosis takes its TOL. *Trends in Ecology & Evolution*, *23*(5), 268–275.

Larkum, A. W. D., Lockhart, P. J., & Howe, C. J. (2007). Shopping for plastids. *Trends in Plant Science*, *12*(5), 189–195.

Lee, J., Cho, C. H., Park, S. I., Choi, J. W., Song, H. S., West, J. A., et al. (2016). Parallel evolution of highly conserved plastid genome architecture in red seaweeds and seed plants. *BMC Biology*, *14*(1), 75.

Li, B., Lopes, J. S., Foster, P. G., Embley, T. M., & Cox, C. J. (2014). Compositional biases among synonymous substitutions cause conflict between gene and protein trees for plastid origins. *Molecular Biology and Evolution*, *31*(7), 1697–1709.

Liu, X., Hempel, F., Stork, S., Bolte, K., Moog, D., Heimerl, T., et al. (2016). Addressing various compartments of the diatom model organism *Phaeodactylum tricornutum* via subcellular marker proteins. *Algal Research*, *20*, 249–257.

Mackiewicz, P., Bodył, A., & Gagat, P. (2012). Protein import into the photosynthetic organelles of Paulinella chromatophora and its implications for primary plastid endosymbiosis. *Symbiosis*, *58*(1–3), 99–107.

Mackiewicz, P., & Gagat, P. (2014). Monophyly of Archaeplastida supergroup and relationships among its lineages in the light of phylogenetic and phylogenomic studies. Are we close to a consensus? *Acta Societatis Botanicorum Poloniae*, *83*(4), 263–280.

Marin, B., Nowack, E. C. M., & Melkonian, M. (2005). A plastid in the making: Evidence for a second primary endosymbiosis. *Protist*, *156*(1), 125–132.

Martin, W. F., Garg, S., & Zimorski, V. (2015). Endosymbiotic theories for eukaryote origin. *Philosophical Transactions of the Royal Society of London. Series B, Biological Sciences*, *370*(1678), 20140330.

Martin, W. F., Rujan, T., Richly, E., Hansen, A., Cornelsen, S., Lins, T., et al. (2002). Evolutionary analysis of *Arabidopsis*, cyanobacterial, and chloroplast genomes reveals plastid phylogeny and thousands of cyanobacterial genes in the nucleus. *Proceedings of the National Academy of Sciences of the United States of America*, *99*(19), 12246–12251.

Massana, R., del Campo, J., Sieracki, M. E., Audic, S., & Logares, R. (2014). Exploring the uncultured microeukaryote majority in the oceans: Reevaluation of ribogroups within stramenopiles. *The ISME Journal*, *8*(4), 854–866.

McFadden, G. I., Reith, M. E., Munholland, J., & Lang-Unnasch, N. (1996). Plastid in human parasites. *Nature, 381*(6582), 482.
Moreira, D., & Deschamps, P. (2014). What was the real contribution of endosymbionts to the eukaryotic nucleus? Insights from photosynthetic eukaryotes. *Cold Spring Harbor Perspectives in Biology, 6*(7), a016014.
Moreira, D., von der Heyden, S., Bass, D., López-García, P., Chao, E. E.-Y., & Cavalier-Smith, T. (2007). Global eukaryote phylogeny: Combined small- and large-subunit ribosomal DNA trees support monophyly of Rhizaria, Retaria and Excavata. *Molecular Phylogenetics and Evolution, 44*(1), 255–266.
Moustafa, A., Beszteri, B., Maier, U. G., Bowler, C., Valentin, K., & Bhattacharya, D. (2009). Genomic footprints of a cryptic plastid endosymbiosis in diatoms. *Science, 324*(5935), 1724–1726.
Nikolaev, S. I., Berney, C., Fahrni, J. F., Bolivar, I., Polet, S., Mylnikov, A. P., et al. (2004). The twilight of Heliozoa and rise of Rhizaria, an emerging supergroup of amoeboid eukaryotes. *Proceedings of the National Academy of Sciences of the United States of America, 101*(21), 8066–8071.
Not, F., Valentin, K., Romari, K., Lovejoy, C., Massana, R., Töbe, K., et al. (2007). Picobiliphytes: A marine picoplanktonic algal group with unknown affinities to other eukaryotes. *Science, 315*(5809), 253–255.
Nowack, E. C. M., & Grossman, A. R. (2012). Trafficking of protein into the recently established photosynthetic organelles of Paulinella chromatophora. *Proceedings of the National Academy of Sciences of the United States of America, 109*(14), 5340–5345.
Nowack, E. C. M., Melkonian, M., & Glöckner, G. (2008). Chromatophore genome sequence of *Paulinella* sheds light on acquisition of photosynthesis by eukaryotes. *Current Biology, 18*(6), 410–418.
Nowack, E. C. M., Price, D. C., Bhattacharya, D., Singer, A., Melkonian, M., & Grossman, A. R. (2016). Gene transfers from diverse bacteria compensate for reductive genome evolution in the chromatophore of *Paulinella chromatophora. Proceedings of the National Academy of Sciences of the United States of America, 113*(43), 12214–12219.
Nowack, E. C. M., Vogel, H., Groth, M., Grossman, A. R., Melkonian, M., & Glöckner, G. (2011). Endosymbiotic gene transfer and transcriptional regulation of transferred genes in *Paulinella chromatophora. Molecular Biology and Evolution, 28*(1), 407–422.
Nozaki, H. (2005). A new scenario of plastid evolution: Plastid primary endosymbiosis before the divergence of the "Plantae," emended. *Journal of Plant Research, 118*(4), 247–255.
Okamoto, N., Chantangsi, C., Horák, A., Leander, B. S., & Keeling, P. J. (2009). Molecular phylogeny and description of the novel katablepharid *Roombia truncata* gen. et sp. nov., and establishment of the Hacrobia taxon nov. *PLoS One, 4*(9), e7080.
Palmer, J. D. (2000). Molecular evolution: A single birth of all plastids? *Nature, 405*(6782), 32–33.
Palmer, J. D. (2003). The symbiotic birth and spread of plastids: How many times and whodunit? *Journal of Phycology, 39*, 4–11.
Parfrey, L. W., Lahr, D. J. G., Knoll, A. H., & Katz, L. A. (2011). Estimating the timing of early eukaryotic diversification with multigene molecular clocks. *Proceedings of the National Academy of Sciences of the United States of America, 108*(33), 13624–13629.
Petersen, J., Ludewig, A.-K., Michael, V., Bunk, B., Jarek, M., Baurain, D., et al. (2014). *Chromera velia*, endosymbioses and the rhodoplex hypothesis—Plastid evolution in cryptophytes, alveolates, stramenopiles, and haptophytes (CASH lineages). *Genome Biology and Evolution, 6*(3), 666–684.
Philippe, H., Brinkmann, H., Lavrov, D. V., Littlewood, D. T. J., Manuel, M., Wörheide, G., et al. (2011). Resolving difficult phylogenetic questions: Why more sequences are not enough. *PLoS Biology, 9*(3), e1000602.

Ponce-Toledo, R. I., Deschamps, P., López-García, P., Zivanovic, Y., Benzerara, K., & Moreira, D. (2017). An early-branching freshwater cyanobacterium at the origin of plastids. *Current Biology, 27*(3), 386–391.

Price, D. C., Chan, C. X., Yoon, H. S., Yang, E. C., Qiu, H., Weber, A. P. M., et al. (2012). *Cyanophora paradoxa* genome elucidates origin of photosynthesis in algae and plants. *Science, 335*(6070), 843–847.

Qiu, H., Lee, J. M., Yoon, H. S., & Bhattacharya, D. (2017). Hypothesis: Gene-rich plastid genomes in red algae may be an outcome of nuclear genome reduction. *Journal of Phycology, 53*(3), 715–719.

Reyes-Prieto, A., Weber, A. P. M., & Bhattacharya, D. (2007). The origin and establishment of the plastid in algae and plants. *Annual Review of Genetics, 41*, 147–168.

Rodríguez-Ezpeleta, N., Brinkmann, H., Burey, S. C., Roure, B., Burger, G., Löffelhardt, W., et al. (2005). Monophyly of primary photosynthetic eukaryotes: Green plants, red algae, and glaucophytes. *Current Biology, 15*(14), 1325–1330.

Rodríguez-Ezpeleta, N., Brinkmann, H., Burger, G., Roger, A. J., Gray, M. W., Philippe, H., et al. (2007). Toward resolving the eukaryotic tree: The phylogenetic positions of jakobids and cercozoans. *Current Biology, 17*(16), 1420–1425.

Rogers, M. B., Gilson, P. R., Su, V., McFadden, G. I., & Keeling, P. J. (2007). The complete chloroplast genome of the chlorarachniophyte *Bigelowiella natans*: Evidence for independent origins of chlorarachniophyte and euglenid secondary endosymbionts. *Molecular Biology and Evolution, 24*(1), 54–62.

Sanchez Puerta, M. V., & Delwiche, C. F. (2008). A hypothesis for plastid evolution in chromalveolates. *Journal of Phycology, 44*(5), 1097–1107.

Schönknecht, G., Chen, W.-H., Ternes, C. M., Barbier, G. G., Shrestha, R. P., Stanke, M., et al. (2013). Gene transfer from bacteria and archaea facilitated evolution of an extremophilic eukaryote. *Science, 339*(6124), 1207–1210.

Seenivasan, R., Sausen, N., Medlin, L. K., & Melkonian, M. (2013). *Picomonas judraskeda* gen. et sp. nov.: The first identified member of the Picozoa phylum nov., a widespread group of picoeukaryotes, formerly known as "picobiliphytes". *PLoS One, 8*(3), e59565.

Ševčíková, T., Horák, A., Klimeš, V., Zbránková, V., Demir-Hilton, E., Sudek, S., et al. (2015). Updating algal evolutionary relationships through plastid genome sequencing: Did alveolate plastids emerge through endosymbiosis of an ochrophyte? *Scientific Reports, 5*(1), 10134.

Shalchian-Tabrizi, K., Eikrem, W., Klaveness, D., Vaulot, D., Minge, M. A., Le Gall, F., et al. (2006). Telonemia, a new protist phylum with affinity to chromist lineages. *Proceedings of the Royal Society B: Biological Sciences, 273*(1595), 1833–1842.

Shi, L.-X., & Theg, S. M. (2013). The chloroplast protein import system: From algae to trees. *Biochimica et Biophysica Acta, 1833*(2), 314–331.

Sibbald, S. J., & Archibald, J. M. (2017). More protist genomes needed. *Nature Ecology & Evolution, 1*(5), 1–3.

Sierra, R., Cañas-Duarte, S. J., Burki, F., Schwelm, A., Fogelqvist, J., Dixelius, C., et al. (2016). Evolutionary origins of Rhizarian parasites. *Molecular Biology and Evolution, 33*(4), 980–983.

Simpson, A. G. B. (2003). Cytoskeletal organization, phylogenetic affinities and systematics in the contentious taxon Excavata (Eukaryota). *International Journal of Systematic and Evolutionary Microbiology, 53*(Pt. 6), 1759–1777.

Stiller, J. W. (2014). Toward an empirical framework for interpreting plastid evolution. *Journal of Phycology, 50*(3), 462–471.

Stiller, J. W., & Hall, B. D. (1997). The origin of red algae: Implications for plastid evolution. *Proceedings of the National Academy of Sciences of the United States of America, 94*(9), 4520–4525.

Stiller, J. W., Reel, D. C., & Johnson, J. C. (2003). A single origin of plastids revisited: Convergent evolution in organellar genome content. *Journal of Phycology*, *39*(1), 95–105.

Stiller, J. W., Schreiber, J., Yue, J., Guo, H., Ding, Q., & Huang, J. (2014). The evolution of photosynthesis in chromist algae through serial endosymbioses. *Nature Communications*, *5*, 5764.

Stork, S., Lau, J., Moog, D., & Maier, U. G. (2013). Three old and one new: Protein import into red algal-derived plastids surrounded by four membranes. *Protoplasma*, *250*(5), 1013–1023.

Taylor, F. J. R. (1974). Implications and extensions of the serial endosymbiosis theory of the origin of eukaryotes. *Taxon*, *23*(2/3), 229.

Tengs, T., Dahlberg, O. J., Shalchian-Tabrizi, K., Klaveness, D., Rudi, K., Delwiche, C. F., et al. (2000). Phylogenetic analyses indicate that the 19'hexanoyloxy-fucoxanthin-containing dinoflagellates have tertiary plastids of haptophyte origin. *Molecular Biology and Evolution*, *17*(5), 718–729.

Timmis, J. N., Ayliffe, M. A., Huang, C. Y., & Martin, W. F. (2004). Endosymbiotic gene transfer: Organelle genomes forge eukaryotic chromosomes. *Nature Reviews Genetics*, *5*(2), 123–135.

Waller, R. F., & McFadden, G. I. (2005). The apicoplast: A review of the derived plastid of apicomplexan parasites. *Current Issues in Molecular Biology*, 7(1), 57–79.

Woehle, C., Dagan, T., Martin, W. F., & Gould, S. B. (2011). Red and problematic green phylogenetic signals among thousands of nuclear genes from the photosynthetic and Apicomplexa-related *Chromera velia*. *Genome Biology and Evolution*, *3*, 1220–1230.

Yabuki, A., Kamikawa, R., Ishikawa, S. A., Kolisko, M., Kim, E., Tanabe, A. S., et al. (2014). *Palpitomonas bilix* represents a basal cryptist lineage: Insight into the character evolution in Cryptista. *Scientific Reports*, *4*, 4641.

Yoon, H. S., Hackett, J. D., Ciniglia, C., Pinto, G., & Bhattacharya, D. (2004). A molecular timeline for the origin of photosynthetic eukaryotes. *Molecular Biology and Evolution*, *21*(5), 809–818.

Yoon, H. S., Price, D. C., Stepanauskas, R. R., Rajah, V. D., Sieracki, M. E., Wilson, W. H., et al. (2011). Single-cell genomics reveals organismal interactions in uncultivated marine protists. *Science*, *332*(6030), 714–717.

Zimorski, V., Ku, C., Martin, W. F., & Gould, S. B. (2014). Endosymbiotic theory for organelle origins. *Current Opinion in Microbiology*, *22*, 38–48.

CHAPTER TWO

Let There Be Light: A Contemporary Primer on Primary Plastid Endosymbiosis

David R. Smith[1]
University of Western Ontario, London, ON, Canada
[1]Corresponding author: e-mail address: dsmit242@uwo.ca

Contents

Abstract

Endosymbiosis, more than any other process, perhaps, is the leading narrative upon which the history of eukaryotic evolution has been written. Primary endosymbiosis, which is the uptake of a prokaryote by another living cell, has arguably been *the* driving force for the origins and diversification of complex life on Earth. The genetic integration of, first, a nonphotosynthetic alphaproteobacterium and, later, a photosynthetic cyanobacterium into a eukaryotic cellular framework have shaped and altered the planet's biodiversity and biogeochemistry in countless ways, from the land, to the water, to the atmosphere. If you are alive today and reading these words, it is in no small part because of endosymbiosis. Like all eukaryotes, we are the product of an ancient endosymbiotic love affair, and for plants and algae the endosymbiotic romance was a complicated triangle. Here, I recount my own passions for the topic of endosymbiosis, highlighting past and present breakthroughs as well as some of the controversies and unanswered questions that have plagued the field. I focus on the evolution of primary plastids, their genomes, and the supergroup to which they are found (the Archaeplastida), including members that have lost photosynthetic capabilities but still retain a colourless plastid.

Advances in Botanical Research, Volume 84
ISSN 0065-2296
http://dx.doi.org/10.1016/bs.abr.2017.06.002

1. INTRODUCTION

When you think about the complexity of our natural world—plants using quantum mechanics for photosynthesis, for example—a smartphone begins to look like a pretty dumb object.

Jeff VanderMeer

Stroll through a flowered park, dip into a cool summer lake, or hike in a dense forest and you will find yourself in good company. I am talking about plants and algae, of course, but more specifically about chloroplasts—the factories of photosynthesis and hubs of countless crucial biochemical reactions. Yes, chloroplasts (and plastids as a whole) are one awesome eukaryotic organelle. Their pigments alone provide the world with much of its beautiful and mellifluous colours, not to mention their clever and life-sustaining ability to convert carbon dioxide and water into oxygen and sugar. Plastids are also the engines primarily responsible for generating the fruits, vegetables, and grains we eat, and they form the basis for the wine and beer many of us imbibe, some of us too often.

For me, the most intriguing thing about plastids is how they came to be. Their journey from a free-living, bacterial existence to integral and inalienable components of plants and algae involved countless winding, diverging, and colliding roads and is replete with whimsy, mystery, death, and disease. The story of plastids has many plots, many characters, and is still ongoing, as emphasized throughout the various chapters in this collection. And it is what first made me fall in love with biology.

Here, I recount some of what is currently known about the origins and early evolution of plastids, focusing first on the process by which they arose: primary endosymbiosis, which is when a prokaryote is taken up by another living cell. I discuss the immediate and downstream consequences of this momentous endosymbiotic event, and how it ultimately gave rise to the eukaryotic supergroup Archaeplastida, which is made up of red algae, glaucophyte algae, green algae, and land plants (Adl et al., 2012). I describe aspects of plastid genetics, genomics, and gene expression, topics which at first glance can appear straightforward and elementary as compared to nuclear systems but which have proven to be far from ordinary (Smith & Keeling, 2015). Although plastids are undoubtedly highly specialized organelles, in certain lineages they have drifted down peculiar evolutionary roads, including ones that have led to the restructuring of the photosynthetic apparatus (Morgan-Kiss, Priscu, Pocock, Gudynaite-Savitch, & Hüner, 2006) or the outright loss of photosynthesis (Figueroa-Martinez, Nedelcu, Smith, & Reyes-Prieto, 2015). In other instances, they have adopted costly and bureaucracy-ridden infrastructures for gene editing (Knoop, 2011), acquired massive genomes made up almost entirely of noncoding DNA (Smith & Lee, 2010), or even forfeited their DNA and gene expression systems altogether (Molina et al., 2014; Smith & Asmail, 2014).

Despite the near-universal acceptance of the endosymbiotic theory, certain questions about plastids and plastid genome evolution remain unanswered and are mired in debate, confusion, and controversy. This is especially true for the steps leading up to and the nuances involved in the genetic merger between a cyanobacterium and a nonphotosynthetic eukaryote. I have witnessed many heated conversations and the occasional all-out shouting match surrounding the topic of whether, for example, the host was "primed" before the primary endosymbiotic event that produced plastids and the role that pathogenic bacteria might have played in this process. But I am getting ahead of myself. Before describing the subtleties of plastid evolution and primary endosymbiosis, one must first cover the basics.

2. THE ABC's OF PRIMARY ENDOSYMBIOSIS

You fit into me
like a hook into an eye
a fish hook
an open eye

Margaret Atwood

When I was an undergraduate student, I detested plant biology classes and anything to do with algae. I still cringe at having to memorize the life cycle of a fern, the encyclopaedic vascular system of a tree, and the multifarious flagellar apparatus of *Chlamydomonas*—and don't get me started on the electron transport chain of photosynthesis. But my interest in plants and algae grew immensely when I took advanced genetics and learned in detail about the endosymbiotic theory for the origins of mitochondria and plastids, which was made popular by Lynn Margulis in the late 1960s (Sagan, 1967), but articulated in various forms much earlier (Wallin, 1927), including by Russian biologist Mereschkowsky (1910). The endosymbiotic theory (or symbiogenesis) holds that the mitochondrion and the plastid were once free-living bacteria taken up via primary endosymbiosis by a host cell (Archibald, 2014). These two endosymbiotic events appear to have occurred about 1.8 and 1.5 billion years ago, respectively (Gray, 2012; Yoon, Hackett, Ciniglia, Pinto, & Bhattacharya, 2004). And as every freshman biology student can tell you, mitochondria are closely related to present-day alphaproteobacteria (Gray, 2012), whereas chloroplasts show strong phylogenetic affinity to contemporary nitrogen-fixing filamentous cyanobacteria (Deusch et al., 2008)—but see de Vries and Archibald (2017).

I can still picture my advanced genetics professor writing on the blackboard the various steps involved in a primary endosymbiosis: feeding, entrapment, sharing, endosymbiotic gene transfer and gene loss, relinquishing control, and eventually complete genetic integration. "Most of you are fixated on procreation", she said, staring down the boys in the class. "You are obsessed with the act of (and the act of avoiding) two becoming two plus one. But look closely, with symbiogenesis we're talking about the opposite: two distinct lifeforms becoming one, and that is why it is so extraordinary". Indeed, in a single 50-min lecture, this talented teacher ignited in me a deep passion for endosymbiosis, eukaryogenesis, and organellogenesis that persists to this day.

Up to that point in my undergraduate education, endosymbiotic theory had been an afterthought in my various biology courses—a subtext to an image of a eukaryotic cell, a single slide of a Pac-Man-esque organism eating a green dot, two simple circles labelled mitochondrial and plastid DNA (mtDNA and ptDNA), end of story. But in advanced genetics I learnt that primary endosymbiosis is not an afterthought or footnote to eukaryotic evolution, rather, it is a fundamental topic in biology, a driving force for the origins and diversification of complex life on Earth, and a leading narrative upon which the history of eukaryotes has been written (Archibald, 2014).

The endosymbiotic theory has been discussed, debated, and lauded in hundreds of research articles, review papers, textbooks, and commentaries over the past half century. There are now so much data in support of the theory that it has arguably completed the journey from hypothesis to scientific fact. Today's scientists, with few exceptions (Kurland, Collins, & Penny, 2006), whole-heartedly accept as truth the endosymbiotic origins of mitochondria and plastids (Archibald, 2015). Consequently, contemporary evolutionary biologists are no longer concerned with proving or disproving symbiogenesis, but are instead focused on understanding the key events leading up to and following the primary endosymbioses that generated mitochondria and plastids, as well as the various eukaryote–eukaryote endosymbioses responsible for the horizontal spread of plastids across the eukaryotic domain (Zimorski, Ku, Martin, & Gould, 2014).

What was the nature of the host cells that acquired the mitochondrial and plastid endosymbionts? What precise alphaproteobacterial and cyanobacterial lineages were the progenitors of the mitochondrion and the plastid? What were the driving evolutionary forces that led to the success of these two ancient endosymbioses? Were they adaptive or nonadaptive? Driven by sharing or, more provocatively, parasitism? Did other earlier endosymbiotic events and/or waves of horizontal gene transfer play a role in the successful integration of mitochondria and plastids? Breakthroughs in molecular sequencing technologies and bioinformatics (Metzker, 2010), including single-cell sequencing, metagenomics, and phylogenomics (Kolisko, Boscaro, Burki, Lynn, & Keeling, 2014; Moran, 2009), are helping scientists address these and other important questions, are providing unexpected and occasionally paradigm-shifting answers, and are ultimately leading to a profound understanding of endosymbiosis at its deepest levels.

Take, for instance, the mitochondrial endosymbiotic event. In just the past 3 years our perception of the host cell and host lineage has improved considerably, and many of these advancements can be summarized in a single word: Loki. The Lokiarchaeota (Loki for short) is a newly discovered archaeal phylum that is more closely related to eukaryotes than any other prokaryotic lineage sampled to date (Spang et al., 2015). Metagenomic analyses of Loki uncovered genes that were previously thought to be unique to eukaryotes, including genes for endocytosis and/or phagocytosis (Embley & Williams, 2015). In other words, Loki likely resembles and descends directly from the archaeal ancestor of all eukaryotes, an ancestor that appears to have harboured the machinery needed to carry out primary endosymbiosis. The discovery of Loki further supports to the notion that the evolutionary transition from prokaryotic to eukaryotic life was predominantly driven by the primary engulfment of an alphaproteobacterium by an archaeal host possessing some eukaryotic-like features. But this does not tell the whole story.

Shortly after the discovery of Loki, Pittis and Gabaldón (2016) using phylogenomics traced the evolutionary histories of proteins believed to be present in the last eukaryotic common ancestor (LECA). In addition to identifying archaeal- and alphaproteobacterial-related proteins, the authors uncovered a third class of bacterial LECA proteins, predating those acquired from the mitochondrial endosymbiont, and some which function in the endoplasmic reticulum and Golgi apparatus. Pittis and Gabaldón (2016) believed that the genes for this third class of proteins came from an earlier bacterial endosymbiont, one that existed before that which gave rise to the mitochondrion, or from waves of horizontal gene transfer. Whatever their origin, these genes might have endowed the host lineage with the complexity needed for the eventual genetic merger with an alphaproteobacterium.

The notion that genes obtained via endosymbiotic and/or horizontal gene transfer can provide the foundation for future endosymbiotic events is a reoccurring theme in evolutionary biology, including plastid evolution. But it is also a controversial and hotly debated idea. The suggestion that the mitochondrial endosymbiotic event occurred later in the evolution of complex cells than previously thought and proceeded an earlier endosymbiosis was embraced by many in the scientific community but vehemently rejected by others. One research team went so far as to say that the paper of Pittis and Gabaldón (2016) had "multiple fatal flaws founded in inappropriate statistical methods and analysis, in addition to erroneous interpretations" (Martin et al., 2016). This debate echoes a continuous thread throughout the study of evolutionary endosymbioses: researchers can be heavily invested in specific views and hypotheses and not always accepting of alternative perspectives, often to the detriment of science. The legend goes that Lynn Margulis' seminal paper on endosymbiotic theory (Sagan, 1967) was not only highly criticized but also rejected 15 times before finally being accepted by the *Journal of Theoretical Biology*. As discussed later, the soap opera surrounding the primary acquisition of plastids within the eukaryotic domain is no less controversial, multifaceted, and debated as its mitochondrial counterpart.

3. THE POLYCHROMATIC PUZZLE PIECES OF PLASTID PRIMARY ENDOSYMBIOSIS

Watermelons:
Green Buddhas
On the fruit stand.
We eat the smile
And spit out the teeth.

Charles Simic

It makes intuitive sense that a heterotrophic organism would want to enslave a bacterium capable of converting sunlight into usable energy. Thus, it is not surprising that around 1.5 billion years ago a unicellular, nonphotosynthetic protist did just that, starting down the road towards eukaryotic phototrophy. The onramp to this road could have been quite simple: for example, a phagotrophic predatory lifestyle involving cyanobacteria as the engulfed prey. However, the evolutionary leap from a predator–prey relationship to a symbiotic photosynthetic one is not so clear. Textbooks like to depict this transition as a single step—a jump from a free-living cyanobacterium to an entrapped and integrated endosymbiont. But, in fact, the segue to a full-fledged host–endosymbiont partnership was unquestionably more drawn out than most textbooks would have us believe, occurring at a population level and on an evolutionary timescale, and likely involving multiple contributing partners rather than being a singular event with only two players.

The early stages of this long-winded metamorphosis from food to endosymbiont to photosynthetic organelle are best depicted by the "shopping bag model" (Larkum, Lockhart, & Howe, 2007) of primary plastid evolution:

> *It seems unlikely that the stable [cyanobacterial] symbiont ultimately acquired by the host cell would be the first one it had ever acquired. The acquisition would almost certainly have been preceded by the uptake of other photosynthetic organisms. … early rounds of failed endosymbiosis, with some would-be endosymbionts eventually lysing and liberating DNA into the cytosol, would result in integration of endosymbiont DNA into the nuclear genome. This DNA would have persisted in the nucleus for a period of time, even if there were no longer functional symbionts in the host cytoplasm. If, finally, a symbiont [was] able to establish a balanced relationship with the host, the reservoir of sequences in the host nucleus that were derived from previous photosynthetic organisms would have provided a pool of sequences to encode proteins to be imported into the newly established plastid.*
>
> ***Howe, Barbrook, Nisbet, Lockhart, and Larkum (2008)***

In short, the shopping bag scenario implies that the protein machinery of plastids has a mixed origin with most proteins originating from the successful cyanobacterial endosymbiont but at least some coming from earlier unsuccessful endosymbionts.

It was long assumed that the preliminary events of primary plastid endosymbiosis, including those described in the shopping bag model, occurred in salt water—somewhere in the saline oceans of the mid-Proterozoic era (Deusch et al., 2008). Just recently, however, this assumption has been laid to rest, and it is now thought that eukaryotic phototrophy first evolved in freshwater and is directly linked to a newly uncovered clade of cyanobacteria called *Gloeomargarita* (Ponce-Toledo et al., 2017).

Like with Loki, it is thanks to the explorations of undersampled environments and the characterization of new organisms that we now know about *Gloeomargarita lithophora*. This recently discovered, deep-branching, and biofilm-forming cyanobacterium might hold countless untold secrets to the beginnings of primary plastids. *G. lithophora*, which is the only member of the *Gloeomargarita* clade currently in culture, was first isolated in 2012 from a freshwater alkaline lake in the Oriental Basin of Mexico (Couradeau et al., 2012). But its claim to fame is its affiliation to plastids. Using a comprehensive phylogenomic data set, including plastid-encoded proteins, nucleus-encoded proteins of plastid origin, and extensive genomic data from cyanobacteria, Ponce-Toledo et al. (2017) showed that *G. lithophora* is the closest prokaryotic relative of primary plastids yet found. This implies that plastids evolved from an ancestor of the *Gloeomargarita*. What's more, the entire *Gloeomargarita* group appears to be restricted to freshwater environments, suggesting that eukaryotic photosynthesis first emerged in a terrestrial freshwater setting. This hypothesis is further supported by the fact that the earliest diverging archaeplastid lineage—the Glaucophyta—is exclusive to freshwater ecosystems (Delwiche & Cooper, 2015).

The notion that primary plastid endosymbiosis began as an assorted shopping bag with the central participant being an ancestral member of the *Gloeomargarita* is compelling. But a more provocative suggestion is that a *Chlamydia* pathogen had a helping hand in primary plastid evolution. A major hurdle en route to a successful endosymbiosis is the ability to overcome host defences. This presents a slight problem when pondering the genesis of primary plastids: present-day cyanobacteria, including *G. lithophora*, do not have the genetic capacity to evade host defences or interact with the host cellular machinery (Ball, Bhattacharya, & Weber, 2016a). So then how did the cyanobacterial progenitor of the plastid endure the initial stages of endosymbiosis? Well, as the shopping bag model posits, many early would-be cyanobacterial endosymbionts probably did not survive but over time gained ground with each subsequent failed attempt (Larkum, Lockhart, & Howe, 2007). An alternative view comes from the observation that the genomes of various archaeplastid plants and algae harbour several dozen genes of seemingly chlamydial origin (Huang, & Gogarten, 2007). These finding spurred the hypothesis "that a chlamydial bacterium entered the host cell together with a cyanobacterium ... [allowing] the cyanobacterium to escape host defenses and establish a tripartite symbiosis through the help of chlamydial-encoded effector proteins and transporters" (Ball, Bhattacharya, & Weber, 2016a). Simply put: eukaryotic photosynthesis might have resulted from a ménage à trois between a heterotrophic protist, a cyanobacterium, and a chlamydial bacterium, whereby key metabolic genes from the latter enabled the symbiotic capture of the cyanobacterium. It is argued that these key genes were ultimately horizontally transferred to the host, thus permitting the eventual loss of the chlamydial partner from the relationship (Ball et al., 2013).

Unsurprisingly, not everyone supports such a salacious view of plastid evolution (Gould, 2016). A subsequent phylogenetic study found a mosaic origin for the putative chlamydial-derived enzymes and no strong evidence for *Chlamydia*-to-host gene transfer (Domman, Horn, Embley, & Williams, 2015). The authors of this study concluded that *Chlamydia* did not have any role in establishing the primary plastid endosymbiosis and noted the following:

> *There is a deeper problem with inferring a special explanation for the presence of putative chlamydial genes on plant genomes ... [Recent] studies have demonstrated that in addition to organellar genes shared with* Cyanobacteria *and* Alphaproteobacteria, *the Archaeplastida share more genes with* Gammaproteobacteria, Actinobacteria, Deltaproteobacteria, Bacilli, Bacteroidetes *and* Betaproteobacteria *than with* Chlamydiae. *Given the extent of HGT, particularly of metabolic genes, among major cellular groups ... these patterns of gene sharing—including those involving* Chlamydiae*—are most simply explained as a mixture of genuine [horizontal gene transfer] and tree reconstruction artefacts.*
>
> ***Domman, Horn, Embley, and Williams (2015)***

Despite these data and that there are no known cases of *Chlamydia* infecting a species from the Archaeplastida, the authors of the "ménage à trois" (MAT) hypothesis—as it has become known—are, for now, sticking with their story: "We disagree that the phylogenetic analyses of Domman et al. (2015) reject our hypothesis ... [their trees] are very similar to ours and emphasize uncertainties that we have accounted for through straightforward interpretations of carbohydrate biochemistry" (Ball, Bhattacharya, & Weber, 2016b). The major challenge facing these hypotheses—and any hypothesis—about primary plastid endosymbiosis is that the events in question happened so long ago. If only there was a more recent primary endosymbiosis between a heterotrophic protist and a cyanobacterium that could be scrutinized. As anyone who works in this field will know, there is, and it is found within the rhizarian genus *Paulinella*.

4. *PAULINELLA CHROMATOPHORA*: A PRIMARY PLASTID REVIVAL TOUR

Outside around the side
Form a circle forward I'm an
Outsider on the side

Formerly a farm boy
Inside I'm on the side
I'm divided undecided
Back then around again
Second time's a charm, boy

Enter the Haggis—Canadian Celtic rock band

On the first day of my 4-year undergraduate genome evolution and endosymbiosis course, I always ask the class the same question: "Raise your hand if you've ever heard the word *Paulinella*?" Not once in my years of teaching this course have I seen a single hand go up. "Come on", I say, "*Paulinella chromatophora*, anybody? I'll even make it easy for you: it is unicellular, likes sunlight, lives in freshwater, belongs to the supergroup Rhizaria, the phylum Cercozoa ..." Still no hands. "It's an alga, for Pete's sake!" Even more blank faces.

It never ceases to amaze me how little undergraduate biology students are taught about eukaryotic microbes, and how excited they get when given the opportunity to learn about this broad assemblage of life called protists. When I finally tell the class that the primary endosymbiotic event that spawned the astonishing diversity of algae and land plants we see today is being played out again in *P. chromatophora*, their eyes go big and half of them are hooked. The other half are hooked when I go to the whiteboard and draw a long horizontal arrow from the words "distinct lifeform" to "organelle" and ask them to tell me at what precise point does the former become the latter. "Wherever that point falls on this arrow", I exclaim, "it is where the photosynthetic endosymbiont of *P. chromatophora* is found".

Yes, *P. chromatophora* deserves to be at the forefront of any course or review on endosymbiosis. Between 60 and 200 million years ago, the ancestor of this little-known amoeboid alga transitioned from a heterotrophic bacterivorous existence, sustained in part by feeding on cyanobacteria, to a phototrophic one, dependent on a cyanobacterial endosymbiont called a chromatophore (Nowack, 2014). *P. chromatophora* is truly one of a kind, being the only known example of primary acquisition of a photosynthetic organelle outside of that which generated the Archaeplastida, and about a billion years more recent, allowing researchers to study photosynthetic organellogenesis in real time. Unlike with primary plastid endosymbiosis, the shift to a phototrophic existence in *Paulinella* likely occurred in a marine environment through the uptake of an α-cyanobacterium related to *Synechococcus* and *Cyanobium* species (Nowack, 2014) (note: *Gloeomargarita* species belong to the β-cyanobacterial clade). Two additional photosynthetic *Paulinella* algae have been identified, including the marine *P. longichromatophora* (Kim & Park, 2016), but there is little doubt that the chromatophores of this genus have a single common origin (Yoon et al., 2009). Presently, there is no evidence of a chlamydial pathogen having any part in the chromatophore endosymbiotic event, but, as described later, endosymbiont gene loss might be driving the fixation of horizontally acquired bacterial genes in the host nuclear genome (Nowack et al., 2016).

What's been learned from examining the burgeoning *Paulinella* primary endosymbiosis? First, many of the concepts and conclusions formed from studying the procurement of primary plastids are echoed in the data from *P. chromatophora*, including, for example, the vital roles of intracellular gene loss, gene transfer, and protein targeting in endosymbiosis and organellogenesis (Gagat, Bodył, & Mackiewicz, 2016). Measuring around one million base pairs, the chromatophore genome is two-thirds smaller than that of its cyanobacterial ancestor (Reyes-Prieto et al., 2010). This severe genomic reduction is largely due to endosymbiotic gene loss, including the disposal of entire biosynthetic pathways for various amino acids and cofactors (Nowack, 2014), thus permanently securing the chromatophore to its host. There are also at least 30 chromatophore genes that have successfully relocated to the host nuclear genome, some of which were shown to be expressed on cytoplasmic ribosomes and posttranslationally targeted to the chromatophore where they assembled with endosymbiont-encoded proteins (Nowack & Grossman, 2012). The trafficking and import of host-encoded proteins into an endosymbiont are considered an early but obligatory step in organellogenesis, leading some to suggest that the chromatophore has officially matured into a bona fide photosynthetic organelle (Nakayama & Archibald, 2012). But new data have shown that the host–chromatophore integration is even further along than previously thought.

For *P. chromatophora*, endosymbiotic gene transfer alone cannot account for the missing genes in the chromatophore genome—or, more importantly, the resulting gaps in the corresponding biosynthetic pathways. Nowack et al. (2016) recently showed that these metabolic cracks have been filled in through horizontal gene transfer from various bacteria to the host nuclear genome, providing a new take on the rules governing primary endosymbiosis:

> *[Chromatophore] genome reduction seems to drive the fixation of horizontally acquired 'compensatory' bacterial genes in the host genome. Thus, similar to endosymbiotic gene transfer, HGT-derived genes may facilitate integration of the*

> *endosymbiont by providing the host with transcriptional/translational control over chromatophore metabolic functions, metabolite fluxes between the cytoplasm and chromatophore, and the processing of genetic information. Therefore, like endosymbiotic gene transfer, HGT establishes key connections that enable the host to coordinate host–chromatophore metabolism, growth, and proliferation.*
>
> ***Nowack et al. (2016)***

The authors of the study hypothesized that these horizontally acquired genes are the outcome of a phagotrophic lifestyle, which was maintained alongside phototrophy in the initial stages of chromatophore integration and eventually lost once the host–chromatophore relationship was firmly established. More broadly, these data underline the fundamental importance of horizontal gene transfer in the establishment of primary endosymbioses.

Paulinella is an excellent case study for the transition from heterotrophy to phototrophy. But how about the reverse direction—the switch from a phototrophic to a heterotrophic existence? As touched upon in the preceding section, there are many primary plastid-containing lineages that have forsaken photosynthesis, and some have even jettisoned their plastid genomes.

5. LOST IN THE LIGHT: UNEXPECTED INSIGHTS FROM NONPHOTOSYNTHETIC PRIMARY PLASTIDS

> *When the dark comes down, oh, the wind is on the sea*
> *With lisping laugh and whimper to the red reef's threnody,*
> *The boats are sailing homeward now across the harbor bar*
> *With many a jest and many a shout from fishing grounds afar.*
> *So furl your sails and take your rest, ye fisher folk so brown,*
> *For task and quest are ended when the dark comes down.*
>
> ***Lucy Maud Montgomery***

When asked to present my research field to a general audience, I love to talk about the eclectic and unusual suspects that have ditched photosynthesis but still retain a colourless plastid (Figueroa-Martinez, Nedelcu, Smith, & Reyes-Prieto, 2015). "Let's get high on some photosynthetic burnouts", I say to the audience, showing them high-definition photos of chlorophyll-lacking flowers and algae, such as the ghostly white corpse plant *Monotropa uniflora* and the nonphotosynthetic green alga *Polytomella magna*, the β-carotene gleaming from its eroding eyespot (Asmail & Smith, 2016). These images usually illicit a few oohs and aahs from the audience, but by far the greatest reaction I get is when I show pictures of disease-causing nonphotosynthetic algae: a close-up of a hand with lesions caused by the parasitic trebouxiophyte *Prototheca wickerhamii*, the worm-like gut contents of an insect infected with the pathogenic green alga *Helicosporidium* sp. … a red blood cell bursting with the malaria parasite *Plasmodium falciparum*, which has a secondary, red algal-derived plastid (Keeling, 2013). After such gruesome shots, I normally see surprised, inquisitive faces looking back at me. "What does a parasite do with a chloroplast anyway?" shouts someone from the back row. "Why hold on to a broken lightbulb?" mumbles another. "Seems like a stupid strategy to me", says a third, referring to an obligate heterotrophic existence.

Indeed, after such a long, hard-fought battle to acquire a plastid, how could any sound-minded species give up photosynthesis? The short answer is that evolution is not always straightforward or adaptive. For instance, a nonlethal mutation that knocks out photosynthetic capabilities could be fixed in a population of algae or plants through random genetic drift, especially if the effective size of the population is small. Why would such a mutation not be lethal? Because many photosynthetic eukaryotes are mixotrophic, meaning that they can survive using inorganic (phototrophy) and organic (chemoheterotrophy) carbon sources—the latter involves prey consumption via phagocytosis or endocytosis, or the intake of small organic compounds through osmosis (Tittel et al., 2003).

It is no great surprise, therefore, that mixotrophs can sometimes find themselves going down the generally one-way road towards a mandatory heterotrophic existence. In spite of the obvious advantages of mixotrophy, it is metabolically expensive to preserve the molecular machineries needed for both trophic strategies. Mixotrophic algae spend about five times more energy on maintaining the photosynthetic apparatus than on heterotrophic abilities (Raven, 1997). Thus, under some conditions, like when the cost of maintaining photosynthesis outweighs its benefits, losing phototrophic capabilities should be advantageous (de Castro, Gaedke, & Boenigk, 2009). In fact, the large number of distinct nonphotosynthetic plant and algal lineages that exist today underscores just how dispensable photosynthesis can be (Stoecker, 1998).

Although it is possible to do away with photosynthesis, it is not so easy to dump a plastid—all known nonphotosynthetic members of the Archaeplastida retain one (Keeling, 2013). This is because as the primary plastid endosymbiosis took hold, the host became dependent upon its cyanobacterial partner for much more than photosynthesis. In plants and algae, many vital biochemical pathways unrelated to photosynthesis are outsourced entirely or partly to the plastid, such as the biosynthesis of aromatic and hydrophobic side-chain amino acids, tetrapyrroles, and terpenoids (Gould, Waller, & McFadden, 2008). Although nearly all the enzymes involved in these pathways are nuclear encoded, most nonphotosynthetic plastids still retain a genome, albeit one that is highly reduced with a much smaller gene content than that in photosynthetic taxa (Figueroa-Martinez, Nedelcu, Smith, & Reyes-Prieto, 2015). The genes that remain in nonphotosynthetic plastids are almost entirely dedicated to plastid gene expression—a complicated process involving both plastid- and nuclear-encoded machinery (Gould, Waller, & McFadden, 2008). In the absence of photosynthesis, this complex infrastructure, spanning two genetic compartments, exists just so that a few key metabolic genes from the ptDNA can be expressed. Consequently, it was long believed that nonphotosynthetic plastids were irreversibly tied to their genomes (Barbrook et al., 2006; Nair & Striepen, 2011). But in 2014 it was shown that genome-less plastids do exist. One of these examples came from my and my colleague's work on *Polytomella*—a genus of free-living, unicellular nonphotosynthetic green algae closely related to *Chlamydomonas reinhardtii* (Smith & Lee, 2014).

After a decade of working on the organelle genetics of *Polytomella*, we were finally able to provide sufficient data to support outright plastid genome loss in this colourless genus. On the day that the paper was accepted, we had champagne on ice ready to celebrate the first example of a primary plastid-bearing lineage with no ptDNA only to discover that another team had beaten us to the summit by only a few weeks. Like *Polytomella*, the nonphotosynthetic and parasitic angiosperm *Rafflesia lagascae* (known for having the largest single flower of any plant) appears to have entirely shed its ptDNA (Molina et al., 2014). It is still not clear how these two lineages have managed to free themselves of a plastid genome and an associated gene expression system when so many other nonphotosynthetic species have not.

Detailed microscopy work and bioinformatic analyses of nuclear-encoded plastid-targeted proteins clearly show that *Polytomella* has a plastid (Moore, Cantor, Sheeler, & Kahn, 1970; Smith & Lee, 2014). But as pointed out by Krause (2015): "One concern with the interpretation of the data from *R. lagascae* may, for some, be the current lack of physical evidence for the existence of a plastid compartment at all". Krauss goes on to suggest that something sneaky may be going on in *Rafflesia*:

> *It is feasible that the intimate association between* Rafflesia *and its host has led to parasite cells being populated with host plastids. The sequestration of host plastids could have relieved the parasite of the selective pressure to keep its own plastid*

genome. Thus, the phylogenetic loss of the plastid genome may be tolerable for the parasite because it can ontogenetically 'hijack' host organelles.

Krause (2015)

The appropriation of plastids by nonphotosynthetic organisms (kleptoplasty) is a well-documented phenomenon, performed by some dinoflagellates (Gast, Moran, Dennett, & Caron, 2007) and even animals, such as the sea slug *Elysia chlorotica*, which steals plastids from the heterokont alga *Vaucheria litorea* (Pelletreau et al., 2011). However, there are currently no confirmed examples of kleptoplasty being performed by an archaeplastid species.

If forfeiting photosynthesis results in severe reduction or outright loss of the plastid genome, then the green alga *Polytoma uvella* is the exception to the rule. This nonphotosynthetic chlamydomonadalean has the largest plastid genome (>230 kb, 75% noncoding DNA) observed in a colourless plant or alga, eclipsing that of its close photosynthetic relative *Chlamydomonas leiostraca* by more than 60 kb (Figueroa-Martinez, Nedelcu, Smith, & Reyes-Prieto, 2017). *P. uvella* is also closely related to *Polytomella*, but each of these two lineages lost photosynthesis independently of one another and, at first glance, their plastid genomic architectures could not be more different: expansion and complete deletion, respectively. Although large, the *P. uvella* plastid genome has, like other nonphotosynthetic ptDNAs, undergone significant gene loss, shedding all coding regions for photosynthetic pathways, and most of its excess genomic baggage is in the form of noncoding DNA. Moreover, plastid genomic inflation and genome loss are not necessarily mutually exclusive—for all we know, the ancestor of *Polytomella* might have also had a bloated plastid genome before discarding its ptDNA (Figueroa-Martinez, Nedelcu, Reyes-Prieto, & Smith, 2017). As highlighted in the following section, the group to which both *Polytomella* spp. and *P. uvella* belong—the Chlamydomonadales—has an unparalleled propensity for plastid genomic inflation, with at least six members known to have ptDNAs exceeding 250 kb (Featherston, Arakaki, Nozaki, Durand, & Smith, 2016).

6. THREE DECADES OF PRIMARY PLASTID GENOMICS

My soul is a hidden orchestra; I know not what instruments, what fiddlestrings and harps, drums and tamboura I sound and clash inside myself. All I hear is the symphony.

Fernando Pessoa

One cannot emphasize enough how much the field of genomics has improved our understanding of primary endosymbiosis and primary plastid evolution (Daniell, Lin, Yu, Chang, 2016). Each advancement in molecular sequencing has brought with it a leap forward in plastid research—from the first ptDNA gene sequences (McIntosh, Poulsen, & Bogorad, 1980), to the first completely sequenced plastid genomes (tobacco and the common liverwort) (Ohyama et al., 1986; Shinozaki et al., 1986) to today's array of mind-boggling metagenomic techniques, which can generate dozens of ptDNAs in a few hours (Li et al., 2014; Stull et al., 2013; Twyford & Ness, 2017). I have benefitted professionally from these advancements: my PhD and postdoctoral work were largely built on plastid genomics, and so were those of many of my friends and colleagues. Personally, I think there is nothing more enjoyable than sitting down at a computer and piecing together an interesting genome, and plastids have more than their fair share of them.

Over the past 30 years, ptDNA has been—and will likely continue to be—among the most highly sequenced kinds of eukaryotic chromosome (Sanitá Lima, Woods, Cartwright, & Smith, 2016). As of February 2017, there are >1400 complete plastid genome sequences from land plants, and at least 125 for green algae, red algae, and glaucophytes, not to mention those from secondary plastids. These sequence data have provided the raw material for countless phylogenies, helping scientists trace the origin, diversification, and lateral spread of plastids (Keeling, 2013), and they have greatly enhanced our understanding of plastid biochemistry and physiology (Gould, Waller, & McFadden, 2008) as well as other sometimes puzzling cellular processes, such as intracellular gene transfer (Kleine, Maier, & Leister, 2009; Smith, Crosby, & Lee, 2011). Plastid DNA is also among the most widely used genetic markers in plant and algal research, both for population-level studies (Parks, Cronn, & Liston, 2009) and for broad-scale comparative analyses, like resolving the relationships among major groups on the eukaryotic tree of life (Ruhfel, Gitzendanner, Soltis, Soltis, & Burleigh, 2014). Some have even argued that plastid genes, such as *rbcL* and *matK*, should be used as universal genetic barcodes for characterizing plant and algal biodiversity (CBOL Plant Working Group, 2009), which seems appropriate given that such sequences have consistently helped researchers identify and/or characterize poorly studied lineages (Worden et al., 2012).

If all of that weren't enough, primary plastids have a become a popular canvas for bioengineering: transgenic ptDNA is aiding in the mass production of food, biofuels, vaccines, and other biopharmaceuticals, such as human blood proteins (Bock, 2001; Daniell, Lin, Yu, Chang, 2016; Su et al., 2015). Archaeologists and palaeobotanists, too, have long depended on plastid genomes for their research, extracting and sequencing ptDNA from fossils that are thousands—even millions—of years old (Hagelberg, Hofreiter, & Keyser, 2015). Using ptDNA isolated and sequenced from permafrost, for example, scientists pieced together the history of plant species populating the Arctic over the past 50,000 years (Willerslev et al., 2014). The results suggest that there was a surprisingly large amount of plant diversity in the ancient Arctic, more diversity, in fact, than the contemporary Arctic. What's more, the types of plants found in the Arctic have changed considerably over time, shifting from primarily wildflowers and other herbaceous vascular plants to more woody plants and grasses, which might have contributed to the decline of large mammals like the woolly mammoth (Willerslev et al., 2014). Similar kinds of studies have also been done on algae (Lejzerowicz et al., 2013; Stoof-Leichsenring et al., 2014).

More than anything, perhaps, the sequencing and characterization of ptDNAs have provided researchers with a treasure trove of interesting and unconventional genomes (Smith & Keeling, 2015), some of which have redefined well-established rules in genetics (Bendich, 2004; Knoop, 2011), helped generate leading hypotheses on molecular evolution (Lynch, 2007), and stimulated important discussions about the roles of adaptive vs nonadaptive processes in shaping genomic complexity (Smith, 2016).

Although not always as extreme as their mitochondrial counterparts (Smith & Keeling, 2015), plastid genomes are, by any measure, remarkably diverse. Within the Archaeplastida alone, ptDNAs can range in size by two orders of magnitude: from 11 kb in the parasitic land plant *Pilostyles aethiopica* (Bellot & Renner, 2016) to over 1000 kb in the beautiful and enormous unicellular green alga *Acetabularia acetabulum* (de Vries et al., 2013). This size variation is primarily a reflection of differences in noncoding DNA content, which has led to various hypotheses about why some plastids are so prone to genomic inflation while others are not (Smith, 2016). That said, plastid gene content can be highly variable: certain red algae have 250 unique genes encrypted in their ptDNAs (Janouškovec, Liu, et al., 2013), whereas heterotrophic archaeplastids can have as few as five (Bellot & Renner, 2016)—or none at all if you consider the ptDNA-lacking *Polytomella* and *Rafflesia*. In some species, particularly prasinophytes, the plastid genes are entirely devoid of introns (Lemieux, Otis, & Turmel, 2014), but in others, like the chlamydomonadalean *Dunaliella salina*, introns have proliferated throughout nearly every coding region (Del Vasto et al., 2015). And the genes and introns themselves can be odd: in *C. reinhardtii*, and its close relatives, the exons and introns of *psaA* are fragmented and scrambled throughout the genome and sewed back together again at the RNA level through *trans*-splicing (Kück, Choquet, Schneider, Dron, & Bennoun, 1987); equally complex exon–intron arrangements exist in other primary plastids as well (Glanz & Kück, 2009).

Posttranscriptional processing is even more elaborate in land plant plastids where cytosine-to-uracil and uracil-to-cytosine substitutional RNA editing can be rampant (Knoop, 2011)—a trait not yet found in any other primary plastid-bearing lineages. For example, in the ptDNA of the seedless vascular plant *Selaginella*, 3415 C-to-U RNA editing sites were identified, including ones in noncoding regions (Oldenkott, Yamaguchi, Tsuji-Tsukinoki, Knie, & Knoop, 2014), paralleling the situation in the mitochondrial compartment (Hecht, Grewe, & Knoop, 2011). *Selaginella* is also one of the few known lineages (along with certain trebouxiophytes) with GC-rich ptDNA (Smith, 2012), contrasting with the typically high AT content of most other plastids (e.g., the green algae *Helicosporidium* sp. and *Pedinomonas minor*). Remarkably, all known species with GC-biased ptDNA also have GC-rich mtDNA, but, strangely, the reverse is not true (Smith, 2012).

The evolutionary roots of these ptDNA architectural features, including GC-richness, RNA editing, and genomic expansion, are poorly understood, but they are thought to have arisen through nonadaptive processes (Gray, Lukeš, Archibald, Keeling, & Doolittle, 2010), and might be connected to mutation rate, recombination, and/or aspects of the DNA maintenance machinery (Smith & Keeling, 2015). Unfortunately, we still have a relatively narrow understanding of plastid mutation rates, but recent studies, including mutation accumulation experiments and substitution rate data, suggest that they might be lower than once though, especially in algae (Ness, Kraemer, Colegrave, & Keightley, 2016; Smith, 2015).

But it is the structure and conformation of plastid genomes that remain one of their most misunderstood and misrepresented features (Bendich, 2004). Pick up a recent review on plastid genetics or a ptDNA genome paper published in the past year and you are likely to find the words "circular chromosome". This is because, with few exceptions (Janouškovec, Sobotka, et al., 2013), plastid genomes assemble and map as circular molecules. Their true conformation, however, is usually more complicated:

Circular forms of [organelle] DNA from plants appear to have exerted a profound influence on 40 years of research, despite the weakness of the data in support of the notion that most or all functions of organellar chromosomes are served by circular DNA molecules (Williamson, 2002; Bendich, 2010). When in-gel methods are employed, chromosomal DNA molecules in the plastids and mitochondria of plants appear as linear and branched-linear forms of various sizes, are found in meristematic tissues, and are typically larger than the size of the genome. In maize, tobacco, and Medicago truncatula, *restriction digest analysis showed that the linear molecules have ends at defined regions of the plastid genome and isomers with three to six distinct ends ... The circular forms account for a few percent or less of total [organelle] DNA ...*

Oldenburg and Bendich (2015)

Overall, the paucity of detailed data on plastid chromosome structure has contributed to the popular misconception that plastid genomes exist as intact, unit-sized circular molecules, when in fact some are clearly made up of complex multigenomic branched structures. Understanding and characterizing this level of chromosomal complexity requires much more than DNA sequencing and bioinformatics analyses: it demands a wide assortment of experiments, from pulsed-field gel electrophoresis to Southern blots—experiments that are often lacking from contemporary organelle genome studies (Sanitá Lima et al., 2016).

7. CONCLUDING THOUGHTS: TOO MUCH SEQUENCING, NOT ENOUGH EXPERIMENTING?

I was taught that the way of progress was neither swift nor easy.

Marie Currie

One of my first projects as a graduate student was to sequence and assemble the plastid genome of *Volvox carteri*. Using PCR, cloning, and Sanger sequencing, I slowly pieced together the ptDNA of this model multicellular green alga. But it was slow going, let me tell you. This was partly because of the old-school PCR/cloning-based approach I was using, but also because the genome was big (>500 kb) and full of repeats (Smith & Lee, 2010), which is a recurring theme among chlamydomonadalean ptDNAs (Featherston, Arakaki, Nozaki, Durand, & Smith, 2016). About halfway through the project, I stumbled upon a useful resource: the United States Department of Energy Joint Genome Institute (DOE JGI) was in the process of sequencing the nuclear genome of *Volvox* (Prochnik et al., 2010) and had deposited in GenBank all their raw Sanger-sequencing data, which were teeming with ptDNA-derived reads. By syphoning off these publically available plastid reads, I quickly completed my *Volvox* plastid assembly. And, as I was told, I did not step on anybody's toes in doing so because the DOE JGI was not even interested in the plastid data, treating it more as a contamination rather than an asset—a mindset that persists today among many in the plant genomics community.

I've since gone on to make a modest research living out of mining sequencing databases, like GenBank's Sequence Read Archive (SRA), for organelle genomes and transcriptomes. With thousands of plant and algal high-throughput DNA and RNA sequencing data sets freely available in the SRA (Smith, 2013), and the recent development of open-source software for assembling organelle genomes from these kinds of data (Castandet, Hotto, Strickler, & Stern, 2016), there's never been a better time to be studying plastid genomics and transcriptomics. However, it is also hard not to get disheartened and disillusioned by this prevailing "sequence first, ask questions later" approach to contemporary organelle research.

Over 2600 organelle genome sequences were published in the last 5 years, and their rate of publication increases with each new year (Sanitá Lima et al., 2016). There are now journals entirely devoted to plastid and mitochondrial genome papers, many of which specialize in "genome announcements"— short 500-word articles that present a new chromosome sequence, its GenBank accession number, and little else (Smith, 2017). Often fast-tracked and sometimes not formally peer-reviewed, these pint-sized genome studies are being used as quick and easy routes to a publication. Approximately 75% of all organelle genome papers released in the past half-decade were genome announcements, and only 3% contained any "complementary" analyses to accompany the sequence data, such as gel electrophoresis, Southern or Northern blotting, or transcriptional profiling (Sanitá Lima et al., 2016). In other words, the plastid research community is generating more and more ptDNA sequences, but data on other crucial aspects of plastid genomes are going undescribed. This cannot be a good thing. What's the use of having thousands of plastid genomes if we don't understand how that information is stored, organized, and expressed, and how it is connected to deeper aspects of the genomic architecture, such as mutation rate, recombination, and DNA repair?

But I shouldn't complain too much. Compared to some other fields, we are lucky to have such large amounts of data to work with. And, again, there are hundreds, if not thousands, of plastid genomes and transcriptomes just waiting to be assembled from publically available data in the SRA, not to mention how new sequencing technologies, like single molecule real-time sequencing, are making it possible to recover entire ptDNAs with a handful of reads. It is, without a doubt, exciting to think about how these and other advancements will change the field of plastid biology and endosymbiosis over the coming years. I, for one, will be watching closely, so keep an eye on your sequencing reads.

ACKNOWLEDGEMENTS

D.R.S. is supported by a Discovery Grant from the Natural Sciences and Engineering Research Council (NSERC) of Canada. He can be found online at www.arrogantgenome.com.

REFERENCES

Adl, S. M., Simpson, A. G., Lane, C. E., Lukeš, J., Bass, D., Bowser, S. S., et al. (2012). The revised classification of eukaryotes. *Journal of Eukaryotic Microbiology*, *59*, 429–514.

Archibald, J. (2014). *One plus one equals one: Symbiosis and the evolution of complex life*. USA: Oxford University Press.

Archibald, J. M. (2015). Endosymbiosis and eukaryotic cell evolution. *Current Biology*, *25*, R911–R921.

Asmail, S. R., & Smith, D. R. (2016). Retention, erosion, and loss of the carotenoid biosynthetic pathway in the nonphotosynthetic green algal genus Polytomella. *New Phytologist*, *209*, 899–903.

Ball, S. G., Bhattacharya, D., & Weber, A. P. (2016a). Pathogen to powerhouse. *Science*, *351*, 659–660.

Ball, S. G., Bhattacharya, D., & Weber, A. P. (2016b). Infection and the first eukaryotes—Response. *Science*, *352*, 1065–1066.

Ball, S. G., Subtil, A., Bhattacharya, D., Moustafa, A., Weber, A. P., Gehre, L., et al. (2013). Metabolic effectors secreted by bacterial pathogens: Essential facilitators of plastid endosymbiosis? *Plant Cell*, *25*, 7–21.

Barbrook, A. C., Howe, C. J., & Purton, S. (2006). Why are plastid genomes retained in non-photosynthetic organisms? *Trends in Plant Science*, *11*, 101–108.

Bellot, S., & Renner, S. S. (2016). The plastomes of two species in the endoparasite genus *Pilostyles* (Apodanthaceae) each retain just five or six possibly functional genes. *Genome Biology and Evolution*, *8*, 189–201.

Bendich, A. J. (2004). Circular chloroplast chromosomes: The grand illusion. *Plant Cell*, *16*, 1661–1666.

Bendich, A. J. (2010). The end of the circle for yeast mitochondrial DNA. *Molecular Cell*, *39*, 831–832.

Bock, R. (2001). Transgenic plastids in basic research and plant biotechnology. *Journal of Molecular Biology*, *312*, 425–438.

Castandet, B., Hotto, A. M., Strickler, S. R., & Stern, D. B. (2016). ChloroSeq, an optimized chloroplast RNA-Seq bioinformatic pipeline, reveals remodeling of the organellar transcriptome under heat stress. *G3: Genes, Genomes, Genetics*, *6*, 2817–2827.

CBOL Plant Working Group. (2009). A DNA barcode for land plants. *Proceedings of the National Academy of Sciences of the United States of America*, *106*, 12794–12797.

Couradeau, E., Benzerara, K., Gerard, E., Moreira, D., Bernard, S., Brown, G. E., et al. (2012). An early-branching microbialite cyanobacterium forms intracellular carbonates. *Science*, *336*, 459–462.

Daniell, H., Lin, C. S., Yu, M., & Chang, W. J. (2016). Chloroplast genomes: Diversity, evolution, and applications in genetic engineering. *Genome Biology*, *17*, 134.

de Castro, F., Gaedke, U., & Boenigk, J. (2009). Reverse evolution: Driving forces behind the loss of acquired photosynthetic traits. *PLoS One*, *4*, e8465.

Del Vasto, M., Figueroa-Martinez, F., Featherston, J., Gonzalez, M. A., Reyes-Prieto, A., Durand, P. M., et al. (2015). Massive and widespread organelle genomic expansion in the green algal genus *Dunaliella*. *Genome Biology and Evolution*, *7*, 656–663.

Delwiche, C. F., & Cooper, E. D. (2015). The evolutionary origin of a terrestrial flora. *Current Biology*, *25*, R899–R910.

Deusch, O., Landan, G., Roettger, M., Gruenheit, N., Kowallik, K. V., Allen, J. F., et al. (2008). Genes of cyanobacterial origin in plant nuclear genomes point to a heterocyst-forming plastid ancestor. *Molecular Biology and Evolution*, *25*, 748–761.

de Vries, J., & Archibald, J. M. (2017). Endosymbiosis: Did plastids evolve from a freshwater cyanobacterium? *Current Biology*, *27*, R103–R105.

de Vries, J., Habicht, J., Woehle, C., Huang, C., Christa, G., Wägele, H., et al. (2013). Is *ftsH* the key to plastid longevity in sacoglossan slugs? *Genome Biology and Evolution*, *5*, 2540–2548.

Domman, D., Horn, M., Embley, T. M., & Williams, T. A. (2015). Plastid establishment did not require a chlamydial partner. *Nature Communications*, *6*, 6421.

Embley, T. M., & Williams, T. A. (2015). Steps on the road to eukaryotes. *Nature*, *521*, 169–170.

Featherston, J., Arakaki, Y., Nozaki, H., Durand, P. M., & Smith, D. R. (2016). Inflated organelle genomes and a circular-mapping mtDNA probably existed at the origin of coloniality in volvocine green algae. *European Journal of Phycology*, *51*, 369–377.

Figueroa-Martinez, F., Nedelcu, A. M., Smith, D. R., & Reyes-Prieto, A. (2015). When the lights go out: The evolutionary fate of free-living colorless green algae. *New Phytologist*, *206*, 972–982.

Figueroa-Martinez, F., Nedelcu, A. M., Reyes-Prieto, A., & Smith, D. R. (2017). The plastid genomes of nonphotosynthetic algae are not so small after all. *Communicative & Integrative Biology*, *10*, e1283080.

Figueroa-Martinez, F., Nedelcu, A. M., Smith, D. R., & Reyes-Prieto, A. (2017). The plastid genome of *Polytoma uvella* is the largest known among colorless algae and plants and reflects contrasting evolutionary paths to nonphotosynthetic lifestyles. *Plant Physiology*, *173*(2), 932–943.

Gagat, P., Bodył, A., & Mackiewicz, P. (2016). Breaking the endosymbiont-organelle border: The case of *Paulinella chromatophora*. *Endocytobiosis and Cell Research*, *27*, 45–51.

Gast, R. J., Moran, D. M., Dennett, M. R., & Caron, D. A. (2007). Kleptoplasty in an Antarctic dinoflagellate: Caught in evolutionary transition? *Environmental Microbiology*, *9*, 39–45.

Glanz, S., & Kück, U. (2009). Trans-splicing of organelle introns—A detour to continuous RNAs. *BioEssays*, *31*, 921–934.

Gould, S. B. (2016). Infection and the first eukaryotes. *Science*, *352*, 1065.

Gould, S. B., Waller, R. F., & McFadden, G. I. (2008). Plastid evolution. *Annual Review of Plant Biology*, *59*, 491–517.

Gray, M. W. (2012). Mitochondrial evolution. *Cold Spring Harbor Perspectives in Biology*, *4*, a011403.

Gray, M. W., Lukeš, J., Archibald, J. M., Keeling, P. J., & Doolittle, W. F. (2010). Irremediable complexity? *Science*, *330*, 920–921.

Hagelberg, E., Hofreiter, M., & Keyser, C. (2015). Introduction. Ancient DNA: The first three decades. *Philosophical Transactions of the Royal Society B: Biological Sciences*, *370*, 201303711.

Hecht, J., Grewe, F., & Knoop, V. (2011). Extreme RNA editing in coding islands and abundant microsatellites in repeat sequences of *Selaginella moellendorffii* mitochondria: The root of frequent plant mtDNA recombination in early tracheophytes. *Genome Biology and Evolution, 3*, 344–358.

Howe, C. J., Barbrook, A. C., Nisbet, R. E. R., Lockhart, P. J., & Larkum, A. W. D. (2008). The origin of plastids. *Philosophical Transactions of the Royal Society B: Biological Sciences, 363*, 2675–2685.

Huang, J., & Gogarten, J. P. (2007). Did an ancient chlamydial endosymbiosis facilitate the establishment of primary plastids? *Genome Biology, 8*, R99.

Janouškovec, J., Liu, S. L., Martone, P. T., Carré, W., Leblanc, C., Collén, J., et al. (2013). Evolution of red algal plastid genomes: Ancient architectures, introns, horizontal gene transfer, and taxonomic utility of plastid markers. *PLoS One, 8*, e59001.

Janouškovec, J., Sobotka, R., Lai, D. H., Flegontov, P., Koník, P., Komenda, J., et al. (2013). Split photosystem protein, linear-mapping topology, and growth of structural complexity in the plastid genome of *Chromera velia*. *Molecular Biology and Evolution, 30*, 2447–2462.

Keeling, P. J. (2013). The number, speed, and impact of plastid endosymbioses in eukaryotic evolution. *Annual Review of Plant Biology, 64*, 583–607.

Kim, S., & Park, M. G. (2016). *Paulinella longichromatophora* sp. nov., a new marine photosynthetic testate amoeba containing a chromatophore. *Protist, 167*, 1–12.

Kleine, T., Maier, U. G., & Leister, D. (2009). DNA transfer from organelles to the nucleus: The idiosyncratic genetics of endosymbiosis. *Annual Review of Plant Biology, 60*, 115–138.

Knoop, V. (2011). When you can't trust the DNA: RNA editing changes transcript sequences. *Cellular and Molecular Life Sciences, 68*, 567–586.

Kolisko, M., Boscaro, V., Burki, F., Lynn, D. H., & Keeling, P. J. (2014). Single-cell transcriptomics for microbial eukaryotes. *Current Biology, 24*, R1081–R1082.

Krause, K. (2015). Grand-scale theft: Kleptoplasty in parasitic plants? *Trends in Plant Science, 20*, 196–198.

Kück, U., Choquet, Y., Schneider, M., Dron, M., & Bennoun, P. (1987). Structural and transcription analysis of two homologous genes for the P700 chlorophyll α-apoproteins in *Chlamydomonas reinhardtii*: Evidence for in vivo trans-splicing. *The EMBO Journal, 6*, 2185–2195.

Kurland, C. G., Collins, L. J., & Penny, D. (2006). Genomics and the irreducible nature of eukaryote cells. *Science, 312*, 1011–1014.

Larkum, A. W. D., Lockhart, P. J., & Howe, C. J. (2007). Shopping for plastids. *Trends in Plant Science, 12*, 189–195.

Lejzerowicz, F., Esling, P., Majewski, W., Szczuciński, W., Decelle, J., Obadia, C., et al. (2013). Ancient DNA complements microfossil record in deep-sea subsurface sediments. *Biology Letters, 9*, 20130283.

Lemieux, C., Otis, C., & Turmel, M. (2014). Six newly sequenced chloroplast genomes from prasinophyte green algae provide insights into the relationships among prasinophyte lineages and the diversity of streamlined genome architecture in picoplanktonic species. *BMC Genomics, 15*, 857.

Li, Q., Li, Y., Song, J., Xu, H., Xu, J., Zhu, Y., et al. (2014). High-accuracy de novo assembly and SNP detection of chloroplast genomes using a SMRT circular consensus sequencing strategy. *New Phytologist, 204*, 1041–1049.

Lynch, M. (2007). *The origins of genome architecture*. Sunderland, MA: Sinauer Associates.

Martin, W. F., Roettger, M., Ku, C., Garg, S. G., Nelson-Sathi, S., & Landan, G. (2016). Late mitochondrial origin is pure artefact. *bioRxiv*. http://dx.doi.org/10.1101/055368.

McIntosh, L., Poulsen, C., & Bogorad, L. (1980). Chloroplast gene sequence for the large subunit of ribulose bisphosphatecarboxylase of maize. *Nature, 288*, 556–560.

Mereschkowsky, C. (1910). Theorie der zwei Plasmaarten als Grundlage der Symbiogenesis, einer neuen Lehre von der Entstehung der Organismen. *Biologisches Centrablatt*, *30*, 278–288.

Metzker, M. L. (2010). Sequencing technologies—The next generation. *Nature Reviews Genetics*, *11*, 31–46.

Molina, J., Hazzouri, K. M., Nickrent, D., Geisler, M., Meyer, R. S., Pentony, M. M., et al. (2014). Possible loss of the chloroplast genome in the parasitic flowering plant *Rafflesia lagascae* (Rafflesiaceae). *Molecular Biology and Evolution*, *31*, 793–803.

Moore, J., Cantor, M. H., Sheeler, P., & Kahn, W. (1970). The ultrastructure of *Polytomella agilis*. *The Journal of Protozoology*, *17*, 671–676.

Moran, M. A. (2009). Metatranscriptomics: Eavesdropping on complex microbial communities: Large-scale sequencing of mRNAs retrieved from natural communities provides insights into microbial activities and how they are regulated. *Microbe*, *4*, 329–335.

Morgan-Kiss, R. M., Priscu, J. C., Pocock, T., Gudynaite-Savitch, L., & Hüner, N. P. (2006). Adaptation and acclimation of photosynthetic microorganisms to permanently cold environments. *Microbiology and Molecular Biology Reviews*, *70*, 222–252.

Nair, S. C., & Striepen, B. (2011). What do human parasites do with a chloroplast anyway? *PLoS Biology*, *9*, e1001137.

Nakayama, T., & Archibald, J. M. (2012). Evolving a photosynthetic organelle. *BMC Biology*, *10*, 35.

Ness, R. W., Kraemer, S. A., Colegrave, N., & Keightley, P. D. (2016). Direct estimate of the spontaneous mutation rate uncovers the effects of drift and recombination in the *Chlamydomonas reinhardtii* plastid genome. *Molecular Biology and Evolution*, *33*, 800–808.

Nowack, E. C. (2014). *Paulinella chromatophora*-rethinking the transition from endosymbiont to organelle. *Acta Societatis Botanicorum Poloniae*, *83*, 387–397.

Nowack, E. C., & Grossman, A. R. (2012). Trafficking of protein into the recently established photosynthetic organelles of *Paulinella chromatophora*. *Proceedings of the National Academy of Sciences of the United States of America*, *109*, 5340–5345.

Nowack, E. C., Price, D. C., Bhattacharya, D., Singer, A., Melkonian, M., & Grossman, A. R. (2016). Gene transfers from diverse bacteria compensate for reductive genome evolution in the chromatophore of *Paulinella chromatophora*. *Proceedings of the National Academy of Sciences of the United States of America*, *113*, 12214–12219.

Ohyama, K., Fukuzawa, H., Kohchi, T., Shirai, H., Sano, T., Sano, S., et al. (1986). Chloroplast gene organization deduced from complete sequence of liverwort *Marchantia polymorpha* chloroplast DNA. *Nature*, *322*, 572–574.

Oldenburg, D. J., & Bendich, A. J. (2015). DNA maintenance in plastids and mitochondria of plants. *Frontiers in Plant Science*, *6*, 883.

Oldenkott, B., Yamaguchi, K., Tsuji-Tsukinoki, S., Knie, N., & Knoop, V. (2014). Chloroplast RNA editing going extreme: More than 3400 events of C-to-U editing in the chloroplast transcriptome of the lycophyte *Selaginella uncinata*. *RNA*, *20*, 1499–1506.

Parks, M., Cronn, R., & Liston, A. (2009). Increasing phylogenetic resolution at low taxonomic levels using massively parallel sequencing of chloroplast genomes. *BMC Biology*, *7*, 84.

Pelletreau, K. N., Bhattacharya, D., Price, D. C., Worful, J. M., Moustafa, A., & Rumpho, M. E. (2011). Sea slug kleptoplasty and plastid maintenance in a metazoan. *Plant Physiology*, *155*, 1561–1565.

Pittis, A. A., & Gabaldón, T. (2016). Late acquisition of mitochondria by a host with chimaeric prokaryotic ancestry. *Nature*, *531*, 101–104.

Ponce-Toledo, R. I., Deschamps, P., López-García, P., Zivanovic, Y., Benzerara, K., & Moreira, D. (2017). An early-branching freshwater cyanobacterium at the origin of plastids. *Current Biology*, *27*, 386–391.

Prochnik, S. E., Umen, J., Nedelcu, A. M., Hallmann, A., Miller, S. M., Nishii, I., et al. (2010). Genomic analysis of organismal complexity in the multicellular green alga *Volvox carteri*. *Science, 329*, 223–226.

Raven, J. A. (1997). Phagotrophy in phototrophs. *Limnology and Oceanography, 42*, 198–205.

Reyes-Prieto, A., Yoon, H. S., Moustafa, A., Yang, E. C., Andersen, R. A., Boo, S. M., et al. (2010). Differential gene retention in plastids of common recent origin. *Molecular Biology and Evolution, 27*, 1530–1537.

Ruhfel, B. R., Gitzendanner, M. A., Soltis, P. S., Soltis, D. E., & Burleigh, J. G. (2014). From algae to angiosperms-inferring the phylogeny of green plants (Viridiplantae) from 360 plastid genomes. *BMC Evolutionary Biology, 14*, 23.

Sagan, L. (1967). On the origin of mitosing cells. *Journal of Theoretical Biology, 14*, 225–274.

Sanitá Lima, M., Woods, L. C., Cartwright, M. W., & Smith, D. R. (2016). The (in)complete organelle genome: Exploring the use and nonuse of available technologies for characterizing mitochondrial and plastid chromosomes. *Molecular Ecology Resources, 16*, 1279–1286.

Shinozaki, K., Ohme, M., Tanaka, M., Wakasugi, T., Hayashida, N., Matsubayashi, T., et al. (1986). The complete nucleotide sequence of the tobacco chloroplast genome: Its gene organization and expression. *The EMBO Journal, 5*, 2043–2049.

Smith, D. R. (2012). Updating our view of organelle genome nucleotide landscape. *Frontiers in Genetics, 3*, 175.

Smith, D. R. (2013). RNA-Seq data: A goldmine for organelle research. *Briefings in Functional Genomics, 12*, 454–456.

Smith, D. R. (2015). Mutation rates in plastid genomes: They are lower than you might think. *Genome Biology and Evolution, 7*, 1227–1234.

Smith, D. R. (2016). The mutational hazard hypothesis of organelle genome evolution: 10 years on. *Molecular Ecology, 25*, 3769–3775.

Smith, D. R. (2017). Goodbye genome paper, hello genome report: The increasing popularity of "genome announcements" and their impact on science. *Briefings in Functional Genomics, 16*, 156–162.

Smith, D. R., & Asmail, S. R. (2014). Next-generation sequencing data suggest that certain nonphotosynthetic green plants have lost their plastid genomes. *New Phytologist, 204*, 7–11.

Smith, D. R., Crosby, K., & Lee, R. W. (2011). Correlation between nuclear plastid DNA abundance and plastid number supports the limited transfer window hypothesis. *Genome Biology and Evolution, 3*, 365–371.

Smith, D. R., & Keeling, P. J. (2015). Mitochondrial and plastid genome architecture: Reoccurring themes, but significant differences at the extremes. *Proceedings of the National Academy of Sciences of Sciences of the United States of America, 112*, 10177–10184.

Smith, D. R., & Lee, R. W. (2010). Low nucleotide diversity for the expanded organelle and nuclear genomes of *Volvox carteri* supports the mutational-hazard hypothesis. *Molecular Biology and Evolution, 27*, 2244–2256.

Smith, D. R., & Lee, R. W. (2014). A plastid without a genome: Evidence from the nonphotosynthetic green algal genus *Polytomella*. *Plant Physiology, 164*, 1812–1819.

Spang, A., Saw, J. H., Jørgensen, S. L., Zaremba-Niedzwiedzka, K., Martijn, J., Lind, A. E., et al. (2015). Complex archaea that bridge the gap between prokaryotes and eukaryotes. *Nature, 521*, 173–179.

Stoecker, D. K. (1998). Conceptual models of mixotrophy in planktonic protists and some ecological and evolutionary implications. *European Journal of Protistology, 34*, 281–290.

Stoof-Leichsenring, K. R., Herzschuh, U., Pestryakova, L. A., Klemm, J., Epp, L. S., & Tiedemann, R. (2014). Genetic data from algae sedimentary DNA reflect the influence of environment over geography. *Scientific Reports, 5*, 12924.

Stull, G. W., Moore, M. J., Mandala, V. S., Douglas, N. A., Kates, H. R., Qi, X., et al. (2013). A targeted enrichment strategy for massively parallel sequencing of angiosperm plastid genomes. *Applications in Plant Sciences, 1*, 1200497.

Su, J., Zhu, L., Sherman, A., Wang, X., Lin, S., Kamesh, A., et al. (2015). Low cost industrial production of coagulation factor IX bioencapsulated in lettuce cells for oral tolerance induction in hemophilia B. *Biomaterials, 70*, 84–93.

Tittel, J., Bissinger, V., Zippel, B., Gaedke, U., Bell, E., Lorke, A., et al. (2003). Mixotrophs combine resource use to outcompete specialists: Implications for aquatic food webs. *Proceedings of the National Academy of Sciences of the United States of America, 100*, 12776–12781.

Twyford, A. D., & Ness, R. W. (2017). Strategies for complete plastid genome sequencing. *Molecular Ecology Resources*. in press. http://onlinelibrary.wiley.com/doi/10.1111/1755-0998.12626/abstract.

Wallin, I. E. (1927). *Symbionticism and the origin of species*. London: Bailliere, Tindall and Cox.

Willerslev, E., Davison, J., Moora, M., Zobel, M., Coissac, E., Edwards, M. E., et al. (2014). Fifty thousand years of Arctic vegetation and megafaunal diet. *Nature, 506*, 47–51.

Williamson, D. (2002). The curious history of yeast mitochondrial DNA. *Nature Reviews Genetics, 3*, 475–481.

Worden, A. Z., Janouskovec, J., McRose, D., Engman, A., Welsh, R. M., Malfatti, S., et al. (2012). Global distribution of a wild alga revealed by targeted metagenomics. *Current Biology, 22*, R675–R677.

Yoon, H. S., Hackett, J. D., Ciniglia, C., Pinto, G., & Bhattacharya, D. (2004). A molecular timeline for the origin of photosynthetic eukaryotes. *Molecular Biology and Evolution, 21*, 809–818.

Yoon, H. S., Nakayama, T., Reyes-Prieto, A., Andersen, R. A., Boo, S. M., Ishida, K. I., et al. (2009). A single origin of the photosynthetic organelle in different *Paulinella* lineages. *BMC Evolutionary Biology, 9*, 98.

Zimorski, V., Ku, C., Martin, W. F., & Gould, S. B. (2014). Endosymbiotic theory for organelle origins. *Current Opinion in Microbiology, 22*, 38–48.

CHAPTER THREE

Secondary Plastids of Stramenopiles

Richard G. Dorrell[1], Chris Bowler[1]
IBENS, Ecole Normale Supérieure, CNRS, Inserm, PSL Research University, Paris, France
[1]Corresponding authors: e-mail address: dorrell@biologie.ens.fr; cbowler@biologie.ens.fr

Contents

Abstract

The stramenopiles encompass an incredible diversity of organisms, including ecologically fundamental single-celled algae such as diatoms, giant macroalgae such as kelps, as well as photo-mixotrophic and heterotrophic species. The photosynthetic species possess plastids of secondary or higher red algal origin. The diversity of stramenopile species provides an ideal system for exploring the fundamental features underpinning plastid establishment in eukaryotes, and also how plastid metabolism

Advances in Botanical Research, Volume 84
ISSN 0065-2296
http://dx.doi.org/10.1016/bs.abr.2017.06.003

has diversified following endosymbiosis. In this chapter, we present an overview of stramenopile diversity and explore the chimeric origins of the stramenopile plastid, which utilises a combination of pathways derived from red algae and other sources to support its function. Next, we discuss unusual features of stramenopile plastid metabolism, some of which, responses to acute nutrient limitation and metabolic crosstalk with the mitochondria, may be specific to the diatoms and underpin their relative success in the contemporary ocean. Finally, we discuss even more dramatic transitions in the evolutionary history and life strategies of individual stramenopile groups, including evidence that stramenopiles may have given rise to some of the other major plastid lineages observed today, such as those of haptophytes and dinoflagellates, thus majorly contributing to the spread of photosynthesis through the tree of life.

1. INTRODUCTION

1.1 Diversity of Stramenopiles

The stramenopiles are an extraordinarily diverse group of eukaryotes, comprising both photosynthetic and nonphotosynthetic lineages (Walker, Dorrell, Schlacht, & Dacks, 2011). While it is unambiguously clear that the stramenopiles are monophyletic, they carry no single unifying synapomorphy (Andersen, 2004; Derelle, López-García, Timpano, & Moreira, 2016; Walker et al., 2011). The most frequently observed and distinctive feature of stramenopiles are flagella (if present) of uneven length (hence the alternative name "heterokont", which may be applied only to the photosynthetic species or to stramenopiles as a whole), and the presence of tripartite tubular hairs on the long flagellum (hence stramenopiles, or "straw-like pili") (Cavalier-Smith, 1989; Patterson, 1989; Taylor, 1976; Walker et al., 2011). More broadly, the stramenopiles form part of the "SAR clade", that is to say the monophyletic group also encompassing rhizarians (including chlorarachniophytes) and alveolates (including dinoflagellates, apicomplexans, and chromerids), although the exact branching order between these groups remains debated (Burki et al., 2016, 2007; He, Sierra, Pawlowski, & Baldauf, 2016).

The photosynthetic stramenopile groups possess a single type of plastid, which originated through the secondary or more complex endosymbiosis of a red alga (Cavalier-Smith, 1989; Le Corguillé et al., 2009). The morphological and physiological diversity of the photosynthetic stramenopiles is huge, perhaps even rivalling that observed within the green algae and plants (Fig. 1).

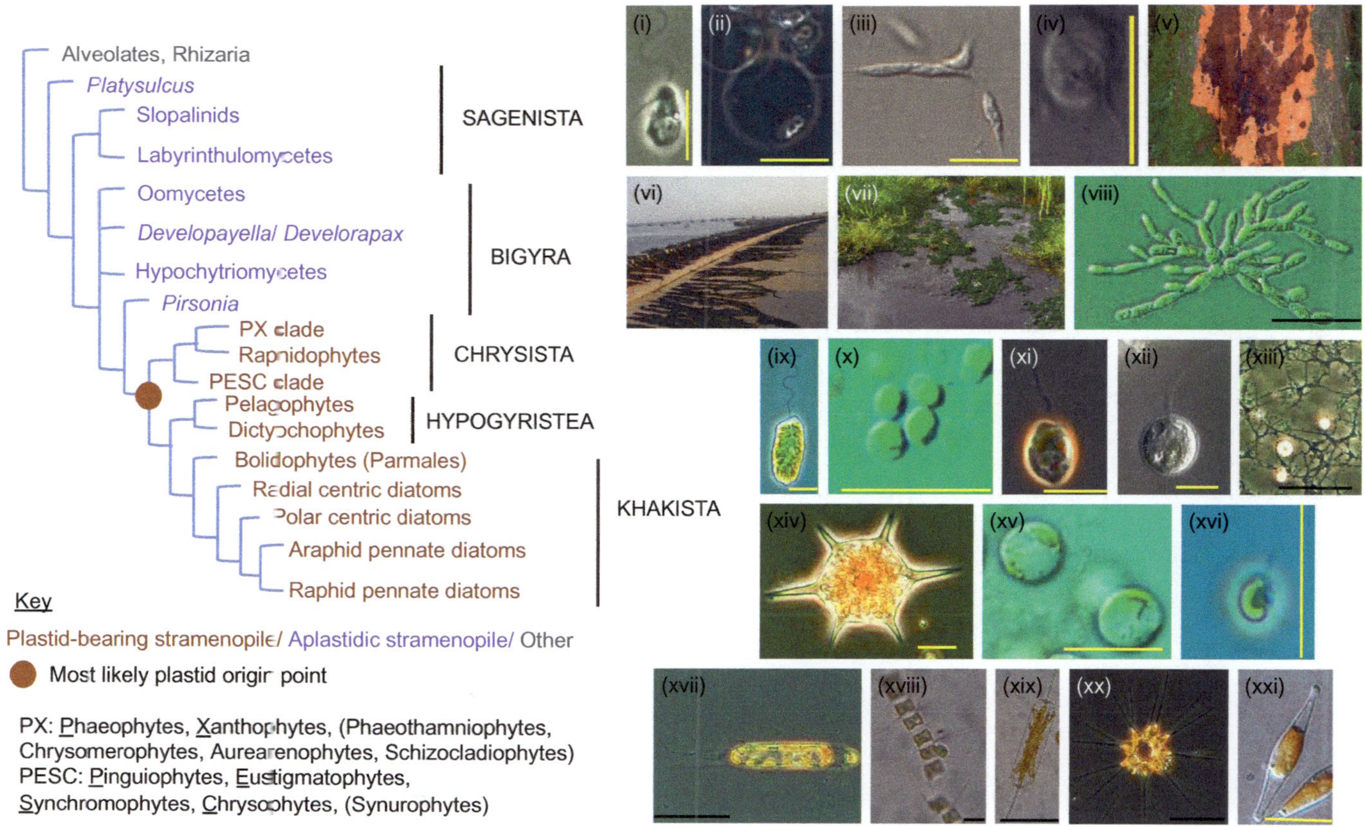

Fig. 1 See legend on next page.

Amongst these lineages are the diatoms, which are the most globally abundant photosynthetic eukaryotes in marine communities (Bowler, Vardi, & Allen, 2010; de Vargas et al., 2015), multicellular seaweed lineages such as the phaeophytes (kelps) and xanthophytes (golden algae) (Walker et al., 2011); microscopic single-celled algae such as the pelagophytes (Gobler & Sunda, 2012); and oleaginous lineages (e.g. eustigmatophytes and pinguiophytes) that may represent valuable systems for biofuel production (Pragya, Pandey, & Sahoo, 2013; Sang, Wang, Liu, Zhang, & Li, 2012). Several of the constituent lineages (e.g. aurearenophytes, synchromophytes, and pinguiophytes) have only been formally described in the last 15 years (Kai, Yoshii, Nakayama, &

Fig. 1 Diversity of plastid-bearing and aplastidic stramenopiles. This schematic tree, using taxonomies inferred in recent studies of stramenopile evolution (Aleoshin, Mylnikov, Mirzaeva, Mikhailov, & Karpov, 2016; Derelle et al., 2016; Dorrell et al., 2017; Yang et al., 2012), shows the branching relationships between the main lineages of photosynthetic and aplastidic stramenopiles. Lineages are coloured by the presence or absence of plastids, and a *brown circle* labels the most probable origin of the stramenopile plastid as inferred from recent studies (Dorrell et al., 2017; Stiller, Huang, Ding, Tian, & Goodwillie, 2009; Stiller et al., 2014). Full definitions of two groups with diverse composition, the "PX" and "PESC" clades are provided below the tree. To the *right*, a representative series of images of different stramenopile lineages are provided. (i–v) aplastidic groups: (i) *Cafeteria roenbergensis* (slopalinid); (ii) *Blastocystis hominis* (slopalinid); (iii) unknown labyrinthulomycete; (iv) *Developayella* sp.; (v) beech tree infected by *Phytophthora kernoviae* (oomycete). (vi–xiii) Chrysista: (vi) littoral giant kelps (phaeophytes; PX clade); (vii) *Vaucheria* sp. (xanthophyte; PX clade); (viii) *Phaeothamnion confervicola* (phaeothamniophyte; PX clade); (ix) *Haramonas dimorpha* (raphidophyte); (x) *Nannochloropsis* sp. (eustigmatophyte; PESC clade); (xi) *Ochromonas danica* (chrysophyte; PESC clade); (xii) *Paraphysomonas vestita* (aplastidic synurophyte; PESC clade); (xiii) *Leukarachnion* sp. (aplastidic synchromophyte; PESC clade). (xiv–xvi) Hypogyristea and *Bolidomonas*: (xiv) *Dictyocha speculum* (dictyochophyte); (xv) *Aureoumbra lagunensis* (pelagophyte); (xvi) *Bolidomonas* sp. (xvii–xxi) diatoms: (xvii) *Corethron hystrix* (radial centric); (xviii) *Chaetoceros* sp. (polar centric); (xix) *Ditylum brightwellii* (polar centric); (xx) *Asterionellopsis glacialis* (araphid pennate); (xxi) *Phaeodactylum tricornutum* (raphid pennate). *Yellow scale bars* (on images i–iv, ix–xii, and xiv–xvi) are to 10 μm, and *black scale bars* (on images viii, xiii, and xvii–xx) are to 50 μm; all remaining images are macroscale. *Image (v) is reprinted with permission of the Forestry Commission under an Open Government Licence. Image (vi) is provided with permission of Dr Zhanru Shao (Institute of Oceanology, Chinese Academy of Sciences) and (xviii) and (xix) are provided by Dr Leïla Tirichine (École Normale Supérieure). All remaining images are taken from Encyclopedia of Life/Micro*scope (http://www.eol.org) and reprinted under a Creative Commons Licence.*

Inouye, 2008; Kessenich, Ruck, Schurko, Wickett, & Alverson, 2014; Walker et al., 2011), and the diversity that has been assigned to them is likely to be substantially expanded by environmental sequencing in future years (de Vargas et al., 2015; Mangot et al., 2017). The initial radiation date of the photosynthetic stramenopiles is estimated (from different molecular clock studies) to have been between 330 and 700 million years before the present (Brown & Sorhannus, 2010; Matari & Blair, 2014; Yoon, Hackett, Ciniglia, Pinto, & Bhattacharya, 2004), with all of the major classes being at least 100 million years old (Brown & Sorhannus, 2010; Medlin, Kooistra, Potter, Saunders, & Andersen, 1997; Silberfeld et al., 2010).

The photosynthetic stramenopiles form a largely monophyletic group (Dorrell et al., 2017; Walker et al., 2011), with most of the major non-photosynthetic lineages (e.g. oomycetes, labyrinthulomycetes, and slopalinids, which include the important human pathogen *Blastocystis*) branching paraphyletically at the base of the stramenopile clade (Fig. 1) (Cavalier-Smith & Scoble, 2013; Derelle et al., 2016; Dorrell et al., 2017). Although the exact branching order and phylogenetic utility of specific photosynthetic stramenopile clades remain debated, the general consensus from recent multigene phylogenies is for a basal division between a clade-containing diatoms and *Bolidomonas* (together forming the "Khakista"), and pelagophytes and dictyochophytes (the "Hypogyristea"), and a separate clade containing all remaining lineages (the "Chrysista") (Fig. 1) (Derelle et al., 2016; Dorrell et al., 2017; Riisberg et al., 2009; Yang et al., 2012).

Stramenopiles have formed important model systems for research into algal plastid dynamics for over four decades. Early studies, focusing on pigment composition and plastid ultrastructure and dynamics in different stramenopile lineages (Gibbs, Luttke, & Cattolico, 1976; Jeffrey & Vesk, 1977; Kiefer, 1973), and biochemical assays performed in vitro and with isolated plastid material (Bhaya & Grossman, 1991; Wittpoth, Kroth, Weyrauch, Kowallik, & Strotmann, 1998), have paved the way for detailed experimentation in vivo using transformable model species such as the diatoms *Phaeodactylum tricornutum* and *Thalassiosira pseudonana*, and the eustigmatophyte alga *Nannochloropsis gaditana* (Li, Gao, & Hu, 2014; Siaut et al., 2007). Alongside this, a wealth of sequence information is now available for photosynthetic stramenopiles, with 12 annotated nuclear genomes (Armbrust et al., 2004; Basu et al., 2017; Bowler et al., 2008; Cock et al., 2010; Dorrell et al., 2017), and over 50 plastid genomes (Brembu et al., 2014; Cattolico et al., 2008; Galachyants et al., 2012; Kamikawa,

Tanifuji, et al., 2015; Kowallik, Stoebe, Schaffran, KrothPancic, & Freier, 1995; Le Corguillé et al., 2009; Liu & Pang, 2016; Lommer et al., 2010; Ong et al., 2010; Oudot-Le Secq et al., 2007; Ruck, Nakov, Jansen, Theriot, & Alverson, 2014; Rumpho et al., 2008; Sabir et al., 2014; Ševčíková et al., 2015; Tajima et al., 2016; Wei et al., 2013; Worden et al., 2012; Yurchenko, Sevcikova, Strnad, Butenko, & Elias, 2016; Zhang et al., 2015) available. The diversity of the resources available for stramenopiles constitutes an exceptional system in which to explore plastid function.

In this chapter, we present an overview of the inferred origins and evolutionary dynamics of the stramenopile plastid. We cover its morphological and genomic diversity, alongside unusual features of stramenopile plastid biochemistry, including processes that are specific to individual lineages such as diatoms, and finally discrete changes to stramenopile plastids that have occurred since their endosymbiotic acquisition, including the secondary loss of photosynthesis in some lineages, and the probable endosymbiotic acquisitions of stramenopile plastids by ancestors of plastid-bearing haptophytes and alveolates. Ultimately, stramenopile plastids provide us with insights into the fundamental evolutionary processes that occur alongside endosymbiotic events, and the degree to which plastid metabolism underpins the ecological success of some of the most dominant photosynthetic eukaryotes, such as diatoms, in the contemporary ocean.

2. ENDOSYMBIOTIC ORIGINS

2.1 Endosymbiotic Histories of Stramenopiles

It is well agreed that the stramenopile plastid is of red algal origin, and arose through a secondary or higher endosymbiosis event related to the endosymbioses that have occurred in dinoflagellates (also within the SAR clade) and in haptophytes and cryptomonads (extrinsic to the SAR clade). This idea was originally proposed based on the conserved presence of chlorophyll *c* in all of the above plastid lineages, and the complex membrane ultrastructure (e.g. four membrane-bound plastid of stramenopiles), consistent with a eukaryotic origin (Cavalier-Smith, 1989), and has been well supported by plastid gene trees, which typically place the stramenopile plastid well within the red algal clade (Dorrell et al., 2017; Le Corguillé et al., 2009; Stiller et al., 2014).

Recently, the exact endosymbiotic events that gave rise to the stramenopile plastid have been debated in more depth: in particular, whether the stramenopile plastid arose directly through a secondary endosymbiosis of a red alga, or alternatively through a more complex (e.g. tertiary) endosymbiosis event (Dorrell et al., 2017; Stiller et al., 2014). Typically, in plastid gene trees, stramenopiles form a sister-group to a monophyletic clade of cryptomonads and haptophytes (Dorrell et al., 2017; Muñoz-Gómez et al., 2017; Sanchez-Puerta, Bachvaroff, & Delwiche, 2007), or alternatively form a sister-group to haptophytes with cryptomonads as the earliest diverging lineage (Iida, Takishita, Ohshima, & Inagaki, 2007; Khan et al., 2007; Stiller et al., 2014). Stiller et al. have proposed, based on the presence of a footprint of nucleus-encoded genes that show a close relationship between photosynthetic stramenopiles and the cryptomonad *Guillardia theta*, that these relationships could be resolved via the tertiary endosymbiosis of a cryptomonad via an ancient stramenopile lineage (Stiller et al., 2014), although it remains to be determined whether the stramenopile plastid can be placed as arising within the cryptomonads, as opposed to in a sister-group relationship. Indeed, some plastid gene trees, particularly those lacking haptophyte sequences (Vogl et al., 2003), or those from which highly variable residues and fast-evolving sites have either been removed or recoded (Dorrell et al., 2017; Khan et al., 2007), record alternative topologies in which the stramenopile plastid arises from a deeper node in the red algal tree to the cryptomonads and haptophytes, consistent with independent secondary endosymbiosis events (Baurain et al., 2010). In any event, resolving the precise endosymbiotic events through which stramenopile plastids originated depends on more finely resolving the red algal donor lineage involved, which remains unknown (de Vries & Gould, 2017; Dorrell et al., 2017).

A related question is when the stramenopile plastid was acquired, and specifically whether the basally divergent, nonphotosynthetic groups (such as oomycetes and labyrinthulomycetes) once possessed red algal plastids which were then lost, or alternatively never acquired these plastids in the first place. Some examples of genes of probable red algal, cyanobacterial, or other plastid-type origin have been found in these basal lineages (Andersson & Roger, 2002; Robertson & Tartar, 2006; Tsui et al., 2009), of which a small number (approximately 10) may be shared with the genomes of photosynthetic stramenopiles, consistent with a common ancestry (Armbrust et al., 2004; Bowler et al., 2008). However, quantitative analysis of the red algal signal across entire oomycete genomes has found no evidence for enrichment in genes from such sources compared to genes inferred to have been acquired from other eukaryotes, suggesting that they may be phylogenetic artefacts or the results of isolated lateral gene transfers, rather than widespread gene transfer events consistent with an endosymbiosis event (Stiller et al., 2009, 2014). Recently, we have profiled the deeper origins of 26 plastid-targeted proteins found in photosynthetic stramenopiles that appear to have closest orthologues in basally divergent aplastidic relatives (Dorrell et al., 2017). We found essentially no evidence for a shared red algal origin for the genes within this dataset, and thus we can conclude that they are unlikely to be remnants of an ancient red algal endosymbiosis shared with basal stramenopile lineages (Dorrell et al., 2017). This suggests a much more recent origin for the red algal plastids in a recent ancestor of the photosynthetic stramenopiles (Fig. 1).

2.2 Chimeric Origins of Stramenopile Plastids

The advent of nuclear genome sequencing of photosynthetic stramenopiles, starting with the model diatom species *T. pseudonana* in 2004 (Armbrust et al., 2004), has added a further level of complexity to the story of stramenopile evolution. Alongside genes of red algal origin, which were expected given the ultimate red ancestry of the stramenopile plastid, stramenopile nuclear genomes have been found to contain large numbers of genes of nonred-origin, most notably genes from prokaryotes (Basu et al., 2017; Bowler et al., 2008) and from green algae (Armbrust et al., 2004; Moustafa et al., 2009), some of which are predicted to support the plastid (Fig. 2). The plastid-targeted proteins of prokaryotic origin notably contain large numbers of proteins associated with expression of the plastid genome, such as aminoacyl-tRNA synthetases (Dorrell et al., 2017). The green algal plastid-targeted proteins identified include fundamental components of plastid metabolism, such as components of the CBB (Calvin–Benson–Bassham) cycle (Li et al., 2006; Petersen, Teich, Brinkmann, & Cerff, 2006), chlorophyll synthesis (Cihlar, Fussy, Horak, & Obornik, 2016; Hunsperger, Randhawa, & Cattolico, 2015; Li et al., 2006), and at least five components of carotenoid biosynthesis (Coesel, Obornik, Varela, Falciatore, & Bowler, 2008; Frommolt et al., 2008). The genes of green algal origin have been proposed to form the footprints of a "cryptic plastid" of green origin acquired by the stramenopile host deep in its evolutionary history (Dorrell & Smith, 2011; Moustafa et al., 2009).

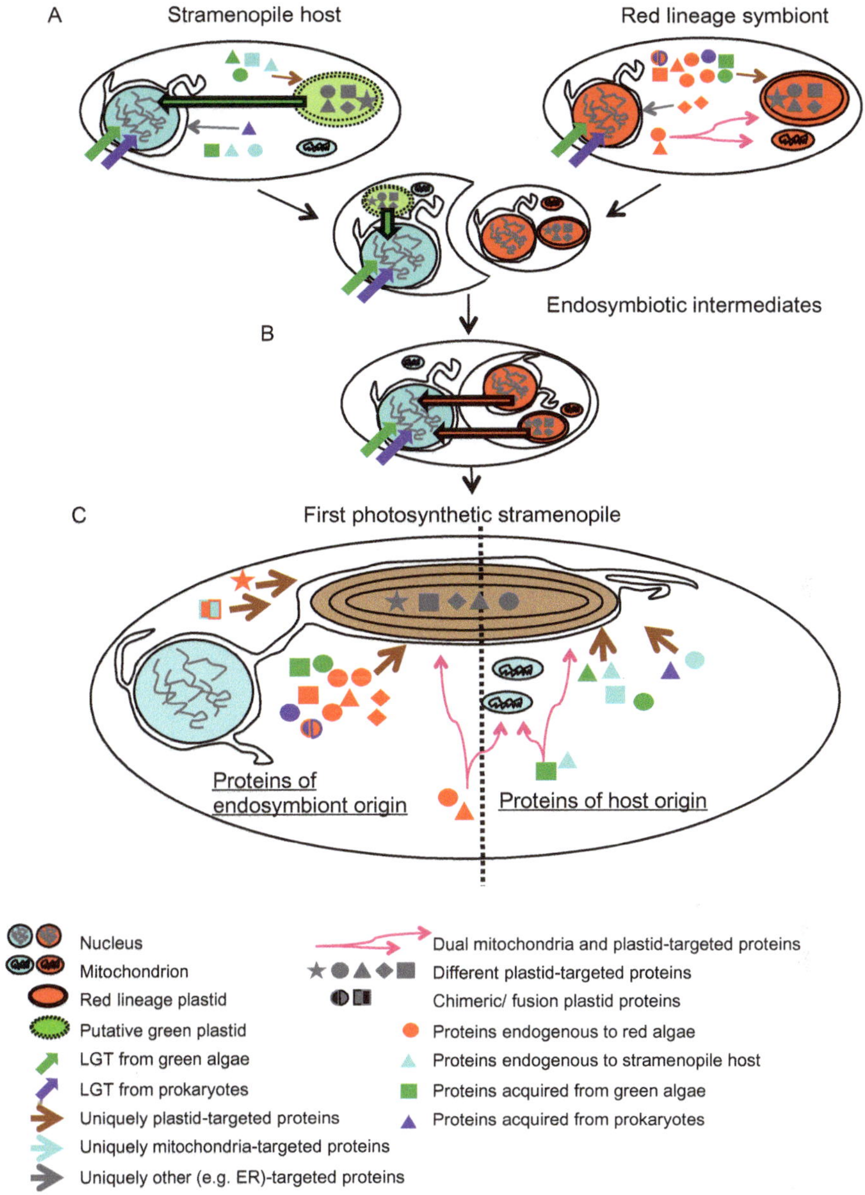

Fig. 2 Mosaic origins of stramenopile plastids. This figure shows the various origins of different plastid-targeted proteins identified that support stramenopile plastids. (A) Shows the ancestral stramenopile host and red-derived plastid lineages involved in the stramenopile plastid endosymbiosis; (B) shows the likely gene transfer events involved at the point of endosymbiotic plastid acquisition; and (C) shows a hypothetical

(Continued)

The total number and evolutionary significance of the green algal genes in stramenopile genomes remain major subjects of debate. In 2009, Moustafa et al. performed a global phylogenomic comparison of the model diatoms *T. pseudonana* and *P. tricornutum*, identifying over 1700 genes of probable green origin, compared to only 442 of red algal origin (Moustafa et al., 2009). This may however be a substantial overestimate of the total number of green gene transfers that have actually occurred. Some of the green genes are likely to be genes of red origin that have been misidentified due to taxonomic undersampling of red algal genomes or phylogenetic artefacts, or genes that were vertically inherited from a common ancestor of the stramenopile and green algal lineages (Burki et al., 2012; Deschamps & Moreira, 2012; Dorrell & Smith, 2011; Woehle, Dagan, Martin, & Gould, 2011). Indeed, a reinspection of the initial data presented by Moustafa et al., using expanded red algal sequence datasets and more stringent criteria for assigning evolutionary origin, radically reduced the number of genes that could be unambiguously assigned a green ancestry (to 89), although the authors did not assess

Fig. 2—Cont'd ancestor of all extant photosynthetic stramenopiles. The *dashed line* in (C) separates plastid-targeted proteins and structures of endosymbiont origin (*left*) and of host origin (*right*). The plastid-targeted proteins acquired from the host lineage include proteins previously targeted to other organelles such as the mitochondria (Dorrell et al., 2017; Gile, Moog, Slamovits, Maier, & Archibald, 2015), the expression products of genes acquired by horizontal transfer from prokaryotes (Bowler et al., 2008), and possibly genes acquired from a cryptic green algal plastid endosymbiosis (Dorrell et al., 2017; Moustafa et al., 2009). In addition, genes encoding plastid-targeted proteins also originated via lateral transfer events from the red algal or more complex (e.g. cryptomonad) ancestor of the stramenopile plastid (Dorrell et al., 2017; Stiller et al., 2014). The endosymbiont-derived proteins probably include genes that were plastid encoded in the endosymbiont, and also genes encoding plastid-targeted and other proteins in the endosymbiont nucleus. Some of the endosymbiont nuclear genes might have been laterally transferred from prokaryotic and green algal sources, providing an alternative mechanism by which these genes eventually were acquired by the stramenopile host (Dorrell et al., 2017; Li, Nosenko, Hackett, & Bhattacharya, 2006). Finally, proteins of chimeric origin, including proteins generated via the fusion of domains acquired from the host and from the endosymbiont, also underpin stramenopile plastid function (Dorrell et al., 2017; Méheust, Zelzion, Bhattacharya, Lopez, & Bapteste, 2016). *Adapted from Dorrell, R. G., Gile, G., McCallum, G., Méheust, R., Bapteste, E. P., Klinger, C. M., et al. (2017). Chimeric origins of ochrophytes and haptophytes revealed through an ancient plastid proteome.* Elife, 6, *23717.*

how these criteria affected their ability to identify genes of red algal origin (Deschamps & Moreira, 2012). Most recently, we have assembled a set of 770 proteins which, based on phylogenetic conservation and the widespread presence of plastid-targeting sequences, are probably ancestral features of stramenopile plastid proteomes, of which 263 could be assigned a clear evolutionary origin (Dorrell et al., 2017). Only 149 (57%) of these 263 proteins resolve with red algae, while 67 (25%) were of unambiguous green algal origin (Dorrell et al., 2017). Thus, the green algal contribution to the stramenopile plastid is somewhat smaller than initially proposed, but still exists (Fig. 2).

The green signal in stramenopile plastids is real: so where did it come from? The available evidence points to gene transfers from deep-branching members of the chlorophyte algae, as opposed to the streptophytes (plants and their closest relatives), which also excludes the possibility of a shared vertical origin, in which case they should show a more global distribution of affinity across the green lineage (Dorrell et al., 2017; Moustafa et al., 2009). These genes might have been acquired through lateral transfer events or through an endosymbiosis: notably, green genes in stramenopile genomes are less likely to encode plastid-targeted proteins than red genes are, suggesting that they are less clearly linked to the functions of a plastid endosymbiont, but are more likely to encode plastid-targeted proteins than other evolutionary categories of genes (e.g. genes of vertical, host origin) (Chan, Reyes-Prieto, & Bhattacharya, 2011; Dorrell et al., 2017), hence may be of partially endosymbiotic origin. We propose that, alongside a cryptic green endosymbiosis in the stramenopile host lineage, transfer of genes from green algae into the red alga, or other endosymbiont, which then gave rise to the current stramenopile plastid, might also be a possible vector for introducing green genes into the stramenopile nucleus (Fig. 2) (Dorrell et al., 2017). Ultimately, these questions will only be resolved once the precise evolutionary origins of the red lineage stramenopile plastid become clear.

3. STRUCTURE OF STRAMENOPILE PLASTIDS

3.1 Stramenopile Plastid Organisation

Stramenopile lineages may possess one or multiple plastids per cell (Fig. 3A); and these plastids have a distinctive ultrastructural organisation (Andersen,

A

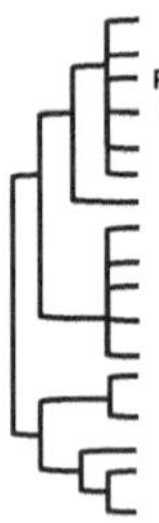

	Number plastids/cell	Pyrenoid	Chlorophyll	Fucoxanthin	Xanthophylls	Girdle lamella	Plastid-nucleus membrane connection	Eyespot	ptDNA
Phaeophytes	Multiple	Some	a, c1, c2	Fuc	V/Z	+	+	Some	R
Aurearenophytes	2	+	a, c?	Fuc	D/D	–	+	+	nd
Phaeothamniophytes	1 to 3	–	a, c1, c2	Fuc	D/D, H	+	+	+	R
Schizocladiophytes	1	–	a, c?	Fuc	nd	+	+	+	R
Chrysomerophytes	1	+	a, c1, c2	Fuc	V/Z	+	+	+	R
Xanthophytes	1	Some	a, c1, c2	–	D/D, H, Va	+	+	Some	R
Raphidophytes	Multiple	+	a, c1, c2	Fuc	V/Z or D/D	Some	–	–	S
Pinguiophytes	1 or 2	+	a, c1, c2	Fuc	V/Z	Some	+	–	R/S
Eustigmatophytes	1 or 2	Some	a	–	V/Z, Va	–	+	+	R
Synchromophytes	Multiple	Some	a, c2	Fuc	V/Z	–	+	–	nd
Synurophytes	1 or 2	Some	a, c1	Fuc	V/Z	+	–	–	R
Chrysophytes	1 or 2	+	a, c1, c2	Fuc, hexFuc	V/Z	+	+	+	R
Pelagophytes	Multiple	Some	a, c1, c2	Fuc, butFuc	D/D	+	+	Some	S
Dictyochophytes	Multiple	+	a, c1, c2	Fuc	D/D	+	+	Some	S
Bolidophytes	1	Some	a, c1, c2, c3	Fuc, hexFuc	D/D	+	+	–	R
Centric diatoms	Multiple	+	a, c1, c2	Fuc, hexFuc	D/D	+	Some	–	R
Pennate diatoms	1 or 2	+	a, c1, c2	Fuc, hexFuc	D/D	+	Some	–	R

Fucoxanthins Fuc: fucoxanthin; hexFuc: 19′-hexanoyloxyfucoxanthin; butFuc: 19′-butanoyloxyfucoxanthin
Xanthophylls V/Z: violaxanthin/zeaxanthin; D/D: diatoxanthin/diadinoxanthin; H: heteroxanthin; Va: vaucherioxanthin
Plastid DNA R: ring genophore; S: scattered genophore

B

(i) (ii)

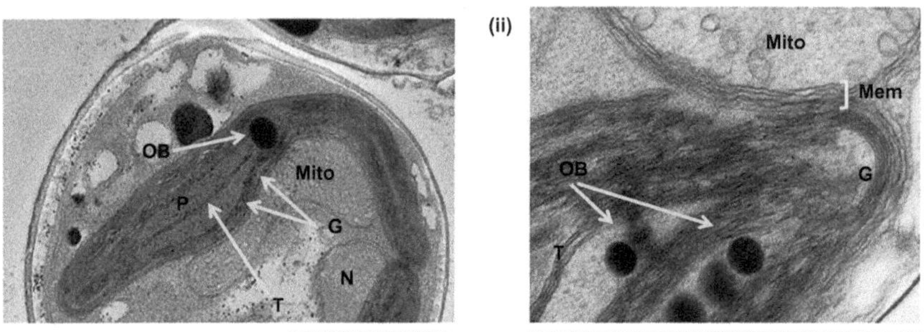

Fig. 3 Stramenopile plastid ultrastructure. The table (A) compares ultrastructural and pigment characteristics for different photosynthetic stramenopile plastids. The figure (B) shows transmission electron micrograph images of plastids from the model diatom species *Phaeodactylum tricornutum*. Image (i) shows a complete plastid containing a visible girdle lamella (labelled G), internal thylakoids (labelled T), pyrenoid (labelled P), and oil body (labelled OB), located in close proximity to the mitochondrion (labelled Mito) and nucleus (labelled N). Image (ii) shows a magnified region of a plastid, showing multiple extraplastidial membranes (labelled Mem), and a close connection between the plastid and the mitochondrion. *Scale bars* in both images are 1 μm. The authors thank Robert A. Andersen (University of Washington) and Christiane Lichtlé and Gérald Zabulon (École Normale Supérieure) for the provision of data and images used for the construction of (A and B), respectively. *Adapted from Andersen, R. A. (2004). Biology and systematics of heterokont and haptophyte algae.* American Journal of Botany, 91, *1508–1522, and using additional information from Bedoshvili, E. D., Popkova, T. P., & Likhoshvaĭ, E. V. (2009). Chloroplast ultrastructure in some diatoms from different classes.* Tsitologiia, 51, *346–357; Horn, S., Ehlers, K., Fritzsch, G., Gil-Rodriguez, M. C., Wilhelm, C., & Schnetter, R. (2007).* Synchroma grande *spec. nov (Synchromophyceae class. nov., Heterokontophyta): An amoeboid marine alga with unique plastid complexes.* Protist, 158, *277–293; Kai, A., Yoshii, Y., Nakayama, T., & Inouye, I. (2008). Aurearenophyceae classis nova, a new class of heterokontophyta based on a new marine unicellular alga* Aurearena cruciata *gen. et sp nov inhabiting sandy beaches.* Protist, 159, *435–457; Walker, G., Dorrell, R. G., Schlacht, A., & Dacks, J. B. (2011). Eukaryotic systematics: A user's guide for cell biologists and parasitologists.* Parasitol, 138, *1638–1663.*

2004; Walker et al., 2011). Typically, plastids are surrounded by four membranes [the inner two corresponding to the ancestral plastid membrane, and the outermost contiguous with the endoplasmic reticulum (ER) and nuclear envelope, forming the chloroplast ER (CER)] (Fig. 2C). The third membrane, which is presumed to correspond to the endosymbiont plasma cell membrane, encloses a separate region of cytoplasm termed the periplastid compartment, or PPC. Nucleomorphs (i.e. vestigial nuclei of endosymbiont origin) have been found within several other lineages with secondary plastids [cryptomonads, chlorarachniophytes] although are unknown within the PPC of stramenopiles (Andersen, 2004; Walker et al., 2011). Nonetheless, vesicles and other membrane structures have been observed inside the PPCs of some species (Flori, Jouneau, Finazzi, Maréchal, & Falconet, 2016; Horn et al., 2007; Ishida, Cavalier-Smith, & Green, 2000). This organisation is similar to the organisation observed in haptophytes, chromerid algae, as well as cryptomonads (barring the presence of a nucleomorph in cryptomonads) (Walker et al., 2011). Within the plastid stroma, a "girdle lamella", consisting of a layer of three stacked thylakoids, encloses all other (stacked and unstacked) thylakoid membranes (Fig. 3B) (Andersen, 2004). There is no clear spatial differentiation between the stacked and unstacked thylakoid regions in the plastid stroma (Prihoda et al., 2012). This thylakoid organisation is unique to stramenopiles, and its biochemical significance remains unknown (Andersen, 2004).

Exceptions to the organisation detailed earlier are known. Girdle lamellae are not found in eustigmatophytes, synchromophytes, aurearenophytes, and some pinguiophyte and raphidophyte species (Fig. 3) (Andersen, 2004). An unconventional outer membrane arrangement is observed in two unrelated lineages (synchromophytes and aurearenophytes), in which a consortium of multiple plastids, each surrounded by two inner membranes (in synchromophytes) or three (in aurearenophytes) is enclosed by a shared outer membrane complex (Horn et al., 2007; Kai et al., 2008). Similarly, there have been lineage-specific innovations in the xanthophyll pigments present in different lineages, with khakistan and hypogyristean lineages typically possessing hexyl- or butyl-fucoxanthins, and diatoxanthin/diadinoxanthin; while chrysistan lineages typically only possess unalkylated fucoxanthin pigments, and use violaxanthin/zeaxanthin in lieu of diatoxanthin (Fig. 3A) (Andersen, 2004). Whether these differences delineate different ecophysiological properties in different stramenopiles remain to be determined.

3.2 Stramenopile Plastid Genomes

Stramenopile plastid genomes typically show similar coding content, organisation, and synteny to one another (Fig. 4). Within sequenced photosynthetic species, the plastid genome ranges from 90 to 166 kb in size, contains between 134 and 197 genes, and has typically low (28%–36%) GC content (Fig. 4). The variability in gene coding content corresponds to varying degrees of plastid gene loss and transfer to the nuclei of different stramenopile lineages, which appears to be an ongoing phenomenon: examples have been found of genes that are encoded in both the plastid and nucleus, and hence might be currently undergoing relocation, in *P. tricornutum* (*acp1*, *tsf*) (Oudot-Le Secq et al., 2007), *T. pseudonana* (*psb28*) (Jiroutova, Koreny, Bowler, & Obornik, 2010), and *Thalassiosira oceanica* (*petF*, in which the plastid-encoded copy of the gene appears to have become pseudogenised) (Lommer et al., 2012). Most dramatically, sequenced pelagophyte plastid genomes are much smaller than those of other photosynthetic stramenopiles, with reduced coding content, and lack obvious inverted repeats as are found in all other studied lineages (Ong et al., 2010; Worden et al., 2012). Other lineages contain highly expanded inverted repeats (the chrysophyte *Ochromonas* and the diatom *Eunotia*) (Ruck et al., 2014; Ševčíková et al., 2015), and protein-coding genes that have been split into multiple open reading frames (*clpC*, in eustigmatophytes and chrysophytes) (Fig. 4) (Ševčíková et al., 2015; Wei et al., 2013; Yurchenko et al., 2016).

Recently, a small number of instances have been documented in which genes have been acquired by a stramenopile plastid lineage in its recent evolutionary history. These include the acquisition of episomal plasmids in diatoms (*Seminavis*, *Cylindrotheca*, and the diatom endosymbionts of the dinoflagellate *Kryptoperidinium*) (Brembu et al., 2014; Imanian et al., 2010), genes encoding predicted serine and/or tyrosine recombinases in diatoms (*Cylindrotheca*, *Kryptoperidinium*, *Fistulifera*) and in raphidophytes (*Heterosigma*) (Imanian et al., 2010), and type I and type II introns in the *rnl* and *atpB* genes of the diatom *Seminavis* (Fig. 4) (Brembu et al., 2014). Some of these genes may represent the remnants of parasitic DNA elements or viruses that infected the plastid genomes of individual stramenopile lineages (Brembu et al., 2014; Imanian et al., 2010).

Other genes appear to have been laterally acquired by stramenopile plastids from other microbial organisms. The plastid genome of the raphidophyte *Heterosigma akashiwo* contains genes encoding a probable G-protein

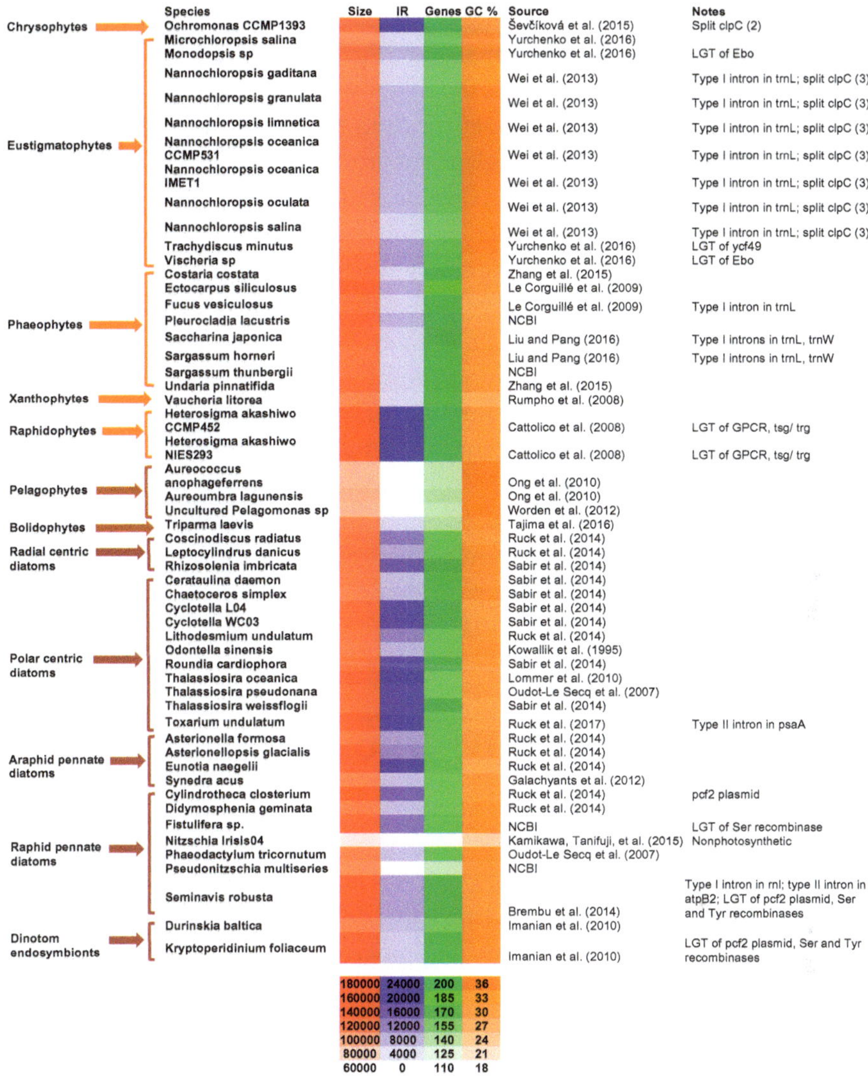

Fig. 4 Characteristics of stramenopile plastid genomes. This figure compares key features of all currently sequenced stramenopile plastid genomes (Brembu et al., 2014; Cattolico et al., 2008; Galachyants et al., 2012; Kamikawa, Tanifuji, et al., 2015; Kowallik et al., 1995; Le Corguillé et al., 2009; Liu & Pang, 2016; Lommer et al., 2010; Ong et al., 2010; Oudot-Le Secq et al., 2007; Ruck et al., 2014; Rumpho et al., 2008; Sabir et al., 2014; Ševčíková et al., 2015; Tajima et al., 2016; Wei et al., 2013; Worden et al., 2012; Yurchenko et al., 2016; Zhang et al., 2015), and the two sequenced plastid genomes from the diatom endosymbionts of dinotom algae (Imanian, Pombert, & Keeling, 2010). Quantitative features [total size in kb, total length of inverted repeat (IR), total number of genes and GC content] are shown as *shaded values*, with a key for each scale provided below the corresponding column. Species are grouped by taxonomic affiliation as per the key on the *left*: chrysistan lineages are grouped under *orange brackets*; hypogyristean lineages and *Bolidomonas* under *brown brackets*; and diatom lineages under *dark red brackets*.

coupled receptor, and a histidine-aspartate signalling circuit (*trg1/tsg1*) which is otherwise only known in red algal plastids and cyanobacteria (Cattolico et al., 2008), while the basally divergent eustigmatophyte *Trachydiscus* contains a plastid open reading frame of unknown function (*ycf49*) otherwise restricted to glaucophytes and cyanidialean red algae (Fig. 4) (Yurchenko et al., 2016). Most dramatically, Yurchenko et al. have recently characterised a six-gene operon (termed *ebo*) that is uniquely found in prokaryotes and in eustigmatophyte plastid lineages, hence might represent an independent lateral gene transfer event into a recent ancestor of the eustigmatophyte plastid (Fig. 4) (Yurchenko et al., 2016). The exact functional properties of the *ebo* operon remain unresolved, although the authors speculate that it may be involved in the synthesis of a novel isoprenoid compound similar to the bacterial photoprotective pigment scytonemin (Yurchenko et al., 2016). Understanding the coding content and metabolic output of this operon may provide valuable insights into the broader physiology of the eustigmatophyte lineage, which can be important in light of the interest of some eustigmatophyte species for biofuel production (Wei et al., 2013).

4. BIOCHEMISTRY AND PHYSIOLOGY

4.1 Plastid Biogenesis

The protein import machineries associated with stramenopile plastids have been studied for over a quarter of a century, and are now well characterised (Bhaya & Grossman, 1991; Gruber et al., 2007; Hempel, Bullmann, Lau, Zauner, & Maier, 2009; Hempel, Felsner, & Maier, 2010; Kilian & Kroth, 2005; Lang, Apt, & Kroth, 1998). Plastid-targeted proteins in stramenopiles must first be translocated from the cytoplasm across the ER surrounding the plastid (CER), and thus possess a bipartite-targeting sequence, consisting of a cleavable signal peptide sequence sufficient for import across the CER, followed by a transit peptide enabling translocation across the remaining three membranes (Fig. 3, panel B) (Bhaya & Grossman, 1991; Lang et al., 1998). Transport across the second outermost membrane (periplastid membrane), into the PPC, involves a protein complex, termed SELMA, that is related to the ER-associated protein degradation machinery. Plastid-targeted proteins generally possess a conserved ASAFAP motif spanning the signal peptide cleavage sequence, with the phenylalanine residue (or less frequently an alternative aromatic or hydrophobic residue) in the +1 position relative to the cleavage site (Gruber et al., 2007; Kilian & Kroth, 2005). Plastid-targeted proteins are sorted out from others by the presence or absence of the +1 phenylalanine residue, which is recognised by a specialist translocon of SELMA (Hempel et al., 2009, 2010). The membrane ultrastructure and import machinery of stramenopile plastids are sufficiently closely related to those of cryptomonads and haptophytes that plastid proteins from these lineages under their native-targeting sequences can be recognised by plastid-targeting predictors trained on stramenopile datasets (Dorrell et al., 2017; Gruber, Rocap, Kroth, Armbrust, & Mock, 2015), and can indeed be imported successfully into stramenopile plastids, using heterologous expression constructs (Felsner et al., 2011; Gould et al., 2006).

Recently, additional levels of complexity have been added to our understanding of stramenopile plastid protein import. The PPC and membranes appear to form the final targeting destination for a number of stramenopile plastid proteins, not least those associated with the translocation of other proteins through the PPC and across the inner two membranes (Dorrell et al., 2017; Gould et al., 2006; Hempel et al., 2010). Moreover, several proteins have been shown to be dually targeted to stramenopile plastids and mitochondria, of which some are likely to be conserved across a broad range of photosynthetic stramenopile lineages (Dorrell et al., 2017; Gile et al., 2015). The additional complexity enabled by the elaborate membrane organisation and targeting pathways of stramenopile plastids might have contributed to the metabolic sophistication of this lineage, for example, by allowing different pathways to operate in different plastid subcompartments, by enabling a global coordination of plastid physiology with that of other organelles, or by providing novel possibilities for the generation of electrochemical gradients from which energy can be generated (Dorrell et al., 2017; Smith, Abbriano, & Hildebrand, 2012).

4.2 Photosynthesis and Photoprotection

Surprising metabolic innovations have occurred in the stramenopile photosynthesis machinery. Proteomic studies of diatom thylakoids have identified an extrinsic PSII subunit (Psb31), which appears (from transcriptome data) to be conserved across all stramenopiles but not found in other lineages, and two PSI subunits, one of which is unique to stramenopiles, and the other is unique to diatoms (Dorrell et al., 2017; Grouneva, Rokka, & Aro, 2011).

Conversely, some subunits of the plastid photosynthesis machinery that are well conserved across other lineages, such as NDH dehydrogenase and Photosystem I subunit K, are absent from stramenopiles (Alboresi et al., 2017; Dorrell et al., 2017; Grouneva et al., 2011). It remains to be determined how many of these lineage-specific innovations have impacted on stramenopile metabolism.

The light harvesting and photoprotective pathways in stramenopiles also differ from those of other algae. In particular, stramenopiles appear to have exploited the chimeric origins of their plastid proteomes to explore physiological capacities that are not known in red algae. Under high light conditions, stramenopiles can dissipate excess light energy from photosystems using a xanthophyll cycle, dependent on the retrograde conversion of violaxanthin to zeaxanthin (in Chrysista) or diatoxanthin to diadinoxanthin (in Khakista and Hypogyristea) (Cao et al., 2013; Lepetit et al., 2017, 2013). This pathway is widespread in green lineages, but is not known in red algae. Consistent with this, the carotenoid biosynthesis pathway in stramenopiles is supplemented with a terminal xanthophyll cycle enzyme pair (violaxanthin deepoxidase and zeaxanthin epoxidase) of green origin (Fig. 5A) (Coesel et al., 2008; Dorrell et al., 2017; Frommolt et al., 2008). Thus, the acquisition of a plastid xanthophyll cycle from green algae might have permitted stramenopiles to survive in a wider range of light regimes than might have been possible purely with a complement of plastid proteins of red origin (Dorrell & Smith, 2011; Dorrell et al., 2017).

A similarly chimeric situation is observed for fucoxanthin/chlorophyll pigment-binding proteins, for which there are at least 10 conserved and phylogenetically distinct gene families in stramenopiles, of both red and green origin (Fig. 5B) (Dittami, Michel, Collén, Boyen, & Tonon, 2010; Dorrell et al., 2017). Notably, amongst the fucoxanthin-/chlorophyll-binding proteins of green origin are Lhcx/Li818/LHCSR-type proteins (Dittami et al., 2010; Dorrell et al., 2017). These proteins have been implicated in inducing nonphotochemical quenching via the coordination of the xanthophyll cycle (Lepetit et al., 2013; Taddei et al., 2016). The genes encoding these proteins are variously upregulated under high and variable light conditions (Lepetit et al., 2017, 2013; Taddei et al., 2016), or in some cases are highly expressed under nonstressful light conditions, but for which mutant lines have diminished nonphotochemical quenching and growth (Bailleul et al., 2010). Thus, Lhcx family proteins have multiple overlapping roles in facilitating tolerance to light stress. In addition, the Lhcx family has undergone considerable expansion in individual diatom species, particularly those (e.g. the polar

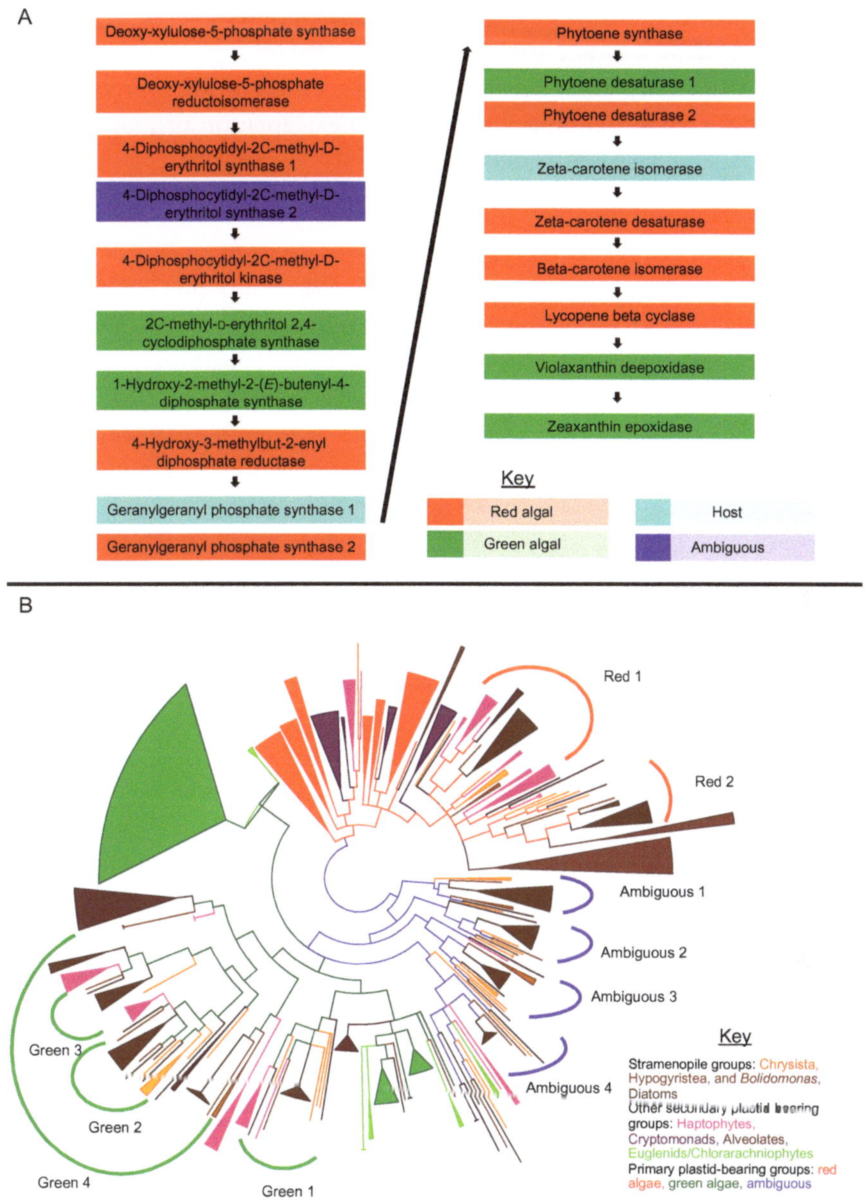

Fig. 5 Chimeric origins of photosynthesis-associated proteins in stramenopile plastids. (A) Shows the most probable evolutionary origins of enzymes associated with carotenoid biosynthesis pathways in stramenopiles, using data integrated from multiple studies (Coesel et al., 2008; Dorrell et al., 2017; Frommolt et al., 2008). While predominantly of red algal origin, several of the components of this pathway appear to have been acquired by stramenopiles from alternative sources, notably including a terminal

(Continued)

species *Fragilariopsis*) that are exposed to unusual light regimes in their native environment (Mock et al., 2017). Ultimately, detailed profiling of mutant phenotypes and expression patterns of a wide range of Lhcx proteins will be essential for understanding their role in the adaptation of stramenopiles to individual environmental niches.

4.3 Central Carbon Metabolism

The carbon metabolism pathways associated with stramenopile plastids also differ somewhat from the canonical plastid metabolism pathways. For example, stramenopile plastids include a complete glycolytic pathway, barring the presence of the initial hexo/glucokinase enzyme (Dorrell et al., 2017; Smith et al., 2012). The phosphofructokinase enzyme used in this pathway, instead of being ATP-dependent, is a pyrophosphate-dependent enzyme of probable cytoplasmic origin, which may facilitate activity of this pathway under low plastidic ATP conditions, which may occur at night (Dorrell et al., 2017; Smith et al., 2012). Flux modelling calculations performed using *P. tricornutum* have proposed that this pathway is predominantly utilised in the dark, under which conditions it can convert fructose- and glucose-6-phosphate imported from the cytoplasm into triose phosphates in the absence of CBB cycle activity (Kim et al., 2016). It remains to be determined what the terminal products of this pathway are used for; for example, whether they enable the synthesis of plastidic amino and fatty acids in the absence of light, or whether they perform an alternative function entirely.

Stramenopiles are probably dependent on active carbon concentration pathways for efficient photosynthesis, as the relative specificity of their RuBisCo for CO_2 over oxygen is somewhat lower than orthologous enzymes in free-living red algae (Badger et al., 1998). Moreover, diatom plastid membranes appear to be highly permeable to CO_2, and it therefore may be necessary to actively concentrate CO_2 around RuBisCo to prevent

Fig. 5—Cont'd xanthophyll cycle of inferred green algal origin. (B) Shows an exemplar tree inferred using combined MrBayes and RAxML analysis under two different substitution matrices (Jones/JTT and WAG), for a 427 taxa × 220 aa alignment of eukaryotic fucoxanthin/chlorophyll-binding proteins. Branches are coloured according to phylogenetic origin; node branches of uncertain origin are shaded by the most probable Archaeplastida affinity (red algal, green algal, or ambiguous). Ten distinct protein isoforms that appear to be conserved across all stramenopiles, of which at least four are of unambiguous green origin, are labelled with coloured arcs, the colour depicting the inferred origin. The underlying trees and alignment data supporting this figure are publicly accessible from https://www.repository.cam.ac.uk/handle/1810/261421.

it from passively diffusing back out of the cell (Roberts, Granum, Leegood, & Raven, 2007; Tsuji et al., 2017). One possible solution is a biophysical carbon-concentrating mechanism, involving the use of a series of carbonic anhydrases located in different plastid subcompartments, to trap CO_2 in bicarbonate anionic form, in which it cannot diffuse out of the plastid. This process, used in conjunction with membrane-bound active bicarbonate transporters, could then enrich plastidic CO_2 around RuBisCo, which is compacted into a pyrenoid (Kroth et al., 2008).

At least one plastidic bicarbonate transporter and multiple carbonic anhydrases essential for this exchange have been characterised from diatom genomes, with nine having been documented in *P. tricornutum* (Bowler et al., 2008; Kikutani et al., 2016; Kroth et al., 2008; Samukawa, Shen, Hopkinson, & Matsuda, 2014; Tanaka, Nakatsuma, Harada, Ishida, & Matsuda, 2005). Most recently, Kikutani et al. have identified a highly unusual θ-carbonic anhydrase targeted to the pyrenoid-penetrating thylakoid lumen of *P. tricornutum*, which is implicated in the final conversion of bicarbonate to CO_2 and delivery of CO_2 to the pyrenoid (Kikutani et al., 2016). Notably, diatoms preferentially assimilate inorganic bicarbonate over CO_2 under CO_2-limiting conditions (Clement, Dimnet, Maberly, & Gontero, 2016); moreover, the expression of several of the diatom plastidic carbonic anhydrases is regulated in response to CO_2 availability (Kikutani et al., 2012; Samukawa et al., 2014), and knockdown mutants of plastidic carbonic anhydrases display reduced CO_2 affinities and retarded growth rates (Kikutani et al., 2016), confirming the importance of this pathway for diatom metabolism. A similar dependence on a plastidic carbonic anhydrase for carbon concentration was recently demonstrated in the eustigmatophyte alga *Nannochloropsis*, suggesting that biophysical carbon concentration may be a widespread feature of stramenopile carbon metabolism (Gee & Niyogi, 2017).

Alongside a biophysical carbon-concentrating mechanism, a biochemical carbon-concentrating mechanism has been proposed to exist in some stramenopile plastids. In 2000, Reinfelder et al. documented the presence of C_4 carboxylase and decarboxylase enzymatic activities in cell lysates of the diatom *Thalassiosira weissflogii*; moreover, *T. weissflogii* grown in medium enriched with radiolabelled bicarbonate initially accumulated a radiolabel on C_4 acids (such as malate) prior to its accumulation in C_3 compounds or sugars (Reinfelder, Kraepiel, & Morel, 2000). Similar results have subsequently been found in other diatom species, such as *P. tricornutum* (Huang, Liu, Zhao, Yang, & Wang, 2016). These findings might be consistent with the presence of a C_4-type carbon-concentrating pathway as found in some higher plants. Genes encoding plastid-targeted proteins required for the fixation of C_4 amino acids (i.e. carbonic anhydrases, pyruvate and phosphoenolpyruvate carboxylase, and phosphoenolpyruvate dikinase) have since been documented in a wide range of diatom genomes (Dorrell et al., 2017; Kroth et al., 2008; Reinfelder et al., 2000). It remains to be determined whether similar pathways operate in other stramenopile lineages, although orthologues of phosphoenolpyruvate carboxylase, pyruvate-phosphate dikinase and pyruvate kinase have all been detected in stramenopile lineages outside of diatoms (Chi et al., 2014).

While there is almost certainly some form of C_4 enzymatic activity in diatoms, it may not necessarily operate in photosynthetic carbon fixation. Notably, the phosphoenolpyruvate carboxykinase and malate dehydrogenase enzymes required for catabolism of C_4 acids and the liberation of the associated CO_2 are not targeted to the diatom plastid, but instead localise to the mitochondria (Davis, Abbriano, Smith, & Hildebrand, 2017; Kroth et al., 2008). Moreover, it is unclear whether the fixation steps of the C_4 pathway are upregulated in diatoms in response to low CO_2 (Clement et al., 2016, 2017; Kustka et al., 2014) and it is clear that the decarboxylation steps are not (Kustka et al., 2014). Finally, while knockdown mutants of the plastid pyruvate-phosphate dikinase and the mitochondrial phosphoenolpyruvate carboxykinase in *P. tricornutum* show impaired photosynthesis (Haimovich-Dayan et al., 2013; Yang, Pan, Bowler, Zhang, & Hu, 2016), the dikinase knockdown does not have an altered affinity for CO_2 (Haimovich-Dayan et al., 2013). Thus, the physiological significance of the diatom C_4 pathway does not appear to be explicitly linked to carbon concentration.

It remains to be determined what role C_4 assimilation plays in diatom plastids. It is possible that the synthesis of C_4 acids has other essential functions in supporting diatom metabolism; for example, generating oxaloacetate for aspartate biosynthesis, or other biosynthetic intermediates for the plastid. Kim et al. have proposed enhanced import of phosphoenolpyruvate into diatom plastids in cells grown under high light, and supplemented with glucose, suggesting a possible role for the C_4 pathway in equilibrating metabolite balance from photosynthesis and respiration (Kim et al., 2016), although their model did not include flux calculations for the activities of phosphoenolpyruvate carboxylase or dikinase. Alternatively, part of the C_4 pathway may act in futile cycles to dispel excess plastidic reducing intermediates or mitochondrial ATP, although the precise events for these interactions have not yet been inferred (Bailleul et al., 2015; Kim et al., 2016). Ultimately, it will be necessary to characterise additional mutant lines, particularly for fundamental C_4-associated enzymes such as phosphoenolpyruvate carboxylase, to more clearly understand the physiological significance of C_4 assimilation in diatoms.

4.4 Primary and Secondary Metabolism Pathways

Previous studies have documented a wide range of biosynthetic processes in stramenopile plastids. Alongside the central carbon metabolism (CBB cycle, gluconeogenesis, and glycolysis) (Kroth et al., 2008; Smith et al., 2012) and carotenoid biosynthesis pathways discussed earlier (Bertrand, 2010; Coesel et al., 2008; Frommolt et al., 2008), stramenopile plastids possess complete pathways for the synthesis of chlorophyll and heme (Cihlar et al., 2016; Dorrell et al., 2017), a wide range of amino acids (Bromke, 2013; Dorrell et al., 2017), fatty acids that are then converted into galactolipids (Dorrell et al., 2017; Shtaida, Khozin-Goldberg, & Boussiba, 2015), and other essential cofactors such as iron–sulphur clusters and riboflavin (Dorrell et al., 2017).

There are some conspicuous absences from stramenopile plastid metabolic pathways. Most significantly, stramenopiles lack plastid-targeted nucleotide biosynthesis pathways, and instead use a conserved range of nucleoside diphosphate (NDP) transporters and NDP kinases to import and charge cytoplasmic nucleosides for use in RNA and DNA synthesis (Ast et al., 2009; Chu et al., 2017; Dorrell et al., 2017). In addition, in some cases, lateral gene transfer or gene replacements have altered the individual metabolic components of specific stramenopile plastids. For example, in the case of 3-dehydroquinate synthase, which performs the second step of the shikimate pathway, chrysistan and diatom plastids possess an enzyme isoform of red or green algal origin, but this has been replaced in pelagophytes and dictyochophytes by an isoform of probable bacterial origin, which is also shared with haptophytes (Dorrell et al., 2017). It remains to be determined how these lineage-specific chimerism events have affected the broader ecophysiology of individual stramenopile lineages.

Several examples of chimeric proteins, in which components encoding adjacent steps or interacting in individual metabolic pathways have fused together, are known in stramenopile metabolism. For example, in the tryptophan biosynthesis pathway, anthranilate synthase and tryptophan synthase, which are typically found as heterodimeric complexes in other organisms, are each encoded by single genes containing fusions of both component domains (Jiroutova, Horak, Bowler, & Obornik, 2007). Recently, we have documented a similar fusion event between two consecutive components of the riboflavin biosynthesis pathway, 3,4-dihydroxy-2-butanone 4-phosphate synthase and GTP cyclohydrolase, which is conserved across all photosynthetic stramenopiles, and arose independently to analogous fusion proteins present in bacteria and plants (Dorrell et al., 2017). These fusion proteins might facilitate enhanced metabolic efficiency in stramenopile plastids, both by ensuring stoichiometric expression of different components, and facilitating the efficient transfer of intermediates through individual metabolism pathways.

4.5 Nutrient Acquisition

Almost all stramenopile lineages are subject to some nutrient limitation in their natural environments. As one case in point, diatoms dominate marine communities in the Southern Ocean where iron is scarce (Bowler et al., 2010; de Vargas et al., 2015). Iron limitation is particularly significant in the context of plastid physiology as it restricts the ability to synthesise photosystem components, therefore, leading to increased vulnerability to photoinhibition, particularly under variable light regimes. Diatom plastids accordingly possess a number of adaptations to facilitate function in iron-poor environments. These include transcriptional responses, such as the upregulation of proteins (Lhcx-type fucoxanthin/chlorophyll proteins) and pathways (e.g. the plastid xanthophyll cycle), implicated in photoprotection (Allen et al., 2008; Smith et al., 2016).

Some stramenopile species, and particularly diatoms, may have acquired specific plastid proteins that allow them to thrive in iron-poor environments. The marine diatom *T. oceanica*, for example, possesses a gene for the photosystem I donor plastocyanin, which is a copper metalloenzyme and may potentially compensate for the iron-containing equivalent cytochrome c_6 under iron limiting conditions (Peers & Price, 2006). Plastocyanin is typically not found in red algal lineages or their endosymbiotic derivatives (Groussman, Parker, & Armbrust, 2015; Raven, 2013). Within the stramenopiles, plastocyanin appears to be restricted to a small proportion of diatom species, which are disparately related to one another (Groussman et al., 2015; Raven, 2013), along with one bolidophyte species (*Bolidomonas* sp. RCC1657; Fig. 6), suggesting multiple lateral acquisition events. Similarly, some diatoms may possess a plastid-targeted isoform of the iron storage protein ferritin of prokaryotic origin (Marchetti et al., 2009; Raven, 2013),

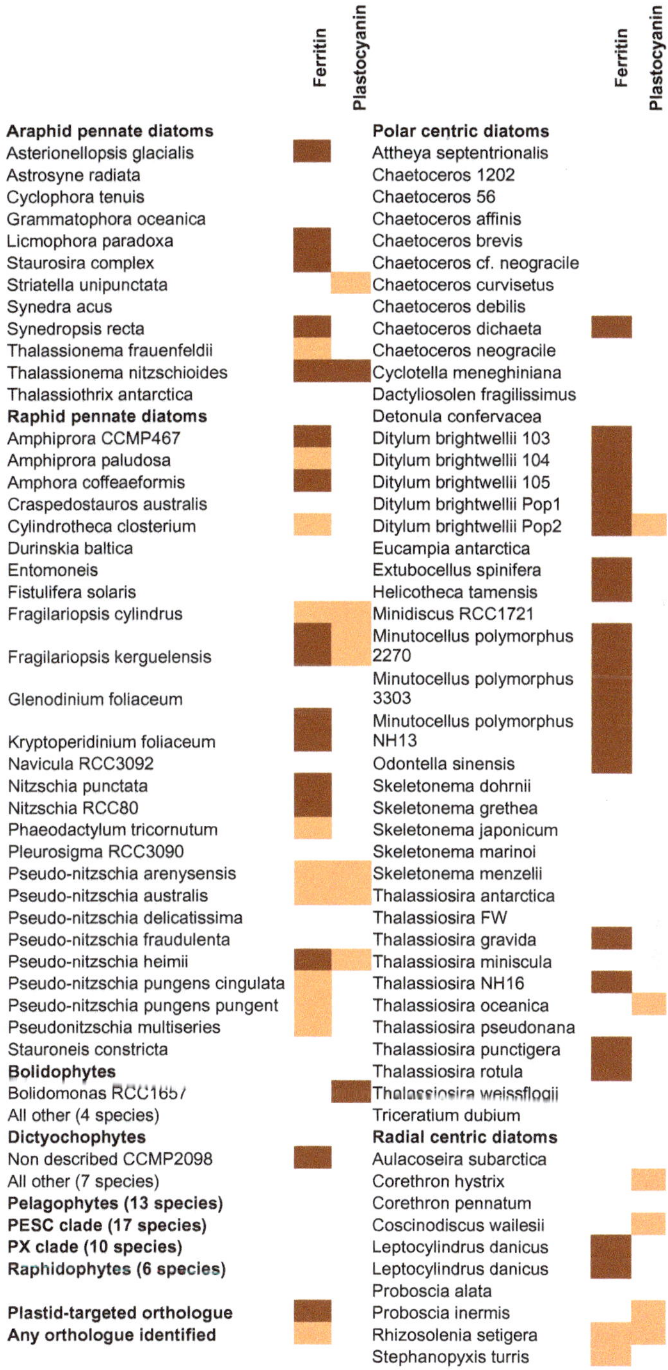

Fig. 6 See legend on next page.

which is expressed in response to iron enrichment and is believed to facilitate the transient storage and remobilisation under transient iron-limiting conditions (Marchetti et al., 2009; Pfaffen et al., 2015). Plastid-targeted ferritin proteins are predicted to be present in some but not all diatom species (Dorrell et al., 2017; Groussman et al., 2015; Marchetti et al., 2009) and are absent from most other stramenopile lineages, although a plastid-targeted isoform may exist in the unclassified arctic dictyochophyte CCMP2098 (Fig. 6) (Dorrell et al., 2017). The plastid localisation of these proteins awaits experimental verification; however, the chimeric nature of diatom plastid proteomes may play a major role in the dominance of this group in iron-poor waters (Bowler et al., 2010; de Vargas et al., 2015).

An alternative strategy that may enable coping with nutrient limitation is photo-mixotrophy, i.e., supplementing the products of photosynthesis with nutrients acquired by feeding on other organisms. Within the stramenopiles, phagocytic consumption is particularly frequently observed in chrysophytes (Beisser et al., 2017; Wilken, Schuurmans, & Matthijs, 2014) and has also been described in raphidophytes (Burkholder, Glibert, & Skelton, 2008; Jeong, 2011). Some pelagophyte (*Aureococcus*) and diatom species have been observed to engage in osmotrophy, taking up external organic nutrients such as urea and glucose in growth media (Burkholder et al., 2008; Kim et al., 2016; Zheng, Quinn, & Sriram, 2013). Comparisons of stramenopiles grown in purely photosynthetic vs mixotrophic conditions (Kim et al., 2016; Wilken, Huisman, Naus-Wiezer, & Van Donk, 2013) have revealed that photo-mixotrophy can substantially change plastid metabolism. These changes include evidence for phosphoenolpyruvate uptake by plastids (Kim et al., 2016) and remodelling of the photosynthesis machinery to reduce both total flux and increase the relative proportions of ATP synthesis via cyclic electron flow compared to inorganic carbon fixation via the CBB

Fig. 6 Evolutionary distribution of iron stress tolerance-associated plastid proteins across stramenopile lineages. This table, integrating information from previous studies of the evolution of iron stress responses in stramenopiles (Groussman et al., 2015) and of stramenopile plastid-targeted proteins (Dorrell et al., 2017), tabulates the inferred evolutionary distribution of ferritin and plastocyanin across genomic and transcriptomic libraries from 151 plastid-bearing stramenopile species. Cells shaded in *brown* correspond to proteins with inferred plastid-targeting sequences as determined by in silico prediction (Gruber et al., 2015; Gschloessl, Guermeur, & Cock, 2008); cells shaded in *orange* correspond to all remaining protein sequences.

cycle (Wilken et al., 2013, 2014). Moreover, photo-mixotrophic stramenopiles express genes associated with photosynthesis to lower relative levels than purely phototrophic relatives (Beisser et al., 2017). It remains to be determined whether photo-mixotrophy also affects nutrient acquisition, and metabolism pathways performed by stramenopiles, and therefore, provide an additional strategy by which their physiology is optimised to respond to nutrient limitation.

4.6 Plastid–Mitochondria Interactions

The mitochondria and plastids of stramenopile cells are typically closely associated with one another (Fig. 3B) (Prihoda et al., 2012). Consistent with this close localisation, extraordinarily complex mitochondrial–plastid interactions are known in diatoms. Allen et al. have documented a complete mitochondrial urea cycle in diatoms, which is upregulated under nitrate limitation, and may have a role in catabolising excess plastidic amino acids and the products of photorespiration (Allen et al., 2011; Prihoda et al., 2012). Recently, Levering et al. have documented a putative plastidic ornithine cycle in *P. tricornutum*, which might plausibly be involved in transporting excess plastidic nitrate via an acetylglutamine shuttle, which could then be dissipated or recycled by the mitochondria (Fig. 7A) (Levering et al., 2016). The plastidic ornithine cycle awaits experimental verification, although notably it appears to be restricted to diatoms and is not found in other stramenopile lineages, including *Bolidomonas* (Dorrell et al., 2017; Levering et al., 2016).

An even more dramatic biophysical connection has recently been demonstrated to occur in diatoms by Bailleul et al. (2015). Excess diatom plastid reducing potential is exported to the mitochondria, where it can be dissipated through respiration, potentially minimising photodamage under high light conditions. Conversely, mitochondrial ATP may be rerouted to the plastid to support photosynthesis under conditions where cyclic electron flow is insufficient to correct the NADPH:ATP ratio essential for the function of the CBB cycle, or to support other plastid biosynthetic processes (Bailleul et al., 2015; Kim et al., 2016). This cycle appears to be highly characteristic of diatoms over other stramenopile lineages. Notably, conventional photoprotective pathways such as cyclic electron flow and the water–water cycle are predicted to only dissipate 15% of the plastid reducing potential in *P. tricornutum* (Bailleul et al., 2015; Levering et al., 2016); in

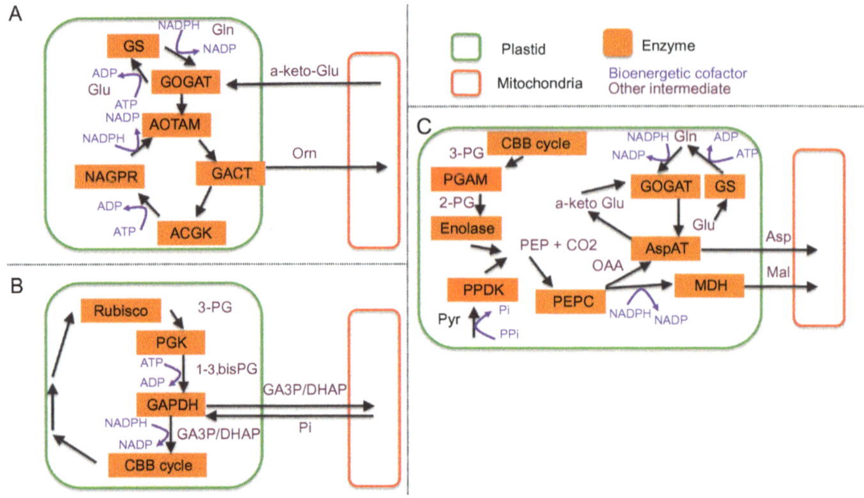

Fig. 7 Models for possible pathways by which excess diatom plastidic reducing potential could be exported to the mitochondria. Three possible scenarios are shown, as proposed by Levering et al. (2016) (A); Smith et al. (2016) and Kim et al. (2016) (B); and positing a role for plastid C_4 synthesis pathways (C) in dissipating excess reducing potential. Abbreviations are used as follows. Cofactors (*blue text*): *ADP*, adenosine diphosphate; *ATP*, adenosine triphosphate; *NADP*, nicotinamide adenine dinucleotide phosphate; *NADPH*, reduced NADP+; *Pi*, phosphate; *Ppi*, pyrophosphate. Metabolites (*plain black text*): *1,3-bisPG*, 1,3-bisphosphoglycerate; *3-PG*, 3-phosphoglycerate; *α-keto-Glu*, α-keto-glutarate; *Asp*, aspartate; *DHAP*, dihydroxyacetone phosphate; *GA3P*, glyceraldehyde-3-phosphate; *Glu*, glutamate; *Gln*, glutamine; *Mal*, malate; *OAA*, oxaloacetate; *Orn*, ornithine; *PEP*, phosphoenolpyruvate; *Pyr*, pyruvate. Enzymes (*orange boxes*): *ACGK*, acetylglutamate kinase; *AOTAM*, acetylornithine aminotransferase; *AspAT*, aspartate aminotransferase; *GACT*, glutamate-*N*-acetyltransferase; *GAPDH*, glyceraldehyde-3-phosphate dehydrogenase; *GS*, glutamine synthetase; *GOGAT*, glutamine oxoglutarate aminotransferase; *MDH*, malate dehydrogenase; *NAGPR*, *N*-acetyl-*δ*-glutamyl-phosphate reductase; *PEPC*, phosphoenolpyruvate carboxylase; *PGAM*, phosphoglycerate mutase; *PGK*, phosphoglycerate kinase; *PPDK*, phosphoenolpyruvate dikinase. In (A), plastidic ornithine, synthesised from α-keto-glutarate and acetylglutamate (generated via previous iterations of the cycle), directly consumes excess plastidic NAPDH and ATP, and also permits the function of a GS/GOGAT cycle that synthesises glutamate at the expense of these two intermediates, prior to being exported to the mitochondria. In (B), triose phosphate, synthesised as part of the CBB cycle using NADPH (GAPDH) at ATP-dependent enzymes (PGK), is directly transported to the mitochondria. In (C), phosphoenolpyruvate, generated either via glycolysis of CBB cycle intermediates or via dikinase activity on plastid pyruvate, is used as a substrate for the synthesis of oxaloacetate, which can then either be directly reduced to malate, or converted into aspartate, indirectly altering the redox poise of the plastid by permitting more NADPH and ATP to be consumed via the GS/GOGAT cycle, prior to being exported to the mitochondria.

contrast, cyclic electron flow appears to form a much larger component of photosynthetic electron transport activity in chrysistan lineages (Fork, Herbert, & Malkin, 1991; Meneghesso et al., 2016). Thus, intricate plastid–mitochondrial interactions may specifically underpin the ecological fitness of diatoms in their natural environments.

It remains to be determined precisely how these connections operate. For example, the exact effectors involved in exporting excess reducing potential from diatom plastids to the mitochondria remain unknown, although different pathways have been proposed (Fig. 7). Smith et al. (2016) and Kim et al. (2016) have suggested that excess reducing potential is dissipated through the export of triose phosphate products of the CBB cycle, which are synthesised at the expense of NADPH and ATP. Once in the mitochondria, they can be directly metabolised by a mitochondrial glycolytic pathway and ultimately the TCA cycle (Fig. 7B). Levering et al. (2016) conversely propose that the plastid ornithine cycle of diatoms could allow the export of excess NADPH and ATP to the mitochondria (Fig. 7A). This efflux could occur both directly, as the cycle contains NADPH-dependent (*N*-acetyl-δ-glutamyl-phosphate reductase) and ATP-dependent (acetylglutamate kinase) steps, and indirectly, as the glutamate required for the perpetuation of the cycle might be synthesised from α-ketoglutarate in an NADPH and ATP-dependent manner, via the activities of glutamine oxoglutarate aminotransferase and glutamine synthetase. Finally (Fig. 7C), excess NADPH might be transferred to the mitochondria via an oxaloacetate derivative, either malate (synthesised by an as yet unknown plastidic malate dehydrogenase) (Prihoda et al., 2012) or aspartate (which could be synthesised by aspartate aminotransferase at the expense of plastidic glutamate, thus enabling dissipation of excess ATP and NADPH via glutamate synthesis, as earlier). In this model, the oxaloacetate might be synthesised via the partial C_4 pathway present in diatom plastids, using plastidic phosphoenolpyruvate carboxylase, and phosphoenolpyruvate generated either via pyruvate-phosphate dikinase activity or via the plastidic glycolytic pathway from the CBB cycle (Fig. 7C). A role for C_4 photosynthesis in the dissipation of excess plastid reducing potential would further explain why mutants of diatom C_4 photosynthesis pathways have partially impaired photosynthetic functions, even though their ability to acquire CO_2 is not compromised (Haimovich-Dayan et al., 2013). Ultimately, more nuanced experimental dissection is required here to identify the precise effector(s) of this unusual pathway.

5. EVOLUTIONARY DYNAMICS

5.1 Loss of Photosynthesis

Several stramenopile lineages have secondarily lost the ability to photosynthesise. In contrast to the situation for oomycetes and other basal lineages, these groups clearly branch with photosynthetic relatives, hence must have once had plastids, and in some of them leucoplasts have been visualised (Kamikawa, Yubuki, et al., 2015; Mylnikov, Mylnikova, & Tikhonenkov, 2008; Sekiguchi, Moriya, Nakayama, & Inouye, 2002). The secondarily nonphotosynthetic lineages include members of the chrysophytes and synurophytes (*Paraphysomonas*, *Apoikia*, *Spumella*) (Kim, Yubuki, Leander, & Graham, 2010; Mangot et al., 2017; Mylnikov et al., 2008); *Leukarachnion*, which forms a sister-group to synchromophytes (Grant, Tekle, Anderson, Patterson, & Katz, 2009); the actinophryids, which resolve with the raphidophytes (Cavalier-Smith & Scoble, 2013); the dictyochophytes *Pteridomonas* and *Ciliophrys* (Sekiguchi et al., 2002); and several pennate diatom species within the *Nitzschia* clade (Kamikawa, Tanifuji, et al., 2015). All of the above lineages are believed to be free-living osmotrophs or phagotrophs, rather than obligate parasites of other lineages, and may constitute major heterotrophic components of freshwater and marine ecosystems (Mangot et al., 2017).

Unravelling the steps by which these lineages have reverted to a purely heterotrophic lifestyle remains a complicated issue. Preliminary evidence indicates a partial reduction or alteration to the photosynthesis machinery in at least some nonphotosynthetic stramenopile lineages, as opposed to an outright loss. The first completed plastid genome of a secondarily nonphotosynthetic stramenopile, the diatom *Nitzschia* sp. NIES-3581 (Kamikawa, Tanifuji, et al., 2015), retains genes encoding components of the ATP synthase complex in the plastid genome, which may have been retained for the purpose of ATP hydrolysis (Kamikawa, Tanifuji, et al., 2015). Plastid genomes from other secondarily nonphotosynthetic stramenopile lineages have yet to be published, although an earlier study documented the retention of *rbcL* genes in nonphotosynthetic dictyochophyte algae, indicating that these species have likewise not eliminated all photosynthesis-associated genes from their plastid genomes either (Sekiguchi et al., 2002). Recently, Beisser et al. have completed a comparative transcriptomic analysis of 18 photosynthetic and nonphotosynthetic chrysophytes, and detected transcribed nucleus-encoded proteins associated with photosynthesis in some nonphotosynthetic genera (*Cornospumella*, *Pedospumella*, and *Poteriospumella*) but not in others (Beisser et al., 2017), suggesting a stepwise or variable reduction in the photosynthesis machinery of different species. Profiling the nucleus-encoded and plastid-targeted proteins associated with other nonphotosynthetic stramenopiles may provide valuable insights into the broader metabolism associated with their plastids, and thus might provide insights into the reasons for the retention of photosynthesis-associated genes in their plastid genomes.

5.2 Transient Symbioses and Kleptoplasty

In some cases, plastids from stramenopile algae may be taken up as transient symbionts, e.g., kleptoplasts, by other eukaryotes. Perhaps the best known example of this phenomenon is the marine sea slug *Elysia chlorotica*, which as a juvenile feeds on the xanthophyte alga *Vaucheria litorea*, and subsists on the sequestered plastids it stores from this association throughout the remainder of its adult life (Dorrell & Howe, 2012; Rumpho, Pelletreau, Moustafa, & Bhattacharya, 2011). Other lineages that may harbour photosynthetic stramenopile symbionts include dinoflagellates (*Podolampas*, which possesses symbionts of dictyochophyte origin) (Park, Kim, & Kim, 2014), foraminiferans (*Elphidium* and *Haynesina*, which possess kleptoplasts derived from a diverse range of diatoms) (Pillet, de Vargas, & Pawlowski, 2011), and fungi (the "marine lichen" *Verrucaria*, which forms a complex symbiosis with the phaeophyte alga *Petroderma maculiforme*) (Sanders, Moe, & Ascaso, 2004).

The exact nature and mechanism by which these nonpermanent symbioses are established and operate remain incompletely understood. Initial studies into the *Elysia*/ *Vaucheria* symbiosis posited gene transfer from the symbiont to the host nucleus as a possible reason for its stability (Pierce, Curtis, & Schwartz, 2010; Rumpho et al., 2009). While isolated gene transfer events may have occurred (Schwartz, Curtis, & Pierce, 2014), there is however no evidence for an extensive suite of endosymbiotically transferred genes that could support the kleptoplasts (Bhattacharya, Pelletreau, Price, Sarver, & Rumpho, 2013; Rauch et al., 2015). More recently, the exact physiological nature of this symbiosis has been called into question: while the sequestered *Vaucheria* plastids may support the host *Elysia* cells through photosynthesis (Cruz et al., 2015; Pierce, Curtis, & Middlebrooks, 2015), they may equally be metabolically inactive, representing thus a purely pseudostable food reserve for the host to exploit during its adult life

(Christa et al., 2014; Rauch et al., 2015). Equivalent studies of other stramenopile photosymbionts remain to be performed, although preliminary evidence indicates the presence of functional photosynthesis but absence of gene transfer in foraminiferan/diatom symbioses (Jauffrais et al., 2016; Pillet & Pawlowski, 2013).

5.3 Tertiary and Higher Endosymbioses

In several instances, stramenopile algae may have been taken up as plastids by other eukaryotic lineages. The best-characterised examples of these events occur in the "dinotoms", an unusual group of dinoflagellate algae that possess permanent plastids of clear diatom origin (Chesnick, Morden, & Schmieg, 1996; Imanian et al., 2010). The most well understood of these lineages, exemplified by the type species *Kryptoperidinium* and *Durinskia*, possess plastids acquired from a pennate diatom probably closely related to *Cylindrotheca* sp. (Dorrell & Howe, 2015; Imanian et al., 2010). Dinotoms with endosymbionts derived from alternative pennate diatom species and from centric diatoms are also known, raising the possibility of serial replacement of diatom endosymbionts within this lineage (Horiguchi & Takano, 2006; Takano, Hansen, Fujita, & Horiguchi, 2008; Yamada, Sym, & Horiguchi, 2017). Studied dinotom endosymbionts are marked by their extreme lack of reduction relative to free-living diatoms, retaining essentially identical plastid genomes (Imanian et al., 2010), as well as functional mitochondria (Imanian, Pombert, Dorrell, Burki, & Keeling, 2012) and a large endosymbiont nucleus, which appears to retain genes for all of the plastid-targeted pathways essential for supporting the dinotom plastid (Hehenberger, Burki, Kolisko, & Keeling, 2016; Hehenberger, Imanian, Burki, & Keeling, 2014). Why the dinotom endosymbiont has undergone essentially no functional integration with its host remains to be determined, although it has been proposed that this reflects the strong interdependence of diatom plastids with other cellular organelles (Dorrell & Howe, 2015).

More recently, two much more ecologically prominent cases of stramenopile endosymbiosis have been proposed. In 2014, Stiller et al. suggested, based on footprints of gene transfer between nuclear genomes, a stramenopile origin for the haptophyte plastid (Stiller et al., 2014). Most recently, we have documented a robust and specific signal consistent with a plastid endosymbiosis of a member of the Hypogyristea (pelagophyte/dictyochophyte clade) by an ancient haptophyte alga (Fig. 8) (Dorrell et al., 2017). This is particularly interesting because multigene trees very clearly

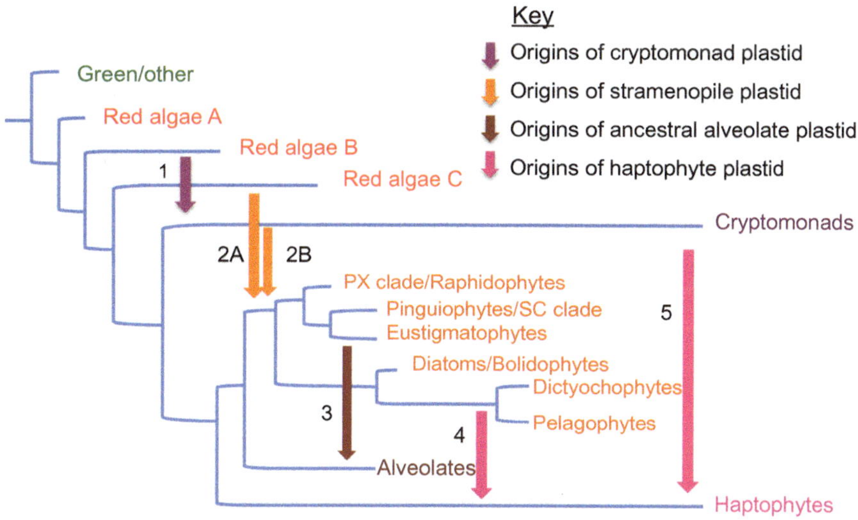

Fig. 8 Possible endosymbiotic connections between stramenopiles and other plastids of red origin. This figure shows a series of endosymbiotic events that may underpin the origins of cryptomonad, stramenopile, alveolate, and haptophyte plastids. The tree diagram reflects the underlying nuclear/mitochondrial relationships between each lineage, and *coloured arrows* indicate the progression of plastid endosymbiosis events observed. For simplicity, only the ancestral alveolate plastid type is shown. An initial secondary endosymbiosis of a red alga occurred in an ancestor of the cryptomonads (1). This plastid was then either transferred to an ancestor of the stramenopiles by tertiary endosymbiosis (2B), or alternatively the stramenopiles may have acquired their own plastid via an independent secondary endosymbiosis of a different red algal lineage (2A). An early ancestor of the eustigmatophytes may have been engulfed by a common ancestor of plastid-bearing alveolates, forming the ancestor of the ancestral dinoflagellate, chromerid, and apicomplexan plastid (3). Finally, an ancestor of the pelagophytes/dictyochophytes was engulfed by an ancestor of the haptophytes (4), yielding a plastid that appears to have been subsequently replaced via a serial endosymbiosis event, possibly from a cryptomonad alga (5). *Adapted from previous studies Dorrell, R. G., Gile, G., McCallum, G., Méheust, R., Bapteste, E. P., Klinger, C. M., et al. (2017). Chimeric origins of ochrophytes and haptophytes revealed through an ancient plastid proteome.* Elife, 6, *23717; Dorrell, R. G., & Smith, A. G. (2011). Do red and green make brown?: Perspectives on plastid acquisitions within chromalveolates.* Eukaryotic Cell, 10, *856–868; Stiller, J. W., Schreiber, J., Yue, J., Guo, H., Ding, Q., & Huang, J. (2014). The evolution of photosynthesis in chromist algae through serial endosymbioses.* Nature Communications, 5, *5764.*

document a nonpelagophyte origin for the haptophyte plastid (Dorrell et al., 2017; Stiller et al., 2014). This discrepancy is supported by numerous discrete features (e.g. a bacterial-type *rpl36* gene, histidine kinase genes, inverted repeats, and the plastid division genes *minD* and *minE*) that are present in haptophyte plastid genomes but are absent from sequenced pelagophyte plastid genomes (de Vries & Gould, 2017; Duplessis et al., 2007; Ong et al., 2010; Rice & Palmer, 2006). The contrasting data could be resolved by invoking a serial endosymbiosis event, in which an ancient haptophyte acquired a plastid of pelagophyte/dictyochophyte origin, which was subsequently replaced by a plastid of nonstramenopile origin (Fig. 8) (Dorrell et al., 2017). Verifying this hypothesis will require the identification of the closest extant relative of the current haptophyte plastid, which remains unknown (Dorrell et al., 2017).

An even less well-characterised stramenopile endosymbiosis may have occurred in an ancestor of plastid-bearing members of the alveolates, i.e., dinoflagellates, chromerids, and apicomplexans. Ševčíková et al. have proposed a sister-group relationship between the plastid genomes of the eustigmatophytes and the chromerid alga *Vitrella* (Fig. 8) (Ševčíková et al., 2015). Notably, both chromerid and eustigmatophyte plastids possess a type I intron in the tRNA-Leu(UAA) gene (Fig. 4) (Janouskovec, Horák, Oborník, Lukes, & Keeling, 2010; Ševčíková et al., 2015), although this feature appears to have arisen independently in several other plastid lineages across the tree of life (Simon, Fewer, Friedl, & Bhattacharya, 2003), and thus its utility as a synapomorphic marker in this case is limited. Similarly, a split *clpC* gene appears to have evolved independently in eustigmatophytes and in specific alveolate lineages (*Chromera*) but not in others (*Vitrella*) (Fig. 4) (Janouskovec et al., 2010; Ševčíková et al., 2015). Thus, the exact evolutionary origins of the alveolate and ochrophyte plastids remain uncertain.

6. CONCLUDING REMARKS

In this chapter, we have explored the extraordinarily broad studied diversity of photosynthetic stramenopiles and their plastids (Figs 1, 3, and 4). We have pinpointed the origin of the stramenopile plastid to a recent common ancestor of the extant photosynthetic lineages. Hence, the majority of basally divergent aplastidic stramenopile groups have never possessed plastids. The late origin of the stramenopile plastid provides a valuable model for exploring the evolutionary transitions associated with plastid acquisition, which is not possible for many other lineages (e.g. Archaeplastida, cryptomonads, and haptophytes) given the relative paucity of sister-groups for each lineage that diverged prior to the plastid endosymbiosis event (Aleoshin et al., 2016; Dorrell et al., 2017). Resolving exactly when this acquisition event occurred will now depend on profiling the sequence content of the very closest relatives of the photosynthetic stramenopiles (e.g. *Pirsonia*, *Developayella*, and hypochytriomycetes), which remain understudied at the genomic level (Fig. 1) (Aleoshin et al., 2016; Massana, del Campo, Sieracki, Audic, & Logares, 2014).

We have further shown that the stramenopile plastid is an evolutionary mosaic, not only utilising proteins of red algal origin, consistent with its endosymbiotic history, but is also supported by plastid-targeted proteins acquired from prokaryotes, green algae, and the stramenopile host lineage (Fig. 2). In some cases, such as the plastid xanthophyll cycle, Lhcx proteins, and iron storage and mobilisation proteins of probable green algal and/or prokaryotic origin, nonvertically inherited plastid proteins have been verified to specifically contribute to the fitness of stramenopiles as a whole, or to the individual stramenopile clades in which they occur, in stressful environments (Fig. 5). Thus, a chimeric plastid proteome may have enabled specific stramenopile clades to rise to their position of environmental dominance. It remains to be determined whether all of the nonvertically acquired plastid proteins in stramenopiles, particularly those with red algal orthologues, have been retained in stramenopile genomes for selective reasons, or through stochastic patterns of acquisition and loss of red orthologues. This will be more clearly understood by profiling the phenotypes of a wider range of mutant lines for stramenopile plastid proteins of nonred origin, or by generating lines in which plastid-targeted proteins of nonred origin have been replaced by orthologues of red derivation, and investigating whether this "artificial reversal" of a lateral gene transfer event has consequences for fitness.

We have profiled the distinctive architecture and relatively conserved genomes of stramenopile plastid lineages, and have explored the unusual biochemical pathways that have arisen in individual stramenopile groups, some of which are broadly conserved (e.g. the use of photoprotective strategies of green algal origin, and a biophysical carbon-concentrating mechanism). Others, such as elaborate iron storage and recycling pathways, a plastidic ornithine cycle, and intricate metabolic connections between the plastids and mitochondria, may be restricted to individual species or to ecologically fundamental groups such as diatoms (Fig. 6). It remains to be determined, however, whether any of these lineage-specific metabolic pathways have contributed to the pronounced ecological presence of diatoms compared to other stramenopile groups. Identifying the key effectors involved in each pathway, for example, the redox enzymes and transporters involved in transferring excess reducing potential and ATP between the chloroplast and mitochondria (Fig. 7) may reveal the elaborations within the diatom plastid proteome that have contributed to the ecological success of this lineage.

Finally, we have discussed the ways in which stramenopile plastids have been modified or transformed following their endosymbiotic acquisition, in some species having been converted into nonphotosynthetic leucoplasts, and in others potentially having been taken up through even more complex endosymbioses by early ancestors of other major eukaryotic lineages (Fig. 8). Most significantly, we describe an ancient endosymbiosis of an ancestor of the pelagophytes/dictyochophytes by an early haptophyte which, based on discrepancies with the plastid genome signal, may represent a serial endosymbiosis event deep in haptophyte history. It will be important to identify the exact origin of the extant haptophyte plastid, specifically whether it can be robustly placed as arising within a separate lineage of algae to the stramenopiles. Similarly, it will be particularly interesting to determine whether the plastid proteomes of photosynthetic alveolates contain proteins that clearly originated from a specific evolutionary point within the stramenopiles, which would help confirm the possible stramenopile origin of the plastid genome of this lineage. Ultimately, understanding both of these ancient and complicated evolutionary stories will require a more precise understanding of stramenopile taxonomy and evolutionary dynamics.

The stramenopiles are a highly diverse, divergent, and ecologically fundamental group of eukaryotic algae, and this is reflected in the convoluted history and unusual biology of their plastids. More nuanced understanding of the metabolic elaboration and evolutionary histories of stramenopile plastids may help reveal the evolutionary and ecophysiological diversification of this major group of photosynthetic eukaryotes.

ACKNOWLEDGEMENTS

Funding is acknowledged from the ERC Advanced Award "Diatomite", the Louis D Foundation of the Institut de France, the French Government "Investissements d'Avenir" programmes MEMO LIFE (ANR-10-LABX-54), OCEANOMICS (ANR-11-BTBR-0008), and PSL* Research University (ANR-11-IDEX-0001-02), and the Gordon and Betty Moore Foundation (all to C.B.). R.G.D. has been supported by the Mairie de Paris and an EMBO early career fellowship (ALTF 1124-2014). C.B. also thanks the Radcliffe Institute of Advanced Study at Harvard University for a scholars' fellowship during the 2016–2017 academic year.

REFERENCES

Alboresi, A., Le Quiniou, C., Yadav, S. K. N., Scholz, M., Meneghesso, A., Gerotto, C., et al. (2017). Conservation of core complex subunits shaped the structure and function of photosystem I in the secondary endosymbiont alga *Nannochloropsis gaditana*. *The New Phytologist, 213*, 714–726.

Aleoshin, V. V., Mylnikov, A. P., Mirzaeva, G. S., Mikhailov, K. V., & Karpov, S. A. (2016). Heterokont predator *Develorapax marinus* gen. et sp. nov.—A model of the ochrophyte ancestor. *Frontiers in Microbiology*, 7, 1194.

Allen, A. E., Dupont, C. L., Obornik, M., Horak, A., Nunes-Nesi, A., McCrow, J. P., et al. (2011). Evolution and metabolic significance of the urea cycle in photosynthetic diatoms. *Nature, 473*, 203–206.

Allen, A. E., Laroche, J., Maheswari, U., Lommer, M., Schauer, N., Lopez, P. J., et al. (2008). Whole-cell response of the pennate diatom *Phaeodactylum tricornutum* to iron starvation. *Proceedings of the National Academy of Sciences of the United States of America, 105*, 10438–10443.

Andersen, R. A. (2004). Biology and systematics of heterokont and haptophyte algae. *American Journal of Botany, 91*, 1508–1522.

Andersson, J. O., & Roger, A. J. (2002). A cyanobacterial gene in nonphotosynthetic protists—An early chloroplast acquisition in eukaryotes? *Current Biology, 12*, 115–119.

Armbrust, E. V., Berges, J. A., Bowler, C., Green, B. R., Martinez, D., Putnam, N. H., et al. (2004). The genome of the diatom *Thalassiosira pseudonana*: Ecology, evolution, and metabolism. *Science, 306*(5693), 79–86.

Ast, M., Gruber, A., Schmitz-Esser, S., Neuhaus, H. E., Kroth, P. G., Horn, M., et al. (2009). Diatom plastids depend on nucleotide import from the cytosol. *Proceedings of the National Academy of Sciences of the United States of America, 106*, 3621–3626.

Badger, M. R., Andrews, T. J., Whitney, S. M., Ludwig, M., Yellowlees, D. C., Leggat, W., et al. (1998). The diversity and coevolution of Rubisco, plastids, pyrenoids, and chloroplast-based CO_2-concentrating mechanisms in algae. *Canadian Journal of Botany, 76*, 1052–1071.

Bailleul, B., Berne, N., Murik, O., Petroutsos, D., Prihoda, J., Tanaka, A., et al. (2015). Energetic coupling between plastids and mitochondria drives CO_2 assimilation in diatoms. *Nature, 524*, 366–369.

Bailleul, B., Rogato, A., de Martino, A., Coesel, S., Cardol, P., Bowler, C., et al. (2010). An atypical member of the light-harvesting complex stress-related protein family modulates diatom responses to light. *Proceedings of the National Academy of Sciences of the United States of America, 107*, 18214–18219.

Basu, S., Patil, S., Mapleson, D., Russo, M. T., Vitale, L., Fevola, C., et al. (2017). Finding a partner in the ocean: Molecular and evolutionary bases of the response to sexual cues in a planktonic diatom. *The New Phytologist, 215*, 140–156.

Baurain, D., Brinkmann, H., Petersen, J., Rodriguez-Ezpeleta, N., Stechmann, A., Demoulin, V., et al. (2010). Phylogenomic evidence for separate acquisition of plastids in cryptophytes, haptophytes, and stramenopiles. *Molecular Biology and Evolution, 27*, 1698–1709.

Beisser, D., Graupner, N., Bock, C., Wodniok, S., Grossmann, L., Vos, M., et al. (2017). Comprehensive transcriptome analysis provides new insights into nutritional strategies and phylogenetic relationships of chrysophytes. *PeerJ, 5*(e), 2832.

Bertrand, M. (2010). Carotenoid biosynthesis in diatoms. *Photosynthesis Research, 106*, 89–102.

Bhattacharya, D., Pelletreau, K. N., Price, D. C., Sarver, K. E., & Rumpho, M. E. (2013). Genome analysis of *Elysia chlorotica* egg DNA provides no evidence for horizontal gene transfer into the germ line of this kleptoplastic mollusc. *Molecular Biology and Evolution, 30*, 1843–1852.

Bhaya, D., & Grossman, A. (1991). Targeting proteins to diatom plastids involves transport through an endoplasmic reticulum. *Molecular & General Genetics*, *229*, 400–404.

Bowler, C., Allen, A. E., Badger, J. H., Grimwood, J., Jabbari, K., Kuo, A., et al. (2008). The *Phaeodactylum* genome reveals the evolutionary history of diatom genomes. *Nature*, *456*, 239–244.

Bowler, C., Vardi, A., & Allen, A. E. (2010). Oceanographic and biogeochemical insights from diatom genomes. *Annual Review of Marine Science*, *2*, 333–365.

Brembu, T., Winge, P., Tooming-Klunderud, A., Nederbragt, A. J., Jakobsen, K. S., & Bones, A. M. (2014). The chloroplast genome of the diatom *Seminavis robusta*: New features introduced through multiple mechanisms of horizontal gene transfer. *Marine Genomics*, *16*, 17–27.

Bromke, M. A. (2013). Amino acid biosynthesis pathways in diatoms. *Metabolites*, *3*, 294–311.

Brown, J. W., & Sorhannus, U. (2010). A molecular genetic timescale for the diversification of autotrophic stramenopiles (Ochrophyta): Substantive underestimation of putative fossil ages. *PloS One*, *5*, 12759.

Burkholder, J. M., Glibert, P. M., & Skelton, H. M. (2008). Mixotrophy, a major mode of nutrition for harmful algal species in eutrophic waters. *Harmful Algae*, *8*, 77–93.

Burki, F., Flegontov, P., Oborník, M., Cihlár, J., Pain, A., Lukes, J., et al. (2012). Re-evaluating the green versus red signal in eukaryotes with secondary plastid of red algal origin. *Genome Biology and Evolution*, *4*, 626–635.

Burki, F., Kaplan, M., Tikhonenkov, D. V., Zlatogursky, V., Minh, B. Q., Radaykina, L. V., et al. (2016). Untangling the early diversification of eukaryotes: A phylogenomic study of the evolutionary origins of Centrohelida, Haptophyta and Cryptista. *Proceedings. Biological Sciences*, *283*(1823).

Burki, F., Shalchian-Tabrizi, K., Minge, M., Skjaeveland, A., Nikolaev, S. I., Jakobsen, K. S., et al. (2007). Phylogenomics reshuffles the eukaryotic supergroups. *PloS One*, *2*, e790.

Cao, S. N., Zhang, X. W., Xu, D., Fan, X., Mou, S. L., Wang, Y. T., et al. (2013). A transthylakoid proton gradient and inhibitors induce a non-photochemical fluorescence quenching in unicellular algae *Nannochloropsis* sp. *FEBS Letters*, *587*, 1310–1315.

Cattolico, R. A., Jacobs, M. A., Zhou, Y., Chang, J., Duplessis, M., Lybrand, T., et al. (2008). Chloroplast genome sequencing analysis of *Heterosigma akashiwo* CCMP452 (West Atlantic) and NIES293 (West Pacific) strains. *BMC Genomics*, *9*, 211.

Cavalier-Smith, T. (1989). The kingdom Chromista. *Chromophyte Algae: Problems and Perspectives*, *38*, 381–407.

Cavalier-Smith, T., & Scoble, J. M. (2013). Phylogeny of Heterokonta: Incisomonas marina, a uniciliate gliding opalozoan related to Solenicola (Nanomonadea), and evidence that Actinophryida evolved from raphidophytes. *European Journal of Protistology*, *49*, 328–353.

Chan, C. X., Reyes-Prieto, A., & Bhattacharya, D. (2011). Red and green algal origin of diatom membrane transporters: Insights into environmental adaptation and cell evolution. *PloS One*, *6*, 29138.

Chesnick, J. M., Morden, C. W., & Schmieg, A. M. (1996). Identity of the endosymbiont of *Peridinium foliaceum* (Pyrrophyta): Analysis of the *rbcLS* operon. *Journal of Phycology*, *32*, 850–857.

Chi, S., Wu, S. X., Yu, J., Wang, X. M., Tang, X. X., & Liu, T. (2014). Phylogeny of C-4-Photosynthesis enzymes based on algal transcriptomic and genomic data supports an archaeal/proteobacterial origin and multiple duplication for most C4-related genes. *PloS One*, *9*, 110154.

Christa, G., Zimorski, V., Woehle, C., Tielens, A. G. M., Wagele, H., Martin, W. F., et al. (2014). Plastid-bearing sea slugs fix CO2 in the light but do not require photosynthesis to survive. *Proceedings. Biological Sciences*, *281*, 1774.

Chu, L., Gruber, A., Ast, M., Schmitz-Esser, S., Altensell, J., Neuhaus, H. E., et al. (2017). Shuttling of (deoxy-) purine nucleotides between compartments of the diatom *Phaeodactylum tricornutum*. *The New Phytologist, 213*, 193–205.

Cihlar, J., Fussy, Z., Horak, A., & Obornik, M. (2016). Evolution of the tetrapyrrole biosynthetic pathway in secondary algae: Conservation, redundancy and replacement. *PloS One, 11*, 166338.

Clement, R., Dimnet, L., Maberly, S. C., & Gontero, B. (2016). The nature of the CO2-concentrating mechanisms in a marine diatom, *Thalassiosira pseudonana*. *The New Phytologist, 209*, 1417–1427.

Clement, R., Lignon, S., Mansuelle, P., Jensen, E., Pophillat, M., Lebrun, R., et al. (2017). Responses of the marine diatom *Thalassiosira pseudonana* to changes in CO_2 concentration: A proteomic approach. *Scientific Reports, 7*, 42333.

Cock, J. M., Sterck, L., Rouze, P., Scornet, D., Allen, A. E., Amoutzias, G., et al. (2010). The *Ectocarpus* genome and the independent evolution of multicellularity in brown algae. *Nature, 465*, 617–621.

Coesel, S., Obornik, M., Varela, J., Falciatore, A., & Bowler, C. (2008). Evolutionary origins and functions of the carotenoid biosynthetic pathway in marine diatoms. *PloS One, 3*, 2896.

Cruz, S., Cartaxana, P., Newcomer, R., Dionisio, G., Calado, R., Serodio, J., et al. (2015). Photoprotection in sequestered plastids of sea slugs and respective algal sources. *Scientific Reports, 5*, 7904.

Davis, A., Abbriano, R., Smith, S. R., & Hildebrand, M. (2017). Clarification of photorespiratory processes and the role of malic enzyme in diatoms. *Protist, 168*, 134–153.

de Vargas, C., Audic, S., Henry, N., Decelle, J., Mahé, F., Logares, R., et al. (2015). Ocean plankton. Eukaryotic plankton diversity in the sunlit ocean. *Science, 348*, 1261605.

de Vries, J., & Gould, S. B. (2017). The monoplastidic bottleneck in algae and plant evolution. *bioRXiv, 2*, 109975.

Derelle, R., López-García, P., Timpano, H., & Moreira, D. (2016). A phylogenomic framework to study the diversity and evolution of stramenopiles (=heterokonts). *Molecular Biology and Evolution, 33*, 2890–2898.

Deschamps, P., & Moreira, D. (2012). Re-evaluating the green contribution to diatom genomes. *Genome Biology and Evolution, 4*, 683–688.

Dittami, S. M., Michel, G., Collén, J., Boyen, C., & Tonon, T. (2010). Chlorophyll-binding proteins revisited-a multigenic family of light-harvesting and stress proteins from a brown algal perspective. *BMC Evolutionary Biology, 10*, 365.

Dorrell, R. G., Gile, G., McCallum, G., Méheust, R., Bapteste, E. P., Klinger, C. M., et al. (2017). Chimeric origins of ochrophytes and haptophytes revealed through an ancient plastid proteome. *Elife, 6*, 23717.

Dorrell, R. G., & Howe, C. J. (2012). What makes a chloroplast? Reconstructing the establishment of photosynthetic symbioses. *Journal of Cell Science, 125*, 1865–1875.

Dorrell, R. G., & Howe, C. J. (2015). Integration of plastids with their hosts: Lessons learnt from dinoflagellates. *Proceedings of the National Academy of Sciences of the United States of America, 112*, 10247–10254.

Dorrell, R. G., & Smith, A. G. (2011). Do red and green make brown? Perspectives on plastid acquisitions within chromalveolates. *Eukaryotic Cell, 10*, 856–868.

Duplessis, M. R., Karol, K. G., Adman, E. T., Choi, L. Y. S., Jacobs, M. A., & Cattolico, R. A. (2007). Chloroplast His-to-Asp signal transduction: A potential mechanism for plastid gene regulation in *Heterosigma akashiwo* (Raphidophyceae). *BMC Evolutionary Biology, 7*, 70.

Felsner, G., Sommer, M. S., Gruenheit, N., Hempel, F., Moog, D., Zauner, S., et al. (2011). ERAD components in organisms with complex red plastids suggest recruitment of a pre-existing protein transport pathway for the periplastid membrane. *Genome Biology and Evolution, 3*, 140–150. http://dx.doi.org/10.1093/gbe/evq074.

Flori, S., Jouneau, P. H., Finazzi, G., Maréchal, E., & Falconet, D. (2016). Ultrastructure of the periplastidial compartment of the diatom Phaeodactylum tricornutum. *Protist, 167*(3), 254–267. http://dx.doi.org/10.1016/j.protis.2016.04.001.

Fork, D. C., Herbert, S. K., & Malkin, S. (1991). Light energy-distribution in the brown alga Macrocystis pyrifera (giant-kelp). *Plant Physiology, 95*(3), 731–739. http://dx.doi.org/10.1104/pp.95.3.731.

Frommolt, R., Werner, S., Paulsen, H., Goss, R., Wilhelm, C., Zauner, S., et al. (2008). Ancient recruitment by chromists of green algal genes encoding enzymes for carotenoid biosynthesis. *Molecular Biology and Evolution, 25*(12), 2653–2667. http://dx.doi.org/10.1093/molbev/msn206.

Galachyants, Y. P., Morozov, A. A., Mardanov, A. V., Beletsky, A. V., Ravin, N. V., Petrova, D. P., et al. (2012). Complete chloroplast genome sequence of freshwater araphid pennate diatom alga Synedra acus from Lake Baikal. *International Journal of Biology, 4*(1), 9.

Gee, C. W., & Niyogi, K. K. (2017). The carbonic anhydrase CAH1 is an essential component of the carbon-concentrating mechanism in Nannochloropsis oceanica. *Proceedings of the National Academy of Sciences of the United States of America, 114*(17), 4537–4542. http://dx.doi.org/10.1073/pnas.1700139114.

Gibbs, S. P., Luttke, A., & Cattolico, R. A. (1976). Synchronous growth and chloroplast replication in chrysophyte alga, olisthodiscusluteus. *Journal of Cell Biology, 70*(2), A215.

Gile, G. H., Moog, D., Slamovits, C. H., Maier, U. G., & Archibald, J. M. (2015). Dual organellar targeting of aminoacyl-tRNA synthetases in diatoms and cryptophytes. *Genome Biology & Evolution,* 7(6), 1728–1742. http://dx.doi.org/10.1093/gbe/evv095.

Gobler, C. J., & Sunda, W. G. (2012). Ecosystem disruptive algal blooms of the brown tide species, Aureococcus anophagefferens and Aureoumbra lagunensis. *Harmful Algae, 14,* 36–45. http://dx.doi.org/10.1016/j.hal.2011.10.013.

Gould, S. B., Sommer, M. S., Kroth, P. G., Gile, G. H., Keeling, P. J., & Maier, U. G. (2006). Nucleus-to-nucleus gene transfer and protein retargeting into a remnant cytoplasm of cryptophytes and diatoms. *Molecular Biology and Evolution, 23*(12), 2413–2422. http://dx.doi.org/10.1093/molbev/msl113.

Grant, J., Tekle, Y. I., Anderson, O. R., Patterson, D. J., & Katz, L. A. (2009). Multigene evidence for the placement of a heterotrophic amoeboid lineage leukarachnion sp among photosynthetic Stramenopiles. *Protist, 160*(3), 376–385. http://dx.doi.org/10.1016/j.protis.2009.01.001.

Grouneva, I., Rokka, A., & Aro, E. M. (2011). The thylakoid membrane proteome of two marine diatoms outlines both diatom-specific and species-specific features of the photosynthetic machinery. *Journal of Proteome Research, 10*(12), 5338–5353. http://dx.doi.org/10.1021/pr200600f.

Groussman, R. D., Parker, M. S., & Armbrust, E. V. (2015). Diversity and evolutionary history of iron metabolism genes in diatoms. *PloS One, 10*(6), 25. http://dx.doi.org/10.1371/J.pone.0129081.

Gruber, A., Rocap, G., Kroth, P. G., Armbrust, E. V., & Mock, T. (2015). Plastid proteome prediction for diatoms and other algae with secondary plastids of the red lineage. *The Plant Journal, 81*(3), 519–528. http://dx.doi.org/10.1111/tpj.12734.

Gruber, A., Vugrinec, S., Hempel, F., Gould, S. B., Maier, U. G., & Kroth, P. G. (2007). Protein targeting into complex diatom plastids: Functional characterisation of a specific targeting motif. *Plant Molecular Biology, 64*(5), 519–530. http://dx.doi.org/10.1007/s11103-007-9171-x.

Gschloessl, B., Guermeur, Y., & Cock, J. M. (2008). HECTAR: A method to predict subcellular targeting in heterokonts. *BMC Bioinformatics, 9,* 393. http://dx.doi.org/10.1186/1471-2105-9-393.

Haimovich-Dayan, M., Garfinkel, N., Ewe, D., Marcus, Y., Gruber, A., Wagner, H., et al. (2013). The role of C4 metabolism in the marine diatom *Phaeodactylum tricornutum*. *The New Phytologist*, *197*(1), 177–185. http://dx.doi.org/10.1111/j.1469-8137.2012.04375.x.

He, D., Sierra, R., Pawlowski, J., & Baldauf, S. L. (2016). Reducing long-branch effects in multi-protein data uncovers a close relationship between Alveolata and Rhizaria. *Molecular Phylogenetics and Evolution*, *101*, 1–7. http://dx.doi.org/10.1016/j.ympev.2016.04.033.

Hehenberger, E., Burki, F., Kolisko, M., & Keeling, P. J. (2016). Functional relationship between a Dinoflagellate host and its diatom endosymbiont. *Molecular Biology and Evolution*, *33*(9), 2376–2390. http://dx.doi.org/10.1093/molbev/msw109.

Hehenberger, E., Imanian, B., Burki, F., & Keeling, P. J. (2014). Evidence for the retention of two evolutionary distinct plastids in dinoflagellates with diatom endosymbionts. *Genome Biology and Evolution*, *6*, 2321–2334.

Hempel, F., Bullmann, L., Lau, J., Zauner, S., & Maier, U. G. (2009). ERAD-derived preprotein transport across the second outermost plastid membrane of diatoms. *Molecular Biology and Evolution*, *26*, 1781–1790.

Hempel, F., Felsner, G., & Maier, U. G. (2010). New mechanistic insights into pre-protein transport across the second outermost plastid membrane of diatoms. *Molecular Microbiology*, *76*, 793–801.

Horiguchi, T., & Takano, Y. (2006). Serial replacement of a diatom endosymbiont in the marine dinoflagellate *Peridinium quinquecorne* (Peridiniales, Dinophyceae). *Phycological Research*, *54*, 193–200.

Horn, S., Ehlers, K., Fritzsch, G., Gil-Rodriguez, M. C., Wilhelm, C., & Schnetter, R. (2007). *Synchroma grande* spec. nov (Synchromophyceae class. nov., Heterokontophyta): An amoeboid marine alga with unique plastid complexes. *Protist*, *158*, 277–293.

Huang, A., Liu, L., Zhao, P., Yang, C., & Wang, G. C. (2016). Metabolic flux ratio analysis and cell staining suggest the existence of C4 photosynthesis in *Phaeodactylum tricornutum*. *Journal of Applied Microbiology*, *120*, 705–713.

Hunsperger, H. M., Randhawa, T., & Cattolico, R. A. (2015). Extensive horizontal gene transfer, duplication, and loss of chlorophyll synthesis genes in the algae. *BMC Evolutionary Biology*, *15*, 4.

Iida, K., Takishita, K., Ohshima, K., & Inagaki, Y. (2007). Assessing the monophyly of chlorophyll-c containing plastids by multi-gene phylogenies under the unlinked model conditions. *Molecular Phylogenetics and Evolution*, *45*, 227–238.

Imanian, B., Pombert, J.-F., Dorrell, R. G., Burki, F., & Keeling, P. J. (2012). Tertiary endosymbiosis in two dinotoms has generated little change in the mitochondrial genomes of their dinoflagellate hosts and diatom endosymbionts. *PloS One*, 7, 43763.

Imanian, B., Pombert, J. F., & Keeling, P. J. (2010). The complete plastid genomes of the two 'dinotoms' *Durinskia baltica* and *Kryptoperidinium foliaceum*. *PloS One*, *5*, 10711.

Ishida, K., Cavalier-Smith, T., & Green, B. R. (2000). Endomembrane structure and the chloroplast protein targeting pathway in *Heterosigma akashiwo* (Raphidophyceae, Chromista). *Journal of Phycology*, *36*, 1135–1144.

Janouskovec, J., Horák, A., Oborník, M., Lukes, J., & Keeling, P. J. (2010). A common red algal origin of the apicomplexan, dinoflagellate, and heterokont plastids. *Proceedings of the National Academy of Sciences of the United States of America*, *107*, 10949–10954.

Jauffrais, T., Jesus, B., Metzger, E., Mouget, J. L., Jorissen, F., & Geslin, E. (2016). Effect of light on photosynthetic efficiency of sequestered chloroplasts in intertidal benthic foraminifera (*Haynesina germanica* and *Ammonia tepida*). *Biogeosciences*, *13*, 2715–2726.

Jeffrey, S. W., & Vesk, M. (1977). Effect of blue-green light on photosynthetic pigments and chloroplast structure in marine diatom *Stephanopyxis turris*. *Journal of Phycology*, *13*, 271–279.

Jeong, H. J. (2011). Mixotrophy in red tide algae raphidophytes. *The Journal of Eukaryotic Microbiology, 58*, 215–222.
Jiroutova, K., Horak, A., Bowler, C., & Obornik, M. (2007). Tryptophan biosynthesis in stramenopiles: Eukaryotic winners in the diatom complex chloroplast. *Journal of Molecular Evolution, 65*, 496–511.
Jiroutova, K., Koreny, L., Bowler, C., & Obornik, M. (2010). A gene in the process of endosymbiotic transfer. *PloS One, 5*, 13234.
Kai, A., Yoshii, Y., Nakayama, T., & Inouye, I. (2008). Aurearenophyceae classis nova, a new class of heterokontophyta based on a new marine unicellular alga *Aurearena cruciata* gen. et sp nov inhabiting sandy beaches. *Protist, 159*, 435–457.
Kamikawa, R., Tanifuji, G., Ishikawa, S. A., Ishii, K. I., Matsuno, Y., Onodera, N. T., et al. (2015). Proposal of a twin arginine translocator system-mediated constraint against loss of ATP synthase genes from nonphotosynthetic plastid genomes. *Molecular Biology and Evolution, 32*, 2598–2604.
Kamikawa, R., Yubuki, N., Yoshida, M., Taira, M., Nakamura, N., Ishida, K., et al. (2015). Multiple losses of photosynthesis in *Nitzschia* (Bacillariophyceae). *Phycological Research, 63*, 19–28.
Kessenich, C. R., Ruck, E. C., Schurko, A. M., Wickett, N. J., & Alverson, A. J. (2014). Transcriptomic insights into the life history of bolidophytes, the sister lineage to diatoms. *Journal of Phycology, 50*, 977–983.
Khan, H., Parks, N., Kozera, C., Curtis, B. A., Parsons, B. J., Bowman, S., et al. (2007). Plastid genome sequence of the cryptophyte alga *Rhodomonas salina* CCMP1319: Lateral transfer of putative DNA replication machinery and a test of chromist plastid phylogeny. *Molecular Biology and Evolution, 24*, 1832–1842.
Kiefer, D. A. (1973). Chlorophyll a fluorescence in marine centric diatoms: Responses of chloroplasts to light and nutrient stress. *Marine Biology, 23*, 39–46.
Kikutani, S., Nakajima, K., Nagasato, C., Tsuji, Y., Miyatake, A., & Matsuda, Y. (2016). Thylakoid luminal theta-carbonic anhydrase critical for growth and photosynthesis in the marine diatom *Phaeodactylum tricornutum. Proceedings of the National Academy of Sciences of the United States of America, 113*(35), 9828–9833.
Kikutani, S., Tanaka, R., Yamazaki, Y., Hara, S., Hisabori, T., Kroth, P. G., et al. (2012). Redox regulation of carbonic anhydrases via thioredoxin in chloroplast of the marine diatom *Phaeodactylum tricornutum. Journal of Biological Chemistry, 287*, 20689–20700.
Kilian, O., & Kroth, P. G. (2005). Identification and characterization of a new conserved motif within the presequence of proteins targeted into complex diatom plastids. *The Plant Journal, 41*, 175–183.
Kim, J., Fabris, M., Baart, G., Kim, M. K., Goossens, A., Vyverman, W., et al. (2016). Flux balance analysis of primary metabolism in the diatom *Phaeodactylum tricornutum. The Plant Journal, 85*, 161–176.
Kim, E., Yubuki, N., Leander, B. S., & Graham, L. E. (2010). Ultrastructure and 18S rDNA phylogeny of *Apoikia lindahlii* comb. nov (Chrysophyceae) and its epibiontic protists, *Filos agilis* gen. et sp nov (Bicosoecida) and *Nanos amicus* gen. et sp nov (Bicosoecida). *Protist, 161*(2), 177–196.
Kowallik, K. V., Stoebe, B., Schaffran, I., KrothPancic, P., & Freier, U. (1995). The chloroplast genome of a chlorophyll a + c-containing alga, *Odontella sinensis. Plant Molecular Biology Reporter, 13*, 336–342.
Kroth, P. G., Chiovitti, A., Gruber, A., Martin-Jezequel, V., Mock, T., Parker, M. S., et al. (2008). A model for carbohydrate metabolism in the diatom *Phaeodactylum tricornutum* deduced from comparative whole genome analysis. *PloS One, 3*, 1426.
Kustka, A. B., Milligan, A. J., Zheng, H. Y., New, A. M., Gates, C., Bidle, K. D., et al. (2014). Low CO2 results in a rearrangement of carbon metabolism to support C4 photosynthetic carbon assimilation in *Thalassiosira pseudonana. New Phytologist, 204*, 507–520.

Lang, M., Apt, K. E., & Kroth, P. G. (1998). Protein transport into "complex" diatom plastids utilizes two different targeting signals. *Journal of Biological Chemistry*, *273*, 30973–30978.

Le Corguillé, G., Pearson, G., Valente, M., Viegas, C., Gschloessl, B., Corre, E., et al. (2009). Plastid genomes of two brown algae, *Ectocarpus siliculosus* and *Fucus vesiculosus*: Further insights on the evolution of red-algal derived plastids. *BMC Evolutionary Biology*, *9*, 253.

Lepetit, B., Gelin, G., Lepetit, M., Sturm, S., Vugrinec, S., Rogato, A., et al. (2017). The diatom *Phaeodactylum tricornutum* adjusts nonphotochemical fluorescence quenching capacity in response to dynamic light via fine-tuned Lhcx and xanthophyll cycle pigment synthesis. *The New Phytologist*, *214*, 205–218.

Lepetit, B., Sturm, S., Rogato, A., Gruber, A., Sachse, M., Falciatore, A., et al. (2013). High light acclimation in the secondary plastids containing diatom *Phaeodactylum tricornutum* is triggered by the redox state of the plastoquinone pool. *Plant Physiology*, *161*, 853–865.

Levering, J., Broddrick, J., Dupont, C. L., Peers, G., Beeri, K., Mayers, J., et al. (2016). Genome-scale model reveals metabolic basis of biomass partitioning in a model diatom. *PloS One*, *11*, 155038.

Li, F., Gao, D., & Hu, H. (2014). High-efficiency nuclear transformation of the oleaginous marine *Nannochloropsis* species using PCR product. *Bioscience, Biotechnology, and Biochemistry*, *78*, 812–817.

Li, S. L., Nosenko, T., Hackett, J. D., & Bhattacharya, D. (2006). Phylogenomic analysis identifies red algal genes of endosymbiotic origin in the chromalveolates. *Molecular Biology and Evolution*, *23*, 663–674.

Liu, F., & Pang, S. J. (2016). Chloroplast genome of *Sargassum horneri* (Sargassaceae, Phaeophyceae): Comparative chloroplast genomics of brown algae. *Journal of Applied Phycology*, *28*, 1419–1426.

Lommer, M., Roy, A. S., Schilhabel, M., Schreiber, S., Rosenstiel, P., & LaRoche, J. (2010). Recent transfer of an iron-regulated gene from the plastid to the nuclear genome in an oceanic diatom adapted to chronic iron limitation. *BMC Genomics*, *11*, 718.

Lommer, M., Specht, M., Roy, A. S., Kraemer, L., Andreson, R., Gutowska, M. A., et al. (2012). Genome and low-iron response of an oceanic diatom adapted to chronic iron limitation. *Genome Biology*, *13*, 66.

Mangot, J. F., Logares, R., Sanchez, P., Latorre, F., Seeleuthner, Y., Mondy, S., et al. (2017). Accessing the genomic information of unculturable oceanic picoeukaryotes by combining multiple single cells. *Scientific Reports*, *7*, 41498.

Marchetti, A., Parker, M. S., Moccia, L. P., Lin, E. O., Arrieta, A. L., Ribalet, F., et al. (2009). Ferritin is used for iron storage in bloom-forming marine pennate diatoms. *Nature*, *457*, 467–470.

Massana, R., del Campo, J., Sieracki, M. E., Audic, S., & Logares, R. (2014). Exploring the uncultured microeukaryote majority in the oceans: Reevaluation of ribogroups within stramenopiles. *The ISME Journal*, *8*, 854–866.

Matari, N. H., & Blair, J. E. (2014). A multilocus timescale for oomycete evolution estimated under three distinct molecular clock models. *BMC Evolutionary Biology*, *14*, 101.

Medlin, L. K., Kooistra, W., Potter, D., Saunders, G. W., & Andersen, R. A. (1997). Phylogenetic relationships of the 'golden algae' (haptophytes, heterokont chromophytes) and their plastids. *Plant Systematics and Evolution*, *11*, 187–219.

Méheust, R., Zelzion, E., Bhattacharya, D., Lopez, P., & Bapteste, E. (2016). Protein networks identify novel symbiogenetic genes resulting from plastid endosymbiosis. *Proceedings of the National Academy of Sciences of the United States of America*, *113*, 3579–3584.

Meneghesso, A., Simionato, D., Gerotto, C., La Rocca, N., Finazzi, G., & Morosinotto, T. (2016). Photoacclimation of photosynthesis in the eustigmatophycean *Nannochloropsis gaditana*. *Photosynthesis Research*, *129*, 291–305.

Mock, T., Otillar, R. P., Strauss, J., McMullan, M., Paajanen, P., Schmutz, J., et al. (2017). Evolutionary genomics of the cold-adapted diatom *Fragilariopsis cylindrus*. *Nature*, *541*, 536–540.

Moustafa, A., Beszteri, B., Maier, U. G., Bowler, C., Valentin, K., & Bhattacharya, D. (2009). Genomic footprints of a cryptic plastid endosymbiosis in diatoms. *Science*, *324*, 1724–1726.

Muñoz-Gómez, S. A., Mejía-Franco, F. G., Durnin, K., Colp, M., Grisdale, C. J., Archibald, J. M., et al. (2017). The new red algal subphylum Proteorhodophytina comprises the largest and most divergent plastid genomes known. *Current Biology*, *27*, 1677–1684.

Mylnikov, A. P., Mylnikova, Z. M., & Tikhonenkov, D. V. (2008). The main cell morphology of the freshwater colorless chrysomonad *Spumella sp* (Ochromonadales, Chrysophyceae). *Inland Water Biology*, *1*, 32–36. http://dx.doi.org/10.1134/s1995082908010069.

Ong, H. C., Wilhelm, S. W., Gobler, C. J., Bullerjahn, G., Jacobs, M. A., McKay, J., et al. (2010). Analyses of the complete chloroplast genome sequences of two members of the pelagophyceae: *Aureococcus anophageferrens* CCMP1984 and *Aureoumbra lagunensis* CCMP 1507. *Journal of Phycology*, *46*, 602–615.

Oudot-Le Secq, M. P., Grimwood, J., Shapiro, H., Armbrust, E. V., Bowler, C., & Green, B. R. (2007). Chloroplast genomes of the diatoms *Phaeodactylum tricornutum* and *Thalassiosira pseudonana*: Comparison with other plastid genomes of the red lineage. *Molecular Genetics and Genomics*, *277*, 427–439.

Park, M. G., Kim, M., & Kim, S. (2014). The acquisition of plastids/phototrophy in heterotrophic dinoflagellates. *Acta Protozoologica*, *53*, 39–50.

Patterson, D. J. (1989). Stramenopiles: Chromophytes from a protistan perspective. *Chromophyte Algae: Problems and Perspectives*, *38*, 357–379.

Peers, G., & Price, N. M. (2006). Copper-containing plastocyanin used for electron transport by an oceanic diatom. *Nature*, *441*, 341–344.

Petersen, J., Teich, R., Brinkmann, H., & Cerff, R. (2006). A "green" phosphoribulokinase in complex algae with red plastids: Evidence for a single secondary endosymbiosis leading to haptophytes, cryptophytes, heterokonts, and dinoflagellates. *Journal of Molecular Evolution*, *62*, 143–157.

Pfaffen, S., Bradley, J. M., Abdulqadir, R., Firme, M. R., Moore, G. R., Le Brun, N. E., et al. (2015). A diatom ferritin optimized for iron oxidation but not iron storage. *Journal of Biological Chemistry*, *290*, 28416–28427.

Pierce, S. K., Curtis, N. E., & Middlebrooks, M. L. (2015). Sacoglossan sea slugs make routine use of photosynthesis by a variety of species-specific adaptations. *Invertebrate Biology*, *134*(2), 103–115.

Pierce, S. K., Curtis, N. E., & Schwartz, J. A. (2010). Chlorophyll synthesis by a sea slug (*Elysia chlorotica*). *Integrative and Comparative Biology*, *50*, 137.

Pillet, L., de Vargas, C., & Pawlowski, J. (2011). Molecular identification of sequestered diatom chloroplasts and kleptoplastidy in foraminifera. *Protist*, *162*(3), 394–404.

Pillet, L., & Pawlowski, J. (2013). Transcriptome analysis of foraminiferan *Elphidium margaritaceum* auestions the role of gene transfer in kleptoplastidy. *Molecular Biology and Evolution*, *30*, 66–69.

Pragya, N., Pandey, K. K., & Sahoo, P. K. (2013). A review on harvesting, oil extraction and biofuels production technologies from microalgae. *Renewable and Sustainable Energy Reviews*, *24*, 159–171.

Prihoda, J., Tanaka, A., de Paula, W. B. M., Allen, J. F., Tirichine, L., & Bowler, C. (2012). Chloroplast-mitochondria cross-talk in diatoms. *Journal of Experimental Botany*, *63*, 1543–1557.

Rauch, C., de Vries, J., Rommel, S., Rose, L. E., Woehle, C., Christa, G., et al. (2015). Why it is time to look beyond algal genes in photosynthetic slugs. *Genome Biology and Evolution*, 7, 2602–2607.

Raven, J. A. (2013). Iron acquisition and allocation in stramenopile algae. *Journal of Experimental Botany*, *64*, 2119–2127.
Reinfelder, J. R., Kraepiel, A. M., & Morel, F. M. (2000). Unicellular C4 photosynthesis in a marine diatom. *Nature*, *407*, 996–999.
Rice, D. W., & Palmer, J. D. (2006). An exceptional horizontal gene transfer in plastids: Gene replacement by a distant bacterial paralog and evidence that haptophyte and cryptophyte plastids are sisters. *BMC Biology*, *4*, 31.
Riisberg, I., Orr, R. J., Kluge, R., Shalchian-Tabrizi, K., Bowers, H. A., Patil, V., et al. (2009). Seven gene phylogeny of heterokonts. *Protist*, *160*, 191–204.
Roberts, K., Granum, E., Leegood, R. C., & Raven, J. A. (2007). Carbon acquisition by diatoms. *Photosynthesis Research*, *93*, 79–88.
Robertson, D. L., & Tartar, A. (2006). Evolution of glutamine synthetase in heterokonts: Evidence for endosymbiotic gene transfer and the early evolution of photosynthesis. *Molecular Biology and Evolution*, *23*, 1048–1055.
Ruck, E. C., Linard, S. R., Nakov, T., Theriot, E. C., & Alverson, A. J. (2017). Hoarding and horizontal transfer led to an expanded gene and intron repertoire in the plastid genome of the diatom, Toxarium undulatum (Bacillariophyta). *Current Genetics*, *63*(3), 499–507.
Ruck, E. C., Nakov, T., Jansen, R. K., Theriot, E. C., & Alverson, A. J. (2014). Serial gene losses and foreign DNA underlie size and sequence variation in the plastid genomes of diatoms. *Genome Biology and Evolution*, *6*, 634–644.
Rumpho, M. E., Pelletreau, K. N., Moustafa, A., & Bhattacharya, D. (2011). The making of a photosynthetic animal. *Journal of Experimental Biology*, *214*, 303–311. http://dx.doi.org/10.1242/jeb.046540.
Rumpho, M. E., Pochareddy, S., Worful, J. M., Summer, E. J., Bhattacharya, D., Pelletreau, K. N., et al. (2009). Molecular characterization of the Calvin cycle enzyme phosphoribulokinase in the stramenopile alga *Vaucheria litorea* and the plastid hosting mollusc *Elysia chlorotica*. *Molecular Plant*, *2*, 1384–1396.
Rumpho, M. E., Worful, J. M., Lee, J., Kannan, K., Tyler, M. S., Bhattacharya, D., et al. (2008). Horizontal gene transfer of the algal nuclear gene *psbO* to the photosynthetic sea slug *Elysia chlorotica*. *Proceedings of the National Academy of Sciences of the United States of America*, *105*, 17867–17871.
Sabir, J. S. M., Yu, M. J., Ashworth, M. P., Baeshen, N. A., Baeshen, M. N., Bahieldin, A., et al. (2014). Conserved gene order and expanded inverted repeats characterize plastid genomes of Thalassiosirales. *PloS One*, *9*, 107854.
Samukawa, M., Shen, C., Hopkinson, B. M., & Matsuda, Y. (2014). Localization of putative carbonic anhydrases in the marine diatom, *Thalassiosira pseudonana*. *Photosynthesis Research*, *121*, 235–249.
Sanchez-Puerta, M. V., Bachvaroff, T. R., & Delwiche, C. F. (2007). Sorting wheat from chaff in multi-gene analyses of chlorophyll c-containing plastids. *Molecular Phylogenetics and Evolution*, *44*, 885–897.
Sanders, W. B., Moe, R. L., & Ascaso, C. (2004). The intertidal marine lichen formed by the pyrenomycete fungus *Verrucaria tavaresiae* (Ascomycotina) and the brown alga *Petroderma maculiforme* (Phaeophyceae): Thallus organization and symbiont interaction. *American Journal of Botany*, *91*, 511–522.
Sang, M., Wang, M., Liu, J. H., Zhang, C. W., & Li, A. F. (2012). Effects of temperature, salinity, light intensity, and pH on the eicosapentaenoic acid production of *Pinguiococcus pyrenoidosus*. *Journal of Ocean University of China*, *11*, 181–186.
Schwartz, J. A., Curtis, N. E., & Pierce, S. K. (2014). FISH labeling reveals a horizontally transferred algal (*Vaucheria litorea*) nuclear gene on a sea slug (*Elysia chlorotica*) chromosome. *The Biological Bulletin*, *227*, 300–312.

Sekiguchi, H., Moriya, M., Nakayama, T., & Inouye, I. (2002). Vestigial chloroplasts in heterotrophic stramenopiles *Pteridomonas danica* and *Ciliophrys infusionum* (Dictyochophyceae). *Protist, 153*, 157–167.

Ševčíková, T., Horák, A., Klimeš, V., Zbránková, V., Demir-Hilton, E., Sudek, S., et al. (2015). Updating algal evolutionary relationships through plastid genome sequencing: Did alveolate plastids emerge through endosymbiosis of an ochrophyte? *Scientific Reports, 5*, 10134.

Shtaida, N., Khozin-Goldberg, I., & Boussiba, S. (2015). The role of pyruvate hub enzymes in supplying carbon precursors for fatty acid synthesis in photosynthetic microalgae. *Photosynthesis Research, 125*, 407–422.

Siaut, M., Heijde, M., Mangogna, M., Montsant, A., Coesel, S., Allen, A., et al. (2007). Molecular toolbox for studying diatom biology in *Phaeodactylum tricornutum*. *Gene, 406*, 23–35.

Silberfeld, T., Leigh, J. W., Verbruggen, H., Cruaud, C., de Reviers, B., & Rousseau, F. (2010). A multi-locus time-calibrated phylogeny of the brown algae (Heterokonta, Ochrophyta, Phaeophyceae): Investigating the evolutionary nature of the "brown algal crown radiation" *Molecular Phylogenetics and Evolution, 56*, 659–674.

Simon, D., Fewer, D., Friedl, T., & Bhattacharya, D. (2003). Phylogeny and self-splicing ability of the plastid tRNA-Leu group I intron. *Journal of Molecular Evolution, 57*, 710–720.

Smith, S. R., Abbriano, R. M., & Hildebrand, M. (2012). Comparative analysis of diatom genomes reveals substantial differences in the organization of carbon partitioning pathways. *Algal Research, 1*, 2–16.

Smith, S. R., Gillard, J. T., Kustka, A. B., McCrow, J. P., Badger, J. H., Zheng, H., et al. (2016). Transcriptional orchestration of the global cellular response of a model pennate diatom to diel light cycling under iron limitation. *PLoS Genetics, 12*, 1006490.

Stiller, J. W., Huang, J. L., Ding, Q., Tian, J., & Goodwillie, C. (2009). Are algal genes in nonphotosynthetic protists evidence of historical plastid endosymbioses? *BMC Genomics, 10*, 484.

Stiller, J. W., Schreiber, J., Yue, J., Guo, H., Ding, Q., & Huang, J. (2014). The evolution of photosynthesis in chromist algae through serial endosymbioses. *Nature Communications, 5*, 5764.

Taddei, L., Stella, G. R., Rogato, A., Bailleul, B., Fortunato, A. E., Annunziata, R., et al. (2016). Multisignal control of expression of the LHCX protein family in the marine diatom *Phaeodactylum tricornutum*. *Journal of Experimental Botany, 67*, 3939–3951.

Tajima, N., Saitoh, K., Sato, S., Maruyama, F., Ichinomiya, M., Yoshikawa, S., et al. (2016). Sequencing and analysis of the complete organellar genomes of Parmales, a closely related group to Bacillariophyta (diatoms). *Current Genetics, 62*, 887–896.

Takano, Y., Hansen, G., Fujita, D., & Horiguchi, T. (2008). Serial replacement of diatom endosymbionts in two freshwater dinoflagenates, *Peridiniopsis* spp. (Peridiniales, Dinophyceae). *Phycologia, 47*, 41–53.

Tanaka, Y., Nakatsuma, D., Harada, H., Ishida, M., & Matsuda, Y. (2005). Localization of soluble beta-carbonic anhydrase in the marine diatom *Phaeodactylum tricornutum*. Sorting to the chloroplast and cluster formation on the girdle lamellae. *Plant Physiology, 138*, 207–217.

Taylor, F. J. R. (1976). Flagellate phylogeny—Study in conflicts. *The Journal of Protozoology, 23*, 28–40.

Tsui, C. K. M., Marshall, W., Yokoyama, R., Honda, D., Lippmeier, J. C., Craven, K. D., et al. (2009). Labyrinthulomycetes phylogeny and its implications for the evolutionary loss of chloroplasts and gain of ectoplasmic gliding. *Molecular Phylogenetics and Evolution, 50*, 129–140.

Tsuji, et al. (2017). Molecular aspects of the biophysical CO_2-concentration mechanism and its regulation in marine diatoms. *Journal of Experimental Botany*. http://dx.doi.org/10.1093/jxb/erx173.

Vogl, C., Badger, J., Kearney, P., Li, M., Clegg, M., & Jiang, T. (2003). Probabilistic analysis indicates discordant gene trees in chloroplast evolution. *Journal of Molecular Evolution*, *56*, 330–340.

Walker, G., Dorrell, R. G., Schlacht, A., & Dacks, J. B. (2011). Eukaryotic systematics: A user's guide for cell biologists and parasitologists. *Parasitology*, *138*, 1638–1663.

Wei, L., Xin, Y., Wang, D. M., Jing, X. Y., Zhou, Q., Su, X. Q., et al. (2013). *Nannochloropsis* plastid and mitochondrial phylogenomes reveal organelle diversification mechanism and intragenus phylotyping strategy in microalgae. *BMC Genomics*, *14*, 534.

Wilken, S., Huisman, J., Naus-Wiezer, S., & Van Donk, E. (2013). Mixotrophic organisms become more heterotrophic with rising temperature. *Ecology Letters*, *16*, 225–233.

Wilken, S., Schuurmans, J. M., & Matthijs, H. C. P. (2014). Do mixotrophs grow as photoheterotrophs? Photophysiological acclimation of the chrysophyte *Ochromonas danica* after feeding. *The New Phytologist*, *204*, 882–889.

Wittpoth, C., Kroth, P. G., Weyrauch, K., Kowallik, K. V., & Strotmann, H. (1998). Functional characterization of isolated plastids from two marine diatoms. *Planta*, *206*, 79–85.

Woehle, C., Dagan, T., Martin, W. F., & Gould, S. B. (2011). Red and problematic green phylogenetic signals among thousands of nuclear genes from the photosynthetic and apicomplexa-related *Chromera velia*. *Genome Biology and Evolution*, *3*, 1220–1230.

Worden, A. Z., Janouskovec, J., McRose, D., Engman, A., Welsh, R. M., Malfatti, S., et al. (2012). Global distribution of a wild alga revealed by targeted metagenomics. *Current Biology*, *22*, 675–677.

Yamada, N., Sym, S. D., & Horiguchi, T. (2017). Identification of highly divergent diatom-derived chloroplasts in dinoflagellates, including a description of *Durinskia kwazulunatalensis* sp. nov. (Peridiniales, Dinophyceae). *Molecular Biology and Evolution*, *34*, 1335–1351.

Yang, E. C., Boo, G. H., Kim, H. J., Cho, S. M., Boo, S. M., Andersen, R. A., et al. (2012). Supermatrix data highlight the phylogenetic relationships of photosynthetic stramenopiles. *Protist*, *163*, 217–231.

Yang, J., Pan, Y. F., Bowler, C., Zhang, L. X., & Hu, H. H. (2016). Knockdown of phosphoenolpyruvate carboxykinase increases carbon flux to lipid synthesis in *Phaeodactylum tricornutum*. *Algal Research*, *15*, 50–58.

Yoon, H. S., Hackett, J. D., Ciniglia, C., Pinto, G., & Bhattacharya, D. (2004). A molecular timeline for the origin of photosynthetic eukaryotes. *Molecular Biology and Evolution*, *21*, 809–818.

Yurchenko, T., Sevcikova, T., Strnad, H., Butenko, A., & Elias, M. (2016). The plastid genome of some eustigmatophyte algae harbours a bacteria-derived six-gene cluster for biosynthesis of a novel secondary metabolite. *Open Biology*, *6*, 160249.

Zhang, L., Wang, X. M., Liu, T., Wang, H. Y., Wang, G. L., Chi, S., et al. (2015). Complete plastid genome of the brown alga *Costaria costata* (Laminariales, Phaeophyceae). *PloS One*, *10*(10), e0140144.

Zheng, Y. T., Quinn, A. H., & Sriram, G. (2013). Experimental evidence and isotopomer analysis of mixotrophic glucose metabolism in the marine diatom *Phaeodactylum tricornutum*. *Microbial Cell Factories*, *12*, 16.

CHAPTER FOUR

Plastid Complexity in Dinoflagellates: A Picture of Gains, Losses, Replacements and Revisions

Ross F. Waller[1], Luděk Kořený
University of Cambridge, Cambridge, United Kingdom
[1]Corresponding author: e-mail address: rfw26@cam.ac.uk

Contents

Abstract

Dinoflagellates are exemplars of plastid complexity and evolutionary possibility. Their ordinary plastids are extraordinary, and their extraordinary plastids provide a window into the processes of plastid gain and integration. No other plastid-bearing eukaryotic group possesses so much diversity or deviance from the basic traits of this cyanobacteria-derived endosymbiont. Although dinoflagellate plastids provide a major

Advances in Botanical Research, Volume 84
ISSN 0065-2296
http://dx.doi.org/10.1016/bs.abr.2017.06.004

contribution to global carbon fixation and energy cycles, they show a remarkable willingness to tinker, modify and dispense with canonical function. The archetype dinoflagellate plastid, the peridinin plastid, has lost photosynthesis many times, has the most divergent organelle genomes of any plastid, is bounded by an atypical plastid membrane number and uses unusual protein trafficking routes. Moreover, dinoflagellates have gained new endosymbionts many times, representing multiple different stages of the processes of organelle formation. New insights into dinoflagellate plastid biology and diversity also suggest that it is timely to revise notions of the origin of the peridinin plastid.

1. INTRODUCTION

Dinoflagellates represent a major plastid-bearing protist lineage that diverged from a common ancestor shared with apicomplexan parasites (Fig. 1). Since this separation dinoflagellates have come to exploit a wide range of marine and aquatic niches, providing critical environmental services

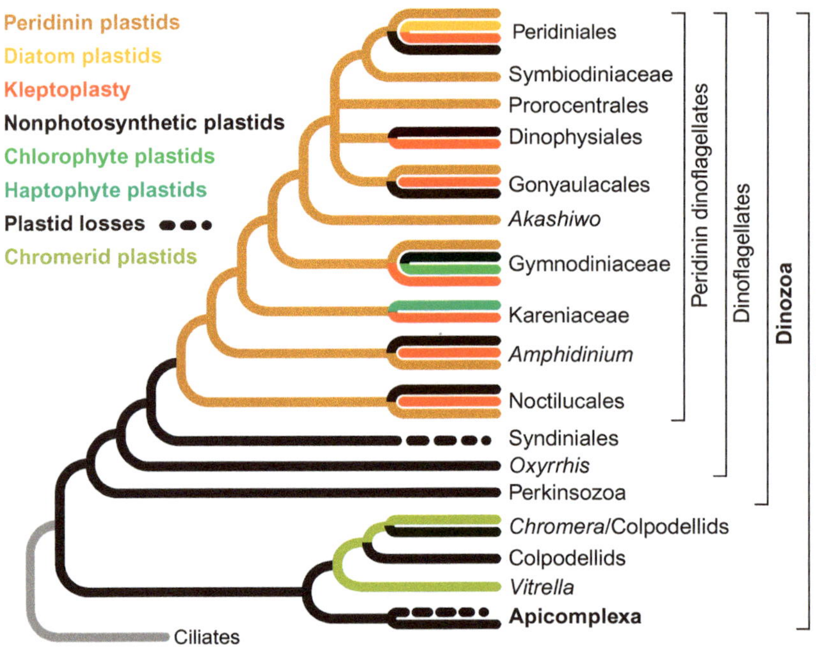

Fig. 1 Schematic of dinoflagellate plastid diversity within the context of alveolate phylogeny. For each clade, the range of plastid types documented is indicated by *colour*. Conservative ancestral plastid state is indicated by *clade branch colour* where multiple lineages have a common plastid type. Phylogeny is based on Janouškovec et al. (2017).

at a global level, as well as having a significant negative impact on some habitats and communities.

As photosynthetic organisms, dinoflagellates contribute a substantial fraction of global carbon fixation that drives food webs as well as capture of some anthropogenic CO_2. Their plastids are also critical to symbiotic associations with a wide range of marine animals. These include symbioses with reef-building corals that are essential for building and maintaining diverse tropical habitats, and that in turn provide barrier protection of vast coastlines from wave erosion and inundation. The loss of the plastid-bearing symbiont leads to coral bleaching, ultimately death of the animal partners and degradation of the reef systems. Plastids of dinoflagellates also generate dimethyl sulphide emissions that play an important role in cloud formation and drive major climatic processes, including those over terrestrial habitats (Charlson, Lovelock, Andreae, & Warren, 1987). The success of dinoflagellates in exploiting nutrient sources, however, can also result in vast blooms up to hundreds of kilometres long that can have major negative impacts. Blooms of taxa that synthesise toxins can lead to accumulation of lethal or injurious toxin doses resulting in widespread death of fish and marine mammals, as well as fatalities in humans. In other cases, the density of dinoflagellate blooms is simply enough to deplete available oxygen resulting in animal mortality through suffocation.

Not all dinoflagellates are photosynthetic, in fact only approximately half of described taxa are (Gómez, 2012). Those that are obligate heterotrophs exploit a wide range of heterotrophic strategies, including micropredation, osmotrophy and parasitism (Coats, 1999; Gaines & Elbrächter, 1987; Hansen & Calado, 1999; Stoecker, 1999). Dinoflagellates, therefore, perform important roles in nutrient recycling and control of prey organism populations. But as parasites they can cause disease out-breaks and even collapses of fisheries stocks (Stentiford & Shields, 2005). Interestingly, most, if not all, of these nonphotosynthetic taxa retain relicts of a photosynthetic plastid, typically as colourless organelles but sometimes merely enzymes or biochemical pathways derived from a former resident plastid. These plastid relicts, therefore, continue to provide metabolic capacity and at least some independence from heterotrophy.

The very wide range of lifestyles that dinoflagellates have adopted have allowed wide exploitation of environmental niches. These include planktonic and sessile niches, as well as those within other organisms in either symbiotic or parasitic partnerships (Gómez, 2012). Recent global ocean sampling of microeukaryotic abundance and diversity places dinoflagellate taxa as one of the most species rich and abundant groups, a testimony to their tremendous success and importance (de Vargas et al., 2015). Notably, these data indicate that approximately three quarters of dinoflagellate taxa detected remain undescribed morphologically, begging the question of what diversity in this group remains to be explored and understood.

The pursuit of a robust understanding of the evolutionary relationships that unite and explain dinoflagellate diversity has long tormented systematists and phylogeneticists. This has been due to both convergent evolution of some traits and high rates of molecular evolution that confounds single-gene phylogenies. Large data sets and phylogenomics, however, are now providing some clearer resolution that is concordant with evolutionary trends such as thecal development and the divergent molecular biology that exists in dinoflagellates (Fig. 1; Bachvaroff, Handy, Place, & Delwiche, 2011; Dorrell et al., 2017; Janouškovec et al., 2017). A clearer picture has also formed of the deep-branching dinoflagellate lineages and their relationships to the nearest nondinoflagellate neighbours (Janouškovec et al., 2015). *Spathulodinium* of the Noctilucales represents the deepest branching photosynthetic dinoflagellate (Gómez, Moreira, & López-García, 2010; Janouškovec et al., 2017), and beneath that branches the chiefly parasitic Syndiniales. *Oxyrrhis* is the deepest 'true' dinoflagellate, and the nearest major group to dinoflagellates are the Perkinsozoa which share some traits characteristic of dinoflagellates but also lacks many (Zhang, Campbell, Sturm, Dungan, & Lin, 2011). Collectively this lineage, referred to as Dinozoa, is sister to the apicomplexan lineage that is best known for its wide range of parasites including human pathogens *Plasmodium* spp. that cause malaria. However, the apicomplexan lineage also has diverse roots including paraphyletic photosynthetic clades *Chromera* and *Vitrella* and a range of colpodellid predators (Fig. 1; Janouškovec et al., 2015).

In the context of the diversity of dinoflagellates and our current understanding of the phylogeny of this group, this chapter addresses the presence and nature of dinoflagellate plastids. Dinoflagellate plastids possess some of the most unusual character states for any plastid-containing group. Further, there have been multiple independent endosymbiotic gains of plastids within dinoflagellates, and this provides remarkable opportunities to consider the modifications to both host and symbiont following such seminal creative events that have been so important in the formation of eukaryotes. In turn, these insights provide a perspective on plastid gain and evolution that is instrumental in considering plastid origins throughout eukaryotes.

2. UNUSUAL FEATURES OF DINOFLAGELLATES THAT MIGHT IMPACT THEIR ENDOSYMBIONTS

Before exploring dinoflagellate plastids in detail, and in particular the features that show variance from those characters found more widely in plastid-containing eukaryotes, it is worth considering some more general features of dinoflagellate biology that might have had impact on the establishment and development of their symbionts.

Most, if not all, photosynthetic dinoflagellates are mixotrophic (Stoecker, 1999). They gain part of their nutrition through a wide range of heterotrophic strategies that provide an alternative source of reduced carbon for energy, as well as necessary elements, minerals and/or vitamins (Gaines & Elbrächter, 1987). Growth studies for *Karlodinium veneficum* (= *Gyrodinium galatheanum*), for example, have shown that by combining feeding and photosynthesis, not only are growth rates increased, but photosynthetic efficiency is also increased, thus a synergy of these two modes of nutrition (Li, Stoecker, & Adolf, 1999). Presumably photosynthetic capacity is often limited by some key nutrients, and heterotrophy overcomes or relaxes these limitations. A further example is the coral symbiont, *Symbiodinium*, that is shown to be an effective free-living predator in order to survive in nutrient replete coral reef waters when outside of its symbiont (Jeong, Yoo, & Kang, 2012). Conversely, the copepod parasite *Dissodinium pseudolunula* retains photosynthesis which it is thought to rely upon during its dispersal stage (Stoecker, 1999). Thus, dinoflagellates show that maintaining multiple nutritional modes is a versatile strategy that can be stably maintained.

Dinoflagellates have developed some ingenious forms of food capture that enable them to hunt, track and ingest a wide range of food items. In some taxa, swimming behaviours are observed to change in response to proximity to prey, and ejectile organelles (trichocysts and nematocysts) can be discharged to immobilise and ensnare their victims (Sheng et al., 2007). Upon capture, prey can be ingested through a number of means (Gaines & Elbrächter, 1987; Hansen & Calado, 1999; Jeong, Yoo, Kim, Seong, & Kang, 2010; Stoecker, 1999). One strategy employs a specialised appendage known as a peduncle to puncture the prey and extract cell contents through this specialised straw. Both liquid cell contents and whole organelles are internalised into food vacuoles for digestion. This form of feeding is called myzocytosis and is seen widely in dinoflagellates as well as lineages related to apicomplexan parasites (e.g. *Colpodella*). Indeed, the peduncle shares ultrastructural similarity to the apical complex of Apicomplexa, which in this group is the major instrument of host cell entry for parasitism (Dodge & Crawford, 1970; Norén, Moestrup, & Rehnstam-Holm, 1999; Siddall, Reece, Graves, & Burreson, 1997). This suggests that an ancestral feeding behaviour has enabled both parasitism in Apicomplexa and ongoing heterotrophy in dinoflagellates. An alternative prey ingestion mode is phagocytosis, again seen widely in dinoflagellates. Prey cells are internalised intact and subsequently processed for digestion in food vacuoles. Interestingly, digestion might not be immediate, with factors such as temperature contributing to delay of digestion by up to a week or more in some examined cases (Li et al., 1999). Dinoflagellates have even achieved predation on cells and cell colonies many times their own cell size. A concertinaed palladium membrane can be issued to surround large prey items into which digestive enzymes are then released. Dinoflagellates with this capacity can recover digested prey nutrients without internalising such prey. Finally, osmotrophy is utilised widely by dinoflagellates, including some that specialise in saprophytic exploitation of food sources.

The ubiquity of heterotrophic behaviours in both photosynthetic and nonphotosynthetic dinoflagellates likely provides options for plastid modification, including loss of function. If heterotrophy is consistently successful, loss or reduction of autotrophic capacity is possible. Indeed, many dinoflagellate groups have independently lost photosynthesis (see below and Fig. 1). Furthermore, there are also multiple cases of plastid recapture via these feeding behaviours and subsequent endosymbioses. These events have resulted in plastid replacements and even coexistence of two evolutionarily distinct plastids in one lineage (Hehenberger, Imanian, Burki, & Keeling, 2014). Thus, there is evidence of an ongoing shifting of balance between autotrophic and nonautotrophic modes of nutrition in dinoflagellates in an equilibrium that is not so obviously seen in other plastid-containing groups. This propensity for ongoing plastid evolution is likely to be attributed, at least in part, to the persistence of mixotrophy.

A further aspect of the biology of dinoflagellates that likely impacts their plastid evolution, although in ways that might be less obvious than heterotrophy, is the long list of very bizarre dinoflagellate genomic traits (Wisecaver & Hackett, 2011). Dinoflagellate nuclear, mitochondrial, and plastid genomes all possess unusual genomic architectures and processes of maintenance and expression. The nuclear genomes of dinoflagellates are extraordinarily large, in the order of ~1.5–190 Gb (Holm-Hansen, 1969;

LaJeunesse, Lambert, Andersen, Coffroth, & Galbraith, 2005). These genomes are maintained in a state of permanent condensation, and in a chromatin form that is apparently near devoid of histones (Chow, Yan, Bennett, & Wong, 2010; Gornik et al., 2012; Rizzo, 1987). When genes are expressed, most if not all mRNAs are processed through *trans*-splicing to receive a short 5′-common leader sequence (Zhang et al., 2007). The mitochondrial genome is similarly unusual, although in very different ways (Nash, Nisbet, Barbrook, & Howe, 2008; Waller & Jackson, 2009). It contains very few genes, but these are fragmented and scrambled in the genomes, with at least one requiring *trans*-splicing to reassemble. In most dinoflagellate mitochondria, mRNAs are extensively posttranscriptionally edited, and translation evidently proceeds without conventional start or stop signals. In this environment of deviance, it is perhaps unsurprising that plastid DNAs share unusual traits in gene organisation and expression also (see below). While it is unclear what has driven the changes seen in the plastid genomes of many dinoflagellates, it is possible that processes directing divergence in the nucleus and mitochondria have affected the plastids too.

3. THE PERIDININ PLASTID

The ‘peridinin plastid’ is the archetypical plastid of dinoflagellates (Fig. 2). It is the one plastid type that occurs broadly in dinoflagellate orders and is, thus, presumed to be an ancestral feature of dinoflagellate radiation. It has for some time been considered to be the product of secondary endosymbiosis of a red alga; however, the simplicity of this hypothesis is discussed further in Section 6. This plastid type has several definitive features, one of the most conspicuous being its pigment composition. The dominant secondary pigment is the carotenoid peridinin that is unique to dinoflagellates. This pigment is unusual for a carotenoid in containing three closed ring structures, and this is thought to contribute to its strong blue-light-absorbing properties (Rapoport et al., 2002). Furthermore, the peridinin plastid contains only a single chlorophyll *c* type (c_2) in addition to chlorophyll *a* (Prezelin, 1987). A second biochemically distinguishing feature of the peridinin plastid is its form II RuBisCO, atypical of eukaryotic types of this central photosynthetic enzyme (Morse, Salois, Markovic, & Hastings, 1995; Whitney, Shaw, & Yellowlees, 1995). Form II RuBisCO assembles as homodimers, rather than the higher order heterodimers of the large and small subunits of form I RuBisCO found through other eukaryotic plastids and their cyanobacterial forebears

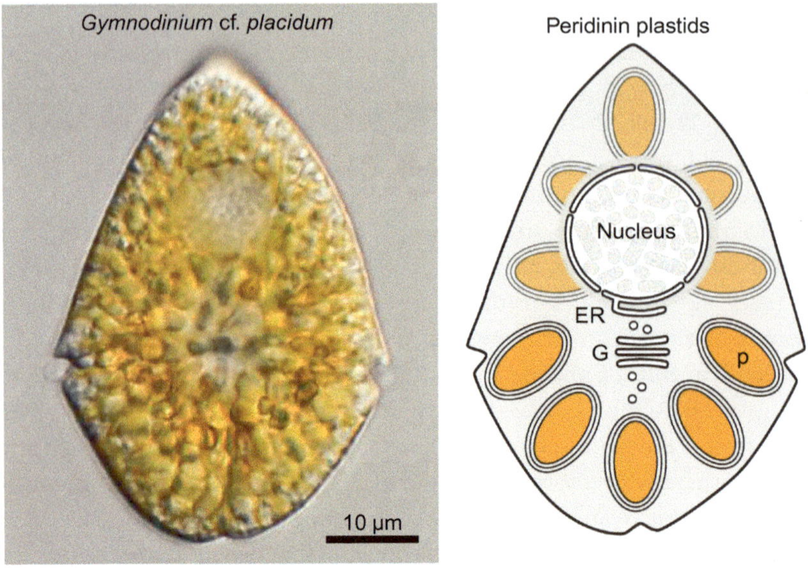

Fig. 2 *Gymnodinium* cf. *placidum* containing peridinin plastids. Plastids (p) are surrounded by three membranes and nucleus-encoded proteins are trafficked to these plastids from the ER via vesicular transport through the Golgi (G). *Micrograph by Gert Hansen.*

(Tabita, Hanson, Satagopan, Witte, & Kreel, 2008). The presence of this unusual RuBisCO suggests protein replacement via lateral gene transfer, most likely from proteobacteria where this RuBisCO form is also prevalent. The only other eukaryotes known to contain this RuBisCO form are the related 'chromerids' of the Alveolata, *Chromera* and *Vitrella* (Fig. 1; Janouškovec, Horák, Oborník, Lukeš, & Keeling, 2010).

Another distinctive trait of the peridinin plastid is the state of the organelle-encoded genome. Most photosynthetic plastid genomes are encoded on single molecules and contain ~90–250 genes (Dorrell & Howe, 2015). The peridinin plastid genome, on the other hand, has been highly reduced to less than 20 genes and fragmented into a population of minicircles containing only one to a few genes each (Howe, Nisbet, & Barbrook, 2008; Zhang, Green, & Cavalier-Smith, 1999). These minicircles are typically 2–3 kbp in size. In each taxon, they contain a common noncoding core sequence, although these core regions do not share sequence identity between taxa. The function of the core domain has not been confirmed, but it contains direct and inverted repeats and is thought to promote the polycistronic transcription of the circles (Moore, Ferguson, Loh,

Hoegh-Guldberg, & Carter, 2003; Nelson & Green, 2005; Nisbet et al., 2008). Some minicircles are thought to lack any genes, others contain gene fragments and some tiny 'microcircles' exist, collectively pointing to an ongoing process of recombination of these genetic elements (Barbrook, Symington, Nisbet, Larkum, & Howe, 2001; Nisbet, Koumandou, Barbrook, & Howe, 2004). Transcribed plastid genes are unusual in receiving a polyuridine 3′-tail, although the significance of this type of posttranscriptional modification is not known (Wang & Morse, 2006). Substitutional editing of plastid transcripts has also been reported in several taxa, adding further to the complexities of these unusual genomes (Dang & Green, 2009; Mungpakdee et al., 2014; Wang & Morse, 2006; Zauner, Greilinger, Laatsch, Kowallik, & Maier, 2004).

With an organelle genome dispersed on many different molecules, these genomes are more difficult to characterise to completion than for single chromosome organelle genomes. Nevertheless, following wide sampling among several photosynthetic taxa, and determination of plastid genes transferred to the nucleus, a clear view of the minicircle gene content has been gained (Dorrell et al., 2017; Mungpakdee et al., 2014; Wang & Morse, 2006). This content is principally for components of the photosynthetic apparatus—photosystem II (PsbA, PsbB, PsbC, PsbD, PsbE, PsbI), photosystem I (PsaA, PsaB), cytochrome *b6/f* (PetB, PetD) and plastid ATP synthase (AtpA, AtpB)—and select genes for their expression (LSU and SSU rRNAs, one to three tRNAs and one to few ORFs). This makes the peridinin plastid's gene complement the most reduced of any known photosynthetic plastid, with the greatest portion of plastid genes relocated to the nucleus. It is possible that minicircles are more easily transferred from organelle to nucleus than components of single plastid chromosomes, making the fragmentation to minicircles a potential accelerant for transfer and contributing to this highly reduced state. Or perhaps segregation during plastid division of equal and/or complete representation of the multiple different minicircle types becomes a limiting factor. Copy number of individual minicircles has been estimated at as little as five copies during exponential growth, which would present challenges to complete sorting of multiple types of minicircles during organelle division (Koumandou & Howe, 2007). This might have further selected for reduced gene number through relocation to the nucleus.

Genes for plastid proteins in the nucleus require their nascent proteins to be trafficked back to these organelles and across their bounding membranes. The peridinin plastid is surrounded by three membranes (Fig. 2), which is a further unusual state for plastids and otherwise only known for the green plastids of euglenids (Schnepf & Elbrächter, 1999). Targeted proteins possess N-terminal bipartite leader sequences consisting of a Sec61-directing signal peptide followed by a plastid-targeting transit peptide (Nassoury, Cappadocia, & Morse, 2003; Patron & Waller, 2007; Patron, Waller, Archibald, & Keeling, 2005). Unlike some other chlorophyll *c*-containing algae, the outermost membrane lacks continuity with the nuclear envelope or endoplasmic reticulum (ER), or obvious ribosomes bound to it. This suggests that vesicular traffic is required for delivery of proteins from the ER to the outer plastid membrane and, indeed, some nucleus-encoded plastid proteins have been detected en route to the plastid in the Golgi apparatus (Nassoury et al., 2003). Curiously, the transit peptide that follows the signal peptide has a predicted stop membrane transfer anchor in a large portion of known plastid-targeted proteins (Nassoury et al., 2003; Patron & Waller, 2007; Patron et al., 2005). This is a plastid protein feature also only known from euglenids and is thought to maintain these proteins in a transmembrane state during their delivery to the plastid (Durnford & Gray, 2006; Durnford & Schwartzbach, 2017). The reason for such a topology is not known, nor why only a portion of plastid proteins show this apparent behaviour, but is a feature that is shared in a protein-specific manner across different dinoflagellate taxa (Patron et al., 2005). Upon fusion to the outer plastid membrane of the Golgi-derived vesicles it is thought that the transit peptide is recognised with equivalents of the translocons of the outer and inner envelope membranes of chloroplasts (TOC and TIC, respectively) to facilitate passage across the last two membranes. While TIC components Tic110 (Hehenberger et al., 2014) and Tic20 (L. Kořený, S. Lam, & R. F. Waller, unpublished) are apparently present in peridinin plastids, no conserved TOC components have been identified and other typically conserved TIC components, such as Tic22, are not detected providing further evidence of an unusually derived plastid.

Plastid membrane number greater than two is widely interpreted as evidence that such plastids were gained through endosymbiosis of a eukaryote that itself contained a plastid, forming a so-called complex plastid, as opposed to a primary endosymbiosis of a prokaryote forming a primary plastid (Gould, Waller, & McFadden, 2008). A complex plastid in dinoflagellates is consistent with molecular phylogenies generally grouping peridinin plastid genes in clades with red algal and other red algal-derived complex plastids (Dorrell et al., 2017; Janouškovec et al., 2010; Waller, Patron, & Keeling, 2006). Most complex plastids, included those in apicomplexans and chromerids, are surrounded by four membranes, and it is unclear what process might generate a three-membrane plastid. It is of note that the deepest lineage of the Dinozoa, the Perkinsozoa, has been suggested to possess plastids surrounded by four membranes (see below). If this is true, and it represents an ancestral peridinin plastid, then it suggests that the three membranes of dinoflagellate plastids arose through loss of one of an original four, well after endosymbiotic gain and integration.

Plastids derived from red algae that are surrounded by four membranes—i.e., in cryptophytes, haptophytes, heterokonts and apicomplexans—share a derived version of the ER-associated degradation (ERAD) machinery, that in most eukaryotes is responsible for exporting misfolded proteins from the ER to the cytosol for proteasomal degradation (Agrawal, van Dooren, Beatty, & Striepen, 2009; Felsner et al., 2011; Stork et al., 2012). The so-called symbiont-specific ERAD-like machinery (SELMA) is an endosymbiont-derived form of this machinery present in all of these organisms' plastids. It has been repurposed for trafficking plastid proteins inward across the second outermost membrane, effectively from the Sec61 translocon on to the TOC and TIC (see Fig. 3). With only three membranes surrounding the peridinin plastid rather than four, it is logical that one translocon would be missing. Peridinin plastid-containing dinoflagellates lack any evidence for a SELMA (L. Kořený, S. Lam, & R. F. Waller, unpublished), consistent with the equivalent of this second outermost membrane being missing in these plastids.

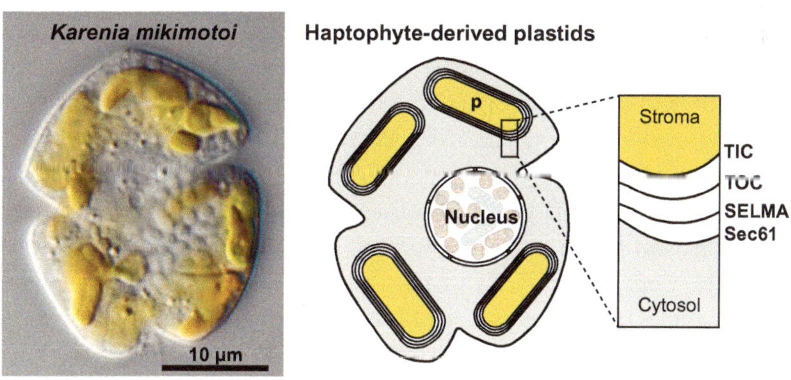

Fig. 3 *Karenia mikimotoi* containing haptophyte-derived plastids (p). Protein import into these plastids is predicted to use four translocons: Sec61, SELMA, TOC and TIC. *Micrograph by Gert Hansen.*

Overall, the peridinin plastid remains relatively poorly understood. This is in part due to the paucity of nuclear genomic data for dinoflagellates, with the exception of the coral symbiont *Symbiodinium* (Aranda et al., 2016; Lin et al., 2015; Shoguchi et al., 2013). Further, dinoflagellate biologists still lack reliable experimental tools such as genetic transformation. Nevertheless, the litany of unusual traits, including form II RuBisCO, plastid minicircular chromosomes, polyuridylated and edited gene transcripts, loss of a bounding membrane, and novel protein targeting, means that this plastid stands out as an exception among other complex plastids. Quite why such divergence might have occurred is unclear.

4. PLASTID REDUCTION AND LOSS

While dinoflagellates are often first thought of as pigmented photosynthetic algae, many, perhaps even most, lack photosynthesis (de Vargas et al., 2015; Gómez, 2012). Such obligate heterotrophs are represented throughout dinoflagellate radiation and all of the major basal members of this lineage are also nonphotosynthetic (Fig. 1). Multiple parallel losses indicate that mixotrophy provides ongoing opportunities for lifestyle changes in dinoflagellates. Loss of photosynthesis was interpreted initially as loss of the plastid, and this was believed to have occurred multiple times independently in dinoflagellates (Saldarriaga, Taylor, Keeling, & Cavalier-Smith, 2001). However, careful scrutiny of molecular data from colourless taxa, for example, *Crypthecodinium*, *Noctiluca* and *Oxyrrhis*, has consistently revealed evidence of persistent plastids, typically by way of genes for plastid-targeted proteins (Janouškovec et al., 2017; Sánchez Puerta, Lippmeier, Apt, & Delwiche, 2007; Slamovits & Keeling, 2008). Even taxa that capture kleptoplasts (*Dinophysis*) or have acquired new photosynthetic plastids ('dinotoms', see below) appear to have retained a colourless plastid (Hehenberger et al., 2014; Janouškovec et al., 2017).

The deepest branching dinozoans, Perkinsozoa, provide the earliest view of nonphotosynthetic plastids in this lineage. *Perkinsus* spp. encode proteins for typical plastid functions with bipartite leader sequences consistent with targeting to complex plastids (Matsuzaki, Kuroiwa, Kuroiwa, Kita, & Nozaki, 2008; Robledo et al., 2011). *Perkinsus* plastids are one of the few dinoflagellate colourless plastids for which there are ultrastructural observations reported in the literature (Robledo et al., 2011; Teles-Grilo et al., 2007). These observations suggest either four membranes surrounding this plastid, or at least three, although direct confirmation that these structures are plastids is currently outstanding. If four, *Perkinsus* plastids could represent an ancestral state of the peridinin plastid, before one membrane was lost. We find no evidence of SELMA in *Perkinsus* genomic data, however, suggesting either that only three membranes are present or that a novel way for proteins to translocate the extra membrane exists in *Perkinsus* spp. (L. Kořený, S. Lam, & R. F. Waller, unpublished). No evidence of a plastid genome has been reported for *Perkinsus* spp., leading to the suggestion that it has been lost (Janouškovec et al., 2015; Matsuzaki et al., 2008). If the *Perkinsus* plastid had previously contained plastid minicircles primarily encoding photosynthetic genes, as for peridinin plastids, then reversion to heterotrophy would be predicted to eliminate these last vestiges of the organelle genome. This prediction extends to all other colourless dinoflagellate plastids derived from the peridinin plastid.

Throughout dinoflagellates, the biochemical functions of the colourless plastids have been inferred from the proteins predicted to be targeted to them. These indicate various synthetic roles in isoprenoid precursors, tetrapyrroles and Fe–S clusters (Bentlage, Rogers, Bachvaroff, & Delwiche, 2016; Gornik et al., 2015; Janouškovec et al., 2017; Matsuzaki et al., 2008; Sánchez Puerta et al., 2007; Slamovits & Keeling, 2008). In eukaryotes, these pathways can all occur in either plastids or the host cytoplasm (and/or mitochondrion for tetrapyrroles and Fe–S clusters), but the presence of plastids often results in loss of at least some of the host-based pathways, and thus reliance on the plastid-located ones (Waller, Gornik, Kořený, & Pain, 2016). To date, the only localisation study for any of these genes was for IspC of the isoprenoid 1-deoxy-D-xylulose 5-phosphate (DOXP) pathway performed in *Perkinsus* (Matsuzaki et al., 2008). IspC was shown to locate to a small number of puncta, suggesting multiple small plastids, although it was not verified that these represented the multimembrane-bound organelles previously seen (Teles-Grilo et al., 2007). Until more complete genomic data are available, and further proteomic studies of these plastids are undertaken, it is unknown if all dinoflagellate colourless plastids will perform the same functions. Nevertheless, these data suggest ongoing dependencies upon plastid metabolism in *Perkinsus* and dinoflagellates that are not fulfilled by heterotrophy, similar to the situation seen for the more thoroughly characterised apicoplasts of apicomplexan parasites (Ralph et al., 2004). In alveolates, an early loss of the host cell pathway for isoprenoid precursor synthesis (the mevalonate pathway) in dinoflagellates and apicomplexans has been suggested to be one factor that has driven retention of the plastid pathway (the DOXP pathway), and thus the organelle, throughout nonphotosynthetic members of both groups (Janouškovec et al., 2017, 2015; Waller et al., 2016).

Despite widespread plastid retention in heterotrophic dinoflagellates, at least one taxon has apparently lost its plastid. *Hematodinium* sp. is a syndinian parasite of crustaceans with no trace of a plastid organelle (Gornik et al., 2015). Its phylogenetic position suggests that ancestrally it would have had a plastid (Fig. 1; Bachvaroff et al., 2011; Janouškovec et al., 2017), and this is corroborated by the presence of one plastid-derived protein (HemD) for the tetrapyrrole pathway that now functions in the cytosol (Gornik et al., 2015). Furthermore, seven enzymes of the plastid diaminopimelate pathway for lysine synthesis are also predicted to be present in the cytosol in *Hematodinium* sp. Although molecular phylogenies of these enzymes were unable to unambiguously identify their origin, this pathway is otherwise only known to be plastidic in eukaryotes, suggesting relocation and retention of the pathway but not the organelle (Gornik et al., 2015; Hudson, Singh, Leustek, & Gilvarg, 2006). Thus, for *Hematodinium* sp. the chance retention of some cytosol-located pathways, relocation of some plastid metabolic proteins to the cytosol and an ability to scavenge metabolites from its host animal have serendipitously alleviated the need for the plastid, and it has been lost. This plastid loss might be a common ancestral feature of the Syndiniales group, but appears to represent a rare evolutionarily event and is otherwise only known from the apicomplexan genus *Cryptosporidium* (Fig. 1).

5. PLASTID REPLACEMENT

Dinoflagellates are uniquely interesting for the study of plastid endosymbiosis as there are multiple verified cases where a new photosynthetic endosymbiont has been gained and replaced or supplemented the role of the original peridinin plastid. These replacements include secondary as well as higher order (tertiary or beyond) endosymbioses, and they provide insight into the processes of plastid capture, integration and modification post gain.

5.1 Complex Plastids Derived From Haptophytes

Within the Gymnodiniales, the Kareniaceae genera *Karenia*, *Karlodinium* and *Takayama* contain endosymbionts of haptophyte origin that now serve as their photosynthetic plastids (Figs 1 and 3). These dinoflagellates still maintain mixotrophy through algivorous phagocytosis (Li et al., 1999; Sheng et al., 2007), so it is plausible that this behaviour was the route to new endosymbiont capture. The foreign source of these plastids was first recognised by pigment analyses that identified a lack of peridinin but the presence of signature accessory pigments of haptophytes, most notably 19′-hexanoyloxyfucoxanthin and/or 19′-butanoyloxyfucoxanthin as well as chlorophylls c_1 and c_2 (Bjørnland & Tangen, 1979; Carreto, Seguel, & Montoya, 2001). This conclusion was substantiated by occurrence of genes for plastid proteins with clear haptophyte affinities in molecular phylogenies (Burki et al., 2014; Ishida & Green, 2002; Patron, Waller, & Keeling, 2006; Takishita, Ishida, & Maruyama, 2004; Tengs et al., 2000; Yoon et al., 2005). Despite the presence of these genes, no trace of the haptophyte nucleus is retained, so all such genes have been relocated a second time to the dinoflagellate nucleus (Schnepf & Elbrächter, 1999). Indeed, there is no obvious relict of the haptophyte cytosol or cytosol-contained structures, with all bounding membranes tightly appressed. The haptophyte-derived plastid, therefore, represents an intimately integrated organelle equivalent to plastids in other algae and plants.

While the plastid compartment in the Kareniaceae is undoubtedly derived from haptophytes, biochemically this plastid is now something of a chimera. Phylogenies of plastid-targeted proteins show that many genes for the peridinin plastid were retained and retargeted as proteins to the new symbiont, either replacing or adding to the haptophyte-derived proteins (Bentlage et al., 2016; Nosenko et al., 2006; Patron et al., 2006; Waller, Slamovits, & Keeling, 2006). This has the further implication that during the period of haptophyte endosymbiont gain and integration, the peridinin plastid was present in some form, likely as a colourless plastid as found in many other dinoflagellates today. Presumably the new photosynthetic plastid ultimately replaced the metabolic roles of this old plastid, and it was eventually lost. However, not before peridinin plastid genes replaced some of those for the haptophyte plastid, perhaps facilitated by them being already expressed from the dinoflagellate nucleus and with targeting signals approximately appropriate for uptake into the haptophyte.

The haptophyte-derived plastid of dinoflagellates provides potentially far-reaching new insight into the establishment of protein import pathways following gain of new complex plastids. Dinoflagellates do not typically use the SELMA machinery for protein import, but haptophytes do (Stork et al., 2012). In the case of dinoflagellates with haptophyte plastids, this machinery has been maintained from the haptophyte and reemployed in the new context of the dinoflagellate. Proteins from all SELMA subcomplexes are present in Kareniaceae, have clear phylogenetic affinities with haptophytes and have bipartite-targeting sequences for the plastid that are otherwise lacking in the cytosolic ERAD paralogues (L. Kořený, S. Lam, & R. F. Waller, unpublished). Thus, these haptophyte-derived plastids show that multiple elements of the plastid machineries are transmissible during establishment of new complex plastids (haptophyte TOC and TIC components are also present). Presence of SELMA also implies that these plastids are likely surrounded by four membranes (Fig. 3), although poor ultrastructural preservation of the plastid membranes has otherwise frustrated attempts to count them directly. Curiously, while the use of bipartite-sorting signals of plastid-targeted proteins is maintained in this reengineered haptophyte plastid, these proteins show subtle but distinct differences in the properties of the transit peptides from those for either peridinin plastids or free-living haptophytes. The new transit peptides for the haptophyte lack the stop-transfer membrane anchors and are also uncharacteristically acidic in nature (Patron & Waller, 2007; Patron et al., 2006). It is possible that the period of coexistence of two different plastid types necessitated some divergence of these transit peptide properties, and these have been maintained since loss of the peridinin plastid.

The haptophyte plastid of dinoflagellates still maintains an organelle genome and, in gene content, this is similar to that of other haptophytes, although with some losses compared to *Emiliania huxleyi* (Gabrielsen et al., 2011). There is evidence, however, that the sequence, architecture and expression machinery for this plastid have undergone modification since this new endosymbiosis. While Kareniaceae plastid-encoded genes branch with haptophytes in phylogenies, they show long branch lengths compared to haptophytes indicating accelerated evolution (Tengs et al., 2000; Yoon, Hackett, & Bhattacharya, 2002; Yoon et al., 2005). All genes are contained on a single organelle chromosome, but there is evidence that some additional gene copies occur on smaller extrachromosomal DNAs, potentially mimicking development of the peridinin minicircles (Espelund et al., 2012). Moreover, two processes for mature transcript production—polyuridylation and substitutional editing—are both now found in these dinoflagellate plastids, whereas they do not occur in the plastids of haptophytes (Dorrell, Hinksman, & Howe, 2016; Dorrell & Howe, 2012; Richardson, Dorrell, & Howe, 2014). These changes all reflect unusual properties of the peridinin plastid and suggest that the factors, presumably targeted proteins, that facilitated these changes in the peridinin plastid persist and are similarly remodelling the genetics of the new haptophyte plastid.

5.2 Complex Plastids Derived From Green Algae

The presence of green pigments, notably chlorophyll *b*, within some members of Gymnodiniaceae demonstrates a second case of plastid replacement in dinoflagellates (Figs 1 and 4; Watanabe, Takeda, Sasa, & Inouye, 1987). *Lepidodinium chlorophorum* (= *Gymnodinium chlorophorum*) and *Lepidodinium viridae* contain plastids surrounded by four membranes in total and containing plastid genomes with clear phylogenetic affinity to green algae (Kamikawa et al., 2015; Matsumoto et al., 2011; Takishita et al., 2008). An early suggestion of the presence of prasinoxanthin, a characteristic pigment of prasinophyte green algae, has not been substantiated, and phylogenetic analysis suggests that the source of this new endosymbiont is within the Pedinophyceae green algae (Matsumoto, Kawachi, Miyashita, & Inagaki, 2012; Matsumoto et al., 2011). *Lepidodinium* spp., therefore, contains green algal plastids presumably gained through secondary endosymbiosis. Uncertain monophyly of this genus might indicate that multiple similar gains have occurred, although all in closely related host taxa (Matsumoto et al., 2012).

Ultrastructural observations of *Lepidodinium* plastids show that the four bounding membranes are not all closely appressed as seen in the Kareniaceae,

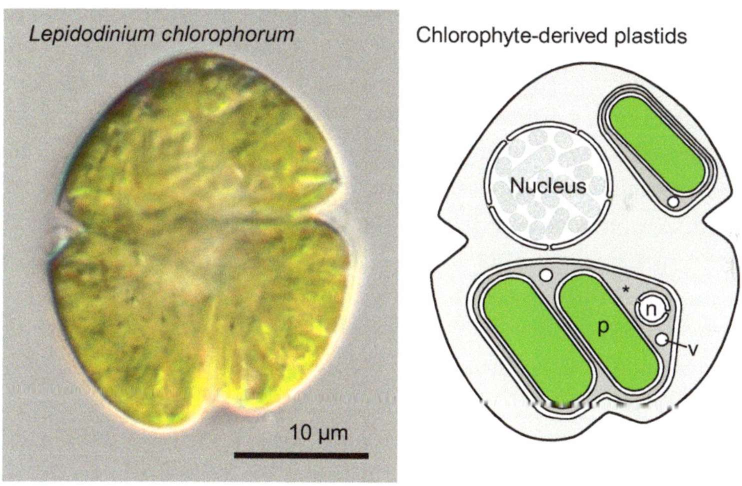

Fig. 4 *Lepidodinium chlorophorum* containing chlorophyte-derived plastids. The plastids (p) are surrounded by four membranes, with a double membrane around each plastid, and a further membrane pair separating the endosymbiont from the host. Between the membrane pairs are relict structures of the endosymbiont cytoplasm including apparent ribosomes (*), vesicles (v) and a possible nucleomorph (n). *Micrograph by Kazuya Takahashi.*

but rather a space occurs between the inner plastid pair containing the thylakoids and the outer pair that separates the host from endosymbiont (Fig. 4). This space contains granules consistent in size with ribosomes and membranous compartments including a double membrane-bound space with pores that is reminiscent of a small relict nucleus, known as a nucleomorph-like structure (Schnepf & Elbrächter, 1999; Watanabe et al., 1987). These observations are highly suggestive of mRNAs being translated in the symbiont cytoplasm and encoded in this small relic nucleomorph-like structure. While the presence of a relict green algal nucleus has not been formally demonstrated, *Lepidodinium* would appear to be at a state of organelle development equivalent to cryptophytes and chlorarachniophytes that possess bona fide nucleomorphs derived from endosymbiont nuclei (Moore & Archibald, 2009).

Genes from the green algal endosymbiont have been transferred to the dinoflagellate nucleus, as is evident by plastid proteins encoded on transcripts possessing the signature transcriptional element of dinoflagellate, the 5′-spliced leader (Minge et al., 2010; Zhang et al., 2007). Several of these are of green algal origin, consistent with nucleus-to-nucleus transfer. Some, however, are peridinin plastid-derived proteins, indicating that, as for the Kareniaceae, a peridinin plastid was present when the green alga was gained, and that some of its proteins have been retargeted to the new plastid. This indicates a common principle, and perhaps propensity, for dinoflagellates to support the gain of new plastids by molecular–genetic adaptations for pre-existing ones (Howe, Barbrook, Nisbet, Lockhart, & Larkum, 2008).

The *Lepidodinium* nucleus-encoded plastid proteins all possess targeting presequences consistent with trafficking to complex plastids—that is, an ER-targeting signal peptide followed by a further extension that might act as a transit peptide. Similar to the Kareniaceae, the properties of the putative transit peptide are somewhat deviant from those of typical plastid transit peptides (Minge et al., 2010), and this might again reflect a need to be distinguished from concurrent targeting to a peridinin plastid while they co-occurred. While we currently have no insight into the translocons that operate to transport across four plastid membranes in *Lepidodinium*, it is of note that neither the dinoflagellate host nor the pedinophyte symbiont would have possessed a SELMA-type apparatus. Therefore, it is unknown what solution has been achieved for protein transport across the third membrane. A redeployment of SELMA such as occurred in the Kareniaceae would not be possible, and a de novo solution would have likely been required in this case.

5.3 Complex Plastid Endosymbionts Derived From Diatoms

A further group of dinoflagellates with stable new plastid endosymbionts occur within the Peridiniales, the best studied of which are *Kryptoperidinium foliaceum* and *Durinskia baltica* (Figs 1 and 5). These dinoflagellates possess endosymbionts derived from diatoms, which is evident by both the presence of signature diatom pigments fucoxanthin as well as chlorophyll c_2 (Tamura, Shimada, & Horiguchi, 2005), and copious molecular–genetic evidence including organelle genomes and extensive transcriptomes of the diatom symbiont (Burki et al., 2014; Imanian & Keeling, 2007; Imanian, Pombert, Dorrell, Burki, & Keeling, 2012; Imanian, Pombert, & Keeling, 2010). These so-called dinotoms are remarkable as stable endosymbiont systems because the symbiont remains relatively intact with a high degree of autonomy. In addition to the diatom plastid, the symbiont maintains a nucleus, mitochondrion, ER and other membranous structures within its cytoplasm (Dodge, 1971; Tamura et al., 2005; Tippit & Pickett-Heaps, 1976; Tomas & Cox, 1973). The plastid and mitochondrion maintain organelle genomes that are seemingly unchanged from free-living diatoms, and the nucleus apparently encodes all genes necessary for plastid and symbiont function (Burki et al., 2014; Imanian & Keeling, 2007; Imanian et al., 2010). In fact, there is no evidence of relocation of functional

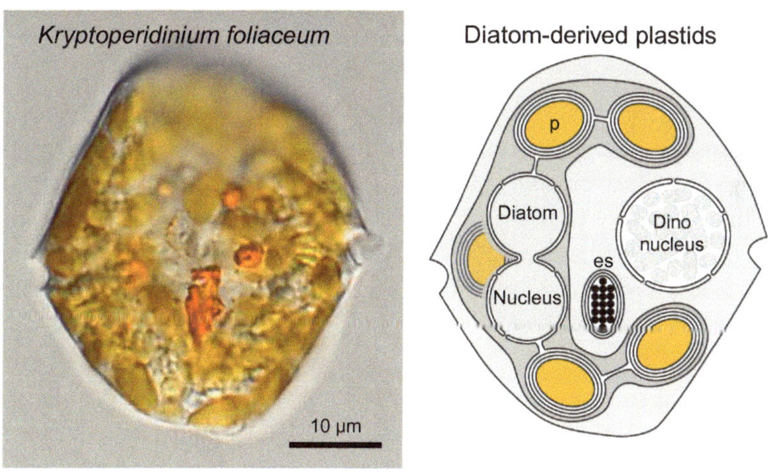

Fig. 5 *Kryptoperidinium foliaceum* containing plastids (p) within a diatom endosymbiont. A single membrane separates the endosymbiont from the host, and a further four membranes surround the plastids. A three membrane-bound eyespot (es) is predicted to be a *colourless* peridinin plastid, implying that these 'dinotoms' contain two types of plastids. *Micrograph by Gert Hansen.*

genes from the symbiont to the dinoflagellate nucleus, suggesting that protein targeting back to this new symbiont has not been developed. This sets it apart from almost all other stable plastid organelles, which are heavily dependent on host nucleus-encoded genes for their protein content.

The membranes surrounding this complex plastid might contribute to its apparent state of arrested development as an organelle. In addition to four membranes immediately surrounding the diatom plastid, an additional membrane then separates the diatom cytosol and organelles from the host dinoflagellate cytosol (Dodge, 1971; Jeffrey & Vesk, 1976; Schnepf & Elbrächter, 1999; Tamura et al., 2005; Tomas & Cox, 1973). Thus, five membranes separate the host nucleus from the plastid stroma (Fig. 5). Many plastid genes remain in the diatom nucleus and are targeted to the plastid via the conventional diatom pathway, including use of the diatom SELMA (Burki et al., 2014; L. Kořený, S. Lam, & R. F. Waller, unpublished). If a protein was to be targeted from the dinoflagellate cytosol, a mechanism for translocating this fifth, outermost, membrane would be required that still leaves the protein competent for translocation across the subsequent four membranes. This complexity might have created a barrier for protein targeting, and the few cases of diatom genes observed in the dinoflagellate nucleus likely represent unproductive random DNA fragment transfers (Burki et al., 2014). The outer membrane even appears to be a barrier to cooperativity between the host and symbiont mitochondria, as both appear to maintain a full suite of independent functions (Imanian, Carpenter, & Keeling, 2007; Imanian & Keeling, 2007; Imanian et al., 2012).

Despite the seemingly stalled nature of this endosymbiont's development as an organelle, change is seen within this symbiont compared to free-living diatoms. The silica frustule is lost allowing an extensive lobed morphology to fill the periphery of the host cytoplasm (Dodge, 1971; Figueroa, Bravo, Fraga, Garcés, & Llaveria, 2009; Tamura et al., 2005; Tippit & Pickett-Heaps, 1976; Tomas & Cox, 1973). Further, the diatom nucleus has apparently relaxed into a permanent state of dispersed chromatin, with no evidence of the chromosome condensation or separation on a mitotic spindle as is so conspicuous in most other diatoms by microscopy (Dodge, 1971; Tippit & Pickett-Heaps, 1976). During cell division, this nucleus is pinched in half through an apparent amitotic process, although if so it is unclear how a viable segregation of genes occurs. It is possible that gene duplication has occurred on short chromosomal elements, akin to the macronucleus of ciliates, but limited details of this unusual symbiont nucleus are currently available.

Dinotoms typically have an eyespot, a conspicuous structure containing carotenoid-rich lipid globules that has long been speculated to be a relict of the peridinin plastid (Fig. 5; Dodge & Crawford, 1969; Schnepf & Elbrächter, 1999; Tamura et al., 2005). Initial evidence for this was its three bounding membranes, as seen for the peridinin plastid, and that eyespot structures are often found within the stroma of peridinin plastids. More recently, genes for plastid proteins have been found in the dinoflagellate nucleus with predicted plastid-targeting bipartite signals (Hehenberger et al., 2014). Given that the diatom plastid is apparently unable to receive proteins from the host cytosol, it is more likely that these are delivered to a plastid directly in the dinoflagellate cytosolic compartment, and the eyespot is the most likely candidate. Moreover, these genes are identified as peridinin plastid-derived and are for metabolic pathways found in other nonphotosynthetic peridinin plastids (isoprenoid precursor synthesis and tetrapyrrole synthesis) consistent with the rationale for colourless plastid retention in other dinoflagellates. The diatom symbiont nucleus also encodes genes for these pathways, and this is further evidence of redundancy and poor cooperativity between this symbiont and its host.

Dinotoms provide further insight into endosymbiosis because while the host cells form a clear monophyletic clade within the Peridiniales, the symbionts are not all derived from a single endosymbiotic event. Molecular phylogenies of plastid genes indicate that the source diatoms differ in different dinotoms—from the pennate genus *Nitzschia*, to centric genera *Chaetoceros*, *Cyclotella* and *Discostella* (Chesnick, Kooistra, Wellbrock, & Medlin, 1997; Horiguchi & Takano, 2006; Tamura et al., 2005; Yamada, Sym, & Horiguchi, 2017; You, Luo, Su, Gu, & Gu, 2015; Zhang, Liu, & Hu, 2014). Moreover, recent detailed sampling of diatoms indicates that within dinotoms with pennate symbionts, six *Nitzschia* species have been independently taken up in at least eight endosymbiotic events (Yamada et al., 2017). These data point to a process of relatively frequent endosymbiont replacements from a population of related diatoms. It is currently unclear what the longevity of each symbiont is; however, in culture these cells are known to maintain the symbiont for decades. It is possible that the apparent degeneration of the diatom mitotic process might gradually undermine the viability of the symbiont, although these observations might be misinterpreted. Nevertheless, these dinoflagellates demonstrate a propensity to establish new diatom endosymbiotic relationships, and this suggests that adaptation to forming specific symbiotic relationships likely happens before long-term establishment of symbionts. There is some evidence of a similar scenario in

Lepidodinium spp. (above), and this might indicate a broad principle relevant to the process of endosymbiosis in general.

5.4 Kleptoplasts

At the furthest extreme of dinoflagellate plastid biology are taxa that exploit temporary photosynthetic plastids gained by feeding on other plastid-containing organisms. These 'stolen' plastids, or kleptoplasts, are maintained from days up to years, and during their tenure the host dinoflagellate derives metabolic benefit from the ongoing photosynthetic activity of these plastids. Some plastids can be maintained through host cell division and some are even seen within dinoflagellate cysts (Onuma & Horiguchi, 2015). However, in all cases, ongoing maintenance and division of the kleptoplast have not been achieved, so the host must continue to predate on a source of these plastids to replenish its stock. Kleptoplasty is seen in a wide range of dinoflagellates. While the Dinophysiales demonstrate a particular proclivity for this behaviour (e.g. multiple *Dinophysis* spp., *Amphisolenia*, *Triposolenia*), kleptoplasts have been seen across several orders including Gymnodiniales (e.g. *Nusuttodinium, Amphidinium*), Gonyaulacales (e.g. *Amylax*), Peridiniales (e.g. *Pfiesteria* and *Cryptoperidiniopsis*) and Noctilucales (e.g. *Noctiluca*) (Daugbjerg, Jensen, & Hansen, 2013; Eriksen, Hayes, & Lewitus, 2002; Gast, Moran, Dennett, & Caron, 2007; Hansen, Miranda, & Azanza, 2004; Kim et al., 2014; Kim, Nam, Shin, & Coats, 2012; Larsen, 1988; Lewitus & Glasgow, 1999; Sweeney, 1976; Takano, Yamaguchi, Inouye, Moestrup, & Horiguchi, 2014; Tarangkoon, Hansen, & Hansen, 2010; Wilcox & Wedemayer, 1985). The sources of kleptoplasts can also be diverse, with many favouring cryptophytes, but others exploiting haptophytes, pelagophytes, chlorophytes and even cyanobacteria (Daugbjerg et al., 2013; Gast et al., 2007; Sweeney, 1976; Tarangkoon et al., 2010; Wilcox & Wedemayer, 1985).

Seemingly just as diverse as the source kleptoplasts is the mode of their gain, and this has consequences for the potential maintenance and longevity of these temporary organelles. Some prey organisms are engulfed whole, accumulate and persist relatively intact (Hansen et al., 2004; Li et al., 1999; Onuma & Horiguchi, 2015; Yamaguchi, Nakayama, Kai, & Inouye, 2011). Some of these symbionts undergo morphological changes within their captors and can lose surface structures, such as periplasts and ejectosomes, and increase in size by up to 20-fold filling a substantial volume of the host cell (Onuma & Horiguchi, 2015; Yamaguchi et al., 2011). Other kleptoplasts are gained by processes that actively disrupt the source cell, including myzocytosis where organelles are harvested directly from the prey cytoplasm. Plastids can even lose surrounding membranes in this process; for example, cryptophyte four membrane plastids can be reduced to double membrane-bound organelles (Schnepf, Winter, & Mollenhauer, 1989). Even more remarkable, both *Dinophysis* and *Amylax* spp. acquire their kleptoplasts second hand by preying on ciliates that themselves are utilising cryptophyte-derived kleptoplasts (Kim et al., 2014, 2012).

Evidence that kleptoplasts provide metabolic gain over their simple catabolic value as food has been provided for several examples and demonstrates further that different cases of kleptoplasty represent a continuum of host benefits. For example, *Dinophysis acuminata* obtains its cryptophyte kleptoplasts indirectly from the ciliate *Mesodinium rubrum*, and with abundant prey consumption heterotrophy is the major carbon source (Riisgaard & Hansen, 2009). However, with reduced prey density, photosynthesis from the sustained kleptoplasts becomes the dominant carbon source and light-dependent growth is seen. *Cryptoperidiniopsis* grazing on algae similarly shows light-dependent growth, and in *Pfiesteria piscicida* carbon capture was seen as light-dependant (Eriksen et al., 2002; Lewitus & Glasgow, 1999). Finally, a new but undescribed Kareniaceae species from the Antarctic selectively feeds on and maintains a haptophyte (*Phaeocystis antarctica*) kleptoplast for up to 30 months (Gast et al., 2007; Sellers, Gast, & Sanders, 2014). This dinoflagellate cannot survive on this prey alone in the dark, suggesting that this dinoflagellate has become an obligate phototroph with a requirement to continuously replenish its supply of kleptoplasts.

The wide presence of kleptoplasty in dinoflagellates indicates a general capacity for establishment of extended nutritional relationships with their prey (Fig. 1). Some lineages, however, show greater propensity for this behaviour, indicating that adaptation has occurred to enhance the benefits derived from kleptoplasts. *Nusuttodinium*, for instance, represents a clade of both marine and freshwater species that all engage in kleptoplasty, and always with a preference for cryptophytes (Takano et al., 2014). The inheritance of this trait, even as speciation to different aquatic habitats occurred, implies stable adaptations specialised for acquiring and managing select kleptoplasts. In *Dinophysis acuta* (Fig. 6), only the cryptophyte plastid is acquired, and yet these plastids display photoregulation via pigment production in response to changing light (Hansen et al., 2016). In the absence of the cryptophyte nucleus or nucleomorph, this is suggestive of some role for the host in controlling the physiology of its kleptoplast. Furthermore, in

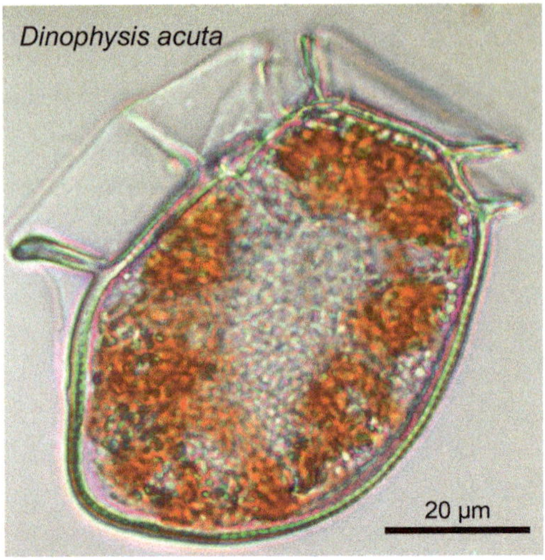

Fig. 6 *Dinophysis acuta* containing pigmented kleptoplasts derived from cryptomonads. *Micrograph by Susanne Busch.*

D. acuminata some plastid genes have been found in the dinoflagellate nucleus, implying that these might target to the cryptophyte kleptoplast to help manage or maintain it (Wisecaver & Hackett, 2010). Interpretation of these findings is confounded by the recent discovery that *D. acuminata* also retains a colourless peridinin plastid that is the likely destination for at least some of these proteins (Janouškovec et al., 2017), and protein location studies are now required to substantiate the potential for protein targeting to kleptoplasts. Finally, *Nusuttodinium aeruginosum* has achieved a limited form of coordinated kleptoplast division with host division, able to pass on a single gained cryptophyte kleptoplast through up to five divisions (and 32 offspring) (Onuma & Horiguchi, 2015). In this case, the cryptophyte nucleus is unable to divide, but, nevertheless, the inheritance of this nucleus correlates with greater kleptoplast size per generation, suggesting that some nucleus-encoded plastid biogenesis processes are maintained for a time. Collectively, these examples suggest that kleptoplasty is more than just delayed prey digestion and exploitation of ongoing photosynthesis, but is a selective, deliberate behaviour with many adaptations akin to endosymbiosis.

6. EVOLUTION OF PLASTIDS IN DINOFLAGELLATES: A CASE TO REVISE CONTEMPORARY NOTIONS

For much of the last one and a half decades, the 'chromalveolate hypothesis' has dominated discussion of the origins of complex plastids derived from red algae—i.e., those found in cryptophytes, haptophytes, heterokonts, apicomplexans and dinoflagellates (Cavalier-Smith, 1999). Central to this discussion was the assertion that minimal endosymbiotic gains of plastids were preferred for accounting for current plastid diversity, and in the hypothesised 'chromalveolate' clade a single ancestral plastid gain would suffice. The monophyly of this clade, however, was ultimately disproven, so separate, alternative explanations for plastid gains in these lineages have been suggested (e.g. Baurain et al., 2010; Bodył, Stiller, & Mackiewicz, 2009; Petersen et al., 2014; Puerta & Delwiche, 2008; Ševčíková et al., 2015; Stiller, 2014; Stiller et al., 2014). While the alveolates (ciliates, apicomplexans and dinoflagellates) are monophyletic, lack of any evidence of plastids in ciliates suggests more recent plastid gain(s) in the apicomplexans and dinoflagellates. This is supported by a metabolic perspective that makes an ancient plastid predating ciliates unlikely (Waller et al., 2016). The discovery of *Chromera* and *Vitrella* that branch at the base of the apicomplexan lineage but, nonetheless, maintain photosynthetic plastids provided a welcome opportunity to compare the apicoplast with the peridinin plastid to consider if these two plastids likely shared a common endosymbiotic origin (Moore et al., 2008; Oborník et al., 2012). Select plastid features were identified as shared between the apicoplast and the 'chromerid' plastid, and the chromerid plastid with the peridinin one (see below), suggesting that a single endosymbiosis might account for these three plastid types. However, it is prudent to review all of the options for plastid gain in these lineages and, in the light of increased knowledge of the biology of dinoflagellate plastids, assess their relative support.

Below, three evolutionarily scenarios are considered that account for the presence of plastids in dinoflagellates, apicomplexans and chromerids (Myzozoa) with emphasis on considering the origin of the peridinin plastid in dinoflagellates.

6.1 Scenario 1: A Common Plastid Was Gained Before the Divergence of Myzozoans

6.1.1 Rationale

A single endosymbiotic gain of a plastid in a common ancestor of all myzozoans means that the obstacles to developing stable host–symbiont partnership need to be overcome only once. In such an event, it would be predicted that shared features of these commonly derived plastid organelles would be represented in the descendent lineages, and in myzozoans, some such features have been observed. All myzozoan plastids have relatively reduced organelle genomes compared to other red algal-derived plastids, with peridinin plastids most reduced in gene content, followed by apicoplasts then chromerid plastids (Janouškovec et al., 2010). Although the apicoplast and peridinin plastids share few overlapping genes due to independent organelle genome reduction, chromerids do share common plastid genes with both groups. Molecular phylogenies with either chromerid and dinoflagellate genes or chromerid with apicomplexan genes show a common alliance of the myzozoan sequences, although always with independent long branches among these taxa (Janouškovec et al., 2010; Moore et al., 2008). Perhaps more significantly, two biochemical features are common to peridinin and chromerid plastids. One is that both use the unusual form II RuBisCO in place of the form I RuBisCO seen throughout other algal groups (Janouškovec et al., 2010). The second is that both polyuridylate at least some of their plastid mRNA transcripts (Janouškovec et al., 2010; Wang & Morse, 2006). Morphologically, a further similarity has been identified as the stacking of thylakoids with a tendency for stacks of three (Janouškovec et al., 2010; Moore et al., 2008; Oborník et al., 2012). The consistency of this character in dinoflagellates, however, is variable (Dodge, 1975).

6.1.2 Caveats

Against these shared similar traits of the myzozoan plastids is a backdrop of tremendous differences also. The pigmentation of peridinin vs chromerid plastids is very different, with peridinin a novel and stable accessory pigment in dinoflagellates along with chlorophyll *c*, but neither are present in chromerids. The peridinin plastids also share highly modified genomes as minicircles, and transfer of many genes to the nucleus which are otherwise still present in the more conventional plastid DNAs of apicomplexans and chromerids (Janouškovec et al., 2013; Wilson & Williamson, 1997). Central to the establishment of stable organelles is the development of protein targeting, yet plastid membrane number and protein-sorting methods are different also (Patron & Waller, 2007). The peridinin plastid is bounded by three membranes and lack the SELMA protein-targeting machinery, but both apicomplexan and chromerid plastids have four membranes and use SELMA (Agrawal et al., 2009; Petersen et al., 2014; Stork et al., 2012). If the *Perkinsus* plastid is also surrounded by four membranes, as some images suggest (Teles-Grilo et al., 2007), then absence of SELMA is independent of membrane loss in the dinoflagellate lineage. Finally, the metabolic division of labour between host cell and plastid after endosymbiosis typically eliminates redundant functions, for example, host vs plastid anabolic pathways for tetrapyrroles. For this pathway, apicomplexans and chromerids utilise the host 'C4' pathway that starts with glycine in the mitochondrion, while dinoflagellates with peridinin plastids maintain the plastid-based 'C5' pathway that starts with glutamyl-tRNA (Janouškovec et al., 2017; Kořený, Sobotka, Janouškovec, Keeling, & Oborník, 2011; Ralph et al., 2004). Basal dinoflagellate lineages Perkinsozoa, *Oxyrrhis* and Syndiniales, however, all exclusively maintain the C4 host pathway (Janouškovec et al., 2017; Waller et al., 2016). In the scenario of a common plastid, maintenance of metabolic redundancy is required for some time during the radiation of dinoflagellates before differential loss of this redundant feature. Thus, if a common myzozoan plastid occurred as hypothesised in Scenario 1, many independent character gains and losses, as well as maintenance of redundancy, are required to account for extant myzozoan plastid diversity.

6.2 Scenario 2: Plastids Were Gained Independently After Divergence From the Common Myzozoan Ancestor

6.2.1 Rationale

An alternative explanation for the many divergent characters that separate the peridinin plastid from those of apicomplexans and chromerids is that these plastids were gained independently. Thus, their differences might be the products of independent processes of integration with their hosts.

The processes necessary to establish a new endosymbiotic organelle are likely to promote a period of accelerated organelle evolution—the 'evolutionary upheavals associated with endosymbiosis' (Stiller, 2014). In particular, the functional relocation of genes from symbiont to host nucleus is likely to drive this change. Organelle DNA insertion into host nuclei is thought to be relatively random (Timmis, Ayliffe, Huang, & Martin, 2004), but from such transfers some symbiont genes will be expressed and translated, and ultimately their proteins will be delivered back to the organelle where they are required. When these imported proteins achieve sufficient function in the organelle, the resident organelle gene is redundant and can be lost through mutational erosion. During this sequence of events, many processes are likely to promote change in both the nucleus and organelle copies of these genes and, thus, the pathways they define. Regulation of the transferred gene will initially occur in the context of a new host environment determined by foreign transcriptional, translational and import control processes acting on the organelle protein. Even during a process of selection for restored function, this varied regulation is likely to have biochemical impact on the organelle. Furthermore, initial selection for function of the transferred gene will be relaxed due to the second gene copy still in the organelle, so accumulation of some mutations is likely prior to fixation of the functional copy. For gene transfers from the symbiont nucleus to the host nucleus, which are the majority of transfers in complex plastid formation, mis-splicing of introns in the foreign nucleus might contribute to further evolution of transferred genes. Finally, during the window of time when two genes provide function, both a host nucleus and organelle copy, partial or full redundancy will relax selection on the organelle-encoded gene also, allowing it to independently evolve. Chance fixation of the organelle gene, that might be subsequently retransferred, will again have promoted evolution of the organelle. It is likely that successful transfer of genes to the host will often requires multiple attempts, and during all of these, there is potential for the biochemistry of the organelle to be progressively, incrementally changed. Thus, the drivers for gene relocation, irrespective of what these are, promote rounds of gene duplication and divergence, and in effect an increased rate of evolution of genes and the processes they define.

Such a period of accelerated evolution during organelle establishment might explain some of the novelty found in the peridinin plastid. For instance, it is easy to envisage how the biochemical pathway for pigment synthesis such as the carotenoids could be impacted by changed regulation, stoichiometries and primary sequence of the synthetic enzymes, allowing potential evolutionary exploration of novel molecular structures such as the peridinin carotenoid. The development of the recombined minicircular organelle DNAs might similarly be the product of transient perturbation of the genome replication and maintenance processes.

6.2.2 Caveats

Two independent gains of plastid organelles in lineages without preexisting plastids are necessarily twice as complex as a single gain and integration. While complexity does not eliminate possibility, in these simple terms it could be argued as less likely. Scenario 2 also requires the two unusual plastid traits, use of form II RuBisCO and polyuridylation of transcripts, to have been acquired independently in chromerids and dinoflagellates. Given that these traits are otherwise apparently only found in these groups, complete independence of their gains seems unlikely also.

6.3 Scenario 3: A Common Plastid Was Present in the Myzozoan Common Ancestor, but the Dinoflagellate Peridinin Plastid Represents a Replacement by a New Endosymbiont

6.3.1 Rationale

Plastid replacement in dinoflagellates has happened many times (Fig. 1). We know from the Kareniaceae, *Lepidodinium* spp. and dinotoms that independent gains of stable plastids have occurred, and in dinotoms a similar endosymbiont type has been gained multiple times within this closely related clade. Moreover, the broad occurrence of kleptoplasty in dinoflagellate orders, including cases of adaptation for symbiont preference and sustained maintenance of the plastid, further indicates the propensity for dinoflagellates to experiment with symbiosis. In this context, an early plastid replacement giving rise to the peridinin plastid is easy to imagine. Nay, one must almost find reason to argue that it has not occurred.

An independent gain of the peridinin plastid would provide opportunity for the accelerated evolution, postulated above during genetic integration of the endosymbiont, to lead to the many novel-derived characters seen today in dinoflagellates. Furthermore, the difference in plastid protein-targeting signals, membrane number and import machinery (lack of SELMA) could be explained by these basic organelle biogenesis processes evolving anew independently. If a former myzozoan plastid was present and replaced after concurrent occupancy, this might further drive the new import processes to be distinct from those seen in apicomplexans and chromerids, as is also observed in plastid replacements in the Kareniaceae and *Lepidodinium* (Minge et al., 2010; Patron et al., 2006).

If a plastid replacement did occur an obvious question is when. If it occurred after the Syndiniales diverged, then this new peridinin plastid might have restored photosynthesis in dinoflagellates. Independent loss of autotrophy in all multiple basal lineages would not be required to explain its conspicuous absence in early dinozoans (Fig. 1). Similarly, the maintenance of redundancy in C4 and C5 pathways for tetrapyrrole synthesis would not be required for this span of dinoflagellate evolution. Instead the myzozoan C4 pathway would simply have been lost in favour of the plastid C5 pathway upon new plastid gain. This hypothesised late replacement implies that the relict plastids in Perkinsozoa and *Oxyrrhis* do not represent derivatives of the peridinin plastid. This implication is difficult to assess in these highly reduced organelles for which there is currently relatively little direct data. However, if *Perkinsus* and *Oxyrrhis* plastids are derived from the peridinin plastid, then an earlier replacement event, before their divergence, must be considered.

6.3.2 Caveats

The use of form II RuBisCO and mRNA polyuridylation in both peridinin and chromerid plastids would seemingly require independent gain of these unusual traits in the scenario of different endosymbiotic events. However, the process of plastid replacement is known to utilise proteins and processes from preexisting resident plastids. In the Kareniaceae, several peridinin plastid proteins have been maintained and retargeted to the haptophyte plastid, and both substitutional RNA editing and polyuridylation have been gained (Dorrell & Howe, 2012; Patron et al., 2006). Therefore, transfer of polyuridylation following plastid replacement has biological precedent, and the adoption of the preexisting alternative RuBisCO protein is also plausible. So, neither of these features present a serious obstacle to Scenario 3. If the peridinin plastid's form II RuBisCO was inherited from an earlier plastid, then it might seem unlikely that a long nonphotosynthetic period would have preceded the new plastid gain. This might favour an early plastid replacement event, rather than one after the Syndiniales. It is of note, however, that there is precedent also for maintenance of RuBisCO in nonphotosynthetic algae such as *Euglena longa* (Záhonová, Füssy, Oborník, Eliáš, & Yurchenko, 2016). An alternative explanation is that a chromerid-type cell, already in possession of these two traits, was itself the source of a new plastid in dinoflagellates that ultimately resulted in the peridinin plastid.

Each of the three scenarios presented here has both strengths and weaknesses, and there is no unambiguous consensus to which best accounts for the evolution of the peridinin plastid. The discussion of plastid origins, however, has been dominated by schemes that seek to minimise the number of endosymbiotic events. For dinoflagellates in particular, the empirical evidence indicates that this is poorly justified, and equal consideration should be given to possible plastid replacement and/or independent origin.

7. CONCLUSION

Dinoflagellate biology provides countless subjects of fascination. Be they: highly complex cell structures for sight, defence and prey capture (Gavelis et al., 2015, 2017); novel nuclear and chromatin conformations; symbioses with animals; or toxic bloom formation: this group of organisms show a penchant for the diverse and unusual, although not at the expense of their broad global impact and importance. As such they are potent subjects for studying organism evolution and function. The biology of their plastids, in particular, offers many insights into endosymbiosis and organellogenesis. Dinoflagellates are seen to develop taxon-selective feeding behaviours and exploitation of temporary plastids over a wide range of timescales. These behaviours have led to repeated cases of stable endosymbioses within closely related dinoflagellate and prey taxa, as is particularly evident in dinotoms. Further, we see new cases of genetically integrated organelles in the Kareniaceae and *Lepidodinium* that show both recycling of complex protein import machineries and development of novel mechanisms for organelle biogenesis. In these plastid replacement events the impact of former plastids on the new ones can be seen with adoption of genes and genetic processes from the old to the new, in doing so forming an amalgam of plastid biochemistries and histories. These somewhat more recent and, indeed, ongoing cases of endosymbiosis in dinoflagellates bring into sharper focus the processes of organelle formation that otherwise lose resolution with antiquity. Many of the principles displayed in plastid formation in dinoflagellates, therefore, might just as likely have contributed to the formation of plastids throughout eukaryotic diversity, be they primary or complex.

ACKNOWLEDGEMENTS

We are very grateful to Jan Janouškovec for useful comments and discussions on this chapter and Gert Hansen, Kazuya Takahashi and Susanne Busch for beautiful micrographs.

REFERENCES

Agrawal, S., van Dooren, G. G., Beatty, W. L., & Striepen, B. (2009). Genetic evidence that an endosymbiont-derived endoplasmic reticulum-associated protein degradation (ERAD) system functions in import of apicoplast proteins. *Journal of Biological Chemistry, 284*(48), 33683–33691.

Aranda, M., Li, Y., Liew, Y. J., Baumgarten, S., Simakov, O., Wilson, M. C., et al. (2016). Genomes of coral dinoflagellate symbionts highlight evolutionary adaptations conducive to a symbiotic lifestyle. *Scientific Reports, 6*(1), 39734.

Bachvaroff, T. R., Handy, S. M., Place, A. R., & Delwiche, C. F. (2011). Alveolate phylogeny inferred using concatenated ribosomal proteins. *Journal of Eukaryotic Microbiology, 58*(3), 223–233.

Barbrook, A. C., Symington, H., Nisbet, R. E., Larkum, A., & Howe, C. J. (2001). Organisation and expression of the plastid genome of the dinoflagellate *Amphidinium operculatum*. *Molecular Genetics and Genomics, 266*(4), 632–638.

Baurain, D., Brinkmann, H., Petersen, J., Rodríguez-Ezpeleta, N., Stechmann, A., Demoulin, V., et al. (2010). Phylogenomic evidence for separate acquisition of plastids in cryptophytes, haptophytes, and stramenopiles. *Molecular Biology and Evolution, 27*(7), 1698–1709.

Bentlage, B., Rogers, T. S., Bachvaroff, T. R., & Delwiche, C. F. (2016). Complex ancestries of isoprenoid synthesis in dinoflagellates. *Journal of Eukaryotic Microbiology, 63*, 123–137.

Bjørnland, T., & Tangen, K. (1979). Pigmentation and morphology of a marine *Gyrodinium* (dinophyceae) with a major carotenoid different from peridinin and fucoxanthin. *Journal of Phycology, 15*(4), 457–463.

Bodył, A., Stiller, J. W., & Mackiewicz, P. (2009). Chromalveolate plastids: Direct descent or multiple endosymbioses? *Trends in Ecology & Evolution, 24*(3), 119–121, author reply 121–122.

Burki, F., Imanian, B., Hehenberger, E., Hirakawa, Y., Maruyama, S., & Keeling, P. J. (2014). Endosymbiotic gene transfer in tertiary plastid-containing dinoflagellates. *Eukaryotic Cell, 13*(2), 246–255.

Carreto, J. I., Seguel, M., & Montoya, N. G. (2001). Pigment profile of the ichthyotoxic dinoflagellate *Gymnodinium* sp. from a massive bloom in southern Chile. *Journal of Plankton Research, 23*, 1171–1175.

Cavalier-Smith, T. (1999). Principles of protein and lipid targeting in secondary symbiogenesis: Euglenoid, dinoflagellate, and sporozoan plastid origins and the eukaryote family tree. *Journal of Eukaryotic Microbiology, 46*(4), 347–366.

Charlson, R. J., Lovelock, J. E., Andreae, M. O., & Warren, S. G. (1987). Oceanic phytoplankton, atmospheric sulphur, cloud albedo and climate. *Nature, 326*, 655–661.

Chesnick, J. M., Kooistra, W. H., Wellbrock, U., & Medlin, L. K. (1997). Ribosomal RNA analysis indicates a benthic pennate diatom ancestry for the endosymbionts of the dinoflagellates *Peridinium foliaceum* and *Peridinium balticum* (Pyrrhophyta). *Journal of Eukaryotic Microbiology, 44*(4), 314–320.

Chow, M. H., Yan, K. T. H., Bennett, M. J., & Wong, J. T. Y. (2010). Birefringence and DNA condensation of liquid crystalline chromosomes. *Eukaryotic Cell, 9*(10), 1577–1587.

Coats, D. W. (1999). Parasitic life styles of marine dinoflagellates. *Journal of Eukaryotic Microbiology, 46*(4), 402–409.

Dang, Y., & Green, B. R. (2009). Substitutional editing of *Heterocapsa triquetra* chloroplast transcripts and a folding model for its divergent chloroplast 16S rRNA. *Gene, 442*(1–2), 73–80.

Daugbjerg, N., Jensen, M. H., & Hansen, P. J. (2013). Using nuclear-encoded LSU and SSU rDNA sequences to identify the eukaryotic endosymbiont in *Amphisolenia bidentata* (Dinophyceae). *Protist, 164*(3), 411–422.

de Vargas, C., Audic, S., Henry, N., Decelle, J., Mahé, F., Logares, R., et al. (2015). Ocean plankton. Eukaryotic plankton diversity in the sunlit ocean. *Science, 348*(6237), 1261605.

Dodge, J. D. (1971). A dinoflagellate with both a mesokaryotic and a eukaryotic nucleus. I. Fine structure of the nuclei. *Protoplasma, 73*(2), 145–157.

Dodge, J. D. (1975). A survey of chloroplast ultrastructure in the Dinophyceae. *Phycologia, 14*, 253–263.

Dodge, J. D., & Crawford, R. M. (1969). Observations on the fine structure of the eyespot and associated organelles in the dinoflagellate *Glenodinium foliaceum*. *Journal of Cell Science, 5*, 479–493.

Dodge, J. D., & Crawford, R. M. (1970). The morphology and fine structure of *Ceratium hirundinella* (Dinophyceae). *Journal of Phycology, 6*, 137–149.

Dorrell, R. G., Hinksman, G. A., & Howe, C. J. (2016). Diversity of transcripts and transcript processing forms in plastids of the dinoflagellate alga *Karenia mikimotoi*. *Plant Molecular Biology, 90*, 233–247.

Dorrell, R. G., & Howe, C. J. (2012). Functional remodeling of RNA processing in replacement chloroplasts by pathways retained from their predecessors. *Proceedings of the National Academy of Sciences of the United States of America, 109*, 18879–18884.

Dorrell, R. G., & Howe, C. J. (2015). Integration of plastids with their hosts: Lessons learned from dinoflagellates. *Proceedings of the National Academy of Sciences of the United States of America*, *112*, 10247–10254.

Dorrell, R. G., Klinger, C. M., Newby, R. J., Butterfield, E. R., Richardson, E., Dacks, J. B., et al. (2017). Progressive and biased divergent evolution underpins the origin and diversification of peridinin dinoflagellate plastids. *Molecular Biology and Evolution*, *34*(2), 361–379.

Durnford, D. G., & Gray, M. W. (2006). Analysis of *Euglena gracilis* plastid-targeted proteins reveals different classes of transit sequences. *Eukaryotic Cell*, *5*(12), 2079–2091.

Durnford, D. G., & Schwartzbach, S. D. (2017). Protein targeting to the plastid of *Euglena*. *Advances in Experimental Medicine and Biology*, *979*(48), 183–205.

Eriksen, N. T., Hayes, K. C., & Lewitus, A. J. (2002). Growth responses of the mixotrophic dinoflagellates, *Cryptoperidiniopsis* sp. and *Pfiesteria piscicida*, to light under prey-saturated conditions. *Harmful Algae*, *1*, 191–203.

Espelund, M., Minge, M. A., Gabrielsen, T. M., Nederbragt, A. J., Shalchian-Tabrizi, K., Otis, C., et al. (2012). Genome fragmentation is not confined to the peridinin plastid in dinoflagellates. *PLoS One*, 7(6), e38809.

Felsner, G., Sommer, M. S., Gruenheit, N., Hempel, F., Moog, D., Zauner, S., et al. (2011). ERAD components in organisms with complex red plastids suggest recruitment of a preexisting protein transport pathway for the periplastid membrane. *Genome Biology and Evolution*, *3*, 140–150.

Figueroa, R. I., Bravo, I., Fraga, S., Garcés, E., & Llaveria, G. (2009). The life history and cell cycle of *Kryptoperidinium foliaceum*, a dinoflagellate with two eukaryotic nuclei. *Protist*, *160*, 285–300.

Gabrielsen, T. M., Minge, M. A., Espelund, M., Tooming-Klunderud, A., Patil, V., Nederbragt, A. J., et al. (2011). Genome evolution of a tertiary dinoflagellate plastid. *PLoS One*, *6*(4), e19132.

Gaines, G., & Elbrächter, M. (1987). Heterotrophic nutrition. In F. Taylor (Ed.), *The biology of dinoflagellates* (pp. 224–269). Oxford, UK: Blackwell Scientific Publications.

Gast, R. J., Moran, D. M., Dennett, M. R., & Caron, D. A. (2007). Kleptoplasty in an Antarctic dinoflagellate: Caught in evolutionary transition? *Environmental Microbiology*, *9*(1), 39–45.

Gavelis, G. S., Hayakawa, S., White, R. A., III, Gojobori, T., Suttle, C. A., Keeling, P. J., et al. (2015). Eye-like ocelloids are built from different endosymbiotically acquired components. *Nature*, *523*, 204–207.

Gavelis, G. S., Wakeman, K. C., Tillmann, U., Ripken, C., Mitarai, S., Herranz, M., et al. (2017). Microbial arms race: Ballistic "nematocysts" in dinoflagellates represent a new extreme in organelle complexity. *Science Advances*, *3*(3), e1602552.

Gómez, F. (2012). A quantitative review of the lifestyle, habitat and trophic diversity of dinoflagellates (Dinoflagellata, Alveolata). *Systematics and Biodiversity*, *10*, 267–275.

Gómez, F., Moreira, D., & López-García, P. (2010). Molecular phylogeny of noctilucoid dinoflagellates (Noctilucales, Dinophyceae). *Protist*, *161*(3), 466–478.

Gornik, S. G., Febrimarsa, Cassin, A. M., Macrae, J. I., Ramaprasad, A., Rchiad, Z., et al. (2015). Endosymbiosis undone by stepwise elimination of the plastid in a parasitic dinoflagellate. *Proceedings of the National Academy of Sciences of the United States of America*, *112*(18), 5767–5772.

Gornik, S. G., Ford, K. L., Mulhern, T. D., Bacic, A., McFadden, G. I., & Waller, R. F. (2012). Loss of nucleosomal DNA condensation coincides with appearance of a novel nuclear protein in dinoflagellates. *Current Biology*, *22*(24), 2303–2312.

Gould, S. B., Waller, R. F., & McFadden, G. I. (2008). Plastid evolution. *Annual Review of Plant Biology*, *59*, 491–517.

Hansen, P. J., & Calado, A. J. (1999). Phagotrophic mechanisms and prey selection in free-living dinoflagellates. *Journal of Eukaryotic Microbiology*, *46*, 382–389.

Hansen, P. J., Miranda, L., & Azanza, R. (2004). Green *Noctiluca scintillans*: A dinoflagellate with its own greenhouse. *Marine Ecology Progress Series*, *275*, 79–87.

Hansen, P. J., Ojamäe, K., Berge, T., Trampe, E. C. L., Nielsen, L. T., Lips, I., et al. (2016). Photoregulation in a kleptochloroplastidic dinoflagellate, *Dinophysis acuta*. *Frontiers in Microbiology*, 7, 785–796.

Hehenberger, E., Imanian, B., Burki, F., & Keeling, P. J. (2014). Evidence for the retention of two evolutionary distinct plastids in dinoflagellates with diatom endosymbionts. *Genome Biology and Evolution*, *6*(9), 2321–2334.

Holm-Hansen, O. (1969). Algae: Amounts of DNA and organic carbon in single cells. *Science*, *163*(3862), 87–88.

Horiguchi, T., & Takano, Y. (2006). Serial replacement of a diatom endosymbiont in the marine dinoflagellate *Peridinium quinquecorne* (Peridiniales, Dinophyceae). *Phycological Research*, *54*, 193–200.

Howe, C. J., Barbrook, A. C., Nisbet, R. E. R., Lockhart, P. J., & Larkum, A. W. D. (2008). The origin of plastids. *Philosophical Transactions of the Royal Society of London Series B, Biological Sciences*, *363*(1504), 2675–2685.

Howe, C. J., Nisbet, R. E. R., & Barbrook, A. C. (2008). The remarkable chloroplast genome of dinoflagellates. *Journal of Experimental Botany*, *59*(5), 1035–1045.

Hudson, A. O., Singh, B. K., Leustek, T., & Gilvarg, C. (2006). An LL-diaminopimelate aminotransferase defines a novel variant of the lysine biosynthesis pathway in plants. *Plant Physiology*, *140*(1), 292–301.

Imanian, B., Carpenter, K. J., & Keeling, P. J. (2007). Mitochondrial genome of a tertiary endosymbiont retains genes for electron transport proteins. *Journal of Eukaryotic Microbiology*, *54*(2), 146–153.

Imanian, B., & Keeling, P. J. (2007). The dinoflagellates *Durinskia baltica* and *Kryptoperidinium foliaceum* retain functionally overlapping mitochondria from two evolutionarily distinct lineages. *BMC Evolutionary Biology*, 7(1), 172–182.

Imanian, B., Pombert, J.-F., Dorrell, R. G., Burki, F., & Keeling, P. J. (2012). Tertiary endosymbiosis in two dinotoms has generated little change in the mitochondrial genomes of their dinoflagellate hosts and diatom endosymbionts. *PLoS One*, 7, e43763.

Imanian, B., Pombert, J.-F., & Keeling, P. J. (2010). The complete plastid genomes of the two "dinotoms" *Durinskia baltica* and *Kryptoperidinium foliaceum*. *PLoS One*, *5*(5), e10711.

Ishida, K.-I., & Green, B. R. (2002). Second- and third-hand chloroplasts in dinoflagellates: Phylogeny of oxygen-evolving enhancer 1 (PsbO) protein reveals replacement of a nuclear-encoded plastid gene by that of a haptophyte tertiary endosymbiont. *Proceedings of the National Academy of Sciences of the United States of America*, *99*(14), 9294–9299.

Janouškovec, J., Gavelis, G. S., Burki, F., Dinh, D., Bachvaroff, T. R., Gornik, S. G., et al. (2017). Major transitions in dinoflagellate evolution unveiled by phylotranscriptomics. *Proceedings of the National Academy of Sciences of the United States of America*, *114*, e171–e180.

Janouškovec, J., Horák, A., Oborník, M., Lukeš, J., & Keeling, P. J. (2010). A common red algal origin of the apicomplexan, dinoflagellate, and heterokont plastids. *Proceedings of the National Academy of Sciences of the United States of America*, *107*(24), 10949–10954.

Janouškovec, J., Sobotka, R., Lai, D.-H., Flegontov, P., Koník, P., Komenda, J., et al. (2013). Split photosystem protein, linear-mapping topology, and growth of structural complexity in the plastid genome of *Chromera velia*. *Molecular Biology and Evolution*, *30*(11), 2447–2462.

Janouškovec, J., Tikhonenkov, D. V., Burki, F., Howe, A. T., Kolisko, M., Mylnikov, A. P., et al. (2015). Factors mediating plastid dependency and the origins of parasitism in

apicomplexans and their close relatives. *Proceedings of the National Academy of Sciences of the United States of America, 112*(33), 10200–10207.
Jeffrey, S. W., & Vesk, M. (1976). Further evidence for a membrane-bound endosymbiont within the dinoflagellate *Peridinium foliaceum. Journal of Phycology, 12*, 450–455.
Jeong, H. J., Yoo, D. Y., & Kang, N. S. (2012). Heterotrophic feeding as a newly identified survival strategy of the dinoflagellate *Symbiodinium. Proceedings of the National Academy of Sciences of the United States of America, 109*, 12604–12609.
Jeong, H. J., Yoo, D. Y., Kim, J. S., Seong, K. A., & Kang, N. S. (2010). Growth, feeding and ecological roles of the mixotrophic and heterotrophic dinoflagellates in marine planktonic food webs. *Ocean Science Journal, 45*, 65–91.
Kamikawa, R., Tanifuji, G., Kawachi, M., Miyashita, H., Hashimoto, T., & Inagaki, Y. (2015). Plastid genome-based phylogeny pinpointed the origin of the green-colored plastid in the dinoflagellate *Lepidodinium chlorophorum. Genome Biology and Evolution, 7*, 1133–1140.
Kim, M., Kim, K. Y., Nam, S. W., Shin, W., Yih, W., & Park, M. G. (2014). The effect of starvation on plastid number and photosynthetic performance in the kleptoplastidic dinoflagellate *Amylax triacantha. Journal of Eukaryotic Microbiology, 61*(4), 354–363.
Kim, M., Nam, S. W., Shin, W., & Coats, D. W. (2012). *Dinophysis caudata* (dinophyceae) sequesters and retains plastids from the mixotrophic ciliate prey *Mesodinium rubrum. Journal of Phycology, 48*, 569–579.
Kořený, L., Sobotka, R., Janouškovec, J., Keeling, P. J., & Oborník, M. (2011). Tetrapyrrole synthesis of photosynthetic chromerids is likely homologous to the unusual pathway of apicomplexan parasites. *Plant Cell, 23*(9), 3454–3462.
Koumandou, V. L., & Howe, C. J. (2007). The copy number of chloroplast gene minicircles changes dramatically with growth phase in the dinoflagellate *Amphidinium operculatum. Protist, 158*(1), 89–103.
LaJeunesse, T. C., Lambert, G., Andersen, R. A., Coffroth, M. A., & Galbraith, D. W. (2005). *Symbiodinium* (Pyrrhophyta) genome sizes (DNA content) are smallest among dinoflagellates. *Journal of Phycology, 41*(4), 880–886.
Larsen, J. (1988). An ultrastructural study of *Amphidinium poecilochroum* (Dinophyceae), a phagotrophic dinoflagellate feeding on small species of cryptophytes. *Phycologia, 27*, 366–377.
Lewitus, A. J., & Glasgow, H. B. (1999). Kleptoplastidy in the toxic dinoflagellate *Pfiesteria piscicida* (Dinophyceae). *Journal of Phycology, 35*, 303–312.
Li, A., Stoecker, D. K., & Adolf, J. E. (1999). Feeding, pigmentation, photosynthesis and growth of the mixotrophic dinoflagellate *Gyrodinium galatheanum. Aquatic Microbial Ecology, 19*, 163–176.
Lin, S., Cheng, S., Song, B., Zhong, X., Lin, X., Li, W., et al. (2015). The Symbiodinium kawagutii genome illuminates dinoflagellate gene expression and coral symbiosis. *Science, 350*(6261), 691–694.
Matsumoto, T., Kawachi, M., Miyashita, H., & Inagaki, Y. (2012). Prasinoxanthin is absent in the green-colored dinoflagellate *Lepidodinium chlorophorum* strain NIES-1868: Pigment composition and 18S rRNA phylogeny. *Journal of Plant Research, 125*(6), 705–711.
Matsumoto, T., Shinozaki, F., Chikuni, T., Yabuki, A., Takishita, K., Kawachi, M., et al. (2011). Green-colored plastids in the dinoflagellate genus *Lepidodinium* are of core chlorophyte origin. *Protist, 162*(2), 268–276.
Matsuzaki, M., Kuroiwa, H., Kuroiwa, T., Kita, K., & Nozaki, H. (2008). A cryptic algal group unveiled: A plastid biosynthesis pathway in the oyster parasite *Perkinsus marinus. Molecular Biology and Evolution, 25*(6), 1167–1179.
Minge, M. A., Shalchian-Tabrizi, K., Tørresen, O. K., Takishita, K., Probert, I., Inagaki, Y., et al. (2010). A phylogenetic mosaic plastid proteome and unusual plastid-targeting signals in the green-colored dinoflagellate *Lepidodinium chlorophorum. BMC Evolutionary Biology, 10*(1), 191–202.

Moore, C. E., & Archibald, J. M. (2009). Nucleomorph genomes. *Annual Review of Genetics*, *43*, 251–264.

Moore, R. B., Ferguson, K. M., Loh, W. K. W., Hoegh-Guldberg, O., & Carter, D. A. (2003). Highly organized structure in the non-coding region of the psbA minicircle from clade C *Symbiodinium*. *International Journal of Systematic and Evolutionary Microbiology*, *53*(Pt. 6), 1725–1734.

Moore, R. B., Oborník, M., Janouškovec, J., Chrudimský, T., Vancová, M., Green, D. H., et al. (2008). A photosynthetic alveolate closely related to apicomplexan parasites. *Nature*, *451*(7181), 959–963.

Morse, D., Salois, P., Markovic, P., & Hastings, J. W. (1995). A nuclear-encoded form II RuBisCO in dinoflagellates. *Science*, *268*(5217), 1622–1624.

Mungpakdee, S., Shinzato, C., Takeuchi, T., Kawashima, T., Koyanagi, R., Hisata, K., et al. (2014). Massive gene transfer and extensive RNA editing of a symbiotic dinoflagellate plastid genome. *Genome Biology and Evolution*, *6*(6), 1408–1422.

Nash, E., Nisbet, R., Barbrook, A., & Howe, C. (2008). Dinoflagellates: A mitochondrial genome all at sea. *Trends in Genetics*, *24*(7), 328–335.

Nassoury, N., Cappadocia, M., & Morse, D. (2003). Plastid ultrastructure defines the protein import pathway in dinoflagellates. *Journal of Cell Science*, *116*(Pt. 14), 2867–2874.

Nelson, M. J., & Green, B. R. (2005). Double hairpin elements and tandem repeats in the non-coding region of *Adenoides eludens* chloroplast gene minicircles. *Gene*, *358*, 102–110.

Nisbet, R. E. R., Hiller, R. G., Barry, E. R., Skene, P., Barbrook, A. C., & Howe, C. J. (2008). Transcript analysis of dinoflagellate plastid gene minicircles. *Protist*, *159*(1), 31–39.

Nisbet, R. E. R., Koumandou, L. V., Barbrook, A. C., & Howe, C. J. (2004). Novel plastid gene minicircles in the dinoflagellate *Amphidinium operculatum*. *Gene*, *331*, 141–147.

Norén, F., Moestrup, Ø., & Rehnstam-Holm, A. S. (1999). *Parvilucifera infectans* Norén et Moestrup gen. et sp. nov. (Perkinsozoa phylum nov.): A parasitic flagellate capable of killing toxic microalgae. *European Journal of Protistology*, *35*, 233–254.

Nosenko, T., Lidie, K. L., Van Dolah, F. M., Lindquist, E., Cheng, J.-F., & Bhattacharya, D. (2006). Chimeric plastid proteome in the Florida "red tide" dinoflagellate *Karenia brevis*. *Molecular Biology and Evolution*, *23*(11), 2026–2038.

Oborník, M., Modrý, D., Lukeš, M., Černotíková-Stříbrná, E., Cihlář, J., Tesařová, M., et al. (2012). Morphology, ultrastructure and life cycle of *Vitrella brassicaformis* n. sp., n. gen., a novel chromerid from the Great Barrier Reef. *Protist*, *163*(2), 306–323.

Onuma, R., & Horiguchi, T. (2015). Kleptochloroplast enlargement, karyoklepty and the distribution of the cryptomonad nucleus in *Nusuttodinium* (= *Gymnodinium*) *aeruginosum* (Dinophyceae). *Protist*, *166*(2), 177–195.

Patron, N. J., & Waller, R. F. (2007). Transit peptide diversity and divergence: A global analysis of plastid targeting signals. *BioEssays*, *29*(10), 1048–1058.

Patron, N. J., Waller, R. F., Archibald, J. M., & Keeling, P. J. (2005). Complex protein targeting to dinoflagellate plastids. *Journal of Molecular Biology*, *348*(4), 1015–1024.

Patron, N. J., Waller, R. F., & Keeling, P. J. (2006). A tertiary plastid uses genes from two endosymbionts. *Journal of Molecular Biology*, *357*(5), 1373–1382.

Petersen, J., Ludewig, A.-K., Michael, V., Bunk, B., Jarek, M., Baurain, D., et al. (2014). *Chromera velia*, endosymbioses and the rhodoplex hypothesis—Plastid evolution in cryptophytes, alveolates, stramenopiles, and haptophytes (CASH lineages). *Genome Biology and Evolution*, *6*(3), 666–684.

Prezelin, B. (1987). Photosynthetic physiology of dinoflagellates. In F. Taylor (Ed.), *The biology of dinoflagellates* (pp. 174–223). Oxford, UK: Blackwell.

Puerta, M. S., & Delwiche, C. F. (2008). A hypothesis for plastid evolution in chromalveolates. *Journal of Phycology*, *44*, 1097–1107.

Ralph, S. A., van Dooren, G. G., Waller, R. F., Crawford, M. J., Fraunholz, M. J., Foth, B. J., et al. (2004). Tropical infectious diseases: Metabolic maps and functions of the *Plasmodium falciparum* apicoplast. *Nature Reviews. Microbiology*, *2*(3), 203–216.

Rapoport, H., Strain, H. H., Svec, W. A., Aitzetmueller, K., Grandolfo, M., Katz, J. J., et al. (2002). Structure of peridinin, the characteristics dinoflagellate carotenoid. *Journal of the American Chemical Society*, *93*(7), 1823–1825.

Richardson, E., Dorrell, R. G., & Howe, C. J. (2014). Genome-wide transcript profiling reveals the coevolution of plastid gene sequences and transcript processing pathways in the fucoxanthin dinoflagellate *Karlodinium veneficum*. *Molecular Biology and Evolution*, *31*(9), 2376–2386.

Riisgaard, K., & Hansen, P. J. (2009). Role of food uptake for photosynthesis, growth and survival of the mixotrophic dinoflagellate *Dinophysis acuminata*. *Marine Ecology Progress Series*, *381*, 51–62.

Rizzo, P. J. (1987). Biochemistry of the dinoflagellate nucleus. In F. Taylor (Ed.), *The biology of dinoflagellates* (pp. 143–173). Oxford: Blackwell Scientific Publications.

Robledo, J. A. F. N., Caler, E., Matsuzaki, M., Keeling, P. J., Shanmugam, D., Roos, D. S., et al. (2011). The search for the missing link: A relic plastid in *Perkinsus*? *International Journal for Parasitology*, *41*(12), 1217–1229.

Saldarriaga, J. F., Taylor, F. J., Keeling, P. J., & Cavalier-Smith, T. (2001). Dinoflagellate nuclear SSU rRNA phylogeny suggests multiple plastid losses and replacements. *Journal of Molecular Evolution*, *53*(3), 204–213.

Sánchez Puerta, M. V., Lippmeier, J. C., Apt, K. E., & Delwiche, C. F. (2007). Plastid genes in a non-photosynthetic dinoflagellate. *Protist*, *158*(1), 105–117.

Schnepf, E., & Elbrächter, M. (1999). Dinophyte chloroplasts and phylogeny—A review. *Grana*, *38*, 81–97.

Schnepf, E., Winter, S., & Mollenhauer, D. (1989). *Gymnodinium aeruginosum* (Dinophyta): A blue-green dinoflagellate with a vestigial, anucleate, cryptophycean endosymbiont. *Plant Systematics and Evolution*, *164*, 75–91.

Sellers, C. G., Gast, R. J., & Sanders, R. W. (2014). Selective feeding and foreign plastid retention in an Antarctic dinoflagellate. *Journal of Phycology*, *50*(6), 1081–1088.

Ševčíková, T., Horák, A., Klimeš, V., Zbránková, V., Demir-Hilton, E., Sudek, S., et al. (2015). Updating algal evolutionary relationships through plastid genome sequencing: Did alveolate plastids emerge through endosymbiosis of an ochrophyte? *Scientific Reports*, *5*, 10134.

Sheng, J., Malkiel, E., Katz, J., Adolf, J., Belas, R., & Place, A. R. (2007). Digital holographic microscopy reveals prey-induced changes in swimming behavior of predatory dinoflagellates. *Proceedings of the National Academy of Sciences of the United States of America*, *104*(44), 17512–17517.

Shoguchi, E., Shinzato, C., Kawashima, T., Gyoja, F., Mungpakdee, S., Koyanagi, R., et al. (2013). Draft assembly of the Symbiodinium minutum nuclear genome reveals dinoflagellate gene structure. *Current Biology*, 1–10.

Siddall, M. E., Reece, K. S., Graves, J. E., & Burreson, E. M. (1997). "Total evidence" refutes the inclusion of *Perkinsus* species in the phylum Apicomplexa. *Parasitology*, *115*(Pt. 2), 165–176.

Slamovits, C. H., & Keeling, P. J. (2008). Plastid-derived genes in the nonphotosynthetic alveolate *Oxyrrhis marina*. *Molecular Biology and Evolution*, *25*(7), 1297–1306.

Stentiford, G. D., & Shields, J. D. (2005). A review of the parasitic dinoflagellates *Hematodinium* species and *Hematodinium*-like infections in marine crustaceans. *Diseases of Aquatic Organisms*, *66*(1), 47–70.

Stiller, J. W. (2014). Toward an empirical framework for interpreting plastid evolution. *Journal of Phycology*, *50*, 462–471.

Stiller, J. W., Schreiber, J., Yue, J., Guo, H., Ding, Q., & Huang, J. (2014). The evolution of photosynthesis in chromist algae through serial endosymbioses. *Nature Communications*, *5*, 5764.

Stoecker, D. K. (1999). Mixotrophy among dinoflagellates. *Journal of Eukaryotic Microbiology*, *46*, 397–401.

Stork, S., Moog, D., Przyborski, J. M., Wilhelmi, I., Zauner, S., & Maier, U.-G. (2012). Distribution of the SELMA translocon in secondary plastids of red algal origin and predicted uncoupling of ubiquitin-dependent translocation from degradation. *Eukaryotic Cell, 11*(12), 1472–1481.

Sweeney, B. M. (1976). *Pedinomonas noctilucae* (Prasinophyceae), the flagellate symbiotic in *Noctiluca* (Dinophyceae) in Southeast Asia. *Journal of Phycology, 12*, 460–464.

Tabita, F. R., Hanson, T. E., Satagopan, S., Witte, B. H., & Kreel, N. E. (2008). Phylogenetic and evolutionary relationships of RubisCO and the RubisCO-like proteins and the functional lessons provided by diverse molecular forms. *Philosophical Transactions of the Royal Society of London. Series B, Biological Sciences, 363*(1504), 2629–2640.

Takano, Y., Yamaguchi, H., Inouye, I., Moestrup, Ø., & Horiguchi, T. (2014). Phylogeny of five species of *Nusuttodinium* gen. nov. (Dinophyceae), a genus of unarmoured kleptoplastidic dinoflagellates. *Protist, 165*(6), 759–778.

Takishita, K., Ishida, K.-I., & Maruyama, T. (2004). Phylogeny of nuclear-encoded plastid-targeted GAPDH gene supports separate origins for the peridinin- and the fucoxanthin derivative-containing plastids of dinoflagellates. *Protist, 155*(4), 447–458.

Takishita, K., Kawachi, M., Noël, M.-H., Matsumoto, T., Kakizoe, N., Watanabe, M. M., et al. (2008). Origins of plastids and glyceraldehyde-3-phosphate dehydrogenase genes in the green-colored dinoflagellate *Lepidodinium chlorophorum*. *Gene, 410*(1), 26–36.

Tamura, M., Shimada, S., & Horiguchi, T. (2005). *Galeidiniium rugatum* gen. et sp. nov. (dinophyceae), a new coccoid dinoflagellate with a diatom endosymbiont1. *Journal of Phycology, 41*(3), 658–671.

Tarangkoon, W., Hansen, G., & Hansen, P. J. (2010). Spatial distribution of symbiont-bearing dinoflagellates in the Indian Ocean in relation to oceanographic regimes. *Aquatic Microbial Ecology, 58*, 197–213.

Teles-Grilo, M. L., Tato-Costa, J., Duarte, S. M., Maia, A., Casal, G., & Azevedo, C. (2007). Is there a plastid in *Perkinsus atlanticus* (phylum Perkinsozoa)? *European Journal of Protistology, 43*(2), 163–167.

Tengs, T., Dahlberg, O. J., Shalchian-Tabrizi, K., Klaveness, D., Rudi, K., Delwiche, C. F., et al. (2000). Phylogenetic analyses indicate that the 19'Hexanoyloxy-fucoxanthin-containing dinoflagellates have tertiary plastids of haptophyte origin. *Molecular Biology and Evolution, 17*(5), 718–729.

Timmis, J. N., Ayliffe, M. A., Huang, C. Y., & Martin, W. (2004). Endosymbiotic gene transfer: Organelle genomes forge eukaryotic chromosomes. *Nature Reviews. Genetics, 5*(2), 123–135.

Tippit, D. H., & Pickett-Heaps, J. D. (1976). Apparent amitosis in the binucleate dinoflagellate *Peridinium balticum*. *Journal of Cell Science, 21*, 273–289.

Tomas, R. N., & Cox, E. R. (1973). Observations on the symbiosis of *Peridinium balticum* and its intracellular alga. I. Ultrastructure. *Journal of Phycology, 9*, 304–323.

Waller, R. F., Gornik, S. G., Kořený, L., & Pain, A. (2016). Metabolic pathway redundancy within the apicomplexan-dinoflagellate radiation argues against an ancient chromalveolate plastid. *Communicative & Integrative Biology, 9*(1), e1116653.

Waller, R. F., & Jackson, C. J. (2009). Dinoflagellate mitochondrial genomes: Stretching the rules of molecular biology. *BioEssays, 31*(2), 237–245.

Waller, R. F., Patron, N. J., & Keeling, P. J. (2006). Phylogenetic history of plastid-targeted proteins in the peridinin-containing dinoflagellate *Heterocapsa triquetra*. *International Journal of Systematic and Evolutionary Microbiology, 56*(Pt 6), 1439–1447.

Waller, R. F., Slamovits, C. H., & Keeling, P. J. (2006). Lateral gene transfer of a multigene region from cyanobacteria to dinoflagellates resulting in a novel plastid-targeted fusion protein. *Molecular Biology and Evolution, 23*(7), 1437–1443.

Wang, Y., & Morse, D. (2006). Rampant polyuridylylation of plastid gene transcripts in the dinoflagellate *Lingulodinium*. *Nucleic Acids Research, 34*(2), 613–619.

Watanabe, M. M., Takeda, Y., Sasa, T., & Inouye, I. (1987). A green dinoflagellate with chlorophylls a and b: Morphology, fine structure of the chloroplast and chlorophyll composition. *Journal of Phycology*, *23*, 382–389.

Whitney, S. M., Shaw, D. C., & Yellowlees, D. (1995). Evidence that some dinoflagellates contain a ribulose-1,5-bisphosphate carboxylase/oxygenase related to that of the alpha-proteobacteria. *Proceedings. Biological Sciences*, *259*(1356), 271–275.

Wilcox, L. W., & Wedemayer, G. J. (1985). Dinoflagellate with blue-green chloroplasts derived from an endosymbiotic eukaryote. *Science*, *227*, 192–194.

Wilson, R. J., & Williamson, D. H. (1997). Extrachromosomal DNA in the Apicomplexa. *Microbiology and Molecular Biology Reviews*, *61*(1), 1–16.

Wisecaver, J. H., & Hackett, J. D. (2010). Transcriptome analysis reveals nuclear-encoded proteins for the maintenance of temporary plastids in the dinoflagellate *Dinophysis acuminata*. *BMC Genomics*, *11*, 366–376.

Wisecaver, J. H., & Hackett, J. D. (2011). Dinoflagellate genome evolution. *Annual Review of Microbiology*, *65*, 369–387.

Yamada, N., Sym, S. D., & Horiguchi, T. (2017). Identification of highly divergent diatom-derived chloroplasts in dinoflagellates, including a description of *Durinskia kwazulunatalensis* sp. nov. (Peridiniales, Dinophyceae). *Molecular Biology and Evolution*, *34*, 1335–1351.

Yamaguchi, H., Nakayama, T., Kai, A., & Inouye, I. (2011). Taxonomy and phylogeny of a new kleptoplastidal dinoflagellate, *Gymnodinium myriopyrenoides* sp. nov. (Gymnodiniales, Dinophyceae), and its cryptophyte symbiont. *Protist*, *162*(4), 650–667.

Yoon, H. S., Hackett, J. D., & Bhattacharya, D. (2002). A single origin of the peridinin- and fucoxanthin-containing plastids in dinoflagellates through tertiary endosymbiosis. *Proceedings of the National Academy of Sciences of the United States of America*, *99*(18), 11724–11729.

Yoon, H. S., Hackett, J. D., Van Dolah, F. M., Nosenko, T., Lidie, K. L., & Bhattacharya, D. (2005). Tertiary endosymbiosis driven genome evolution in dinoflagellate algae. *Molecular Biology and Evolution*, *22*(5), 1299–1308.

You, X., Luo, Z., Su, Y., Gu, L., & Gu, H. (2015). *Peridiniopsis jiulongensis*, a new freshwater dinoflagellate with a diatom endosymbiont from China. *Nova Hedwigia*, *101*(3), 313–326.

Záhonová, K., Füssy, Z., Oborník, M., Eliáš, M., & Yurchenko, V. (2016). RuBisCO in non-photosynthetic alga *Euglena longa*: Divergent features, transcriptomic analysis and regulation of complex formation. *PLoS One*, *11*(7), e0158790.

Zauner, S., Greilinger, D., Laatsch, T., Kowallik, K. V., & Maier, U.-G. (2004). Substitutional editing of transcripts from genes of cyanobacterial origin in the dinoflagellate *Ceratium horridum*. *FEBS Letters*, *577*(3), 535–538.

Zhang, H., Campbell, D. A., Sturm, N. R., Dungan, C. F., & Lin, S. (2011). Spliced leader RNAS, mitochondrial gene frameshifts and multi-protein phylogeny expand support for the genus *Perkinsus* as a unique group of alveolates. *PLoS One*, *6*(5), e19933.

Zhang, Z., Green, B. R., & Cavalier-Smith, T. (1999). Single gene circles in dinoflagellate chloroplast genomes. *Nature*, *400*(6740), 155–159.

Zhang, H., Hou, Y., Miranda, L., Campbell, D. A., Sturm, N. R., Gaasterland, T., et al. (2007). Spliced leader RNA trans-splicing in dinoflagellates. *Proceedings of the National Academy of Sciences of the United States of America*, *104*(11), 4618–4623.

Zhang, Q., Liu, G., & Hu, Z. (2014). Description of a new freshwater bloom-forming dinoflagellate with a diatom endosymbiont, *Peridiniopsis minima* sp. nov. (Peridiniales, Dinophyceae) from China. *Algological Studies*, *145*, 119–133.

CHAPTER FIVE

The Dark Side of the Chloroplast: Biogenesis, Metabolism and Membrane Biology of the Apicoplast

Giel G. van Dooren[1], Sanduni V. Hapuarachchi
Research School of Biology, Australian National University, Acton, ACT, Australia
[1]Corresponding author: e-mail address: giel.vandooren@anu.edu.au

Contents

Abstract

Members of the phylum Apicomplexa contain plastids, termed apicoplasts, that were derived by secondary endosymbiosis. Unlike most of their sun-loving cousins, apicomplexans are parasites that live in the dark recesses of the animal hosts they infect. As a consequence, apicoplasts are not photosynthetic, but nevertheless carry out essential metabolic processes. In this chapter, we examine the evolution, biogenesis and functions of the apicoplast. In particular, we focus on the biology of the membranes that surround this organelle, which play key roles in the biogenesis of the organelle, and link the metabolic functions of the apicoplast with the rest of the cell.

Advances in Botanical Research, Volume 84
ISSN 0065-2296
http://dx.doi.org/10.1016/bs.abr.2017.06.007

1. INTRODUCTION: APICOMPLEXAN PARASITES

Apicomplexans are a sinister group of single-celled, intracellular parasites that include some of the most prevalent and destructive pathogens on Earth. The best known of these include members of the genus *Plasmodium*, the causative agents of malaria. Malaria is estimated to infect ~200 M humans annually, leading to ~400 K deaths, predominantly young children in sub-Saharan Africa (World Health Organization, 2016). *Toxoplasma gondii* infects an estimated one-third of humans around the world, causing disease in immunocompromised individuals and neonates (Montoya & Liesenfeld, 2004). *T. gondii* is also a major pathogen in animals, being a prominent cause of abortions in livestock such as sheep and causing disease and death in many marsupials (Hill, Chirukandoth, & Dubey, 2005; Parameswaran et al., 2010). *Eimeria* species cause coccidiosis in poultry, resulting in an estimated cost of US$2B to the poultry industry worldwide (Peek & Landman, 2011). *Cryptosporidium* is one of the major causes of severe diarrhoea and associated deaths in infants and children in low-income countries (Kotloff et al., 2013; Striepen, 2013).

2. DISCOVERY OF THE APICOPLAST

The first clues to the existence of a plastid in apicomplexan parasites came with the identification of a circular 35-kb extrachromosomal DNA molecule from *Plasmodium* (Fig. 1A; Kilejian, 1975; Wilson et al., 1996). Initially assumed to be the mitochondrial genome, the presence of plastid-like genes on the 35-kb circle suggested instead that this genome was of plastid origin (Gardner, Feagin, et al., 1991; Gardner et al., 1994; Gardner, Williamson, & Wilson, 1991; Howe, 1992; Kohler et al., 1997). Subsequent in situ hybridisation studies localised the 35-kb genome of *T. gondii* to an organelle bound by four membranes that was not the mitochondrion (Kohler et al., 1997; McFadden, Reith, Munholland, & Lang-Unnasch, 1996). Previous electron microscopy studies had identified a multimembrane-bound organelle in apicomplexan parasites, located apical to the nucleus in *T. gondii*, that was variously termed the Golgi adjunct, Hohlzylinder and spherical body, to name but a few (Aikawa, 1966; Hepler, Huff, & Sprinz, 1966; Sheffield & Melton, 1968; Siddall, 1992). Localisation of the plastid-like genome to this structure led to its definitive renaming as the 'apicoplast', for *api*complexan *plast*id (Kohler et al., 1997).

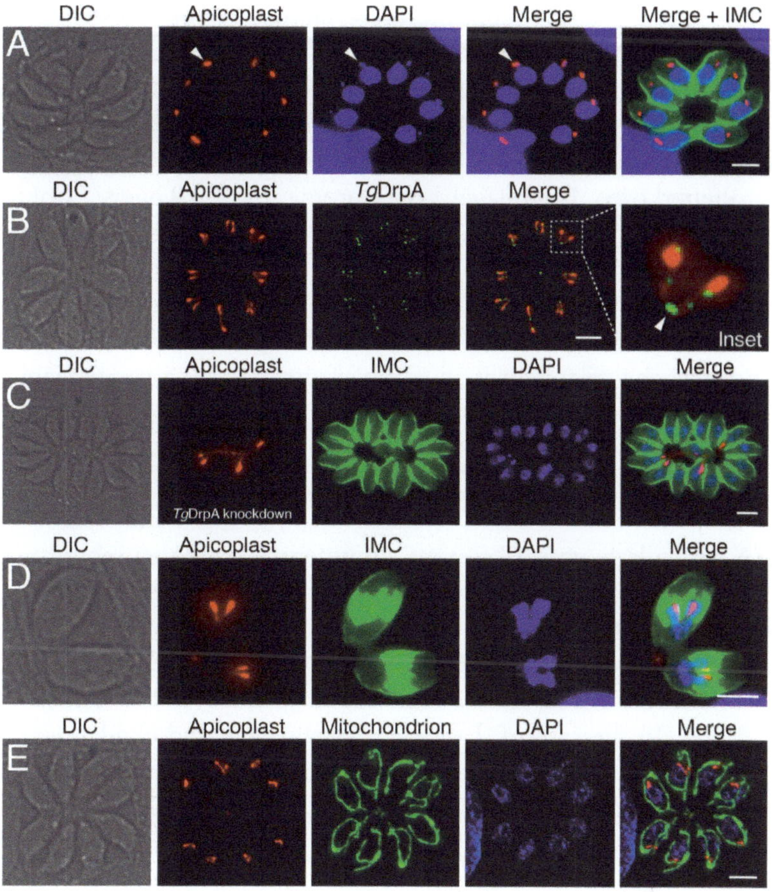

Fig. 1 Apicoplast morphology and biogenesis in *T. gondii*. (A) An eight-cell *T. gondii* vacuole. Each *T. gondii* parasite contains a single apicoplast (*red*). The 35-kb genome of the apicoplast can often be visualised by the DNA-labelling dye DAPI (*blue*; *arrowheads*). The inner membrane complex (IMC, *green*) localises at the periphery of each parasite. (B) An eight-cell *T. gondii* parasite showing typical U-shaped dividing apicoplasts (*red*). The apicoplast division protein *Tg*DrpA (*green*) occurs at the midpoint of dividing apicoplasts (merge and Inset). (C) Defects in apicoplast division are apparent upon the ATc-regulated knockdown of *Tg*DrpA. Most parasites of this 16-cell vacuole have lost their apicoplast. Some parasites retain a branch of the single, connected apicoplast (*red*) that remains in the vacuole. The IMC (*green*) localises at the parasite periphery. (D) Apicoplasts (*red*) divide during the process of daughter cell budding. Each parasite in this two-cell vacuole contains the IMC scaffold (*green*) of two new daughter cells. One branch of the U-shaped apicoplast localises in each of the daughter cells. The dividing nucleus (*blue*) of the parasite is also visible. (E) The relationship of the apicoplast (*red*) and mitochondrion (*green*) of *T. gondii*. Each parasite contains a single apicoplast and mitochondrion. Scale bars are 2 μm.

Antibiotics that target replication (e.g. ciprofloxacin), transcription (e.g. rifampin) or translation (e.g. clindamycin, thiostrepton) of the 35-kb genome inhibit the growth of *T. gondii* and *Plasmodium falciparum* parasites (Clough, Strath, Preiser, Denny, & Wilson, 1997; Fichera, Bhopale, & Roos, 1995; Fichera & Roos, 1997; McConkey, Rogers, & McCutchan, 1997; Pfefferkorn, Nothnagel, & Borotz, 1992; Rogers, Bukhman, McCutchan, & Draper, 1997; Williamson et al., 2002). Treatment with the DNA replication inhibitor ciprofloxacin leads to loss of the apicoplast genome, indicating the essential role of the apicoplast genome in parasite survival (Fichera & Roos, 1997). These studies provided the first definitive evidence that the apicoplast is an essential organelle for parasite viability in *Plasmodium*, *T. gondii* and most other apicomplexans.

3. THE 'DELAYED DEATH' PHENOMENON

A particularly unusual feature of numerous drugs that inhibit replication and translation of the apicoplast genome is that they do not lead to immediate growth arrest of the parasite. This phenomenon was first observed in *T. gondii* and has been termed 'delayed death' (Fichera et al., 1995; Fichera & Roos, 1997; Pfefferkorn & Borotz, 1994; Pfefferkorn et al., 1992). In the case of *T. gondii*, ciprofloxacin and clindamycin have minimal effects on parasite growth when first added to a parasite culture, with parasites capable of replicating inside the host cell, then egressing and invading new host cells (Fichera et al., 1995; Fichera & Roos, 1997; Pfefferkorn et al., 1992). It is in this second intracellular cycle of replication that growth inhibition by ciprofloxacin and clindamycin takes effect, with the potency of these drugs increasing by >1000-fold. Notably, washing out these drugs after the first cycle leads to equivalent defects in parasite growth in the second cycle compared to when the drug remained present throughout the assay (Fichera et al., 1995), suggesting that these drugs are causing an effect in the parasite during the first replication cycle that is important for development during the second replication cycle. Genetic modifications that lead to apicoplast loss in *T. gondii* mirror the effects of chemically induced delayed death (He et al., 2001; Mazumdar, Wilson, Masek, Hunter, & Striepen, 2006; van Dooren et al., 2009). In these examples, parasites lacking apicoplasts are able to egress from existing host cells and can reinvade new host cells, but exhibit limited development in these new host cells. In the case of *Plasmodium*, the phenomenon of delayed death is less clear. Of the drugs proposed to target apicoplast 'housekeeping' functions, only the translation inhibitors clindamycin, chloramphenicol and tetracycline have been shown to cause clear delayed death effects (Dahl & Rosenthal, 2007, 2008; Dahl et al., 2006; Goodman, Su, & McFadden, 2007). Treatments that cause loss of *P. falciparum* apicoplasts lead to subsequent defects in parasite growth following invasion of new host cells (Yeh & DeRisi, 2011), suggesting that, like in *T. gondii*, parasites lacking an apicoplast are unable to develop fully in new host cells.

Despite its characterisation over 20 years ago, the reason for delayed death remains elusive. During invasion, apicomplexan parasites encase themselves in a so-called parasitophorous vacuole that separates them from the host cell they inhabit (Fig. 1). Apicomplexans typically undergo multiple rounds of cell division within a single parasitophorous vacuole, before egressing from the host cell, destroying the host cell in the process. In the case of *T. gondii*, parasites within a vacuole can develop normally as long as one parasite in that vacuole retains an apicoplast (He et al., 2001). One hypothesis for why delayed death occurs, then, is that apicoplasts are producing one or more metabolic products that can be shared among other parasites (including apicoplast-less parasites) within a vacuole. When vacuoles contain no parasites with an apicoplast, growth of parasites within this vacuole is impaired. In support of this, a recent study found that *T. gondii* parasites maintain actomyosin-dependent cytosolic connections with other parasites in the same vacuole that likely facilitate the transfer of metabolites between parasites (Frenal et al., 2017). Apicoplast loss in mutants deficient in these cytosolic connections leads to more rapid impairment of parasite growth than when connections are present (Frenal et al., 2017). A second hypothesis is that apicoplasts generate a metabolic product that is required for proper development of parasites in the next developmental cycle (i.e. following subsequent invasion of new host cells). For example, a metabolic product generated by the apicoplast may be important in establishing a functional parasitophorous vacuole in the new host cell (He et al., 2001).

Both scenarios explaining delayed death posit that apicoplasts contribute one or more metabolic products that are essential for parasite development. Apicoplasts generate numerous metabolic products (see Section 7). In the case of *P. falciparum* blood stages, the only essential product is the isoprenoid isopentenyl pyrophosphate (IPP; Yeh & DeRisi, 2011). This suggests that one or more downstream metabolites of IPP are essential for proper development of apicomplexan parasites in new vacuoles and that loss in production of these metabolites results in delayed death. Consistent with this, genetic ablation of *T. gondii* genes required for IPP synthesis or for the import of precursors of IPP into the apicoplast leads to rapid parasite death (Brooks et al., 2010; Nair et al., 2011). Whether parasite death caused by the loss of IPP synthesis is more rapid than that seen in delayed death has not been examined in detail, although this may hold the key to understanding the delayed death phenomenon. Regardless of the biological reasons, the delayed death caused by antiapicoplast drugs has important implications for how these drugs can be used in a clinical setting. Drugs targeting the apicoplast cannot be used as a monotherapy to treat acute stages of the diseases caused by apicomplexans (Burkhardt et al., 2007; Goodman & McFadden, 2013). However, apicoplast-targeting drugs can be useful in combination therapies with fast-acting drugs that have a short half-life in human serum, or as prophylactic agents. For example, clindamycin can be paired with fast-acting, short-lived artemisinin derivatives to treat acute stages of *P. falciparum* infection (Ramharter et al., 2005), and the tetracycline analogue doxycycline is a common prophylactic agent for tourists who travel in malaria endemic regions.

4. THE EVOLUTION OF THE APICOPLAST

Like all plastids, apicoplasts trace their evolutionary history to cyanobacteria that were acquired during a primary endosymbiotic event. Electron micrographs in *T. gondii* indicated the presence of four membranes around this organelle (Kohler et al., 1997; McFadden & Roos, 1999). In *Plasmodium*, an early electron microscopy study suggested the presence of three membranes (Hopkins et al., 1999), although a more recent tomographic study clearly indicated the presence of four (Lemgruber et al., 2013). The presence of four membranes surrounding the apicoplast is consistent with these organelles having been acquired by secondary endosymbiosis, where a primary plastid-containing eukaryote was incorporated into a non-photosynthetic eukaryote and evolved to become a fully integrated plastid organelle (Gibbs, 1981). In this scenario, the two innermost membranes are equivalent to the two membranes that surround primary plastids, the third innermost membrane (termed the periplastid membrane) corresponds to the plasma membrane of the eukaryotic endosymbiont and the outermost membrane represents the endomembrane system of the eukaryotic host cell (Cavalier-Smith, 2000; Gould, Maier, & Martin, 2015). Numerous studies have localised the apicomplexan homologue of autophagy-related protein 8 (Atg8) to apicoplasts, where it likely localises on the outermost membrane (e.g. Kitamura et al., 2012; Lavine & Arrizabalaga, 2012). In other eukaryotes, Atg8 is a membrane-bound marker of autophagosomes, the organelles that engulf and recycle cellular waste products and dispose of them by fusing with lysosomes. Autophagosomes also function as a cellular defence mechanism by engulfing intracellular pathogens (Huang & Brumell, 2014). The localisation of Atg8 homologues to the outermost apicoplast membrane is consistent with an autophagosomal origin for this membrane.

The identity of the photosynthetic eukaryote that gave rise to the apicoplast has been subject to considerable debate. Phylogenetic analysis of a gene encoded on the 35-kb genome suggested a green algal origin for the apicoplast (Kohler et al., 1997), although the gene order of the 35-kb genome was more consistent with a red algal origin (McFadden & Waller, 1997). Phylogenetic analysis of a nucleus-encoded, plastid-targeted protein termed glyceraldehyde-3-phosphate dehydrogenase argued that apicoplasts were derived in the same endosymbiotic event as other organisms that contain a red algal-derived secondary plastid (Fast, Kissinger, Roos, & Keeling, 2001; Harper & Keeling, 2003). The discovery a decade ago of a phylum of photosynthetic protists termed the chromerids provided the definitive evidence that apicoplasts were derived from red algae (Moore et al., 2008). The chloroplasts of chromerids are surrounded by four membranes, and phylogenetic analyses of chromerid photosynthetic proteins indicated close similarity to the red algal-derived chloroplasts of dinoflagellates, a group of protists that are closely related to apicomplexans (Janouskovec, Horak, Obornik, Lukes, & Keeling, 2010; Moore et al., 2008). Phylogenetic and morphological data revealed that chromerids are closer relatives of apicomplexans than of dinoflagellates (Janouskovec et al., 2010; Moore et al., 2008). This indicates that chromerids are a sister taxon to apicomplexans and that the red algal-derived plastid of apicomplexans, chromerids and dinoflagellates was present in the common ancestor of all three of these organisms.

The acquisition of a red algal-derived plastid was a key moment in the evolution of these groups of organisms. The exact timing of plastid acquisition remains a topic of debate. One scenario, called the chromalveolate hypothesis, posits that the red algal symbiont became incorporated as an organelle early in eukaryotic evolution, in the shared common ancestor of alveolates (which include apicomplexans, chromerids, dinoflagellates and nonplastid-bearing ciliates) and chromists (which include diatoms, brown algae and nonplastid-containing oomycetes) (Cavalier-Smith, 1999). In this scenario, there has been loss of plastids in lineages such as the ciliates. Another scenario posits a later acquisition of the plastid in the alveolates, subsequent to the divergence of ciliates from dinoflagellates and apicomplexans (Waller, Gornik, Koreny, & Pain, 2016) and other scenarios argue for the plastids in this lineage being derived from a tertiary endosymbiosis (Gould et al., 2015).

Regardless of the timing, the evolution of a free-living red algal cell into a fully integrated plastid organelle required three important adaptations (van Dooren & Striepen, 2013). First, unlike many eukaryotic organelles, plastids can only be derived from the division of preexisting plastids. Plastid evolution therefore requires that organellar division and segregation is coupled to cell division. Second, when the apicoplast progenitor first evolved, most of the proteins that functioned in the organelle would have been encoded in the red algal nucleus and posttranslationally imported across the two plastid membranes and into the plastid lumen. During the process of secondary plastid acquisition, many genes encoding plastid-targeted proteins in the symbiont nucleus migrated to the nucleus of the heterotrophic host cell, to the extent that modern apicoplasts retain no vestiges of the red algal nucleus. Secondary plastid integration therefore required the evolution of mechanisms to target the protein products of these host cell nucleus-encoded genes across each of the four membranes that surround the plastid. Finally, and perhaps most importantly, in order to reap the benefits of their newly acquired plastid, host cells had to evolve mechanisms of integrating plastid metabolism with that of their own metabolism. This required the evolution of solute transporters and/or channels to facilitate the exchange of metabolites between host and plastid. In the remainder of this review, we will examine what is known and what remains to be discovered about division and segregation, protein targeting and solute transport in apicoplasts. Central to all three are the integral membrane and membrane-associated proteins that carry out these processes, and this review will focus in particular on these.

5. APICOPLAST DIVISION AND SEGREGATION

The division of the primary plastids of red algae is, in some respects, similar to the division of the cyanobacteria from which they were derived. Red algal plastids divide by binary fission in a process mediated by the bacterial-derived GTPase FtsZ (Miyagishima et al., 2001). Similar to its function in bacteria, FtsZ assembles into filaments that form a ring in the plastid lumen at the midpoint of the organelle. These rings then undergo GTP-dependent contraction to mediate division of the organelle.

A eukaryote-derived dynamin-related protein termed Arc5/DRP5B also has a role in plastid division in red algae. This protein forms spiral filaments on the cytosolic side of the plastid midpoint and, like FtsZ, undergoes GTP-dependent constriction to facilitate plastid division (Miyagishima et al., 2003). Analysis of apicomplexan genomes failed to identify any orthologues of FtsZ or Arc5 (Vaishnava & Striepen, 2006; van Dooren et al., 2009), indicating that apicoplast division is fundamentally different from the division of the red algal plastids from which they evolved. Furthermore, FtsZ and Arc5 orthologues are absent from the genomes of chromerids (Woo et al., 2015), suggesting that the plastid in the common ancestor of apicomplexans and chromerids had already lost the red algal mode of plastid division.

Although lacking orthologues of Arc5, apicomplexan genomes do encode several other dynamin-related proteins (Breinich et al., 2009; van Dooren et al., 2009). Analysis of dynamin-related proteins in *T. gondii* revealed that one such dynamin, termed *Tg*DrpA, localised to the outermost membrane of the apicoplast at the midpoint of dividing organelles (Fig. 1B; van Dooren et al., 2009). Expression of dominant-negative *Tg*DrpA or knockdown of *Tg*DrpA transcripts resulted in specific defects in apicoplast division (Fig. 1C; van Dooren et al., 2009 and G. van Dooren, unpublished). This indicates that *Tg*DrpA has replaced FtsZ and Arc5 in mediating the division of apicoplasts. Analysis of the *Tg*DrpA homologue of *P. falciparum* suggests that it may localise to the apicoplast (Charneau et al., 2007), although the role of *Pf*DrpA has not been functionally determined. Interestingly, *Cryptosporidium* species, apicomplexan parasites that have lost their apicoplasts, retain DrpA homologues, suggesting either that DrpA may have multiple roles in these parasites, or that the *Cryptosporidium* DrpA homologue has acquired novel functions.

Apicomplexan cells form through a process of daughter cell budding, which makes cell division in these parasites quite different from the cytokinesis that occurs in other eukaryotic cells (Francia & Striepen, 2014). Daughter cell formation in these parasites commences with the building of a membranous scaffold consisting of microtubules and a so-called inner membrane complex (IMC), a series of flattened sacs that, in mature cells, lie just beneath the plasma membrane. In developing daughter cells, the IMC and microtubules create a scaffold that resembles the shape of the future cell. Nuclei and organelles, including apicoplasts, divide and segregate into the space enclosed by this scaffold (Fig. 1D). Cytokinesis (i.e. division of the plasma membrane of the parasite) occurs once scaffold formation and organellar division is complete (Francia & Striepen, 2014; Nishi, Hu,

Murray, & Roos, 2008). Although scaffold formation by the IMC and microtubules is generally conserved, there are key differences in cell division between different apicomplexan species, and even between different life cycle stages of the same apicomplexan (Francia & Striepen, 2014; Striepen, Jordan, Reiff, & van Dooren, 2007).

Asexual blood and liver stages of *Plasmodium* parasites divide by a process termed schizogony. Here, parasites undergo repeated rounds of mitosis and karyokinesis (nuclear division) that are not coupled to daughter budding or cytokinesis, producing a multinucleate cell that then undergoes daughter budding and cytokinesis in a single event to produce multiple (1000s in the case of liver-stage schizonts) parasites. During schizogony, apicoplasts, which start of as small, rounded organelles, undergo extensive elongation and branching (Stanway et al., 2011; Stanway, Witt, Zobiak, Aepfelbacher, & Heussler, 2009; van Dooren et al., 2005; Waller, Reed, Cowman, & McFadden, 2000). Division of the elongated apicoplast into multiple small apicoplasts occurs in a seemingly synchronised fashion shortly before cytokinesis, with the number of apicoplasts usually equivalent to the number of daughter cells that are formed at the end of schizogony (Stanway et al., 2011, 2009; van Dooren et al., 2005). The apicomplexan *Sarcocystis neurona* divides by a process termed endopolygeny, where the nucleus undergoes multiple rounds of DNA replication and mitosis that are not coupled to karyokinesis, producing a large polyploid nucleus, with karyokinesis, daughter budding and cytokinesis then occurring in a single event to produce numerous parasites. Similar to *Plasmodium* schizonts, apicoplasts of *S. neurona* become elongated and branched, and then are divided in a synchronous fashion shortly before cytokinesis (Vaishnava et al., 2005). The disease-causing tachyzoite stage of *T. gondii* parasites undergoes a mode of cell division termed endodyogeny. Here, each round of mitosis and karyokinesis is coupled to daughter budding and cytokinesis, with apicoplast division closely coupled to daughter budding (Fig. 1D; Striepen et al., 2000).

The complex modes of cell division in apicomplexans, where apicoplasts must elongate and segregate into multiple daughter cells in one event, are not compatible with the FtsZ-mediated binary fission that occur in chloroplasts of red algal and other eukaryotes. It is indeed conceivable that loss of FtsZ and the adoption of alternative modes of apicoplast division may have preadapted apicomplexans to their complex modes of cell division, or conversely that the selective pressures that led to apicomplexan-style cell divisions may have necessitated the evolution of FtsZ-independent modes of plastid division. Regardless, the evolution of daughter budding in apicomplexans raises interesting questions about the coordination of apicoplast division and segregation.

In *T. gondii*, where a single apicoplast divides to form two apicoplasts, *Tg*DrpA is positioned at the midpoint of the organelle just before division (Figs 1B and 2; van Dooren et al., 2009). The factors that govern *Tg*DrpA positioning are unknown. Dynamin-related proteins have numerous roles in the fission of a range of organelles and vesicles in other eukaryotes (Praefcke & McMahon, 2004). The recruitment and positioning of dynamins at the membranes they are set to divide are coordinated by dynamin-interacting proteins and lipids on the target membrane (Antonny et al., 2016; Koirala et al., 2013). No DrpA-associated proteins have been identified in apicomplexans, and nothing is known about the lipid-binding properties of DrpA. Dynamins in other systems have been shown to bind to phosphatidylinositol phosphates, negatively charged membrane lipids (Klein, Lee, Frank, Marks, & Lemmon, 1998). Intriguingly, recent studies have identified phosphatidylinositol-3-phosphate (PI3P) as a constituent of the outermost apicoplast membrane in both *T. gondii* and *P. falciparum* (Tawk et al., 2010, 2011). In the case of *T. gondii*, PI3P was sometimes observed to localise to the midpoint of apicoplasts prior to division, and depletion of PI3P or disruption of the kinase responsible for PI3P synthesis leads to defects in apicoplast biogenesis. It is conceivable that PI3P

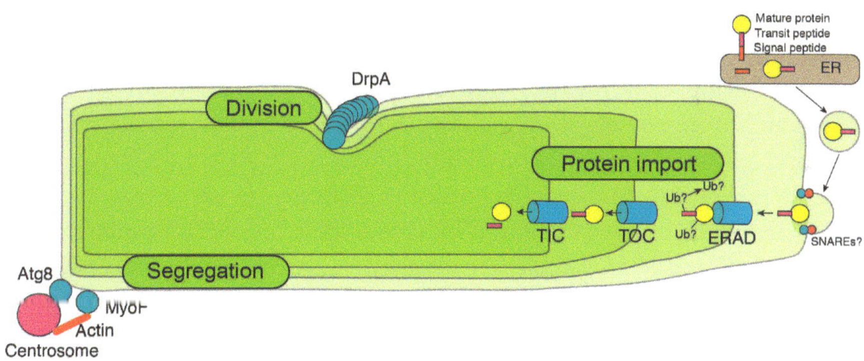

Fig. 2 Schematic of the apicoplast, highlighting the proteins proposed to function in apicoplast division and segregation, and protein import into the organelle. See Sections 5 and 6 for details. Abbreviations: *Atg8*, autophagy-related protein 8; *DrpA*, dynamin-related protein A; *ER*, endoplasmic reticulum; *ERAD*, ER-associated protein degradation machinery; *MyoF*, myosin F; *SNAREs*, soluble *N*-ethylmaleimide-sensitive factor attachment protein α receptor; *TIC*, translocon at the inner envelope membrane of the chloroplast; *TOC*, translocon at the outer envelope membrane of the chloroplast; *Ub*, ubiquitin-like protein.

plays a role in positioning *Tg*DrpA on apicoplasts, although the phenotype associated with impairing PI3P synthesis appears to be enlargement and general loss of the apicoplast, rather than specific defects in organellar division seen upon *Tg*DrpA impairment (Daher et al., 2015; Tawk et al., 2011; Fig. 1C). Mechanical factors may also have a role in apicoplast division. Dynamins are known to form helical filaments around the membranes to which they bind. Notably, the diameter of the organelle or vesicle to which dynamin filaments are bound is critical to whether dynamin can mediate fission of that membrane (Ingerman et al., 2005). During apicoplast division in *T. gondii* tachyzoites, apicoplasts are typically stretched into a U-shape as a result of daughter cell budding (Fig. 1B and D; Striepen et al., 2000; van Dooren et al., 2009). This may result in contraction of the apicoplast diameter at the midpoint of the organelle, thus facilitating dynamin filament assembly around the apicoplast, and subsequent fission of the organelle.

Following division, apicoplasts must correctly segregate into daughter cells. Centrosomes appear to play a key role in mediating daughter cell budding, including the correct segregation of the apicoplast into each daughter bud (Francia & Striepen, 2014). Centrosomes are the microtubule-organising centres for the mitotic spindle. In *T. gondii*, centrosome duplication immediately precedes, and regulates, the onset of daughter budding (Chen & Gubbels, 2013; Nishi et al., 2008). Following centrosome duplication, apicoplasts become attached to the centrosomes (Leveque et al., 2015; Striepen et al., 2000; Fig. 2). Mechanical pressure resulting from the growth of the IMC then results in apicoplasts adopting a U-shape or elongated shape up until the point at which they divide (Striepen et al., 2000; van Dooren et al., 2009). The ends of the apicoplast remain attached to the centrosomes until the completion of cell division, thus ensuring their correct segregation into daughter parasites (Striepen et al., 2000).

Recent studies suggest a central role for actin in apicoplast segregation in both *T. gondii* and *P. falciparum*. Treatment of *T. gondii* and *P. falciparum* parasites with the actin inhibitor cytochalasin D resulted in defects in apicoplast segregation (Jacot, Daher, & Soldati-Favre, 2013). Similarly, knockout of actin or knockdown of actin regulatory factors in *T. gondii* resulted in apicoplast segregation defects (Andenmatten et al., 2013; Jacot et al., 2013). The myosin-related protein *Tg*MyoF localises to the periphery of the apicoplast in *T. gondii*, and *Tg*MyoF knockdown resulted in apicoplast segregation defects, suggesting that *Tg*MyoF is the actin-dependent motor that drives apicoplast segregation (Jacot et al., 2013; Fig. 2). How it does this is not entirely clear. *Tg*MyoF is important for correct positioning of centrosomes at the onset of daughter budding, suggesting that actin/*Tg*MyoF-mediated centrosomal positioning may be crucial for apicoplast segregation (Jacot et al., 2013).

One important question is what mediates the connection between apicoplasts and centrosomes? A recent study revealed that knockdown of the outer membrane-localised *T. gondii* Atg8 (*Tg*Atg8) protein resulted in defects in apicoplast segregation (Leveque et al., 2015). *Tg*Atg8 proteins were enriched at the ends of dividing apicoplasts, and knockdown of *Tg*Atg8 impaired apicoplast association with centrosomes (Leveque et al., 2015). This implies that *Tg*Atg8 has a role in the attachment of apicoplasts to centrosomes, either providing a direct link between apicoplast and centrosome, or functioning to position the apicoplast relative to the centrosome (Fig. 2). It should be noted that the impairment of numerous apicoplast processes, including fatty acid biosynthesis, protein import and translation of the 35-kb genome, led to apicoplast loss and apparent segregation defects (Mazumdar et al., 2006; van Dooren, Tomova, Agrawal, Humbel, & Striepen, 2008; Yeh & DeRisi, 2011). Future studies in this area must therefore be careful to distinguish between processes that directly impair segregation, and those that impair apicoplast biogenesis more generally.

There are numerous unanswered questions when it comes to apicoplast division and segregation. One particularly intriguing question is how the single, elongated apicoplasts of *Plasmodium* and *S. neurona* are divided into multiple apicoplasts in a single event. Equally intriguing is how apicoplasts are correctly segregated into daughter cells during schizogony and endopolygeny. Of note, the apicoplasts of *S. neurona* also appear to associate with centrosomes, suggesting centrosomal association may be key in ensuring correct segregation of apicoplasts in the complex and varied modes of cell division that are observed in apicomplexans (Vaishnava et al., 2005; Vaishnava & Striepen, 2006). The association between apicoplasts and nuclei is less clear in *Plasmodium* parasites (Stanway et al., 2011; van Dooren et al., 2005), and no study has examined whether apicoplasts associate with centrosomes in *Plasmodium*. Another key question is how each of the multiple membranes surrounding the organelle is divided.

6. PROTEIN TARGETING TO THE APICOPLAST

The evolution of the red algal symbiont to a fully integrated plastid saw the transfer of many of the plastid protein-encoding genes in the red algal nucleus to the nucleus of the host cell. Such transfer has the effect of assigning regulatory control of symbiont metabolic process and biogenesis to the host cell. This process enabled the host/symbiont pair to ultimately discard the red algal nucleus, and the energetically costly processes associated with its maintenance. Most nucleus-encoded proteins that are targeted to the primary plastids of red algae, green algae and plants contain an N-terminal 'transit peptide' sequence, a short, positively charged peptide that directs proteins across the two membranes surrounding the plastid (Dobberstein, Blobel, & Chua, 1977; Patron & Waller, 2007). The transit peptide first interacts with receptor proteins on the so-called *t*ranslocon at the *o*uter envelope membrane of the *c*hloroplast (TOC) complex (Jarvis & Soll, 2002). These receptors facilitate the translocation of plastid proteins through a β-barrel pore in the outer membrane termed Toc75 (Schleiff, Jelic, & Soll, 2003). The plastid-targeted protein then crosses a multiprotein *t*ranslocon at the *i*nner envelope membrane of the *c*hloroplast (TIC) complex to enter the plastid lumen (Jarvis & Soll, 2002; Nakai, 2015). Once in the lumen, a protease termed the stromal processing peptidase cleaves the transit peptide from the remainder of the protein, and the plastid protein can then carry out its function.

Analysis of N-terminal domain of numerous nucleus-encoded apicoplast-targeted proteins in both *T. gondii* and *P. falciparum* revealed the presence of a two-part (bipartite) targeting signal that is both necessary and sufficient to direct import into the apicoplast (DeRocher, Hagen, Froehlich, Feagin, & Parsons, 2000; Foth et al., 2003; Waller et al., 1998, 2000). The first part of the signal is thought to function as a signal peptide to direct cotranslational import of the apicoplast-targeted precursor protein into the endoplasmic reticulum (ER; Fig. 2; DeRocher et al., 2000; Waller et al., 2000). The second is a positively charged domain that in many respects resembles the transit peptide of primary plastids, and which facilitates the transport of apicoplast-directed proteins from the bulk secretory pathway into the apicoplast lumen (Fig. 2; DeRocher et al., 2000; Foth et al., 2003; Tonkin, Roos, & McFadden, 2006; Waller et al., 2000). As in primary plastids, the 'transit peptide' region of the apicoplast-targeted protein is cleaved from the rest of the protein once in the lumen (Fig. 2; van Dooren, Su, D'Ombrain, & McFadden, 2002; Waller et al., 1998). Notably, numerous apicoplast-targeted proteins, in particular proteins that localise to the periphery (e.g. outer membranes) of the apicoplast, lack canonical bipartite-targeting signals and instead contain poorly defined alternative-targeting motifs (e.g. DeRocher et al., 2008; DeRocher, Karnataki, Vaney, & Parsons, 2012; Karnataki, Derocher, Coppens, Feagin, &

Parsons, 2007; Karnataki, Derocher, Coppens, Nash, et al., 2007). Aromatic amino acids appear to play critical roles in the sorting of proteins at the periphery of the apicoplast. For example, a conserved tyrosine residue in the triose phosphate transporter of apicoplasts in *T. gondii* and *P. falciparum* is essential for sorting this integral membrane protein from the secretory pathway to the apicoplast periphery (DeRocher et al., 2012; Lim, Sayers, Goodman, & McFadden, 2016). A separate study identified the presence of a phenylalanine residue in the +1 position of the transit peptide of many apicoplast lumen proteins, demonstrating that this residue is important for targeting proteins from the apicoplast periphery to the organelle lumen (Sheiner et al., 2015). This mirrors similar aromatic-targeting signals that are found in other organisms with secondary red algal plastids (e.g. Gould et al., 2006).

Although the signals required for apicoplast targeting are reasonably well understood, the molecular mechanisms by which proteins translocate across each of the four membranes surrounding the organelle are less clear. In the remainder of this section, we will review what is known, starting from translocation across the innermost membranes, which are best understood, and working our way outwards.

Given the homology of the two innermost apicoplast membranes to the two membranes that surround primary plastids, it is not surprising that the mechanisms by which apicoplast proteins translocate across these membranes appear to be conserved from primary plastids. The innermost apicoplast membrane of *T. gondii* contains a homologue of Tic20 (*Tg*Tic20; Fig. 2; van Dooren et al., 2008), an integral membrane protein proposed to function as the TIC complex pore in plant plastids (Kikuchi et al., 2013). Knockdown of *Tg*Tic20 resulted in defects in the import of nucleus-encoded apicoplast proteins into the organelle, providing evidence that *Tg*Tic20 functions in protein import (van Dooren et al., 2008). Interestingly, knockdown of *Tg*Tic20 to below detectable levels did not fully impair protein import, suggesting that *Tg*Tic20 may not function as the pore of this complex (van Dooren et al., 2008). Apicomplexan parasites also contain homologues of Tic22, an intermembrane space protein that functions as part of the TIC complex in plants. Apicomplexan Tic22 is a soluble protein that localises to the apicoplast periphery (Glaser et al., 2012; Kalanon, Tonkin, & McFadden, 2009), although the exact intermembrane space to which it localises is not known. Knockdown of the *T. gondii* homologue of Tic22 (*Tg*Tic22) impaired apicoplast protein import (Glaser et al., 2012). The *P. falciparum* Tic22 homologue (*Pf*Tic22) functions as a chaperone, with two hydrophobic grooves along the surface of the protein critical to this function (Glaser et al., 2012).

The second innermost apicoplast membrane is homologous to the outer membrane of primary plastids. Apicomplexans contain a homologue of Toc75, the pore that facilitates translocation of plastid proteins across the outer membrane (Fig. 2; Bullmann et al., 2010). The Toc75 homologue of *T. gondii* (*Tg*Toc75) localises to the apicoplast, and knockdown of *Tg*Toc75 led to specific defects in apicoplast protein import (Sheiner et al., 2015). Although the membrane to which *Tg*Toc75 localises has not been defined, knockdown of *Tg*Toc75 does not impair the import of proteins that localise to the periphery of the apicoplast, suggesting it mediates import through one of the inner membranes (Sheiner et al., 2015). Taken together, the evidence is consistent with import across the two innermost apicoplast membranes being conserved with protein import into primary plastids (Fig. 2). Transit peptide-like sequences facilitate interaction with the TOC complex of the second innermost membrane, with proteins translocating through the Toc75 pore. Tic22 then functions as a chaperone in the space between the inner two membranes, binding to apicoplast-targeted proteins to maintain them in an unfolded state and facilitating their transfer to the TIC complex of the inner membrane. The TIC complex, which consists of Tic20 and likely other proteins, facilitates translocation across the inner membrane. Once across the inner membrane, the transit peptide is cleaved by a stromal processing peptidase, a homologue of which is present in apicoplast-containing apicomplexans (van Dooren et al., 2002).

Conservation of the translocation mechanisms across the two inner apicoplast membranes between apicomplexans and organisms with primary plastids makes evolutionary sense, since it does not require the evolution of any novel components. The same is not true for translocation across the periplastid membrane (the second outermost membrane) that surrounds the apicoplast. Proposed to be homologous to the plasma membrane of the red algal symbiont, the mechanism of protein translocation across this membrane was long debated (van Dooren, Schwartzbach, Osafune, & McFadden, 2001). The first insights into a possible mechanism came from studies of the nucleomorph (remnant red algal nucleus) of cryptomonad plastids (Sommer et al., 2007). This nucleomorph encodes multiple homologues of proteins involved in the ER-associated degradation (ERAD) pathway, a process by which eukaryotes translocate misfolded ER-resident proteins across the ER membrane and into the cytosol for proteosomal degradation (Sommer et al., 2007). ERAD components were found to localise to numerous red algal-derived secondary plastids, including those of *P. falciparum* and *T. gondii* (Agrawal, van Dooren, Beatty, & Striepen, 2009; Hempel, Bullmann, Lau, Zauner, & Maier, 2009; Kalanon et al., 2009; Sommer et al., 2007; Spork et al., 2009). This led to the proposal that the plastid-localised ERAD system, originally derived from the red algal symbiont, may have been repurposed to enable translocation of plastid-targeted proteins across the periplastid membrane (Fig. 2; Sommer et al., 2007). Consistent with this hypothesis, knockdown of *Tg*Der1, the putative pore of the plastid-localised ERAD complex in *T. gondii*, led to specific defects in protein import into this organelle (Agrawal et al., 2009).

As proteins translocate through the ERAD complex at the ER membrane, they are ubiquitinated, a process that facilitates their subsequent degradation by the proteasome (Hirsch, Gauss, Horn, Neuber, & Sommer, 2009). Interestingly, all the necessary components involved in this ubiquitination, including a putative ubiquitin-like protein, E1, E2 and E3 proteins that facilitate the addition of ubiquitin to proteins, and a deubiquitinating enzyme that removes ubiquitin, localise to the red algal-derived secondary plastids of several organisms, including *T. gondii* and *P. falciparum* (Agrawal et al., 2013; Hempel, Felsner, & Maier, 2010; Sommer et al., 2007; Spork et al., 2009). These ubiquitination components are functional, and knockdown of the ubiquitin-conjugating E2 enzyme of the *T. gondii* apicoplast impaired apicoplast protein import (Agrawal et al., 2013). Together, these data support a model whereby translocation across the periplastid membrane is mediated by the plastid-localised ERAD complex (Fig. 2). The data also imply that proteins are ubiquitinated during or after this translocation process, and deubiquitinated before translocation across the TOC complex (Fig. 2), although direct evidence that ubiquitination occurs is currently lacking.

Apicoplast proteins traffic via the secretory pathway, and it is likely that translocation across the outermost apicoplast membrane is mediated by the fusion of secretory vesicles with this membrane (Fig. 2). There is debate as to whether the sorting of vesicles bearing apicoplast-targeted proteins away from the bulk secretory pathway occurs pre- or post-Golgi. Initial studies found that fusing an ER-retention motif, which is typically recognised by a *cis* Golgi-localised receptor, to apicoplast-targeted proteins did not impair apicoplast targeting (DeRocher, Gilbert, Feagin, & Parsons, 2005; Tonkin, Struck, Mullin, Stimmler, & McFadden, 2006). Furthermore, these studies demonstrated that apicoplast protein targeting is largely insensitive to brefeldin A, an inhibitor of ER-to-Golgi trafficking (DeRocher et al.,

2005; Tonkin, Struck, et al., 2006). These studies suggested that vesicles destined to the apicoplast are derived directly from the ER. A recent study that repeated both the ER-retention motif and brefeldin A experiments concluded, instead, that the Golgi does play a role in apicoplast targeting (Heiny, Pautz, Recker, & Przyborski, 2014), indicating that further studies are required to elucidate the trafficking route of apicoplast proteins post-ER. Accumulating evidence suggests there may be multiple routes for apicoplast protein targeting. Cargo vesicles that contain proteins destined for the apicoplast periphery (e.g. those with 'noncanonical' apicoplast-targeting signals) do not contain proteins destined for the apicoplast lumen (e.g. those with 'canonical' apicoplast-targeting signals) (Bouchut, Geiger, DeRocher, & Parsons, 2014; Chaudhari, Dey, Narayan, Sharma, & Patankar, 2017).

A major unanswered question is how apicoplast proteins are sorted at the ER (or Golgi) compartment for trafficking to the apicoplast? There is likely a receptor that can recognise the transit peptide domain of apicoplast lumen proteins and mediate sorting into apicoplast-bound vesicles. Proteins targeted to the apicoplast periphery that lack the canonical bipartite signal must be recognised and sorted in a different manner, and this might explain why they are trafficked in different vesicles than their lumenal counterparts (Bouchut et al., 2014; Chaudhari et al., 2017). Another important question is what facilitates the targeting and fusion of apicoplast-destined vesicles with the outer membrane of the organelle? PI3P is known to recruit proteins that function in endosomal targeting (Di Paolo & De Camilli, 2006), and it is conceivable that apicoplast-localised PI3P plays a role in directing and fusing apicoplast vesicles to their target organelle. Nevertheless, defects in the import of apicoplast lumen and peripheral proteins upon knockdown of PI3P synthesis were concomitant with apicoplast loss, precluding direct analysis of its role in import (Daher et al., 2015). Apicoplast biogenesis, and possibly apicoplast protein targeting, was impaired in *T. gondii* mutants defective in *s*oluble *N*-ethylmaleimide-sensitive factor *a*ttachment *p*rotein α (αSNAP), a protein that facilitates SNARE (*SNAP re*ceptor)-mediated fusion at target membranes (Stewart et al., 2016). SNARE-mediated vesicle fusion typically depends on a vesicle-localised SNARE and a separate target membrane-localised SNARE, and it is now of particular interest to determine which SNAREs (if any) function in these roles for apicoplasts and their corresponding cargo vesicles (Fig. 2).

There is still much to learn about protein import into apicoplasts. How are apicoplast-targeting signals recognised in the secretory pathway? What is the nature of the vesicles that take apicoplast cargo to the organelle, and how is the fusion of these with the apicoplast mediated (Fig. 2)? Could there be direct contacts between the ER and apicoplast that mediate protein targeting (Vaishnava & Striepen, 2006)? Are apicoplast proteins ubiquitinated (and deubiquitinated) on their journey across the periplastid membrane, and, if so, what is the importance of this process (Fig. 2)? What receptors recognise apicoplast proteins at the TOC and TIC complexes (Fig. 2)? What is the nature of the pore at the TIC complex? Can proteins be selectively targeted to each of the three intermembrane spaces that exist at the apicoplast, or to each of the four apicoplast membranes, and, if so, what mediates this selectivity?

7. INTEGRATING METABOLISM OF THE APICOPLAST WITH THAT OF THE PARASITE

The retention of the apicoplast across hundreds of millions of years of evolution implies that it serves an important role in the cell. Early clues to the possible functions of the apicoplast came with analysis of its 35 kb genome. Although essential, the 35-kb apicoplast genome itself provides few clues to what the essential functions of the apicoplast are. Genes found on the 35-kb genome encode for a variety of housekeeping functions, including proteins and RNA molecules that function in transcription and translation of the 35-kb genome (Wilson et al., 1996). One notable exception is a gene initially annotated as ORF470 (Wilson et al., 1996). This open reading frame encodes a protein with homology to SufB, a protein that functions in the so-called Suf pathway to synthesise iron–sulphur (Fe–S) clusters (Ellis, Clough, Saldanha, & Wilson, 2001), implicating the apicoplast in the synthesis of Fe–S clusters. The availability of nuclear genome sequences for *P. falciparum*, *T. gondii* and many other apicomplexans (Gardner et al., 2002; www.eupathdb.org) provided an essential resource to identify proteins containing the canonical bipartite apicoplast-targeting signal, and thereby uncovering enzymes that carry out metabolic functions in this organelle (Ralph et al., 2004). These studies have found that the apicoplast, like its photosynthetic chloroplast ancestor, functions in a range of biosynthetic pathways. Although no longer photosynthetic, the apicoplast contains proteins that function in the synthesis of Fe–S clusters, fatty acids, isoprenoids and haem. Rather than provide an in-depth review of each of these pathways (which have been reviewed extensively elsewhere; e.g. Ralph et al., 2004; Seeber & Soldati-Favre, 2010), we will consider the metabolic products that must be exchanged between the apicoplast and the parasite to ensure that the apicoplast can fulfil its metabolic obligation to the host cell that enslaved it many generations earlier.

7.1 Isoprenoid Synthesis

Isoprenoids are a diverse and important class of molecules, which include derivatives that function in a range of processes, including in posttranslational protein modification, as secondary messengers and as membrane lipids. Apicoplasts contain the complete set of seven enzymes that encode the so-called nonmevalonate pathway for isoprenoid biosynthesis (Guggisberg, Amthor, & Odom, 2014; Jomaa et al., 1999; Ralph et al., 2004). Isoprenoid biosynthesis enzymes are encoded in all apicomplexans that retain an apicoplast, with the pathway shown to be essential in *T. gondii* tachyzoites and blood stages of *P. falciparum* parasites, and a validated drug target in *P. falciparum* (Jomaa et al., 1999; Nair et al., 2011; Yeh & DeRisi, 2011; Zhang et al., 2011). The pathway requires glyceraldehyde-3-phosphate (G3P) and pyruvate as substrates. Both G3P and pyruvate are synthesised in the apicoplast from the precursors dihydroxyacetone phosphate (DHAP) and phosphoenolpyruvate (PEP), respectively (Table 1; Fig. 3). DHAP and PEP are intermediates in glycolysis and must be imported into the apicoplast from the cytosol. Translocation of DHAP and PEP into apicoplasts requires a transporter, termed *Tg*APT in *T. gondii* and oTPT in *Plasmodium*, that is homologous to a family of so-called plastidic phosphate transporters (pPT) from plants and red algae (Weber, Linka, & Bhattacharya, 2006). This transporter, which localises to the inner membrane in plants, localises to the outermost apicoplast membrane in *Plasmodium*, and possibly to multiple apicoplast membranes in *T. gondii* (Karnataki, Derocher, Coppens, Nash, et al., 2007; Mullin et al., 2006; Fig. 3). Heterologous expression studies revealed that *Tg*APT and *Plasmodium* oTPT translocate both DHAP and PEP in exchange for inorganic phosphate (Brooks et al., 2010; Lim, Linka, Mullin, Weber, & McFadden, 2010). Knockdown of *Tg*APT indicated that this transporter is critical for parasite growth and for metabolic processes that rely on the substrates transported by *Tg*APT (Brooks et al., 2010). In addition to an outer membrane TPT, *Plasmodium* parasites contain a second pPT-family protein, termed iTPT, that is postulated to localise to the innermost apicoplast membrane, and which is also capable of transporting DHAP and PEP in exchange for inorganic phosphate (Lim et al., 2010; Mullin et al., 2006).

Table 1 Substrates, Energy Sources, Cofactors and Products of the Major Apicoplast Metabolic Pathways

Pathway	Substrates	Energy Sources and Electron Donors/ Acceptors	Cofactors	Products
Isoprenoid synthesis	DHAP, PEP, CTP	NADPH, ATP	Fe–S clusters	IPP
Fe–S cluster synthesis	Fe^{2+}, cysteine	ATP, NAD(P)H, $FADH_2$	Pyridoxal phosphate	Fe–S clusters[a]
Fatty acid, lipoic acid and phospholipid synthesis	PEP, DHAP, CTP, SAM, dephospho-CoA	ATP, $NADP^+$/NADPH, NAD^+/NADH, FAD	Fe–S clusters, TPP, biotin, lipoic acid, CoA	Lipoic acid[a], short-chain fatty acids, LPA/PA, CDP-DAG, CoA?[b]
Haem synthesis	ALA			CPIII
Housekeeping	dNTPs, NTPs, amino acids		Mg^{2+}	DNA[c], RNA[c], proteins[c]

[a]Fe–S clusters and lipoic acid are cofactors for apicoplast proteins only and are not exported from the organelle.
[b]CoA is a cofactor required for fatty acid biosynthesis. The final enzyme of the CoA pathway (DPCK) is predicted to localise to the apicoplast, but experimental evidence for this is lacking.
[c]Apicoplast synthesised DNA, RNA and proteins are not thought to be exported from the organelle.
Abbreviations: *ALA*, δ-aminolevulinic acid; *ATP*, adenosine triphosphate; *CDP-DAG*, cytidine diphosphate-diacylglycerol; *CoA*, coenzyme A; *CPIII*, coproporphyrinogen III; *CTP*, cytidine triphosphate; *DHAP*, dihydroxyacetone phosphate; *dNTPs*, deoxynucleotide triphosphates; *FAD/FADH*$_2$, flavin adenine dinucleotide; *Fe–S*, iron–sulphur; *IPP*, isopentenyl pyrophosphate; *LPA*, lysophosphatidic acid; *NAD*$^+$*/NADH*, nicotinamide adenine dinucleotide; *NADP*$^+$*/NADPH*, nicotinamide adenine dinucleotide phosphate; *NTPs*, nucleotide triphosphates; *PA*, phosphatidic acid; *PEP*, phosphoenolpyruvate; *SAM*, *S*-adenosylmethionine; *TPP*, thiamine pyrophosphate.

Together, these results indicate that the apicoplast pPT-family proteins function to translocate the substrates of isoprenoid biosynthesis into the organelle. One outstanding question is whether apicoplast pPTs translocate DHAP and PEP across all four apicoplast membranes or whether other as yet unknown transporters exist to translocate these substrates across some of the membranes (Fig. 3). iTPT appears to be dispensable in blood stages of mouse malaria-causing *Plasmodium berghei* (Banerjee, Jaijyan, Surolia, Singh, & Surolia, 2012) and lacks a homologue in other apicomplexans such as *T. gondii*, suggesting translocation through one or more of the inner membranes of the organelle requires different transporters. Another possibility is

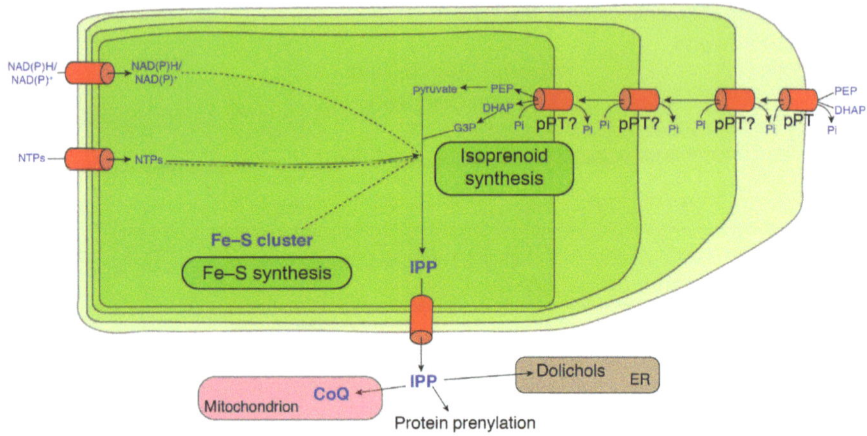

Fig. 3 Schematic of the apicoplast, highlighting the substrates and products of isoprenoid biosynthesis. *Dotted lines* represent the contribution of cofactors and energy sources involved in the pathway; *filled lines* represent substrates and products of the pathway. See Section 7.1 for details. Abbreviations: *CoQ*, coenzyme Q; *DHAP*, dihydroxyacetone phosphate; *dNTPs*, deoxynucleotide triphosphates; *ER*, endoplasmic reticulum; *Fe–S*, iron–sulphur; *G3P*, glyceraldehyde-3-phosphate; *IPP*, isopentenyl pyrophosphate; *NAD$^+$/NADH*, nicotinamide adenine dinucleotide; *NADP$^+$/NADPH*, nicotinamide adenine dinucleotide phosphate; *NTPs*, nucleotide triphosphates; *PEP*, phosphoenolpyruvate; *Pi*, inorganic phosphate; *pPT*, plastidic phosphate transporter.

that apicoplast membranes contain general solute channels, similar to the porins that are found on the outer mitochondrial membrane, and which facilitate the diffusion of small metabolites across the membranes to which they localise (Young, Bay, Hausner, & Court, 2007). Porins are integral membrane β-barrel proteins, much like the Toc75 protein that functions in protein translocation across the second innermost membrane of the apicoplast (see Section 6). β-Barrel membrane proteins are difficult to distinguish by analysis of primary amino acid sequence, which is perhaps why no other apicoplast β-barrel proteins have yet been identified.

The pPT-family proteins that translocate DHAP and PEP into the apicoplast are one of only a few solute transporters that have been shown to localise to the apicoplast, and are the only solute transporters that have been functionally characterised. This highlights the dearth in knowledge around how apicoplast metabolism is integrated into broader cell metabolism and emphasises the importance of identifying and functionally characterising more apicoplast solute transporters. The apicoplast isoprenoid biosynthesis pathway requires ATP as an energy source, NADPH as an electron donor, CTP as a substrate for the third enzyme of the pathway and Fe–S clusters as cofactors for the final two enzymes of the pathway (Table 1; Fig. 3). Nothing is known about how apicoplasts import NADPH or nucleotides such as ATP and CTP, although transporters for these must exist. Fe–S clusters are synthesised in the apicoplast (see Section 7.2). The product of isoprenoid biosynthesis is IPP, which must be exported from the apicoplast. IPP has a range of uses in cells (Fig. 3). One is in facilitating the prenylation and subsequent membrane association of Rabs, proteins with key roles in membrane trafficking events. Impairment of IPP synthesis in blood stages of *P. falciparum* led to defects in Rab prenylation, and subsequent defects in the digestive vacuole of the parasite, the organelle that digests host cell haemoglobin (Howe, Kelly, Jimah, Hodge, & Odom, 2013). IPP is also a precursor for dolichols, which function in protein glycosylation in the ER, and a precursor in the synthesis of coenzyme Q, an electron carrier in the mitochondrial electron transport chain (Fig. 3; Guggisberg et al., 2014; Ralph et al., 2004). The isopentenyl group that contributes to coenzyme Q synthesis is one of several products of apicoplast metabolism that are required by the mitochondrion (Fig. 3). Apicomplexan parasites have a single mitochondrion that, like the apicoplast, is essential for parasite growth and survival (Fig. 1E; van Dooren, Stimmler, & McFadden, 2006). Apicoplasts and mitochondria form a physical connection in blood stages of *Plasmodium* parasites, possibly to facilitate efficient transfer of compounds such as isoprenoids, although such physical connections are not always observed between these two organelles in *T. gondii* tachyzoites or *Plasmodium* liver stages (Fig. 1E; Stanway et al., 2011; van Dooren et al., 2005).

7.2 Fe–S Cluster Synthesis

Fe–S clusters are important prosthetic groups found on a broad range of proteins, where they typically function in electron transfer reactions. There are several known pathways for Fe–S cluster synthesis. The so-called ISC pathway operates in most eukaryotic mitochondria (Balk & Pilon, 2011; Lill & Muehlenhoff, 2005). Eukaryotes also contain a cytosolic Fe–S cluster synthesis pathway, which is dependent on a sulphur-containing product of the mitochondrial ISC pathway (Balk & Pilon, 2011; Lill & Muehlenhoff, 2005). In addition, plant chloroplasts harbour a so-called sulphur mobilisation (Suf) pathway for Fe–S cluster synthesis (Balk & Pilon, 2011). Like plants, apicomplexan parasites contain three separate Fe–S synthesis pathways, including a functional Suf pathway that localises to the apicoplast (Fig. 4; Charan et al., 2014; Ellis et al., 2001; Gisselberg, Dellibovi-Ragheb,

Matthews, Bosch, & Prigge, 2013; Haussig, Matuschewski, & Kooij, 2014; Kumar et al., 2011). Overexpression of a dominant-negative version of the apicoplast Fe–S cluster synthesis protein SufC in *P. falciparum*, where the active site lysine has been converted to alanine, impaired parasite growth (Gisselberg et al., 2013). The SufC dominant-negative strain is only capable of growing in the presence of IPP, which previous studies have demonstrated is the only essential apicoplast product in *P. falciparum* blood stages (Gisselberg et al., 2013; Yeh & DeRisi, 2011; Zhang et al., 2011). This demonstrates that apicoplast Fe–S synthesis is essential for growth of blood-stage *P. falciparum* parasites and that apicoplast-synthesised Fe–S clusters are essential only for the apicoplast and do not, for instance, contribute to cytosolic Fe–S cluster synthesis (Gisselberg et al., 2013).

Numerous apicoplast proteins are known or predicted to contain Fe–S clusters, including the final two enzymes of isoprenoid synthesis (IspG and IspH), the LipA protein that functions in lipoic acid synthesis, a MiaA protein that adds isoprenoids to tRNA molecules, and the electron transfer

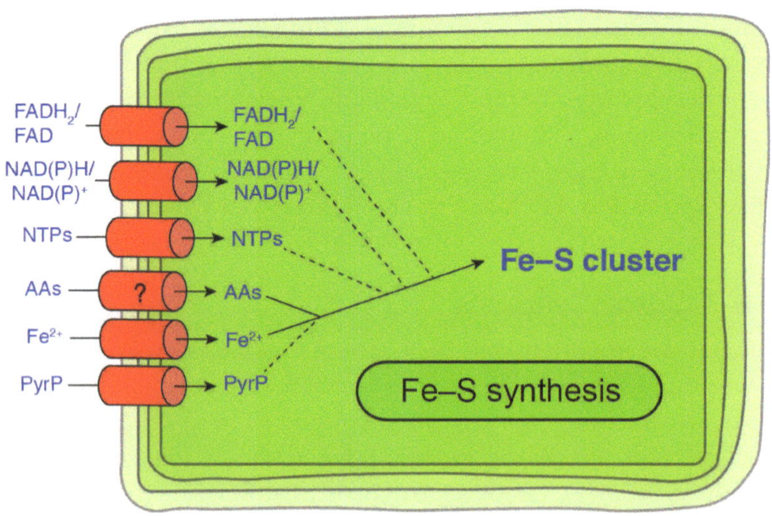

Fig. 4 Schematic of the apicoplast, highlighting the substrates and products of Fe–S cluster biosynthesis. *Dotted lines* represent contribution of cofactors and energy sources involved in the pathway; *filled lines* represent substrates and products of the pathway. See Section 7.2 for details. Abbreviations: *AAs*, amino acids; *FAD/FADH$_2$*, flavin adenine dinucleotide; *Fe–S*, iron–sulphur; *NAD$^+$/NADH*, nicotinamide adenine dinucleotide; *NADP$^+$/NADPH*, nicotinamide adenine dinucleotide phosphate; *NTPs*, nucleotide triphosphates; *PyrP*, pyridoxal phosphate.

protein ferredoxin (Ralph et al., 2004; Rohrich et al., 2005; Vollmer, Thomsen, Wiek, & Seeber, 2001). Expression of dominant-negative SufC leads to defects in apicoplast biogenesis in *P. falciparum*, indicating that one or more of these apicoplast Fe–S proteins are essential for both parasite survival and biogenesis of the organelle (Gisselberg et al., 2013). Interestingly, defects in IPP synthesis (e.g. by treatment with the IPP inhibitor fosmidomycin) do not impair apicoplast biogenesis (Gisselberg et al., 2013). This suggests that, although the Fe–S-containing proteins that function in isoprenoid biosynthesis (IspG and IspH) are essential for *P. falciparum* survival, the defects in apicoplast biogenesis upon impairment of the Suf pathway are the result of defects in other Fe–S-containing proteins in the apicoplast (Gisselberg et al., 2013).

Fe–S cluster synthesis requires sources of iron and sulphur. Sulphur is derived from the amino acid cysteine, in a reaction catalysed by the cysteine desulphurase SufS (Charan et al., 2014; Gisselberg et al., 2013). It is unclear whether apicoplasts contain specific amino acid transporters, or whether they obtain the amino acids for Fe–S synthesis and translation of the 35-kb genome through the degradation of apicoplast-targeted proteins (or of apicoplast transit peptides following their cleavage from mature proteins during import) (Fig. 4). The source of iron is also unknown. Uptake of iron into plant chloroplasts is mediated by an inner membrane transporter termed Pic1 (Duy et al., 2007). There are no clear homologues of Pic1 in apicomplexans, suggesting alternative iron-uptake mechanisms into this organelle (Fig. 4). Assembly of Fe–S clusters via the Suf pathway requires energy in the form of ATP, NAD(P)H and possibly $FADH_2$ as electron sources, as well as pyridoxal phosphate (vitamin B6), which serves as a cofactor for the SufS protein (Table 1; Fig. 4; Charan et al., 2014). Apicomplexans are capable of synthesising pyridoxal phosphate through a pathway that localises to the cytosol (Gengenbacher et al., 2006). Apicoplasts must, therefore, import pyridoxal phosphate from the cytosol (Fig. 4).

7.3 Fatty Acid, Lipoic Acid and Phospholipid Synthesis

Fatty acids are a constituent of major lipid classes in all cells, including the phospholipids that make up the bulk of cellular membranes. Fatty acid synthesis in eukaryotes occurs by two main routes. Many eukaryotes synthesise fatty acids via the so-called Type I pathway, which utilises a large, single-subunit enzyme to catalyse all steps of fatty acid synthesis. Plastid-bearing eukaryotes such as plants can also synthesise fatty acids via the so-called Type II pathway, which consists of multiple subunits that localise to the plastid. Apicoplasts of *T. gondii* and *Plasmodium*, but not of *Babesia* and *Theileria*, contain all the necessary enzymes for Type II fatty acid biosynthesis (Ralph et al., 2004; Seeber & Soldati-Favre, 2010; Waller et al., 1998). The apicoplast-localised fatty acid biosynthesis pathway is important for the growth of *T. gondii* tachyzoites and of liver (but not blood) stages of *Plasmodium* parasites (Mazumdar et al., 2006; Vaughan et al., 2009; Yu et al., 2008).

Apicoplast fatty acid biosynthesis commences with the conversion of pyruvate to acetyl-coenzyme A (CoA), a reaction catalysed by the multi-subunit pyruvate dehydrogenase complex (Fig. 5). Acetyl-CoA is converted

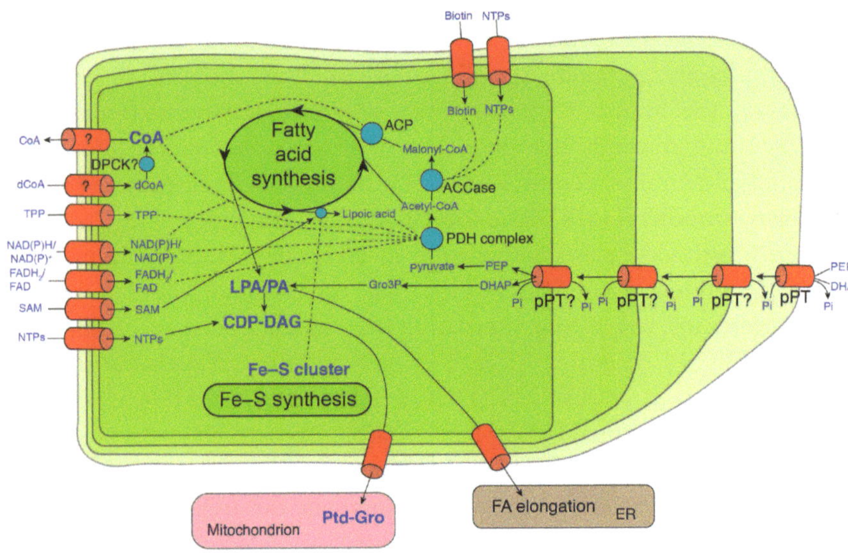

Fig. 5 Schematic of the apicoplast, highlighting the substrates and products of fatty acid, lipoic acid, coenzyme A and phospholipid biosynthesis. *Dotted lines* represent contribution of cofactors and energy sources involved in the pathway; *filled lines* represent substrates and products of the pathway. See Section 7.3 for details. Abbreviations: *CDP-DAG*, cytidine diphosphate-diacylglycerol; *CoA*, coenzyme A; *DHAP*, dihydroxyacetone phosphate; *ER*, endoplasmic reticulum; *FA*, fatty acid; *FAD/FADH₂*, flavin adenine dinucleotide; *Fe–S*, iron–sulphur; *Gro3P*, glycerol-3-phosphate; *LPA*, lysophosphatidic acid; *NAD^+/NADH*, nicotinamide adenine dinucleotide; *$NADP^+$/NADPH*, nicotinamide adenine dinucleotide phosphate; *NTPs*, nucleotide triphosphates; *PA*, phosphatidic acid; *PEP*, phosphoenolpyruvate; *Pi*, inorganic phosphate; *pPT*, plastidic phosphate transporter; *Ptd-Gro*, phosphatidylglycerol; *SAM*, *S*-adenosylmethionine; *TPP*, thiamine pyrophosphate.

to malonyl-CoA in a reaction catalysed by acetyl-CoA carboxylase (ACCase) and is then fed into an elongation cycle that adds two carbons to the growing fatty acid chain with each cycle (Fig. 5; Ralph et al., 2004). The apicoplast fatty acid synthesis pathway produces predominantly short-chain (14 and 16 carbons) fatty acids (Ramakrishnan et al., 2012). Eight carbon-chain fatty acids are also modified in the apicoplast by a two-enzyme pathway to generate lipoic acid, which serves as a cofactor for the E2 subunit of pyruvate dehydrogenase (Fig. 5; Crawford et al., 2006). Within the apicoplast, synthesised fatty acids can be added to glycerol-3-phosphate to form the phospholipid lysophosphatidic acid (LPA) and possibly also phosphatidic acid (PA) (Amiar et al., 2016; Shears et al., 2017). The head group of PA can be further modified to form cytidine diphosphate-diacylglycerol (CDP-DAG) via an apicoplast-localised CDP-DAG synthase (Kong et al., 2017). Free fatty acids and/or LPA/PA and/or CDP-DAG can then be exported from the apicoplast to organelles such as the ER and mitochondrion (Fig. 5). Both the ER and the mitochondrion can form contact sites with apicoplasts, which may facilitate the transfer of these lipids (Tomova et al., 2009; van Dooren et al., 2005). Exported apicoplast-derived fatty acids and phospholipids have several fates in the parasite. Fatty acid chains can be further lengthened in the ER via a fatty acid elongation pathway (Fig. 5; Ramakrishnan et al., 2012, 2015). Phospholipids can also be further modified in the mitochondrion to produce phosphatidylglycerol (Fig. 5; Kong et al., 2017). In *T. gondii*, loss of fatty acid and phospholipid synthesis led to defects in apicoplast biogenesis, as well as defects in the biogenesis of other organelles such as the IMC (Amiar et al., 2016; Martins-Duarte, Carias, Vommaro, Surolia, & de Souza, 2016; Mazumdar et al., 2006), suggesting that apicoplast-derived fatty acids are important both locally within the organelle and globally within the parasite.

Fatty acid biosynthesis requires pyruvate as a substrate, which, as previously noted, is synthesised in the apicoplast from PEP (Fig. 5). PEP, a product of glycolysis, is imported from the cytosol by pPTs (Fig. 5). The pyruvate dehydrogenase complex that synthesises acetyl-CoA requires numerous cofactors. The E1 subunits require thiamine pyrophosphate (TPP), which is synthesised from thiamine (vitamin B1) by a thiamine pyrophosphokinase that localises to the cytosol in *P. falciparum* (Eschbach, Muller, Gilberger, Walter, & Wrenger, 2006). Apicoplasts must therefore import TPP from the parasite cytosol (Fig. 5). The E3 subunit of pyruvate dehydrogenase requires NAD^+ and FAD as electron acceptors, while the E2 subunit requires lipoic acid and CoA as cofactors (Fig. 5; Table 1). Lipoic acid synthesis is mediated by a two-step pathway in the apicoplast, catalysed by the enzymes LipB and LipA, and involves the modification of octanoyl-acyl carrier protein (ACP), an intermediate of the fatty acid synthesis pathway (Crawford et al., 2006). The final step of lipoic acid synthesis is catalysed by LipA, an Fe–S cluster-containing protein, and requires *S*-adenosylmethionine (SAM) as a sulphur donor (Fig. 5). The E2 subunit also requires CoA. CoA is synthesised in a five-step pathway from pantothenate (Spry, Kirk, & Saliba, 2008). The first four enzymes of this pathway are predicted to localise to the cytosol in *P. falciparum*, whereas the last enzyme, termed dephospho-CoA kinase (DPCK), contains an apparent apicoplast-targeting signal (Spry et al., 2008). Localisation of DPCK to the apicoplast requires experimental verification, but if DPCK does indeed localise to the apicoplast, then the organelle must import dephospho-CoA, the substrate of this enzyme (Fig. 5). ~9% of all enzymes require CoA as a cofactor, including enzymes that are found in the parasite cytosol and other parasite organelles (Spry et al., 2008). If DPCK localises solely to the apicoplast, CoA must be exported from the organelle, implying that the apicoplast is the major site of CoA synthesis in these parasites. This would be unexpected, since previous studies in *P. falciparum* blood stages suggest both that CoA biosynthesis is essential and that IPP is the only essential pathway of the apicoplast (Divo, Geary, Davis, & Jensen, 1985; Howe et al., 2013; Yeh & DeRisi, 2011). Experimentally determining the localisation of DPCK is an important priority in addressing this paradox, and in fully understanding the role of the apicoplast in CoA biology in the parasite. If DPCK is not localised to the apicoplast, then the parasite must import CoA from the cytosol (Fig. 5).

After its synthesis by pyruvate dehydrogenase, acetyl-CoA is converted to malonyl-CoA in an ATP-dependent reaction catalysed by ACCase, an enzyme that requires biotin as a cofactor (Jelenska et al., 2001). In the next step of fatty acid synthesis, the malonate from malonyl-CoA is conjugated to ACP, which serves as the scaffold protein on which fatty acids are built. ACP contains a CoA-derived phosphopantetheine cofactor, an addition that is catalysed by apicoplast-localised ACP synthase enzyme (Ralph et al., 2004). The fatty acid synthesis elongation cycle requires NADH and NADPH as electron donors (Fig. 5). Phospholipid synthesis requires glycerol-3-phosphate, which is derived from pPT-imported DHAP, and CDP-DAG synthesis requires CTP (Table 1; Fig. 5; Amiar et al., 2016; Kong et al., 2017).

7.4 Haem Synthesis

Haem biosynthesis occurs in *T. gondii* and *P. falciparum*, but is absent from several other apicoplast-bearing apicomplexans, including *Babesia* and *Theileria*. The haem biosynthetic pathway is essential in *T. gondii* tachyzoites (E.T. Tjhin and G. van Dooren, unpublished) and in insect and liver stages (but not blood stages) of the *Plasmodium* life cycle (Ke et al., 2014; Nagaraj et al., 2013; Smith et al., 2015). Haem synthesis in apicomplexans exists as an unusual 'hybrid' pathway, where the eight enzymes that catalyse the reactions of this pathway are distributed between the mitochondrion, apicoplast and cytosol (Sigala & Goldberg, 2014; van Dooren, Kennedy, & McFadden, 2012). The first committed step of haem biosynthesis occurs with the synthesis of δ-aminolevulinic acid (ALA) in a reaction catalysed by the enzyme ALA synthase (ALAS). ALAS localises to the parasite mitochondrion (Sato, Clough, Coates, & Wilson, 2004; Varadharajan, Dhanasekaran, Bonday, Rangarajan, & Padmanaban, 2002). The next enzyme of the pathway, porphobilinogen synthase, localises to the apicoplast (Dhanasekaran, Chandra, Chandrasekhar Sagar, Rangarajan, & Padmanaban, 2004; Sato et al., 2004; Shanmugam, Wu, Ramirez, Jaffe, & Roos, 2010). This indicates that apicoplasts must take up mitochondrially derived ALA (Fig. 6). The next three enzymes of the pathway, which produce the compound coproporphyrinogen III, all localise to the apicoplast, with the sixth enzyme, coproporphyrinogen III oxidase, localising to the cytosol (Nagaraj, Arumugam, et al., 2009; Nagaraj et al., 2008; Nagaraj, Prasad,

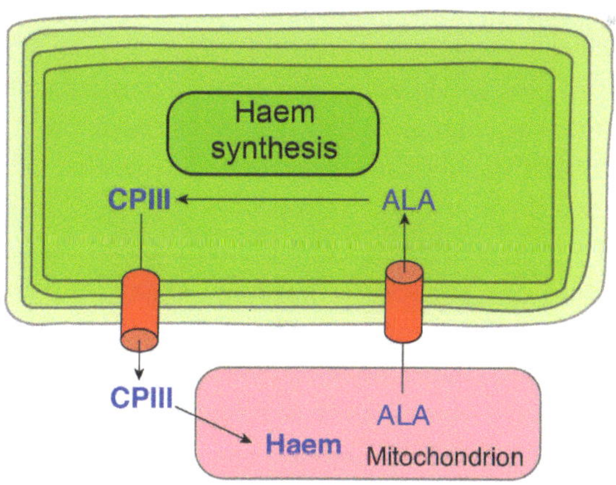

Fig. 6 Schematic of the apicoplast, highlighting the substrates and products of haem biosynthesis. *Filled lines* represent substrates and products of the pathway. See Section 7.4 for details. Abbreviations: *ALA*, δ-aminolevulinic acid; *CPIII*, coproporphyrinogen III.

Arumugam, Rangarajan, & Padmanaban, 2010; Sato et al., 2004). This indicates that coproporphyrinogen III must be exported from the apicoplast into the cytosol (Fig. 6). The transporters that import ALA and export coproporphyrinogen III are unknown. The final two enzymes in haem synthesis localise to the mitochondrion (Nagaraj, Arumugam, Prasad, Rangarajan, & Padmanaban, 2010; Nagaraj, Prasad, Rangarajan, & Padmanaban, 2009; van Dooren et al., 2006), the organelle that requires the bulk of haem in the parasite (van Dooren et al., 2012). The haem pathway, therefore, highlights a further instance where the apicoplast produces organic molecules required by the mitochondrion (Fig. 6).

8. CONCLUSIONS AND PERSPECTIVES

Since its discovery over two decades ago, the apicoplast has emerged as a fascinating and important organelle in apicomplexan parasites. Like its chloroplast ancestors, the apicoplast functions as a biosynthetic organelle that makes substantial contributions to the metabolism of the parasite. This can be exploited through drugs that target essential apicoplast processes. The versatile genetic tools available in apicomplexans such as *T. gondii*, *P. falciparum* and *P. berghei* have facilitated an in-depth understanding of the biology of the apicoplast, with protein tagging, and gene knockout and knockdown approaches providing valuable information on the biogenesis of the organelle, and the importance of apicoplast-localised biochemical pathways. Indeed, apicoplasts are perhaps the most-studied and best-understood of all secondary plastids, and provide a versatile model for understanding how secondary plastids have evolved.

Yet there is still a lot we do not understand, in particular about the membrane biology of this organelle. As highlighted in this review, little is known about how apicoplast metabolism is integrated with that of the rest of the cell. Key in this are the metabolite transporters that facilitate the movement of substrates and products of apicoplast biochemical pathways into and out of the organelle. Only one transporter has been characterised at a molecular level, although many more must exist (Figs 3–6). A challenge in this area is to understand how substrates move across each of the four membranes that surround the organelle. Does each membrane contain selective transporters? Or can solute transport occur through general solute pores in one or more of the membranes? Are these transporters targeted to multiple membranes, or does each membrane require a different transporter to transport a particular substrate? Tied to this are questions around how proteins can be targeted to particular membranes or intermembrane spaces of the apicoplast, and whether intermembrane spaces function to partition particular metabolic processes or biochemical reactions. Developing tools to elucidate the specific membranes/intermembrane spaces to which particular apicoplast proteins localise are key to answering these questions.

Finally, there are the unknown unknowns. Although the wealth of apicomplexan genomic data has provided valuable information on putative apicoplast functions, the vast majority of predicted apicoplast-targeted proteins have no ascribed function (Ralph et al., 2004). Determining the importance and role of these proteins will be of particular interest. Recent genome-wide knockout studies in apicomplexans provide valuable resources for determining which apicoplast proteins to focus on in future research (Bushell et al., 2017; Sidik et al., 2016). Developing more targeted forward genetic screens, for instance utilising the chemical rescue of apicoplast defects by the addition of IPP in *P. falciparum* (Yeh & DeRisi, 2011), also holds particular promise for identifying proteins that function in apicoplast processes. Recent developments in organellar purification and proximity labelling techniques may facilitate proteomic approaches to studying the apicoplast (Botte et al., 2013; Rhee et al., 2013), and sensitive metabolomic approaches can provide insights into the metabolic pathways of the organelle (Amiar et al., 2016; Ramakrishnan et al., 2012). A challenge for the coming 20 years of studying apicoplast biology is to develop new and more sophisticated means of probing apicoplast biogenesis and metabolism, and examining how the biology of this fascinating organelle integrates into the broader biology of the parasite.

ACKNOWLEDGEMENTS

Research in the authors' laboratory is supported by a grant from the Australian Research Council (DP150102883). S.V.H. is supported by an Australian Government Research Training Program Domestic Scholarship.

REFERENCES

Agrawal, S., Chung, D. W., Ponts, N., van Dooren, G. G., Prudhomme, J., Brooks, C. F., et al. (2013). An apicoplast localized ubiquitylation system is required for the import of nuclear-encoded plastid proteins. *PLoS Pathogens*, *9*, e1003426.

Agrawal, S., van Dooren, G. G., Beatty, W. L., & Striepen, B. (2009). Genetic evidence that an endosymbiont-derived endoplasmic reticulum-associated protein degradation (ERAD) system functions in import of apicoplast proteins. *The Journal of Biological Chemistry*, *284*, 33683–33691.

Aikawa, M. (1966). The fine structure of the erythrocytic stages of three avian malarial parasites, *Plasmodium fallax, P. lophurae*, and *P. cathemerium*. *The American Journal of Tropical Medicine and Hygiene, 15*, 449–471.

Amiar, S., MacRae, J. I., Callahan, D. L., Dubois, D., van Dooren, G. G., Shears, M. J., et al. (2016). Apicoplast-localized lysophosphatidic acid precursor assembly is required for bulk phospholipid synthesis in *Toxoplasma gondii* and relies on an algal/plant-like glycerol 3-phosphate acyltransferase. *PLoS Pathogens, 12*, e1005765.

Andenmatten, N., Egarter, S., Jackson, A. J., Jullien, N., Herman, J. P., & Meissner, M. (2013). Conditional genome engineering in *Toxoplasma gondii* uncovers alternative invasion mechanisms. *Nature Methods, 10*, 125–127.

Antonny, B., Burd, C., De Camilli, P., Chen, E., Daumke, O., Faelber, K., et al. (2016). Membrane fission by dynamin: What we know and what we need to know. *The EMBO Journal, 35*, 2270–2284.

Balk, J., & Pilon, M. (2011). Ancient and essential: The assembly of iron-sulfur clusters in plants. *Trends in Plant Science, 16*, 218–226.

Banerjee, T., Jaijyan, D. K., Surolia, N., Singh, A. P., & Surolia, A. (2012). Apicoplast triose phosphate transporter (TPT) gene knockout is lethal for *Plasmodium*. *Molecular and Biochemical Parasitology, 186*, 44–50.

Botte, C. Y., Yamaryo-Botte, Y., Rupasinghe, T. W., Mullin, K. A., MacRae, J. I., Spurck, T. P., et al. (2013). Atypical lipid composition in the purified relict plastid (apicoplast) of malaria parasites. *Proceedings of the National Academy of Sciences of the United States of America, 110*, 7506–7511.

Bouchut, A., Geiger, J. A., DeRocher, A. E., & Parsons, M. (2014). Vesicles bearing *Toxoplasma* apicoplast membrane proteins persist following loss of the relict plastid or Golgi body disruption. *PLoS One, 9*, e112096.

Breinich, M. S., Ferguson, D. J., Foth, B. J., van Dooren, G. G., Lebrun, M., Quon, D. V., et al. (2009). A dynamin is required for the biogenesis of secretory organelles in *Toxoplasma gondii*. *Current Biology, 19*, 277–286.

Brooks, C. F., Johnsen, H., van Dooren, G. G., Muthalagi, M., Lin, S. S., Bohne, W., et al. (2010). The *Toxoplasma* apicoplast phosphate translocator links cytosolic and apicoplast metabolism and is essential for parasite survival. *Cell Host & Microbe, 7*, 62–73.

Bullmann, L., Haarmann, R., Mirus, O., Bredemeier, R., Hempel, F., Maier, U. G., et al. (2010). Filling the gap, evolutionarily conserved Omp85 in plastids of chromalveolates. *The Journal of Biological Chemistry, 285*, 6848–6856.

Burkhardt, D., Wiesner, J., Stoesser, N., Ramharter, M., Uhlemann, A. C., Issifou, S., et al. (2007). Delayed parasite elimination in human infections treated with clindamycin parallels 'delayed death' of *Plasmodium falciparum* in vitro. *International Journal for Parasitology, 37*, 777–785.

Bushell, E., Gomes, A. R., Sanderson, T., Anar, B., Girling, G., Herd, C., et al. (2017). Functional profiling of a plasmodium genome reveals an abundance of essential genes. *Cell, 170*(2), 260–272.

Cavalier-Smith, T. (1999). Principles of protein and lipid targeting in secondary symbiogenesis: Euglenoid, dinoflagellate, and sporozoan plastid origins and the eukaryote family tree. *The Journal of Eukaryotic Microbiology, 46*, 347–366.

Cavalier-Smith, T. (2000). Membrane heredity and early chloroplast evolution. *Trends in Plant Science, 5*, 174–182.

Charan, M., Singh, N., Kumar, B., Srivastava, K., Siddiqi, M. I., & Habib, S. (2014). Sulfur mobilization for Fe-S cluster assembly by the essential SUF pathway in the *Plasmodium falciparum* apicoplast and its inhibition. *Antimicrobial Agents and Chemotherapy, 58*, 3389–3398.

Charneau, S., Bastos, I. M., Mouray, E., Ribeiro, B. M., Santana, J. M., Grellier, P., et al. (2007). Characterization of PfDYN2, a dynamin-like protein of *Plasmodium falciparum* expressed in schizonts. *Microbes and Infection, 9*, 797–805.

Chaudhari, R., Dey, V., Narayan, A., Sharma, S., & Patankar, S. (2017). Membrane and luminal proteins reach the apicoplast by different trafficking pathways in the malaria parasite *Plasmodium falciparum*. *PeerJ*, *5*, e3128.

Chen, C. T., & Gubbels, M. J. (2013). The *Toxoplasma gondii* centrosome is the platform for internal daughter budding as revealed by a Nek1 kinase mutant. *Journal of Cell Science*, *126*, 3344–3355.

Clough, B., Strath, M., Preiser, P., Denny, P., & Wilson, I. R. (1997). Thiostrepton binds to malarial plastid rRNA. *FEBS Letters*, *406*, 123–125.

Crawford, M. J., Thomsen-Zieger, N., Ray, M., Schachtner, J., Roos, D. S., & Seeber, F. (2006). *Toxoplasma gondii* scavenges host-derived lipoic acid despite its de novo synthesis in the apicoplast. *The EMBO Journal*, *25*, 3214–3222.

Daher, W., Morlon-Guyot, J., Sheiner, L., Lentini, G., Berry, L., Tawk, L., et al. (2015). Lipid kinases are essential for apicoplast homeostasis in *Toxoplasma gondii*. *Cellular Microbiology*, *17*, 559–578.

Dahl, E. L., & Rosenthal, P. J. (2007). Multiple antibiotics exert delayed effects against the *Plasmodium falciparum* apicoplast. *Antimicrobial Agents and Chemotherapy*, *51*, 3485–3490.

Dahl, E. L., & Rosenthal, P. J. (2008). Apicoplast translation, transcription and genome replication: Targets for antimalarial antibiotics. *Trends in Parasitology*, *24*, 279–284.

Dahl, E. L., Shock, J. L., Shenai, B. R., Gut, J., DeRisi, J. L., & Rosenthal, P. J. (2006). Tetracyclines specifically target the apicoplast of the malaria parasite *Plasmodium falciparum*. *Antimicrobial Agents and Chemotherapy*, *50*, 3124–3131.

DeRocher, A. E., Coppens, I., Karnataki, A., Gilbert, L. A., Rome, M. E., Feagin, J. E., et al. (2008). A thioredoxin family protein of the apicoplast periphery identifies abundant candidate transport vesicles in *Toxoplasma gondii*. *Eukaryotic Cell*, *7*, 1518–1529.

DeRocher, A., Gilbert, B., Feagin, J. E., & Parsons, M. (2005). Dissection of brefeldin A-sensitive and -insensitive steps in apicoplast protein targeting. *Journal of Cell Science*, *118*, 565–574.

DeRocher, A., Hagen, C. B., Froehlich, J. E., Feagin, J. E., & Parsons, M. (2000). Analysis of targeting sequences demonstrates that trafficking to the *Toxoplasma gondii* plastid branches off the secretory system. *Journal of Cell Science*, *113*, 3969–3977.

DeRocher, A. E., Karnataki, A., Vaney, P., & Parsons, M. (2012). Apicoplast targeting of a *Toxoplasma gondii* transmembrane protein requires a cytosolic tyrosine-based motif. *Traffic*, *13*, 694–704.

Dhanasekaran, S., Chandra, N. R., Chandrasekhar Sagar, B. K., Rangarajan, P. N., & Padmanaban, G. (2004). Delta-aminolevulinic acid dehydratase from *Plasmodium falciparum*: Indigenous versus imported. *The Journal of Biological Chemistry*, *279*, 6934–6942.

Di Paolo, G., & De Camilli, P. (2006). Phosphoinositides in cell regulation and membrane dynamics. *Nature*, *443*, 651–657.

Divo, A. A., Geary, T. G., Davis, N. L., & Jensen, J. B. (1985). Nutritional requirements of *Plasmodium falciparum* in culture. I. Exogenously supplied dialyzable components necessary for continuous growth. *The Journal of Protozoology*, *32*, 59–64.

Dobberstein, B., Blobel, G., & Chua, N. H. (1977). In vitro synthesis and processing of a putative precursor for the small subunit of ribulose-1,5-bisphosphate carboxylase of *Chlamydomonas reinhardtii*. *Proceedings of the National Academy of Sciences of the United States of America*, *74*, 1082–1085.

Duy, D., Wanner, G., Meda, A. R., von Wiren, N., Soll, J., & Philippar, K. (2007). PIC1, an ancient permease in *Arabidopsis* chloroplasts, mediates iron transport. *Plant Cell*, *19*, 986–1006.

Ellis, K. E., Clough, B., Saldanha, J. W., & Wilson, R. J. (2001). Nifs and Sufs in malaria. *Molecular Microbiology*, *41*, 973–981.

Eschbach, M. L., Muller, I. B., Gilberger, T. W., Walter, R. D., & Wrenger, C. (2006). The human malaria parasite *Plasmodium falciparum* expresses an atypical N-terminally extended pyrophosphokinase with specificity for thiamine. *Biological Chemistry*, *387*, 1583–1591.

Fast, N. M., Kissinger, J. C., Roos, D. S., & Keeling, P. J. (2001). Nuclear-encoded, plastid-targeted genes suggest a single common origin for apicomplexan and dinoflagellate plastids. *Molecular Biology and Evolution, 18*, 418–426.
Fichera, M. E., Bhopale, M. K., & Roos, D. S. (1995). In vitro assays elucidate peculiar kinetics of clindamycin action against *Toxoplasma gondii*. *Antimicrobial Agents and Chemotherapy, 39*, 1530–1537.
Fichera, M. E., & Roos, D. S. (1997). A plastid organelle as a drug target in apicomplexan parasites. *Nature, 390*, 407–409.
Foth, B. J., Ralph, S. A., Tonkin, C. J., Struck, N. S., Fraunholz, M., Roos, D. S., et al. (2003). Dissecting apicoplast targeting in the malaria parasite *Plasmodium falciparum*. *Science, 299*, 705–708.
Francia, M. E., & Striepen, B. (2014). Cell division in apicomplexan parasites. *Nature Reviews. Microbiology, 12*, 125–136.
Frenal, K., Jacot, D., Hammoudi, P. M., Graindorge, A., Maco, B., & Soldati-Favre, D. (2017). Myosin-dependent cell-cell communication controls synchronicity of division in acute and chronic stages of *Toxoplasma gondii*. *Nature Communications, 8*, 15710.
Gardner, M. J., Feagin, J. E., Moore, D. J., Spencer, D. F., Gray, M. W., Williamson, D. H., et al. (1991). Organisation and expression of small subunit ribosomal RNA genes encoded by a 35-kilobase circular DNA in *Plasmodium falciparum*. *Molecular and Biochemical Parasitology, 48*, 77–88.
Gardner, M. J., Goldman, N., Barnett, P., Moore, P. W., Rangachari, K., Strath, M., et al. (1994). Phylogenetic analysis of the rpoB gene from the plastid-like DNA of *Plasmodium falciparum*. *Molecular and Biochemical Parasitology, 66*, 221–231.
Gardner, M. J., Hall, N., Fung, E., White, O., Berriman, M., Hyman, R. W., et al. (2002). Genome sequence of the human malaria parasite *Plasmodium falciparum*. *Nature, 419*, 498–511.
Gardner, M. J., Williamson, D. H., & Wilson, R. J. (1991). A circular DNA in malaria parasites encodes an RNA polymerase like that of prokaryotes and chloroplasts. *Molecular and Biochemical Parasitology, 44*, 115–123.
Gengenbacher, M., Fitzpatrick, T. B., Raschle, T., Flicker, K., Sinning, I., Muller, S., et al. (2006). Vitamin B6 biosynthesis by the malaria parasite *Plasmodium falciparum*: Biochemical and structural insights. *The Journal of Biological Chemistry, 281*, 3633–3641.
Gibbs, S. P. (1981). The chloroplasts of some algal groups may have evolved from endosymbiotic eukaryotic algae. *Annals of the New York Academy of Sciences, 361*, 193–208.
Gisselberg, J. E., Dellibovi-Ragheb, T. A., Matthews, K. A., Bosch, G., & Prigge, S. T. (2013). The suf iron-sulfur cluster synthesis pathway is required for apicoplast maintenance in malaria parasites. *PLoS Pathogens, 9*, e1003655.
Glaser, S., van Dooren, G. G., Agrawal, S., Brooks, C. F., McFadden, G. I., Striepen, B., et al. (2012). Tic22 is an essential chaperone required for protein import into the apicoplast. *The Journal of Biological Chemistry, 287*, 39505–39512.
Goodman, C. D., & McFadden, G. I. (2013). Targeting apicoplasts in malaria parasites. *Expert Opinion on Therapeutic Targets, 17*, 167–177.
Goodman, C. D., Su, V., & McFadden, G. I. (2007). The effects of anti-bacterials on the malaria parasite *Plasmodium falciparum*. *Molecular and Biochemical Parasitology, 152*, 181–191.
Gould, S. B., Maier, U. G., & Martin, W. F. (2015). Protein import and the origin of red complex plastids. *Current Biology, 25*, R515–R521.
Gould, S. B., Sommer, M. S., Hadfi, K., Zauner, S., Kroth, P. G., & Maier, U. G. (2006). Protein targeting into the complex plastid of cryptophytes. *Journal of Molecular Evolution, 62*, 674–681.
Guggisberg, A. M., Amthor, R. E., & Odom, A. R. (2014). Isoprenoid biosynthesis in *Plasmodium falciparum*. *Eukaryotic Cell, 13*, 1348–1359.

Harper, J. T., & Keeling, P. J. (2003). Nucleus-encoded, plastid-targeted glyceraldehyde-3-phosphate dehydrogenase (GAPDH) indicates a single origin for chromalveolate plastids. *Molecular Biology and Evolution, 20*, 1730–1735.

Haussig, J. M., Matuschewski, K., & Kooij, T. W. (2014). Identification of vital and dispensable sulfur utilization factors in the *Plasmodium* apicoplast. *PLoS One, 9*, e89718.

He, C. Y., Shaw, M. K., Pletcher, C. H., Striepen, B., Tilney, L. G., & Roos, D. S. (2001). A plastid segregation defect in the protozoan parasite *Toxoplasma gondii. The EMBO Journal, 20*, 330–339.

Heiny, S. R., Pautz, S., Recker, M., & Przyborski, J. M. (2014). Protein traffic to the *Plasmodium falciparum* apicoplast: Evidence for a sorting branch point at the Golgi. *Traffic, 15*, 1290–1304.

Hempel, F., Bullmann, L., Lau, J., Zauner, S., & Maier, U. G. (2009). ERAD-derived preprotein transport across the second outermost plastid membrane of diatoms. *Molecular Biology and Evolution, 26*, 1781–1790.

Hempel, F., Felsner, G., & Maier, U. G. (2010). New mechanistic insights into pre-protein transport across the second outermost plastid membrane of diatoms. *Molecular Microbiology, 76*, 793–801.

Hepler, P. K., Huff, C. G., & Sprinz, H. (1966). The fine structure of the exoerythrocytic stages of *Plasmodium fallax. The Journal of Cell Biology, 30*, 333–358.

Hill, D. E., Chirukandoth, S., & Dubey, J. P. (2005). Biology and epidemiology of *Toxoplasma gondii* in man and animals. *Animal Health Research Reviews, 6*, 41–61.

Hirsch, C., Gauss, R., Horn, S. C., Neuber, O., & Sommer, T. (2009). The ubiquitylation machinery of the endoplasmic reticulum. *Nature, 458*, 453–460.

Hopkins, J., Fowler, R., Krishna, S., Wilson, I., Mitchell, G., & Bannister, L. (1999). The plastid in *Plasmodium falciparum* asexual blood stages: A three-dimensional ultrastructural analysis. *Protist, 150*, 283–295.

Howe, C. J. (1992). Plastid origin of an extrachromosomal DNA molecule from *Plasmodium*, the causative agent of malaria. *Journal of Theoretical Biology, 158*, 199–205.

Howe, R., Kelly, M., Jimah, J., Hodge, D., & Odom, A. R. (2013). Isoprenoid biosynthesis inhibition disrupts Rab5 localization and food vacuolar integrity in *Plasmodium falciparum. Eukaryotic Cell, 12*, 215–223.

Huang, J., & Brumell, J. H. (2014). Bacteria-autophagy interplay: A battle for survival. *Nature Reviews. Microbiology, 12*, 101–114.

Ingerman, E., Perkins, E. M., Marino, M., Mears, J. A., McCaffery, J. M., Hinshaw, J. E., et al. (2005). Dnm1 forms spirals that are structurally tailored to fit mitochondria. *The Journal of Cell Biology, 170*, 1021–1027.

Jacot, D., Daher, W., & Soldati-Favre, D. (2013). *Toxoplasma gondii* myosin F, an essential motor for centrosomes positioning and apicoplast inheritance. *The EMBO Journal, 32*, 1702–1716.

Janouskovec, J., Horak, A., Obornik, M., Lukes, J., & Keeling, P. J. (2010). A common red algal origin of the apicomplexan, dinoflagellate, and heterokont plastids. *Proceedings of the National Academy of Sciences of the United States of America, 107*, 10949–10954.

Jarvis, P., & Soll, J. (2002). Toc, tic, and chloroplast protein import. *Biochimica et Biophysica Acta, 1590*, 177–189.

Jelenska, J., Crawford, M. J., Harb, O. S., Zuther, E., Haselkorn, R., Roos, D. S., et al. (2001). Subcellular localization of acetyl-CoA carboxylase in the apicomplexan parasite *Toxoplasma gondii. Proceedings of the National Academy of Sciences of the United States of America, 98*, 2723–2728.

Jomaa, H., Wiesner, J., Sanderbrand, S., Altincicek, B., Weidemeyer, C., Hintz, M., et al. (1999). Inhibitors of the nonmevalonate pathway of isoprenoid biosynthesis as antimalarial drugs. *Science, 285*, 1573–1576.

Kalanon, M., Tonkin, C. J., & McFadden, G. I. (2009). Characterization of two putative protein translocation components in the apicoplast of *Plasmodium falciparum*. *Eukaryotic Cell*, *8*, 1146–1154.

Karnataki, A., Derocher, A. E., Coppens, I., Feagin, J. E., & Parsons, M. (2007). A membrane protease is targeted to the relict plastid of *Toxoplasma* via an internal signal sequence. *Traffic*, *8*, 1543–1553.

Karnataki, A., Derocher, A., Coppens, I., Nash, C., Feagin, J. E., & Parsons, M. (2007). Cell cycle-regulated vesicular trafficking of *Toxoplasma* APT1, a protein localized to multiple apicoplast membranes. *Molecular Microbiology*, *63*, 1653–1668.

Ke, H., Sigala, P. A., Miura, K., Morrisey, J. M., Mather, M. W., Crowley, J. R., et al. (2014). The heme biosynthesis pathway is essential for *Plasmodium falciparum* development in mosquito stage but not in blood stages. *The Journal of Biological Chemistry*, *289*, 34827–34837.

Kikuchi, S., Bedard, J., Hirano, M., Hirabayashi, Y., Oishi, M., Imai, M., et al. (2013). Uncovering the protein translocon at the chloroplast inner envelope membrane. *Science*, *339*, 571–574.

Kilejian, A. (1975). Circular mitochondrial DNA from the avian malarial parasite *Plasmodium lophurae*. *Biochimica et Biophysica Acta*, *390*, 276–284.

Kitamura, K., Kishi-Itakura, C., Tsuboi, T., Sato, S., Kita, K., Ohta, N., et al. (2012). Autophagy-related Atg8 localizes to the apicoplast of the human malaria parasite *Plasmodium falciparum*. *PLoS One*, 7, e42977.

Klein, D. E., Lee, A., Frank, D. W., Marks, M. S., & Lemmon, M. A. (1998). The pleckstrin homology domains of dynamin isoforms require oligomerization for high affinity phosphoinositide binding. *The Journal of Biological Chemistry*, *273*, 27725–27733.

Kohler, S., Delwiche, C. F., Denny, P. W., Tilney, L. G., Webster, P., Wilson, R. J., et al. (1997). A plastid of probable green algal origin in apicomplexan parasites. *Science*, *275*, 1485–1489.

Koirala, S., Guo, Q., Kalia, R., Bui, H. T., Eckert, D. M., Frost, A., et al. (2013). Interchangeable adaptors regulate mitochondrial dynamin assembly for membrane scission. *Proceedings of the National Academy of Sciences of the United States of America*, *110*, E1342–E1351.

Kong, P., Ufermann, C. M., Zimmermann, D. L. M., Yin, Q., Suo, X., Helms, J. B., et al. (2017). Two phylogenetically and compartmentally distinct CDP-diacylglycerol synthases cooperate for lipid biogenesis in *Toxoplasma gondii*. *The Journal of Biological Chemistry*, *292*, 7145–7159.

Kotloff, K. L., Nataro, J. P., Blackwelder, W. C., Nasrin, D., Farag, T. H., Panchalingam, S., et al. (2013). Burden and aetiology of diarrhoeal disease in infants and young children in developing countries (the Global Enteric Multicenter Study, GEMS): A prospective, case-control study. *Lancet*, *382*, 209–222.

Kumar, B., Chaubey, S., Shah, P., Tanveer, A., Charan, M., Siddiqi, M. I., et al. (2011). Interaction between sulphur mobilisation proteins SufB and SufC: Evidence for an iron-sulphur cluster biogenesis pathway in the apicoplast of *Plasmodium falciparum*. *International Journal for Parasitology*, *41*, 991–999.

Lavine, M. D., & Arrizabalaga, G. (2012). Analysis of monensin sensitivity in *Toxoplasma gondii* reveals autophagy as a mechanism for drug induced death. *PLoS One*, 7, e42107.

Lemgruber, L., Kudryashev, M., Dekiwadia, C., Riglar, D. T., Baum, J., Stahlberg, H., et al. (2013). Cryo-electron tomography reveals four-membrane architecture of the *Plasmodium* apicoplast. *Malaria Journal*, *12*, 25.

Leveque, M. F., Berry, L., Cipriano, M. J., Nguyen, H. M., Striepen, B., & Besteiro, S. (2015). Autophagy-related protein ATG8 has a noncanonical function for apicoplast inheritance in *Toxoplasma gondii*. *MBio*, *6*, e01446–01415.

Lill, R., & Muehlenhoff, U. (2005). Iron-sulfur-protein biogenesis in eukaryotes. *Trends in Biochemical Sciences*, *30*, 133–141.

Lim, L., Linka, M., Mullin, K. A., Weber, A. P., & McFadden, G. I. (2010). The carbon and energy sources of the non-photosynthetic plastid in the malaria parasite. *FEBS Letters*, *584*, 549–554.

Lim, L., Sayers, C. P., Goodman, C. D., & McFadden, G. I. (2016). Targeting of a transporter to the outer apicoplast membrane in the human malaria parasite *Plasmodium falciparum*. *PLoS One*, *11*, e0159603.

Martins-Duarte, E. S., Carias, M., Vommaro, R., Surolia, N., & de Souza, W. (2016). Apicoplast fatty acid synthesis is essential for pellicle formation at the end of cytokinesis in *Toxoplasma gondii*. *Journal of Cell Science*, *129*, 3320–3331.

Mazumdar, J., Wilson, E. H., Masek, K., Hunter, C. A., & Striepen, B. (2006). Apicoplast fatty acid synthesis is essential for organelle biogenesis and parasite survival in *Toxoplasma gondii*. *Proceedings of the National Academy of Sciences of the United States of America*, *103*, 13192–13197.

McConkey, G. A., Rogers, M. J., & McCutchan, T. F. (1997). Inhibition of *Plasmodium falciparum* protein synthesis. Targeting the plastid-like organelle with thiostrepton. *The Journal of Biological Chemistry*, *272*, 2046–2049.

McFadden, G. I., Reith, M. E., Munholland, J., & Lang-Unnasch, N. (1996). Plastid in human parasites. *Nature*, *381*, 482.

McFadden, G. I., & Roos, D. S. (1999). Apicomplexan plastids as drug targets. *Trends in Microbiology*, 7, 328–333.

McFadden, G. I., & Waller, R. F. (1997). Plastids in parasites of humans. *BioEssays*, *19*, 1033–1040.

Miyagishima, S. Y., Nishida, K., Mori, T., Matsuzaki, M., Higashiyama, T., Kuroiwa, H., et al. (2003). A plant-specific dynamin-related protein forms a ring at the chloroplast division site. *Plant Cell*, *15*, 655–665.

Miyagishima, S., Takahara, M., Mori, T., Kuroiwa, H., Higashiyama, T., & Kuroiwa, T. (2001). Plastid division is driven by a complex mechanism that involves differential transition of the bacterial and eukaryotic division rings. *Plant Cell*, *13*, 2257–2268.

Montoya, J. G., & Liesenfeld, O. (2004). Toxoplasmosis. *Lancet*, *363*, 1965–1976.

Moore, R. B., Obornik, M., Janouskovec, J., Chrudimsky, T., Vancova, M., Green, D. H., et al. (2008). A photosynthetic alveolate closely related to apicomplexan parasites. *Nature*, *451*, 959–963.

Mullin, K. A., Lim, L., Ralph, S. A., Spurck, T. P., Handman, E., & McFadden, G. I. (2006). Membrane transporters in the relict plastid of malaria parasites. *Proceedings of the National Academy of Sciences of the United States of America*, *103*, 9572–9577.

Nagaraj, V. A., Arumugam, R., Chandra, N. R., Prasad, D., Rangarajan, P. N., & Padmanaban, G. (2009). Localisation of *Plasmodium falciparum* uroporphyrinogen III decarboxylase of the heme-biosynthetic pathway in the apicoplast and characterisation of its catalytic properties. *International Journal for Parasitology*, *39*, 559–568.

Nagaraj, V. A., Arumugam, R., Gopalakrishnan, B., Jyothsna, Y. S., Rangarajan, P. N., & Padmanaban, G. (2008). Unique properties of *Plasmodium falciparum* porphobilinogen deaminase. *The Journal of Biological Chemistry*, *283*, 437–444.

Nagaraj, V. A., Arumugam, R., Prasad, D., Rangarajan, P. N., & Padmanaban, G. (2010). Protoporphyrinogen IX oxidase from *Plasmodium falciparum* is anaerobic and is localized to the mitochondrion. *Molecular and Biochemical Parasitology*, *174*, 44–52.

Nagaraj, V. A., Prasad, D., Arumugam, R., Rangarajan, P. N., & Padmanaban, G. (2010). Characterization of coproporphyrinogen III oxidase in *Plasmodium falciparum* cytosol. *Parasitology International*, *59*, 121–127.

Nagaraj, V. A., Prasad, D., Rangarajan, P. N., & Padmanaban, G. (2009). Mitochondrial localization of functional ferrochelatase from *Plasmodium falciparum*. *Molecular and Biochemical Parasitology*, *168*, 109–112.

Nagaraj, V. A., Sundaram, B., Varadarajan, N. M., Subramani, P. A., Kalappa, D. M., Ghosh, S. K., et al. (2013). Malaria parasite-synthesized heme is essential in the mosquito and liver stages and complements host heme in the blood stages of infection. *PLoS Pathogens*, *9*, e1003522.

Nair, S. C., Brooks, C. F., Goodman, C. D., Strurm, A., McFadden, G. I., Sundriyal, S., et al. (2011). Apicoplast isoprenoid precursor synthesis and the molecular basis of fosmidomycin resistance in *Toxoplasma gondii*. *The Journal of Experimental Medicine*, *208*, 1547–1559.

Nakai, M. (2015). The TIC complex uncovered: The alternative view on the molecular mechanism of protein translocation across the inner envelope membrane of chloroplasts. *Biochimica et Biophysica Acta*, *1847*, 957–967.

Nishi, M., Hu, K., Murray, J. M., & Roos, D. S. (2008). Organellar dynamics during the cell cycle of *Toxoplasma gondii*. *Journal of Cell Science*, *121*, 1559–1568.

Parameswaran, N., Thompson, R. C., Sundar, N., Pan, S., Johnson, M., Smith, N. C., et al. (2010). Non-archetypal type II-like and atypical strains of *Toxoplasma gondii* infecting marsupials of Australia. *International Journal for Parasitology*, *40*, 635–640.

Patron, N. J., & Waller, R. F. (2007). Transit peptide diversity and divergence: A global analysis of plastid targeting signals. *BioEssays*, *29*, 1048–1058.

Peek, H. W., & Landman, W. J. (2011). Coccidiosis in poultry: Anticoccidial products, vaccines and other prevention strategies. *The Veterinary Quarterly*, *31*, 143–161.

Pfefferkorn, E. R., & Borotz, S. E. (1994). Comparison of mutants of *Toxoplasma gondii* selected for resistance to azithromycin, spiramycin, or clindamycin. *Antimicrobial Agents and Chemotherapy*, *38*, 31–37.

Pfefferkorn, E. R., Nothnagel, R. F., & Borotz, S. E. (1992). Parasiticidal effect of clindamycin on *Toxoplasma gondii* grown in cultured cells and selection of a drug-resistant mutant. *Antimicrobial Agents and Chemotherapy*, *36*, 1091–1096.

Praefcke, G. J., & McMahon, H. T. (2004). The dynamin superfamily: Universal membrane tubulation and fission molecules? *Nature Reviews. Molecular Cell Biology*, *5*, 133–147.

Ralph, S. A., van Dooren, G. G., Waller, R. F., Crawford, M. J., Fraunholz, M. J., Foth, B. J., et al. (2004). Tropical infectious diseases: Metabolic maps and functions of the *Plasmodium falciparum* apicoplast. *Nature Reviews. Microbiology*, *2*, 203–216.

Ramakrishnan, S., Docampo, M. D., Macrae, J. I., Pujol, F. M., Brooks, C. F., van Dooren, G. G., et al. (2012). Apicoplast and endoplasmic reticulum cooperate in fatty acid biosynthesis in apicomplexan parasite *Toxoplasma gondii*. *The Journal of Biological Chemistry*, *287*, 4957–4971.

Ramakrishnan, S., Docampo, M. D., MacRae, J. I., Ralton, J. E., Rupasinghe, T., McConville, M. J., et al. (2015). The intracellular parasite *Toxoplasma gondii* depends on the synthesis of long-chain and very long-chain unsaturated fatty acids not supplied by the host cell. *Molecular Microbiology*, *97*, 64–76.

Ramharter, M., Oyakhirome, S., Klein Klouwenberg, P., Adegnika, A. A., Agnandji, S. T., Missinou, M. A., et al. (2005). Artesunate-clindamycin versus quinine-clindamycin in the treatment of *Plasmodium falciparum* malaria: A randomized controlled trial. *Clinical Infectious Diseases*, *40*, 1777–1784.

Rhee, H. W., Zou, P., Udeshi, N. D., Martell, J. D., Mootha, V. K., Carr, S. A., et al. (2013). Proteomic mapping of mitochondria in living cells via spatially restricted enzymatic tagging. *Science*, *339*, 1328–1331.

Rogers, M. J., Bukhman, Y. V., McCutchan, T. F., & Draper, D. E. (1997). Interaction of thiostrepton with an RNA fragment derived from the plastid-encoded ribosomal RNA of the malaria parasite. *RNA*, *3*, 815–820.

Rohrich, R. C., Englert, N., Troschke, K., Reichenberg, A., Hintz, M., Seeber, F., et al. (2005). Reconstitution of an apicoplast-localised electron transfer pathway involved in the isoprenoid biosynthesis of *Plasmodium falciparum*. *FEBS Letters*, *579*, 6433–6438.

Sato, S., Clough, B., Coates, L., & Wilson, R. J. (2004). Enzymes for heme biosynthesis are found in both the mitochondrion and plastid of the malaria parasite *Plasmodium falciparum*. *Protist, 155*, 117–125.

Schleiff, E., Jelic, M., & Soll, J. (2003). A GTP-driven motor moves proteins across the outer envelope of chloroplasts. *Proceedings of the National Academy of Sciences of the United States of America, 100*, 4604–4609.

Seeber, F., & Soldati-Favre, D. (2010). Metabolic pathways in the apicoplast of apicomplexa. *International Review of Cell and Molecular Biology, 281*, 161–228.

Shanmugam, D., Wu, B., Ramirez, U., Jaffe, E. K., & Roos, D. S. (2010). Plastid-associated-porphobilinogen synthase from *Toxoplasma gondii*: Kinetic and structural properties validate therapeutic potential. *The Journal of Biological Chemistry, 285*, 22122–22131.

Shears, M. J., MacRae, J. I., Mollard, V., Goodman, C. D., Sturm, A., Orchard, L. M., et al. (2017). Characterization of the *Plasmodium falciparum* and *P. berghei* glycerol 3-phosphate acyltransferase involved in FASII fatty acid utilization in the malaria parasite apicoplast. *Cell Microbiology, 19*, e12633.

Sheffield, H. G., & Melton, M. L. (1968). The fine structure and reproduction of *Toxoplasma gondii*. *The Journal of Parasitology, 54*, 209–226.

Sheiner, L., Fellows, J. D., Ovciarikova, J., Brooks, C. F., Agrawal, S., Holmes, Z. C., et al. (2015). *Toxoplasma gondii* Toc75 functions in import of stromal but not peripheral apicoplast proteins. *Traffic, 16*, 1254–1269.

Siddall, M. E. (1992). Hohlzylinders. *Parasitology Today, 8*, 90–91.

Sidik, S. M., Huet, D., Ganesan, S. M., Huynh, M. H., Wang, T., Nasamu, A. S., et al. (2016). A genome-wide CRISPR screen in *Toxoplasma* identifies essential apicomplexan genes. *Cell, 166*, 1423–1435.

Sigala, P. A., & Goldberg, D. E. (2014). The peculiarities and paradoxes of *Plasmodium* heme metabolism. *Annual Review of Microbiology, 68*, 259–278.

Smith, C. M., Jerkovic, A., Puy, H., Winship, I., Deybach, J. C., Gouya, L., et al. (2015). Red cells from ferrochelatase-deficient erythropoietic protoporphyria patients are resistant to growth of malarial parasites. *Blood, 125*, 534–541.

Sommer, M. S., Gould, S. B., Lehmann, P., Gruber, A., Przyborski, J. M., & Maier, U. G. (2007). Der1-mediated preprotein import into the periplastid compartment of chromalveolates? *Molecular Biology and Evolution, 24*, 918–928.

Spork, S., Hiss, J. A., Mandel, K., Sommer, M., Kooij, T. W., Chu, T., et al. (2009). An unusual ERAD-like complex is targeted to the apicoplast of *Plasmodium falciparum*. *Eukaryotic Cell, 8*, 1134–1145.

Spry, C., Kirk, K., & Saliba, K. J. (2008). Coenzyme A biosynthesis: An antimicrobial drug target. *FEMS Microbiology Reviews, 32*, 56–106.

Stanway, R. R., Mueller, N., Zobiak, B., Graewe, S., Froehlke, U., Zessin, P. J., et al. (2011). Organelle segregation into *Plasmodium* liver stage merozoites. *Cellular Microbiology, 13*, 1768–1782.

Stanway, R. R., Witt, T., Zobiak, B., Aepfelbacher, M., & Heussler, V. T. (2009). GFP-targeting allows visualization of the apicoplast throughout the life cycle of live malaria parasites. *Biology of the Cell, 101*, 415–430. 415 p following 430.

Stewart, R. J., Ferguson, D. J., Whitehead, L., Bradin, C. H., Wu, H. J., & Tonkin, C. J. (2016). Phosphorylation of alphaSNAP is required for secretory organelle biogenesis in *Toxoplasma gondii*. *Traffic, 17*, 102–116.

Striepen, B. (2013). Parasitic infections: Time to tackle cryptosporidiosis. *Nature, 503*, 189–191.

Striepen, B., Crawford, M. J., Shaw, M. K., Tilney, L. G., Seeber, F., & Roos, D. S. (2000). The plastid of *Toxoplasma gondii* is divided by association with the centrosomes. *The Journal of Cell Biology, 151*, 1423–1434.

Striepen, B., Jordan, C. N., Reiff, S., & van Dooren, G. G. (2007). Building the perfect parasite: Cell division in apicomplexa. *PLoS Pathogens*, *3*, e78.
Tawk, L., Chicanne, G., Dubremetz, J. F., Richard, V., Payrastre, B., Vial, H. J., et al. (2010). Phosphatidylinositol 3-phosphate, an essential lipid in *Plasmodium*, localizes to the food vacuole membrane and the apicoplast. *Eukaryotic Cell*, *9*, 1519–1530.
Tawk, L., Dubremetz, J. F., Montcourrier, P., Chicanne, G., Merezegue, F., Richard, V., et al. (2011). Phosphatidylinositol 3-monophosphate is involved in *Toxoplasma* apicoplast biogenesis. *PLoS Pathogens*, *7*, e1001286.
Tomova, C., Humbel, B. M., Geerts, W. J., Entzeroth, R., Holthuis, J. C., & Verkleij, A. J. (2009). Membrane contact sites between apicoplast and ER in *Toxoplasma gondii* revealed by electron tomography. *Traffic*, *10*, 1471–1480.
Tonkin, C. J., Roos, D. S., & McFadden, G. I. (2006). N-terminal positively charged amino acids, but not their exact position, are important for apicoplast transit peptide fidelity in *Toxoplasma gondii*. *Molecular and Biochemical Parasitology*, *150*, 192–200.
Tonkin, C. J., Struck, N. S., Mullin, K. A., Stimmler, L. M., & McFadden, G. I. (2006). Evidence for Golgi-independent transport from the early secretory pathway to the plastid in malaria parasites. *Molecular Microbiology*, *61*, 614–630.
Vaishnava, S., Morrison, D. P., Gaji, R. Y., Murray, J. M., Entzeroth, R., Howe, D. K., et al. (2005). Plastid segregation and cell division in the apicomplexan parasite *Sarcocystis neurona*. *Journal of Cell Science*, *118*, 3397–3407.
Vaishnava, S., & Striepen, B. (2006). The cell biology of secondary endosymbiosis—How parasites build, divide and segregate the apicoplast. *Molecular Microbiology*, *61*, 1380–1387.
van Dooren, G. G., Kennedy, A. T., & McFadden, G. I. (2012). The use and abuse of heme in apicomplexan parasites. *Antioxidants & Redox Signaling*, *17*, 634–656.
van Dooren, G. G., Marti, M., Tonkin, C. J., Stimmler, L. M., Cowman, A. F., & McFadden, G. I. (2005). Development of the endoplasmic reticulum, mitochondrion and apicoplast during the asexual life cycle of *Plasmodium falciparum*. *Molecular Microbiology*, *57*, 405–419.
van Dooren, G. G., Reiff, S. B., Tomova, C., Meissner, M., Humbel, B. M., & Striepen, B. (2009). A novel dynamin-related protein has been recruited for apicoplast fission in *Toxoplasma gondii*. *Current Biology*, *19*, 267–276.
van Dooren, G. G., Schwartzbach, S. D., Osafune, T., & McFadden, G. I. (2001). Translocation of proteins across the multiple membranes of complex plastids. *Biochimica et Biophysica Acta*, *1541*, 34–53.
van Dooren, G. G., Stimmler, L. M., & McFadden, G. I. (2006). Metabolic maps and functions of the *Plasmodium* mitochondrion. *FEMS Microbiology Reviews*, *30*, 596–630.
van Dooren, G. G., & Striepen, B. (2013). The algal past and parasite present of the apicoplast. *Annual Review of Microbiology*, *67*, 271–289.
van Dooren, G. G., Su, V., D'Ombrain, M. C., & McFadden, G. I. (2002). Processing of an apicoplast leader sequence in *Plasmodium falciparum* and the identification of a putative leader cleavage enzyme. *The Journal of Biological Chemistry*, *277*, 23612–23619.
van Dooren, G. G., Tomova, C., Agrawal, S., Humbel, B. M., & Striepen, B. (2008). *Toxoplasma gondii* Tic20 is essential for apicoplast protein import. *Proceedings of the National Academy of Sciences of the United States of America*, *105*, 13574–13579.
Varadharajan, S., Dhanasekaran, S., Bonday, Z. Q., Rangarajan, P. N., & Padmanaban, G. (2002). Involvement of delta-aminolaevulinate synthase encoded by the parasite gene in de novo haem synthesis by *Plasmodium falciparum*. *The Biochemical Journal*, *367*, 321–327.
Vaughan, A. M., O'Neill, M. T., Tarun, A. S., Camargo, N., Phuong, T. M., Aly, A. S., et al. (2009). Type II fatty acid synthesis is essential only for malaria parasite late liver stage development. *Cellular Microbiology*, *11*, 506–520.
Vollmer, M., Thomsen, N., Wiek, S., & Seeber, F. (2001). Apicomplexan parasites possess distinct nuclear-encoded, but apicoplast-localized, plant-type ferredoxin-NADP+ reductase and ferredoxin. *The Journal of Biological Chemistry*, *276*, 5483–5490.

Waller, R. F., Gornik, S. G., Koreny, L., & Pain, A. (2016). Metabolic pathway redundancy within the apicomplexan-dinoflagellate radiation argues against an ancient chromalveolate plastid. *Communicative & Integrative Biology*, *9*, e1116653.

Waller, R. F., Keeling, P. J., Donald, R. G. K., Striepen, B., Handman, E., Lang-Unnasch, N., et al. (1998). Nuclear-encoded proteins target to the plastid in *Toxoplasma gondii* and *Plasmodium falciparum*. *Proceedings of the National Academy of Sciences of the United States of America*, *95*, 12352–12357.

Waller, R. F., Reed, M. B., Cowman, A. F., & McFadden, G. I. (2000). Protein trafficking to the plastid of *Plasmodium falciparum* is via the secretory pathway. *The EMBO Journal*, *19*, 1794–1802.

Weber, A. P., Linka, M., & Bhattacharya, D. (2006). Single, ancient origin of a plastid metabolite translocator family in Plantae from an endomembrane-derived ancestor. *Eukaryotic Cell*, *5*, 609–612.

Williamson, D. H., Preiser, P. R., Moore, P. W., McCready, S., Strath, M., & Wilson, R. J. (2002). The plastid DNA of the malaria parasite *Plasmodium falciparum* is replicated by two mechanisms. *Molecular Microbiology*, *45*, 533–542.

Wilson, R. J., Denny, P. W., Preiser, P. R., Rangachari, K., Roberts, K., Roy, A., et al. (1996). Complete gene map of the plastid-like DNA of the malaria parasite *Plasmodium falciparum*. *Journal of Molecular Biology*, *261*, 155–172.

Woo, Y. H., Ansari, H., Otto, T. D., Klinger, C. M., Kolisko, M., Michalek, J., et al. (2015). Chromerid genomes reveal the evolutionary path from photosynthetic algae to obligate intracellular parasites. *eLife*, *4*, e06974.

World Health Organization. (2016). *2016 World Malaria report.* Geneva, Switzerland: World Health Organisation.

Yeh, E., & DeRisi, J. L. (2011). Chemical rescue of malaria parasites lacking an apicoplast defines organelle function in blood-stage *Plasmodium falciparum*. *PLoS Biology*, *9*, e1001138.

Young, M. J., Bay, D. C., Hausner, G., & Court, D. A. (2007). The evolutionary history of mitochondrial porins. *BMC Evolutionary Biology*, 7, 31.

Yu, M., Kumar, T. R., Nkrumah, L. J., Coppi, A., Retzlaff, S., Li, C. D., et al. (2008). The fatty acid biosynthesis enzyme FabI plays a key role in the development of liver-stage malarial parasites. *Cell Host & Microbe*, *4*, 567–578.

Zhang, B., Watts, K. M., Hodge, D., Kemp, L. M., Hunstad, D. A., Hicks, L. M., et al. (2011). A second target of the antimalarial and antibacterial agent fosmidomycin revealed by cellular metabolic profiling. *Biochemistry*, *50*, 3570–3577.

CHAPTER SIX

Chromerids and Their Plastids

Zoltán Füssy*, Miroslav Oborník*,†,‡,1
*Institute of Parasitology, Biology Centre of the Czech Academy of Sciences, České Budějovice, Czech Republic
†Faculty of Science, University of South Bohemia, České Budějovice, Czech Republic
‡Institute of Microbiology of the Czech Academy of Sciences, Centre Algatech, Třeboň, Czech Republic
[1]Corresponding author: e-mail address: obornik@paru.cas.cz

Contents

Abstract

Chrompodellids, a monophyletic group consisting of predatory colpodellids and the chromerid algae *Chromera velia* and *Vitrella brassicaformis*, are the closest known relatives to apicomplexan parasites. The photosynthetic plastids of *Chromera* and *Vitrella* are evolutionarily linked to apicoplasts, nonphotosynthetic plastids of apicomplexans, allowing the reconstruction of processes that shaped these organelles into their current form and function. Chromerid and apicomplexan plastids are derived from a rhodophyte endosymbiont and thus share common origin with the plastids of the "red lineage". Plastids in apicomplexans and chrompodellids play major roles in cellular biochemistry and dependence on plastid pathways underlies the organelle's essentiality in most nonphotosynthetic lineages. The architecture and gene content of plastid genomes differ significantly in the two described chromerids, as do their nuclear and mitochondrial genomes. Life cycle complexity also varies in chromerids, possibly reflecting adaptive mechanisms in their respective ecological niches. Finally, accumulating evidence suggests a tertiary endosymbiosis with an ochrophyte symbiont was the origin of current plastids in chromerid algae and apicomplexan parasites.

Advances in Botanical Research, Volume 84
ISSN 0065-2296
http://dx.doi.org/10.1016/bs.abr.2017.07.001

1. INTRODUCTION

Apicomplexans represent one of the most successful obligate parasites of animals including humans. The presence of a relic nonphotosynthetic plastid (termed the apicoplast) in their cells implies that these deadly parasites evolved from a phototrophic ancestry, which was corroborated by the discovery of the chromerids *Chromera velia* (Moore et al., 2008) and *Vitrella brassicaformis* (Oborník et al., 2012). These coral-associated algae from Australia are the closest known phototrophic relatives to parasitic Apicomplexa. Each known chromerid species possesses multimembrane-bound complex plastids that are inferred to share a common origin with the apicoplast of apicomplexans (Janouškovec, Horák, Oborník, Lukeš, & Keeling, 2010; Moore et al., 2008; Oborník, Janouškovec, Chrudimský, & Lukeš, 2009; Oborník & Lukeš, 2013, 2015). However, *Chromera* and *Vitrella* are polyphyletic and only form a monophyletic clade when grouped with the free-living predatory colpodellids (Fig. 1); members of the resulting clade are referred to as "chrompodellids" (Janouškovec et al., 2015). Chrompodellids are the sister group to all apicomplexans, with *Vitrella* branching at the root and *Chromera* branching near the crown predatory genera *Voromonas* and *Colpodella*, suggesting that photosynthesis was lost at least three times independently in apicomplexans and apicomplexan-related lineages (Fig. 1; Janouškovec et al., 2015). Based on sequence evidence, a nonphotosynthetic plastid is present at least in the colpodellids *Voromonas pontica*, *Alphamonas edax*, and *Colpodella angusta* (Gile & Slamovits, 2014; Janouškovec et al., 2015). While many apicomplexans are known, there is also a potentially prominent group of apicomplexan-related lineages that occupy various environmental niches; some may contain photosynthetic plastids, but others are potentially heterotrophs or parasites (Janouškovec, Horák, Barott, Rohwer, & Keeling, 2012; Janouškovec et al., 2015). Interestingly, only two chromerid algae are available in culture in spite of the fact that almost a decade has passed since the groups' discovery (Moore et al., 2008; Oborník et al., 2012).

Together with dinoflagellates and perkinsids, apicomplexans and chrompodellids constitute the group Myzozoa (Cavalier-Smith & Chao, 2004; Fig. 1) mainly characterized by the presence of a particular feeding structure termed the apical complex (Morrissette & Sibley, 2002; Okamoto & Keeling, 2014) and a rhodophyte-derived plastid. About half of the dinoflagellate species contain a multimembrane-bound photosynthetic plastid, but

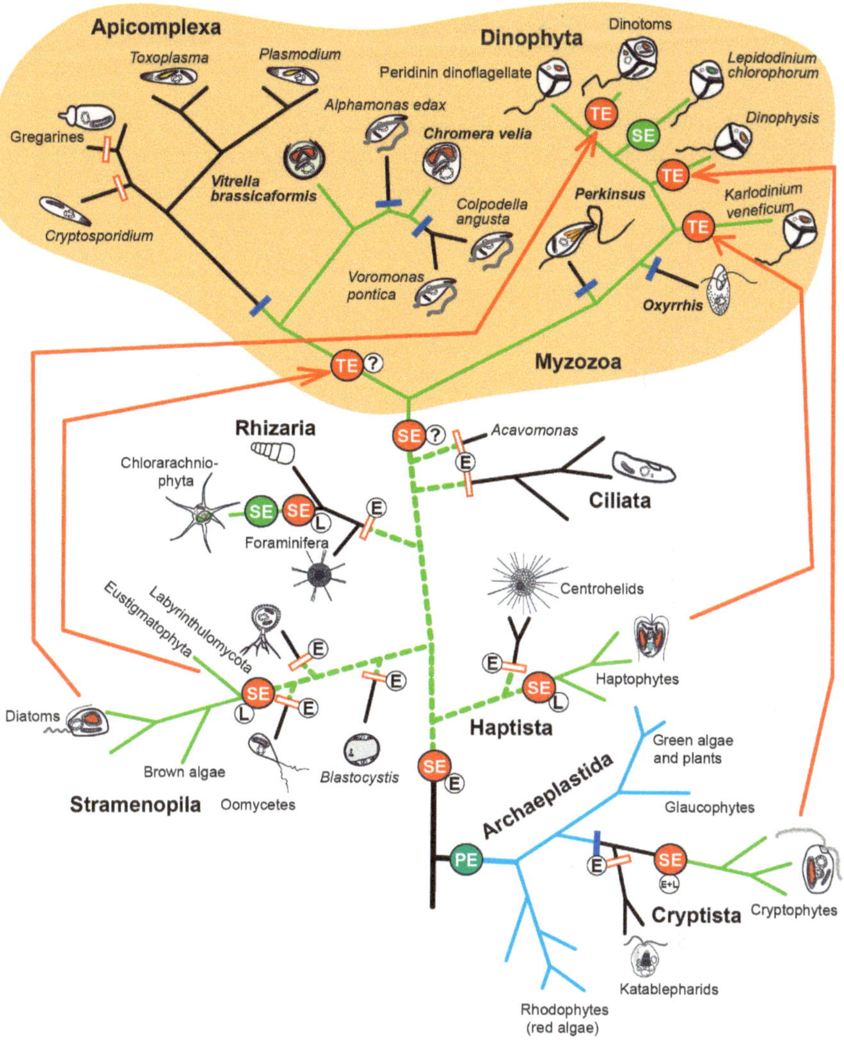

Fig. 1 *Acquisitions of plastids in photoautotrophic eukaryotes based on recent host phylogenomic analyses.* The scheme depicts hypotheses on possible plastid endosymbioses in Archaeplastida and lineages with rhodophyte-derived plastids. The tree is roughly based on current phylogenomic analyses (Burki et al., 2016; He, Sierra, Pawlowski, & Baldauf, 2016). *Coloured lines* indicate photoautotrophic lineages, *full lines* represent late plastid acquisition scenarios (i.e. where only crown groups are phototrophs but early lineages are primary heterotrophs), while *dashed lines* represent early plastid acquisition scenario (i.e. a modified chromalveolates hypothesis with Cryptista branching within Archaeplastida). *Colours of circles* indicate if the plastid was acquired from the *red* or *green* lineage. PE/SE/TE—endosymbiosis order, i.e., primary/secondary/tertiary. Note that the order of endosymbiosis cannot be qualified without knowing the

(Continued)

photosynthesis was lost numerous times in the remaining dinoflagellate lineages (Janouškovec et al., 2017; Nuismer & Otto, 2004; Taylor, 1987). Therefore, Myzozoa represents an important benchmark to study processes linked to plastid acquisitions and losses of plastids from the perspective of evolutionary biology. Additionally, the relict apicomplexan plastid (the apicoplast) is hailed as a potential drug target in these harmful parasites because this organelle contains unique metabolic pathways essential for the parasite survival that have no counterpart in the human host (McFadden & Roos, 1999).

Chromerids are also less evolutionarily divergent than phototrophic dinoflagellates and have been studied in an effort to reconstruct the ancestral state of the entire clade of Myzozoa. In this review, we summarize the current knowledge on the biology and plastids of chromerids, which we extend to colpodellids where data are available. We also discuss the emerging hypothesis that the original myzozoan plastid was replaced in chromerids and apicomplexans via horizontal transfer of a plastid from eustigmatophytes.

2. MORPHOLOGY AND LIFE CYCLES OF CHROMPODELLIDS

As is apparent from the phylogeny of chrompodellids (Fig. 1; Janouškovec et al., 2015), chromerid algae *C. velia* and *V. brassicaformis* are not directly related. Accordingly, their morphology and life cycles substantially differs; the predominant life stage of *Chromera* is represented by the immotile vegetative cells (autospores). These are coccoid, 5–7 μm in diameter, bounded by a thick and resistant cell wall (Fig. 2A). The alga reproduces by simple binary division resulting in autosporangia with 2–4 autospores covered by an additional membrane (Moore et al., 2008; Oborník et al., 2011). In addition to autospores, *Chromera* also forms zoosporangia (up to 15 μm in diameter) containing from 2 to 10 (always an even number)

Fig. 1—Cont'd exact sequence of the plastid acquisitions in the red-derived plastid lineages; our marking essentially reflects a scenario where all red-derived plastids were independently acquired from red algae. *Blue* bars indicate confirmed losses of photosynthesis, *red-lined* bars indicate losses of plastids (no plastid-like organelles have been found in these lineages). E or L indicate evolutionary events necessitated by the early and late plastid acquisition scenarios (E + L under both scenarios, as cryptophytes seem to branch within Archaeplastida with an ancestral primary plastid), question marks indicate events discussed in this chapter.

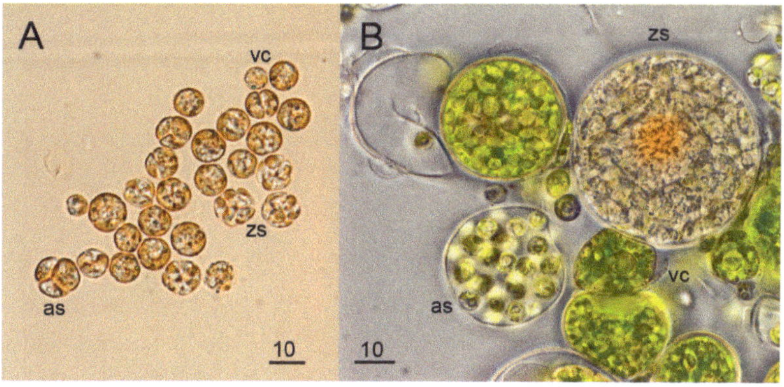

Fig. 2 Light microscopy images of cultures of (A) *Chromera velia* and (B) *Vitrella brassicaformis*. Vegetative cells (vc), autosporangia (as), and zoosporangia (zs) can be recognized in both species. Scale bars in μm.

motile zoospores in a process strongly resembling schizogony, the process of asexual formation of trophozoites in apicomplexan parasites (Oborník, Kručinská, & Esson, 2016), and the intracyst division of *Colpodella vorax* and *Co. tetrahymenae* (see later; Brugerolle, 2002; Cavalier-Smith & Chao, 2004). Zoospores morphologically resemble colpodellids, which also appear as elongated cells with two heterodynamic flagella at the anterior apex (Fig. 3). The zoospores of *C. velia* are equipped with a pseudoconoid, a structure similar to the tubular component of the apical complex (i.e. the conoid) of some apicomplexans and colpodellids. Sexual reproduction has not been observed for *C. velia*, and the zoospores presumably transform into coccoid cells (Oborník et al., 2011). The purpose of the zoospores is unknown, but they might facilitate dispersal into new regions of the biofilm on the coral reef surface. Resistant cysts formed under adverse environmental conditions remain viable for years. *C. velia* apparently uses closed mitosis, keeping the nuclear envelope intact throughout the division cycle (our unpublished data), similarly to other apicomplexans (Francia & Striepen, 2014). *C. velia* cell contains a typical eukaryotic nucleus (in contrast to dinoflagellates, chromosomes are not condensed during the entire life cycle), a single plastid, and several small mitochondria (Figs 3 and 5). Typical apicomplexan ultrastructural characters are also found, such as cortical alveoli supported with a subsheet of microtubules and a micropore (Oborník et al., 2011).

In comparison, the size of vegetative cells of *Vitrella* varies substantially during development. Starting from 3 μm diameter when released from an

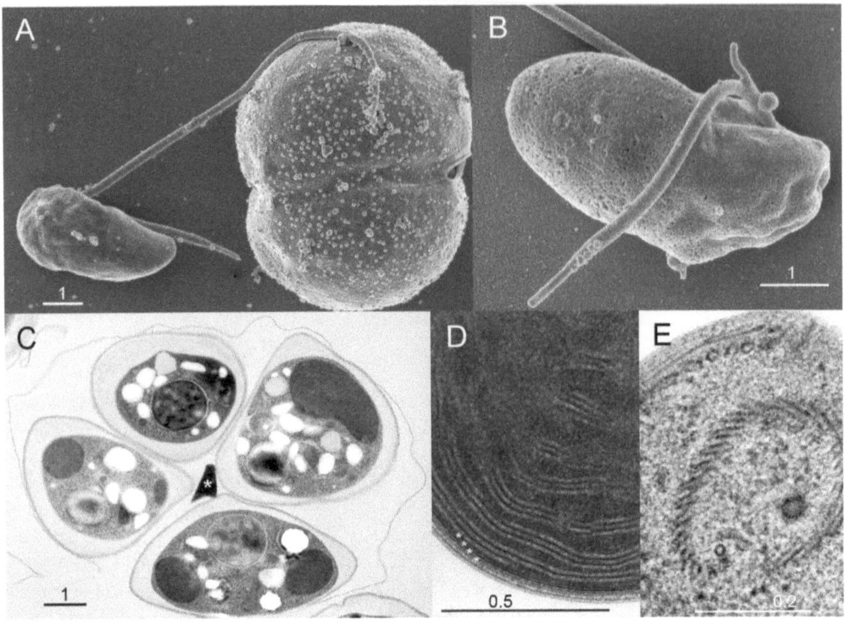

Fig. 3 Scanning electron micrographs (SEM) of *C. velia*: (A) zoospore (*left*) and autosporangium (*right*) containing two autospores and (B) zoospore with two heterodynamic flagella with a typical finger-like projection on the shorter flagellum (C). Transmission electron micrographs (TEM) of *C. velia*: (C) autosporangium with four autospores, asterisk indicates the residual body; (D) plastid surrounded by four membranes (marked by asterisks) and thylakoid organized in stacks of three; and (E) the pseudoconoid. Scale bars in μm.

autosporangium, autospores grow up to the diameter of 40 μm and then transform into sporangia containing dozens of autospores or zoospores (Figs 2B and 4). The vegetative autospores and sporangia are bounded by a very thick laminar cell wall (Oborník et al., 2012). Zoospores in *Vitrella* are of two types; one type is formed by budding from the central mother cell and exhibits extracytoplasmic flagellar assembly (Füssy, Masařová, Kručinská, Esson, & Oborník, 2017), while the second type of zoospores develops intracytoplasmic axonemes and flagella are ejected from the cell upon spore maturation and release (Oborník et al., 2012). Both types apparently lack a pseudoconoid. Numerous fusions of zoospores have been observed in *Vitrella*, possibly representing the sexual stage of the life cycle (Füssy et al., 2017; Oborník et al., 2016). The cells of *V. brassicaformis* contain a typical eukaryotic nucleus, a large and reticulated mitochondrion, and a

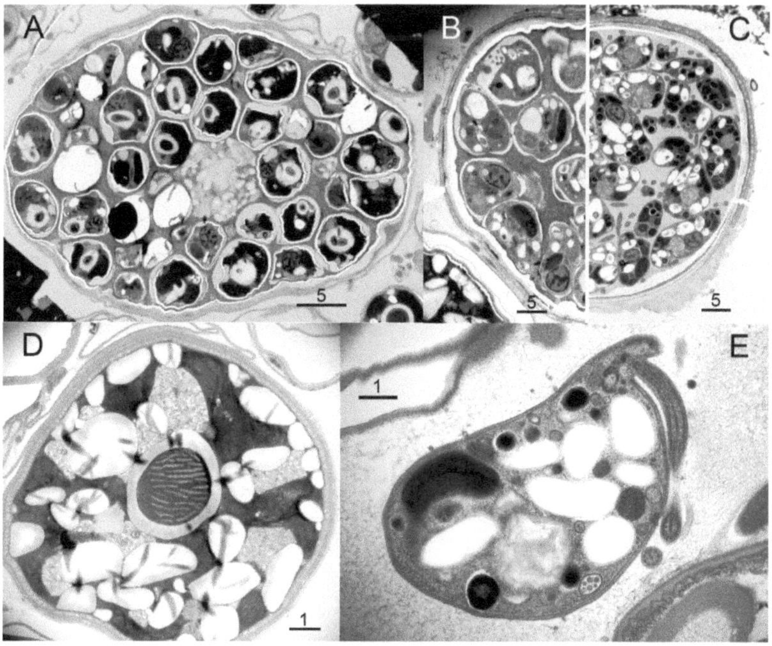

Fig. 4 Transmission electron micrographs of *V. brassicaformis*: (A) autosporangium containing numerous autospores with conspicuous pyrenoids; (B) zoosporangium containing zoospores with intracytoplasmic axonemes; (C) zoosporangium containing zoospores with extracellular formation of flagella; (D) vegetative cell with a central pyrenoid; and (E) zoospore. Scale bars in μm.

single large plastid (Figs 4 and 5). Cortical alveoli and subpellicular microtubules are also present (Oborník et al., 2012).

Although *Chromera* and *Vitrella* were isolated from inside their associated corals by procedures usually used to isolate intracellular symbionts such as dinoflagellates of the genus *Symbiodinium* (Moore et al., 2008), the symbiotic character of chromerids remains unclear. Plastid sequences closely related to *Chromera* are found infrequently in environmental data from coral reefs and apparently reside on coral surfaces (Janouškovec et al., 2012). On the other hand, *Chromera* was found in tissues of coral larvae of the genus *Acropora* (Cumbo et al., 2013). *Vitrella*-related sequences are quite abundant in thrombolites, coral reefs, and calcifying environments around the world (Janouškovec et al., 2012, 2015).

Not much is known about the biology of colpodellids. They are biflagellated, 5–20 μm long, unicellular organisms with an open conoid and rhoptries (extrusive organelles) used to hunt for other eukaryotes. There

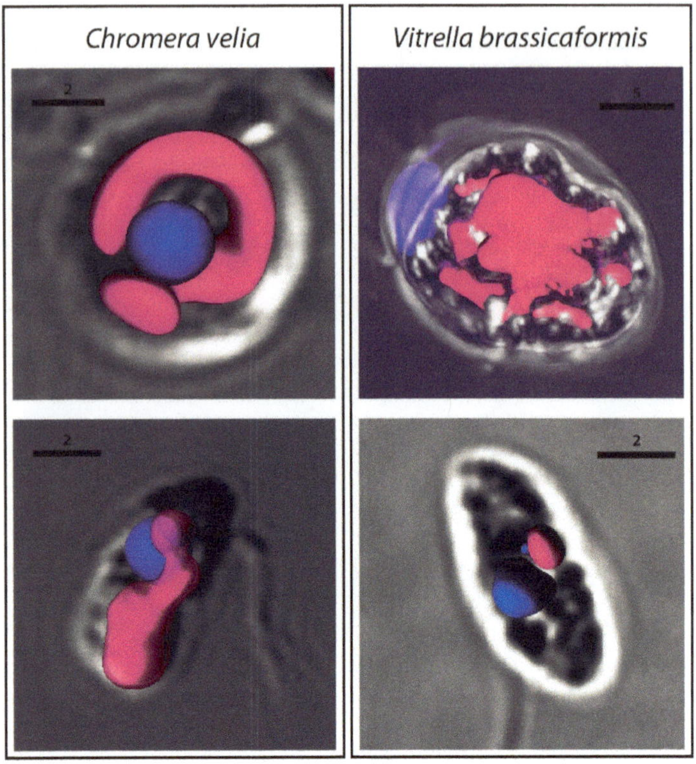

Fig. 5 Plastids in autospores (*upper*) and zoospores (*lower*) of *C. velia* and *V. brassicaformis*. 3D renderings of *Z*-stacked confocal scans, overlaid with visible light pictures to indicate the cells' shape; nuclei in *blue* and plastids in *purple*. Note that DAPI fails to enter *Vitrella* autospores; therefore only the cell wall is stained. Scale bars in μm.

are 11 described species whose morphological diversity generated uncertainty regarding their phylogenetic position (Cavalier-Smith & Chao, 2004; Simpson & Patterson, 1996). Several accessions clustering with colpodellids appeared in environmental data from various environments, including seawater, freshwater, soil, microaerobic, and animal samples (Janouškovec et al., 2012). Transcriptomic data suggest that colpodellids harbour a nonphotosynthetic plastid (Gile & Slamovits, 2014; Janouškovec et al., 2015). After feeding, colpodellids internalize their flagella, encyst, and divide into tetrads, a process similar to zoospore development in *C. velia*. Conjugation of cells occurs after excystment, which suggests sexual behaviour (Brugerolle, 2002; Simpson & Patterson, 1996), but an absence of genomic data prohibits drawing accurate conclusions.

3. CHROMERID PLASTID ULTRASTRUCTURE AND PIGMENTATION

Double-membrane-bound primary plastids of the phylum Archaeplastida, consisting of primary algae (glaucophytes, rhodophytes, and chlorophytes) and streptophytes, arose via endosymbiosis with a cyanobacterium. Plastids are, however, far more widespread in eukaryotes, and most phototrophic lineages emerged upon higher-order endosymbiotic events where both symbiotic partners were eukaryotes, one devouring the other. The algal symbiont and the host can be referred to as endosymbiont and exosymbiont, respectively. In comparison to primary plastids, secondary or tertiary plastids are surrounded by more membranes, reflecting the eukaryotic origin of the endosymbiont (reviewed in Archibald, 2015; Gould, Waller, & McFadden, 2008; Keeling, 2013).

Plastids of both chromerid algae, *C. velia* and *V. brassicaformis*, are surrounded by four-membrane envelopes like apicoplasts (Janouškovec et al., 2010; Moore et al., 2008; Oborník et al., 2011). In *C. velia*, the plastid is cone shaped, held in shape by the internal cilia anchored at the apex of the cell. Thylakoid lamellae are stacked in sets of three or more (Fig. 3D). The single plastid of *V. brassicaformis* has multiple lobes and a conspicuous pyrenoid surrounded by a cup-shaped starch sheath (Fig. 4). Long thylakoids stack in sets of three. Plastids of motile zoospores of both *C. velia* and *V. brassicaformis* are ovoid and reduced in size but still contain thylakoids. Chlorophyll autofluorescence seems less abundant than in plastids of autospores, suggesting photosynthesis is not an effective process at this life stage (Fig. 5).

Principal photosynthetic pigments in *Chromera* and *Vitrella* include chlorophyll *a*, violaxanthin, and β-carotene. Furthermore, *Chromera* produces a novel type of isofucoxanthin, while *Vitrella* relies on vaucheriaxanthin (Moore et al., 2008; Oborník et al., 2012). Both chromerids lack chlorophyll *c*, the hallmark pigment of rhodophyte-derived secondary plastids (Oborník et al., 2012). Nevertheless, it was shown that chlorophyll *c* is also absent from eustigmatophytes (Lubián et al., 2000) and the xanthophyte *Xanthonema debile* (Gardian, Tichý, & Vácha, 2011). *Chromera* accumulates high amounts of violaxanthin that participate in light harvesting (Kotabová, Kaňa, Jarešová, & Prášil, 2011). *C. velia* also exhibits efficient nonphotochemical quenching (NPQ) and de-epoxidation, which might partially be attributable to high violaxanthin content (Kotabová et al., 2011). The roles of *V. brassicaformis* pigments in light harvesting and NPQ are currently unknown.

For a photosynthetically naive host, plastid acquisition has the obvious advantage of permitting the harvest of light energy and its conversion to biomass. Still, even well-established endosymbioses do not always persist as permanent associations; sometimes plastids (or some of their functions) are lost, or the organelle is replaced in a serial manner with another phototrophic endosymbiont (Fig. 1; e.g. Keeling, 2013). Consistent with their non-photosynthetic roles (discussed later), apicoplasts are inconspicuous colourless organelles lacking thylakoids. For similar reasons we have no ultrastructural data on the plastid of colpodellids. Interestingly, the number of membranes of the apicoplast was a subject of passionate debate. Reconstruction of apicoplast ultrastructure in the malaria parasite *Plasmodium falciparum* based on serial section transmission electron microscopy (TEM) proposed a model with three membranes surrounding the apicoplast (Hopkins et al., 1999). Intriguingly, additional membrane layers were found associated with some regions of the apicoplast (Hopkins et al., 1999). Moreover, the apicoplast was shown to form permanent contact with the singular mitochondrion during the entire life cycle of *Plasmodium* (Ralph, Foth, Hall, & McFadden, 2004a). In contrast, other researchers proposed the presence of four membranes surrounding all apicoplasts in Apicomplexa (Lim & McFadden, 2010). This inconsistency likely arose not only due to the ambiguous visibility of membranes in the TEM micrographs of *Plasmodium* but also due to the complex structure of its apicoplast envelope. The analysis of serial ultrathin sections of tachyzoites of *Toxoplasma gondii* showed that the four individual layers might not surround the entire apicoplast in *T. gondii*, and a pocket-like invagination into the four-membrane-bound apicoplast was discovered (Köhler, 2005). The author of this study even suggests a primary origin for the *T. gondii* apicoplast, which we see unlikely considering phylogenetic data (e.g. Denny, Preiser, Williamson, & Wilson, 1998; Janouškovec et al., 2010; Lang-Unnasch, Reith, Munholland, & Barta, 1998; Oborník et al., 2002) as well as the presence of compound targeting presequences for importing the nuclear-encoded proteins into the apicoplast (Roos et al., 1999; Waller et al., 1998). Currently, it is mostly agreed upon that four membranes surround the apicoplast (e.g. Lemgruber et al., 2013), similar to the plastid of chromerids. Locally occurring membrane folds cannot, however, be fully excluded.

4. DIVERGENT PLASTID GENOMES OF CHROMERID ALGAE

C. velia possesses a highly divergent linear-mapping plastid genome of 121 kb with inverted repeats at the ends. Contained in the plastid genome are 74 highly divergent AT-rich protein-coding genes (Table 1), from which two, *psaA* and *atpB*, are split into two cistrons within their coding regions; these cistrons are separately transcribed, translated, and incorporated into the functional photosystem I and the plastid ATPase complex, respectively (Janouškovec et al., 2013). Linear plastid genomes, though abundant in plant meristems, have been mostly considered an intermediate during replication of circular plastid genomes, but this topology is quite common in mitochondria (Bendich, 2007; Leung & Wong, 2009; Smith & Keeling, 2015), and could eventually be observed in other plastids than that of *C. velia*. Light-harvesting complex proteins exhibit a divergent nature and are frequently lost (our unpublished data). Still, *Chromera* is capable of performing highly efficient photosynthesis in various light conditions (Kotabová et al., 2011), suggesting that the evolutionary erosion does not necessarily take its toll on the organelle's functionality. The translation apparatus of the *C. velia* plastid uses a noncanonical code for tryptophan (UGA instead of UGG). The use of UGA for tryptophan is uncommon for plastid-encoded proteins and has not been reported for any apicomplexan lineage other than coccidians (Moore et al., 2008) (but they are known from plants, see Beier & Grimm, 2001); even there the usage of the UGA code varies for different genera and transcripts. For instance, 17 out of 28 of plastid-encoded proteins are translated using noncanonical UGA for tryptophan in *T. gondii*, while it is only 7 in *Eimeria tenella* (Cai, Fuller, McDougald, & Zhu, 2003; Moore et al., 2008; Wilson, 2002).

In contrast, *Vitrella* plastid contains a highly compacted circular-mapping GC rich 85 kb genome with 81 protein coding genes (Table 1); this genome is highly syntenic with that of *P. falciparum* apicoplast, with only four inversions (Janouškovec et al., 2010). The plastid genome of *V. brassicaformis* lacks any genetic curiosity found in *Chromera;* it uses canonical coding, and *psaA* and *atpB* represent single cistrons. Altogether, there are hardly any features typical for chromerid plastid genomes, mirroring the many ways evolutionary forces shape this group of eukaryotes. It seems like the only notable character typical for both plastid genomes of *Vitrella* and

Table 1 Characteristics of Nuclear, Plastid, and Mitochondrial Genomes of *Chromera velia* and *Vitrella brassicaformis*

	Chromera	***Vitrella***
Nuclear genomes		
Genome size (Mb)	193.6	72.7
% GC	49	58
Number of genes	26,112	22,817
Median length of gene	1218	1199
Median number of introns per gene	2	6
% of genes with introns	75.73	94.75
Total intron length (Mb)	60.23	16.41
Number of TE	4366	686
Total TE length (Mb)	30.2	1.5
Total TE length (%)	16	2
Plastid genomes		
Genome size (kb)	121.2	85.5
% GC	36.63	47.74
Noncanonical coding	Trp (W) UGA	—
Number of (protein coding) genes[a]	74	81
Genome architecture	linear	circular
Mitochondrial genomes		
Number of protein coding genes	2	3
Number of rRNA gene fragments	8	unknown
Genome architecture	Heterogeneous linear molecules	Heterogeneous linear molecules

[a]According to Janouškovec et al. (2010).

Chromera is the structure of the ribosomal superoperon resembling that of rhodophyte and rhodophyte-derived plastids.

Plastid genomes of apicomplexan parasites are reduced in connection with the loss of photosynthesis. The genome is circular of 29, 6–39, 6 kb in length with 29–44 protein-coding genes and no traces of genes related to photosynthesis (Arisue & Hashimoto, 2015; Wilson et al., 1996). Secondary peridinin pigmented plastids of dinoflagellates host even more reduced genomes; however, this particular reduction poses no apparent functional limitations even in phototrophic species. Peridinin plastid genomes are composed of small circular molecules (1.8–3.3 kb in *Symbiodinium*), the so-called minicircles, which encode a mere 14 photosynthesis-related proteins, two ribosomal RNAs (rRNAs), and a few transfer RNAs (tRNAs) (Green, 2004; Howe, Nisbet, & Barbrook, 2008; Mungpakdee et al., 2014; Zhang, Green, & Cavalier-Smith, 1999); the remaining plastid proteins need to be imported from the cytoplasm to maintain the organellar functions. Plastid genomes of colpodellids might have been lost (Janouškovec et al., 2015).

5. NUCLEAR AND MITOCHONDRIAL GENOMES OF CHROMERIDS

Similarly to plastid genomes, also nuclear and mitochondrial genomes show gradual deterioration over the course of evolution in apicomplexans and chrompodellids. The nuclear genome of *C. velia* is considerably larger (193.6 Mb) than that of *V. brassicaformis* (72.7 Mb). This difference in sizes is caused by substantially higher incidence of transposable elements and longer introns in *C. velia* (altogether 90.4 Mb and 17.9 Mb in *Chromera* and *Vitrella*, respectively) (Table 1). The numbers of predicted proteins do not differ to such extent: 26,112 and 22,817 proteins were predicted in *Chromera* and *Vitrella*, respectively. Evolutionary analysis of apicomplexan genomes with the genomes of chromerids showed that a massive genes loss occurred since their common ancestor transited from phototrophy to parasitism (over 3800 genes were lost), while only few genes were gained during this transition (about 80) (Woo et al., 2015). In accordance, apicomplexan nuclear genomes are reduced in size compared to their photosynthetic siblings; coccidian genomes range between 55 and 80 Mb, *Plasmodium* genomes range between 20 and 25 Mb, piroplasmid genomes range between 6.5 and 9 Mb, and *Cryptosporidium* genomes are around 9 Mb (Cornillot et al., 2012; Kissinger & DeBarry, 2011). This leads us to think that genes currently functioning in parasitism-related processes of apicomplexans had likely been present in the common, phototrophic, ancestor of the entire lineage (Woo et al., 2015). Following the split of chrompodellids, further gene losses and gains occurred during the adaptation of apicomplexan parasites to specific hosts. Therefore, reductive evolution accompanies the transition to parasitism in Apicomplexa.

Mitochondrial genomes of Myzozoa are generally highly reduced; in apicomplexans they are linear homogeneous molecules of 6–11 kb, in dinoflagellates heterogeneous molecules of various arrangements are found. Myzozoan mitochondrial genomes encode only three proteins, and fragments of rRNA (Feagin, Mericle, Werner, & Morris, 1997; Nash, Nisbet, Barbrook, & Howe, 2008). Two of the proteins represent subunits of Complex II of the respiratory chain (cytochrome oxidase 1, *cox1*; cytochrome oxidase 3, *cox3*) and the third protein is a component of Complex III (cytochrome *b*, *cob*). While *V. brassicaformis* mitochondrial genome encodes an identical gene set, *C. velia* lost *cob*. Notably, mitochondrial genome was entirely lost in *Cryptosporidium*. These reductions in mitochondrial genomes came with concurrent modifications of the mitochondrial metabolism. Complex I is absent in all Myzozoa, and is functionally replaced by alternative NADH dehydrogenases (Flegontov et al., 2015). The respiratory chain of *Vitrella* conform this scheme. *Chromera* lost Complex III and apparently feeds electrons to alternative oxidase instead, and recruits L- and D-lactate: cytochrome *c* oxidoreductases (D-LDH, L-LDH) and L-galactono-1,4-lactone dehydrogenase (G14LDH) to generate membrane potential via Complex IV, effectively splitting the respiratory chain into two subchains (Flegontov et al., 2015; Oborník & Lukeš, 2015). The mitochondrion of *Cryptosporidium* lost its respiratory function and became a mitosome-like organelle with a sole function, to assemble Fe—S clusters (LaGier, Tachezy, Stejskal, Kutisova, & Keithly, 2003).

6. ROLES OF CHROMERID PLASTIDS IN CELLULAR METABOLISM

Plastid proteomes consist of (1) proteins encoded by the plastid genome and (2) proteins that are nuclear encoded; while the former are synthesized in the plastid stroma, the latter need to be imported into the organelle and therefore possess a targeting presequence (reviewed in Patron & Waller, 2007). Plastid-targeting presequences of primary algae (chlorophytes, rhodophytes, and glaucophytes) are termed chloroplast transit peptides and have no conserved primary sequence or length. Instead, they are characteristic by their amino-acid composition; they are richer in hydroxylated residues (serine and threonine), depleted in acidic residues (aspartic acid and glutamic acid), and contain high ratios of nonpolar amino-acids, especially alanine. Transit peptides of rhodophytes and glaucophytes contain an additional feature, phenylalanine located at, or within a few positions from, the N terminus (Patron & Waller, 2007). Nucleus-encoded plastid proteins of all higher-order plastid-bearing organisms possess a bipartite N-terminal presequence, and they are transported through the endomembrane system en route to the plastids, starting with the endoplasmic reticulum import (Bolte et al., 2009; Patron & Waller, 2007; Sheiner & Striepen, 2013). The bipartite presequence consists of a signal peptide followed by a transit peptide and each part of the presequence is cleaved off after the successful translocation into its respective compartment. The mature protein is then released into the plastid stroma where it might be subcompartmented based on other encoded motifs. The signal peptide parts of the bipartite presequences are indistinguishable from signal peptides of secretory proteins (Kilian & Kroth, 2005; Roos et al., 1999); therefore the following transit peptides are crucial to direct proteins further to their destination in the plastid. The features of transit peptides targeting to rhodophyte-derived plastids, including chromerid plastids and apicoplasts, are very similar to those of rhodophytes. Especially the phenylalanine motif is highly conserved in *Toxoplasma* and *Chromera* (Patron & Waller, 2007; Woehle, Dagan, Martin, & Gould, 2011), though none of the four identified full-length plastid proteins of the colpodellid *Voromonas* carried a phenylalanine at the N-terminus of their transit peptide, suggesting a released evolutionary pressure on this motif (Gile & Slamovits, 2014).

In spite the principal role of plastids is the conversion of the energy of light into matter, they also have important functions beyond photosynthesis. Plastids are deeply integrated into the cellular biochemistry, for example by tight coupling with mitochondrial energy metabolism in diatoms (Bailleul et al., 2015). Searching for proteins with plastid targeting presequence features allows a more or less complete picture of what role plastids play in the cellular metabolism in chromerids and apicomplexans. Also, some plastid-located pathways reveal interesting evolutionary stories.

Tetrapyrroles, such as heme and chlorophylls, are essential for life in the vast majority of organisms (Kořený et al., 2012); they facilitate electron transport in photosynthesis and oxidative phosphorylation. Therefore, living creatures must make tetrapyrroles autonomously or sequester them from outer sources. The first common precursor is synthesized differently in phototrophs and heterotrophs. While aplastidic (primary heterotrophic) eukaryotes and α-proteobacteria produce tetrapyrroles from succinyl-CoA and glycine as precursors, other organisms (Archaea, bacteria, and most eukaryotic phototrophs) create tetrapyrroles from glutamate (Oborník & Green, 2005). Localization of the pathway reflects this difference in precursors. Tetrapyrrole synthesis is mitochondrial–cytosolic in aplastidic eukaryotes, and entirely plastid located in phototrophic eukaryotes (Kořený & Oborník, 2011). Plastid tetrapyrrole synthesis appeared convergently in most phototrophic lineages, possibly due to an increased need for tetrapyrroles in this compartment and the possibility to generate glutamate in high amounts using the reductive power of NADPH produced during photosynthesis (glutamine synthase/glutamate synthase cycle). Reliance on plastid tetrapyrrole biosynthesis, in the long run, leads to the loss of mitochondrial–cytosolic pathway and a dependence of the host on plastid tetrapyrrole synthesis. Apicomplexan parasites and chromerids (and likely also colpodellids; Janouškovec et al., 2015) are so far the only known exceptions from this rule. Although they contain plastids, apicomplexans and chromerids synthesize tetrapyrroles from mitochondrial precursors like in aplastidic eukaryotes. In apicomplexans, next three or four steps (depending on the lineage) locate to apicoplast, and the pathway terminates in the mitochondrion. In chromerids, the rest of the reactions are supposedly plastid located (Kořený, Sobotka, Janouškovec, Keeling, & Oborník, 2011). Such unusual arrangement qualifies chromerids to be the only phototrophs on Earth known so far synthesizing chlorophyll from glycine instead of glutamate. At the same time, it is one of the most convincing evidence supporting the common origin of chromerids and apicomplexans and their plastids (Kořený, Oborník, & Lukeš, 2013; Kořený et al., 2011; Oborník & Lukeš, 2013).

Loss of photosynthesis is not directly linked to the loss of plastids. In fact, outright plastid losses might be exceedingly rare, and only three documented examples can be found in myzozoans: apicomplexan parasites of the genus *Cryptosporidium* (Zhu, Marchewka, & Keithly, 2000), eugregarine *Gregarina niphandroides*, an early branching apicomplexan related to *Cryptosporidium* (Toso & Omoto, 2007), and parasitic dinoflagellates of the genus *Hematodinium* (Gornik et al., 2015). This is because, upon integration, several biochemical pathways of the host and the endosymbiont functionally overlap and metabolic streamlining might bring about the selective loss of host or endosymbiont pathways. If in such intertwined metabolic situation the plastid contains exclusive biosynthetic steps, the organelle becomes indispensable even if its other functions are completely lost (Borza, Popescu, & Lee, 2005; Janouškovec et al., 2015; Neuhaus & Emes, 2000; Oborník et al., 2009; Ralph et al., 2004b; Slamovits & Keeling, 2008; Waller, Gornik, Kořený, & Pain, 2016). Further essential compounds that can be produced by plastids include isopentenyl pyrophosphate (IPP), fatty acids, and Fe—S clusters. Interestingly, cytosolic IPP biosynthesis was lost quite early after plastid acquisition in Myzozoa, leading to a plastid dependency universally in all living descendants (except for the earlier mentioned plastid-less Myzozoa) (Janouškovec et al., 2015, 2017; Waller et al., 2016). Plastid fatty acid synthesis complements cytosolic synthesis, and there is no apparent tendency to lose either pathway. Fe—S cluster assembly occurs in apicoplasts as IPP and fatty acid biosynthesis enzymes rely on these catalyst complexes (Janouškovec et al., 2015). Chromerids and colpodellids supposedly rely on all these plastid pathways (Janouškovec et al., 2015; Waller et al., 2016), although redundancy of fatty acid syntheses is not yet clearly understood in chromerids and no enzymes of the plastid fatty acid pathway are known from colpodellids. Apicoplast biosynthesis pathways have been reviewed elsewhere, and to some degree are lineage and stage specific (e.g. McFadden & Yeh, 2017; Pradel & Schlitzer, 2010).

In theory, organisms can resolve the metabolic restriction placed by the dependence on a plastid biosynthesis pathway might by obtaining a "repair set" of enzymes via serial plastid acquisition or horizontal gene transfer from other sources. Alternatively, plastid dependence might diminish if an external source of the essential compounds is ensured through parasitism or predation (Figueroa-Martinez, Nedelcu, Smith, & Reyes-Prieto, 2015; Gornik et al., 2015; Janouškovec et al., 2015; Kořený, Lukeš, & Oborník, 2010; Zhu et al., 2000).

7. AMBIGUITY OF CHROMERID PLASTID SEQUENCE-BASED PHYLOGENIES

High divergence of chromerid, apicomplexan, and dinoflagellate sequences complicates phylogenetic analyses. A dataset containing divergent sequences is prone to cause phylogenetic artefacts such as long-branch attraction. Nucleotide composition bias (e.g. AT-rich sequences) of plastid genomes may also influence the analyses. For instance, a phylogenetic analysis based on several plastid genes suggested that the apicoplast originated from a green lineage because the nonphotosynthetic plastid of the parasitic green alga *Helicosporidium* appeared as the closest relative (Lau, McElwain, Brayton, Knowles, & Roalson, 2009). Since many lines of evidence support a rhodophyte ancestry of apicoplasts (Janouškovec et al., 2010; Moore et al., 2008; Oborník et al., 2012), the reported affiliation with green plastids appears to be a phylogenetic artefact, perhaps resulting from analogous evolutionary forces acting on these genomes (Lukeš, Leander, & Keeling, 2009).

Still, accumulating molecular evidence suggests that the plastid of apicomplexans and chrompodellids could originate in a tertiary endosymbiosis of an ochrophyte (stramenopile alga), presumably a eustigmatophyte or a closely related lineage (Fig. 1; Oborník et al., 2012; Ševčíková et al., 2015). Such endosymbiosis would imply that dinoflagellate and apicomplexan/chrompodellid plastids are not directly related. In an earlier work, Janouškovec et al. (2010) performed phylogenetic analyses (using RAxML, PhyML-CAT, MrBayes, PhyloBayes) with a dataset of 34 plastid genes from two chromerids (*C. velia* and *V. brassicaformis*), ochrophytes (without eustigmatophytes), haptophytes, cryptophytes, and Archaeplastida. Here chromerid plastids appeared as a sister lineage to the ochrophyte plastids with a robust support (Fig. 6A; Janouškovec et al., 2010). After addition of apicomplexan or dinoflagellate sequences to this dataset (which resulted in the reduction of the dataset to 23 and 11 shared genes, respectively), chromerid plastids showed an affiliation with the added group (Janouškovec et al., 2010). However, the plastid genomes of apicomplexans, dinoflagellates, as well as *C. velia* are too divergent to be used for reliable phylogenies.

To reduce phylogenetic artefacts, Ševčíková et al. (2015) omitted these divergent taxa from analyses, and phylogenies of 69 conserved plastid genes showed an affinity of the plastid genomes of *Vitrella,* eustigmatophytes, and other ochrophytes (Fig. 6A; Ševčíková et al., 2015). Unfortunately, eustigmatophyte plastids also exhibited a long branch, and the support of clades including the *Vitrella* plastid changed dependent on the method used for phylogenetic reconstruction and fast-evolving site removal. While the maximum likelihood analysis (RAxML under GTR-GAMMA model) placed the *Vitrella* plastid to the root of eustigmatophyte plastids with the bootstrap support of >80%, the Bayesian approach (PhyloBayes under CAT + GTR model) inferred an unresolved position within ochrophytes. *Vitrella* began to branch sister to eustigmatophytes and *Ochromonas* when analysed in a set of 34 slow-evolving genes (Fig. 6A; Ševčíková et al., 2015). This topology had weak support under PhyloBayes, but the consistency of maximum likelihood suggests there might be phylogenetic signal supporting their

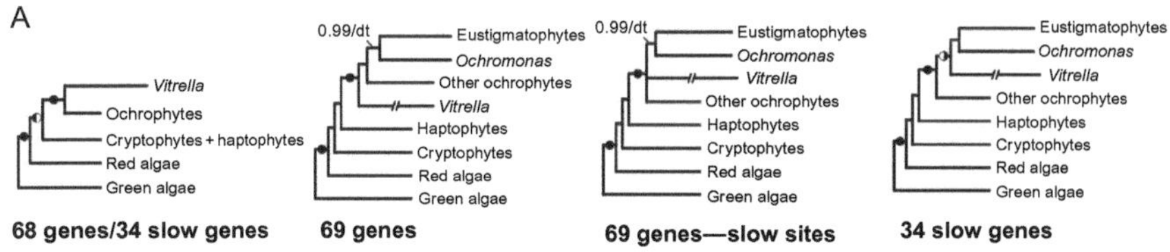

B

	Dinoflagellates*	Chromerids	Eustigmatophytes
Main pigments	Chlorophyll *a* + *c*, diadinoxanthin, peridinin	Chlorophyll *a*, violaxanthin, beta-carotene	Chlorophyll *a*, violaxanthin, beta-carotene
Plastid genome	Minicircles	1 Chromosome	1 Chromosome
Transcript polyuridylation	Yes	Yes	No
No. plastid membranes	3	4	4
Protein transport	SELMA independent	SELMA dependent	SELMA dependent
RuBisCO type	II	II	I

Fig. 6 *Features supporting common origin of chromerid and eustigmatophyte plastids.* (A) Schematics of plastid-encoded protein phylogenies based on Bayesian inferences. *Vitrella* branches sister to ochrophytes in the leftmost tree (fig. 5 in Janouškovec et al., 2010), while addition of eustigmatophytes and removal of fast-evolving sites and genes result in placing of *Vitrella* more robustly inside ochrophytes (fig. S8, fig. S9, fig. 2B, respectively, in Ševčíková et al., 2015). Only with slow-evolving gene set *Vitrella* was recovered sister to *Ochromonas* and eustigmatophytes by the Bayesian approach, though with low support. In contrast, maximum likelihood analysis recovered *Vitrella* sister to eustigmatophytes in all cases (note the "different topology" at the *Ochromonas*/eustigmatophytes node). Full circles mark highest support for nodes, while half circles mark nodes where only PhyloBayes/RAxML had highest support values. Branches are not in scale, // denotes a long branch for *Vitrella* (see Section 7). (B) Table summarizing complex traits supporting monophyly of chromerids with dinoflagellates or eustigmatophytes. *Characteristics shown for peridinin dinoflagellates supposedly bearing the original myzozoan plastid.

relationship. Given the possibility of the long-branch attraction of *Vitrella* and other myzozoans and eustigmatophytes, the results do not permit clear conclusions; screening for apicomplexan-related phototrophic lineages introduce less divergent members of the clade to such analyses (Janouškovec et al., 2012).

Nuclear gene phylogenies were more consistent in retrieving relationships between the myzozoan exosymbionts (Janouškovec et al., 2010, 2015; Moore et al., 2008; Oborník et al., 2009; Woehle et al., 2011; Woo et al., 2015). Nevertheless, nonvertically inherited genes (such as those transferred from the endosymbiont or accidentally from other genetic sources) may bring confusion into the origin of the apicomplexan/chrompodellid plastids. For instance, a handful of chromerid genes shows affinity to chlorophytes (Burki et al., 2012), though a previous work suggested a substantially larger green-algal contribution (Woehle et al.,

2011); the latter result was nevertheless falsified as an outcome of taxon sampling and an automated phylogenomic analysis. This point out that few endosymbiont-derived sequences can provide unequivocal phylogenies. Furthermore, several plastid-located genes of apicomplexans are horizontal acquisitions from bacteria, which might not be isolated cases (Janouškovec et al., 2015; Petersen et al., 2014). Horizontal gene transfer further lowers the number of genes obtained in Myzozoa directly from their endosymbiont. Considering a scenario whereby an original myzozoan plastid was serially replaced by a rhodophyte-derived plastid of eustigmatophytes or their relatives, phylogenomic position of concatenated chromerid nuclear-encoded plastid proteins might exhibit background noise and weak supports. On the other hand, weak support often hinders robust explanations of topologies of single genes as to their origin in either plastid (original or replacement). In the absence of reliable sequence data, investigation of rare evolutionary events and complex traits may bring an alternative to ambiguous topologies.

8. THE ORIGIN OF CHROMERID PLASTIDS

Dinoflagellates and chromerids share some plastid-related features which support the monophyly of their plastids (Fig. 6B). Localization patterns of common plastid pathways, such as IPP, fatty acid, and tetrapyrrole biosynthesis, are conserved among myzozoans, although lineage-specific differences are seen in details (Waller et al., 2016). Both groups perform polyuridylation of some plastid-encoded transcripts, and both groups encode Rubisco type II apparently acquired from a proteobacterium at the base of Myzozoa (Janouškovec et al., 2010). Interestingly, polyuridylation is also present in the tertiary plastid-bearing dinoflagellates, kareniaceans, which have haptophyte-derived plastids, suggesting that factors underlying this process were retargeted from the peridinin plastid during serial endosymbiosis (Dorrell & Howe, 2015). This is much similar to the retargeting of tetrapyrrole biosynthesis in the dinoflagellate *Lepidodinium* (Cihlář, Füssy, Horák, & Oborník, 2016), where the peridinin-pigmented plastid was functionally replaced after serial endosymbiosis with a chlorophyte (Elbrachter & Schnepf, 1996; Matsumoto et al., 2011). In the event of serial plastid replacement in apicomplexan/chrompodellid lineage, retargeting of polyuridylation (and Rubisco II) could plausibly explain why chromerids share these features with dinoflagellates.

Similar pigment composition could be indicative of a common origin of photosynthesis and photoprotection in chromerids and eustigmatophytes (Fig. 6). *Chromera* and *Vitrella* lack chlorophyll *c*, which is also absent from all investigated eustigmatophytes (Lubián et al., 2000). Both *Chromera* and the eustigmatophyte *Nannochloropsis* accumulate high amounts of violaxanthin that participate in light harvesting (Kotabová et al., 2011; Owens, Gallagher, & Alberte, 1987). At the same time, primary pigments in *Vitrella* (chlorophyll *a*, violaxanthin, vaucheriaxanthin, and β-carotene) are virtually identical to those found in *Nannochloropsis limnetica* (Oborník et al., 2012).

The number of plastid membranes in Myzozoa might be considered as an evolutionary trait. How many layers form the plastid envelope is dependent on the mechanism of engulfment, the actual number of membranes of the engulfed plastid, and subsequent membrane losses (Gould, Maier, & Martin, 2015; Petersen et al., 2014; Stiller et al., 2014). In principle, a phototrophic organelle can be acquired either by myzocytosis or phagocytosis, with differing number of plastid membranes (Bodył & Moszczyński, 2006). During myzocytosis, a predatory host attaches a primary algal prey and sucks its contents including plastids into a feeding vacuole (phagosome), leaving the prey cytoplasmic membrane hollow. The engulfed plastid is hence separated from the host cytoplasm only by the phagosomal membrane (Schnepf & Deichgraber, 1984) that could have transformed into the plastid outermost membrane resulting in a three-membrane-bound plastid (Bodył & Moszczyński, 2006). During phagocytosis, a host feeding vacuole engulfs an entire prey cell including its cytoplasmic membrane; in this case, the resulting plastid would be four-membrane-bound. Higher-order endosymbioses were reasoned to give rise to plastids with higher number of plastid membranes, gaining additional membranes depending on the mechanism of engulfment. However, dinoflagellates developed many strategies to maintain a photosynthetic organelle confined to a membrane-bound compartment. These strategies range from delayed phagocytosis of an intact algal cell (*Noctiluca*) to sophisticated extracting a double-membrane chloroplast directly into the cytosol whilst digesting the prey remainder (*Dinophysis*) (reviewed in Park, Kim, & Kim, 2014). Therefore, stripping off the extra plastid membranes during higher-order endosymbioses does not seem so implausible as some researchers argue (Gould et al., 2015). Consistently, the number of envelope layers might differ in dinoflagellates with the ancestral peridinin-pigmented plastids (three; Nassoury, Cappadocia, & Morse, 2003; Patron, Waller, Archibald, & Keeling, 2005) and dinoflagellates that underwent serial endosymbiosis or thrive on kleptoplastidy (e.g. two membranes in *Dinophysis*, four in *Lepidodinium*, five in dinotoms) (Eschbach, Speth, Hansmann, & Sitte, 1990; Hansen, Botes, & De Salas, 2007; Lucas & Vesk, 1990). Notably, fucoxanthin-pigmented plastids in kareniacean dinoflagellates are three-membrane-bound, in spite of its tertiary origin (which should yield five- to six-membrane-bound plastids) (Kite & Dodge, 1988; Steidinger, Truby, & Dawes, 1978). The exact mechanism of plastid sequestering is, therefore, crucial for the determination of its envelope's structure. The additional membrane of the apicoplast compared to the dinoflagellate peridinin plastids might be a result of different endosymbiont acquisition mechanisms, differential membrane loss in the apicomplexan vs dinoflagellate lineage, or different origin of their plastids, the apicoplast arising from a higher-order endosymbiosis (tertiary rather than secondary as in dinoflagellates). Although we cannot reliably conclude on independent origins of the apicoplast and the peridinin-pigmented plastid based solely on ultrastructural features, it is worth taking into account.

Differences in plastid membrane numbers seem to be of relevance here, as they codetermine the transport route of proteins. The endoplasmic reticulum in chromerids, dinoflagellates, and apicomplexans is not continuous with the outermost membrane of their plastids, which is the case in ochrophytes, cryptophytes, and haptophytes (reviewed in Gould et al., 2008). The membrane architecture of plastid envelopes and the protein translocation machinery must be established soon after an endosymbiotic event because any later changes would interfere with the functionality of the transport of plastid proteins (Bodył & Moszczyński, 2006). Consistently, in *Lepidodinium* and *Karlodinium* that underwent plastid replacement, plastid targeting presequences are deviant from those of peridinin dinoflagellates, suggesting an adaptation occurred to suit the needs of protein transport into the new compartment (Minge et al., 2010; Patron, Waller, & Keeling, 2006). Notably, all diatom symbionts of dinotoms are enclosed by a single-unit membrane (Eschbach et al., 1990; Pienaar, Sakai, & Horiguchi, 2007). Such envelope architecture possibly prevents transport of host nuclear-encoded proteins using the same secretory pathway as peridinin plastids and hinders a deeper integration of this endosymbiont (Hehenberger, Imanian, Burki, & Keeling, 2014; Imanian & Keeling, 2007).

Protein transport mechanism in dinoflagellates also differs from that in apicomplexans/chromerids, certainly reflecting dissimilarity of plastid envelopes in these lineages. Plastid-targeted proteins of peridinin dinoflagellates carry two major classes of targeting presequences (Patron et al., 2005). Both classes have the bipartite signal + transit peptide structure typical for rhodophyte-derived plastids (see earlier), but one class contains a hydrophobic domain downstream of the transit peptide that possibly serves as a membrane anchor (Patron et al., 2005). This architecture of plastid protein presequences is also present in the early branching myzozoans, *Perkinsus* and *Oxyrrhis* (Matsuzaki, Kuroiwa, Kuroiwa, Kita, & Nozaki, 2008; Slamovits & Keeling, 2008). In comparison, the anchoring transmembrane domain is missing in apicomplexans, diatoms, cryptophytes, and *Chromera* (Apt et al., 2002; Gould et al., 2006; Gruber et al., 2007; Patron & Waller, 2007; Waller et al., 1998; Woehle et al., 2011), as well as in eustigmatophytes (e.g. *Trachydiscus minutus*, our unpublished data). Furthermore, recent analyses failed to find in dinoflagellates components of the protein import complex (symbiont-specific ER-associated degradation (ERAD)-like machinery: SELMA) universally found in other organisms with rhodophyte-derived plastids, i.e., ochrophytes, cryptophytes, haptophytes, apicomplexans and *Chromera* (Petersen et al., 2014; Stork et al., 2012). Some of SELMA components, Der1, Ufd1, Hrd1, and Cdc48, are encoded in the nucleomorph of cryptophytes (Douglas et al., 2001; Lane et al., 2007; Sommer et al., 2007), suggesting they genuinely derived from the engulfed red alga. SELMA is essential for protein transport across the second outermost plastid membrane, so its absence from dinoflagellates could coincide with the loss of one of the plastid membranes. Also, Petersen et al. (2014) found robust discrepancies in plastid marker origins of *Chromera* and peridinin-pigmented dinoflagellates, and therefore propose separate origins of apicomplexan/chrompodellid and dinoflagellate plastids. This proposal is indeed in line with differences seen in organellar genome organization, morphology, and protein import.

There are, however, shared features of dinoflagellate and apicomplexan/chrompodellid plastids that we cannot neglect, such as shared horizontal gene transfers, the patterns of metabolic pathways localization, and the unusual polyuridylation of plastid transcripts. We suggest an alternative endosymbiotic scenario whereby a plastid was acquired at the base of Myzozoa, but due to plastid wearing down, a replacement of the ancestral plastid occurred in the apicomplexan/chrompodellid lineage. As illustrated earlier, organellar genomes of myzozoans indeed show signs of a high erosion rate (Hackett et al., 2004; Zhang et al., 1999). Several mechanisms, e.g., RNA editing in plastids, have been suggested to alleviate the deleterious effects of mutations (Dorrell & Howe, 2015). While in some dinoflagellates plastid erosion could lead to a trophic switch from autotrophy to heterotrophy or parasitism, it might also be the cause for the plastid promiscuity frequently observed in dinoflagellate lineages experiencing serial endosymbioses and plastid sequestration. Indeed, differential erosion seen in dinoflagellates and apicomplexans might just reflect the lineage-specific capacity to maintain the integrity of organellar genomes. Nevertheless, a hypothetical plastid replacement would also have alleviated the effect of organellar deterioration in the apicomplexan/chrompodellid lineage. The observed pigmentation and a phylogenetic affinity of plastid genomes of chromerids to eustigmatophytes suggest this replacement plastid might have come from this or a related ochrophyte lineage.

In conclusion, chromerids have been crucial in understanding the evolutionary origin and fate of the apicoplasts, and have helped to recognize their close relationship with peridinin-pigmented plastids of dinoflagellates. The big picture is still missing a lot of pieces, though. How complex was the life cycle in first Myzozoa? How are chromerid plastids integrated into cellular biochemistry? Was there need and opportunity to replace the ancestral plastid in the apicomplexan/chrompodellid branch of Myzozoa? A closer look at these questions might bring further light into processes and limitations that accompany evolutionary transitions of phototrophs to deadly parasites.

ACKNOWLEDGEMENTS

Support from the Czech Science Foundation (Project 16-24027S to M.O.) is greatly acknowledged. The authors thank Marek Eliáš for sharing *Trachydiscus minutus* data and Heather Esson for language correction.

Declaration of conflict of interest: Authors declare no conflict of interest.

REFERENCES

Apt, K. E., Zaslavkaia, L., Lippmeier, J. C., Lang, M., Kilian, O., Wetherbee, R., et al. (2002). In vivo characterization of diatom multipartite plastid targeting signals. *Journal of Cell Science, 115*(Pt. 21), 4061–4069. http://dx.doi.org/10.1242/jcs.00092.

Archibald, J. M. (2015). Genomic perspectives on the birth and spread of plastids. *Proceedings of the National Academy of Sciences of the United States of America, 112*(33), 10147–10153. http://dx.doi.org/10.1073/pnas.1421374112.

Arisue, N., & Hashimoto, T. (2015). Phylogeny and evolution of apicoplasts and apicomplexan parasites. *Parasitology International, 64*(3), 254–259. http://dx.doi.org/10.1016/j.parint.2014.10.005.

Bailleul, B., Berne, N., Murik, O., Petroutsos, D., Prihoda, J., Tanaka, A., et al. (2015). Energetic coupling between plastids and mitochondria drives CO2 assimilation in diatoms. *Nature, 524*(7565), 366–369. http://dx.doi.org/10.1038/nature14599.

Beier, H., & Grimm, M. (2001). Misreading of termination codons in eukaryotes by natural nonsense suppressor tRNAs. *Nucleic Acids Research, 29*(23), 4767–4782. http://dx.doi.org/10.1093/nar/29.23.4767.

Bendich, A. J. (2007). The size and form of chromosomes are constant in the nucleus, but highly variable in bacteria, mitochondria and chloroplasts. *BioEssays*, *29*(5), 474–483. http://dx.doi.org/10.1002/bies.20576.

Bodył, A., & Moszczyński, K. (2006). Did the peridinin plastid evolve through tertiary endosymbiosis? A hypothesis. *European Journal of Phycology*, *41*(4), 435–448. http://dx.doi.org/10.1080/09670260600961080.

Bolte, K., Bullmann, L., Hempel, F., Bozarth, A., Zauner, S., & Maier, U. (2009). Protein targeting into secondary plastids. *Journal of Eukaryotic Microbiology*, *56*(1), 9–15. http://dx.doi.org/10.1111/j.1550-7408.2008.00370.x.

Borza, T., Popescu, C. E., & Lee, R. W. (2005). Multiple metabolic roles for the non-photosynthetic plastid of the green alga *Prototheca wickerhamii*. *Eukaryotic Cell*, *4*(2), 253–261. http://dx.doi.org/10.1128/EC.4.2.253-261.2005.

Brugerolle, G. (2002). *Colpodella vorax*: Ultrastructure, predation, life-cycle, mitosis, and phylogenetic relationships. *European Journal of Protistology*, *38*(2), 113–125. http://dx.doi.org/10.1078/0932-4739-00864.

Burki, F., Flegontov, P., Oborník, M., Cihlář, J., Pain, A., Lukeš, J., et al. (2012). Re-evaluating the green versus red signal in eukaryotes with secondary plastid of red algal origin. *Genome Biology and Evolution*, *4*(6), 626–635. http://dx.doi.org/10.1093/gbe/evs049.

Burki, F., Kaplan, M., Tikhonenkov, D. V., Zlatogursky, V., Minh, B. Q., Radaykina, L. V., et al. (2016). Untangling the early diversification of eukaryotes: A phylogenomic study of the evolutionary origins of Centrohelida, Haptophyta and Cryptista. *Proceedings of the Royal Society B: Biological Sciences*, *283*(1823), 20152802. http://dx.doi.org/10.1098/rspb.2015.2802.

Cai, X., Fuller, A. L., McDougald, L. R., & Zhu, G. (2003). Apicoplast genome of the coccidian *Eimeria tenella*. *Gene*, *321*(1–2), 39–46. http://dx.doi.org/10.1016/j.gene.2003.08.008.

Cavalier-Smith, T., & Chao, E. E. (2004). Protalveolate phylogeny and systematics and the origins of Sporozoa and dinoflagellates (phylum Myzozoa nom. nov.). *European Journal of Protistology*, *40*(3), 185–212. http://dx.doi.org/10.1016/j.ejop.2004.01.002.

Cihlář, J., Füssy, Z., Horák, A., & Oborník, M. (2016). Evolution of the tetrapyrrole biosynthetic pathway in secondary algae: Conservation, redundancy and replacement. *PloS One*, *11*(11), e0166338. http://dx.doi.org/10.1371/journal.pone.0166338.

Cornillot, E., Hadj-Kaddour, K., Dassouli, A., Noel, B., Ranwez, V., Vacherie, B., et al. (2012). Sequencing of the smallest Apicomplexan genome from the human pathogen *Babesia microti*. *Nucleic Acids Research*, *40*(18), 9102–9114. http://dx.doi.org/10.1093/nar/gks700.

Cumbo, V. R., Baird, A. H., Moore, R. B., Negri, A. P., Neilan, B. A., Salih, A., et al. (2013). *Chromera velia* is endosymbiotic in larvae of the reef corals *Acropora digitifera* and A. tenuis. *Protist*, *164*(2), 237–244. http://dx.doi.org/10.1016/j.protis.2012.08.003.

Denny, P., Preiser, P., Williamson, D., & Wilson, I. (1998). Evidence for a single origin of the 35 kb plastid DNA in apicomplexans. *Protist*, *149*, 51–59. http://dx.doi.org/10.1016/S1434-4610(98)70009-4.

Dorrell, R., & Howe, C. (2015). Integration of plastids with their hosts: Lessons learned from dinoflagellates. *Proceedings of the National Academy of Sciences of the United States of America*, *112*(33), 201421380. http://dx.doi.org/10.1073/pnas.1421380112.

Douglas, S., Zauner, S., Fraunholz, M., Beaton, M., Penny, S., Deng, L.-T., et al. (2001). The highly reduced genome of an enslaved algal nucleus. *Nature*, *410*(6832), 1091–1096. http://dx.doi.org/10.1038/35074092.

Elbrachter, M., & Schnepf, E. (1996). *Gymnodinium chlorophorum*, a new, green, bloom-forming dinoflagellate (Gymnodiniales, Dinophyceae) with a vestigial prasinophyte endosymbiont. *Phycologia*, *35*(5), 381–393. http://dx.doi.org/10.2216/I0031-8884-35-5-381.1.

Eschbach, S., Speth, V., Hansmann, P., & Sitte, P. (1990). Freeze-fracture study of the single membrane between host cell and endocytobiont in the dinoflagellates *Glenodinium foliaceum* and *Peridinium balticum*. *Journal of Phycology*, *26*(2), 324–328. http://dx.doi.org/10.1111/j.0022-3646.1990.00324.x.
Feagin, J. E., Mericle, B. L., Werner, E., & Morris, M. (1997). Identification of additional rRNA fragments encoded by the *Plasmodium falciparum* 6 kb element. *Nucleic Acids Research*, *25*(2), 438–446.
Figueroa-Martinez, F., Nedelcu, A. M., Smith, D. R., & Reyes-Prieto, A. (2015). When the lights go out: The evolutionary fate of free-living colorless green algae. *New Phytologist*, *206*, 972–982.
Flegontov, P., Michálek, J., Janouškovec, J., Lai, D. H., Jirků, M., Hajdušková, E., et al. (2015). Divergent mitochondrial respiratory chains in phototrophic relatives of apicomplexan parasites. *Molecular Biology and Evolution*, *32*(5), 1115–1131. http://dx.doi.org/10.1093/molbev/msv021.
Francia, M. E., & Striepen, B. (2014). Cell division in apicomplexan parasites. *Nature Reviews Microbiology*, *12*(2), 125–136. http://dx.doi.org/10.1038/nrmicro3184.
Füssy, Z., Masařová, P., Kručinská, J., Esson, H. J., & Oborník, M. (2017). Budding of the alveolate alga *Vitrella brassicaformis* resembles sexual and asexual processes in apicomplexan parasites. *Protist*, *168*(1), 80–91. http://dx.doi.org/10.1016/j.protis.2016.12.001.
Gardian, Z., Tichý, J., & Vácha, F. (2011). Structure of PSI, PSII and antennae complexes from yellow-green alga *Xanthonema debile*. *Photosynthesis Research*, *108*(1), 25–32. http://dx.doi.org/10.1007/s11120-011-9647-z.
Gile, G. H., & Slamovits, C. H. (2014). Transcriptomic analysis reveals evidence for a cryptic plastid in the colpodellid *Voromonas pontica*, a close relative of chromerids and apicomplexan parasites. *PLoS One*, *9*(5), e96258. http://dx.doi.org/10.1371/journal.pone.0096258.
Gornik, S. G., Febrimarsa, Cassin, A. M., MacRae, J. I., Ramaprasad, A., Rchiad, Z., et al. (2015). Endosymbiosis undone by stepwise elimination of the plastid in a parasitic dinoflagellate. *Proceedings of the National Academy of Sciences of the United States of America*, *112*(18), 5767–5772. http://dx.doi.org/10.1073/pnas.1423400112.
Gould, S. B., Maier, U.-G., & Martin, W. F. (2015). Protein import and the origin of red complex plastids. *Current Biology*, *25*(12), R515–R521. http://dx.doi.org/10.1016/j.cub.2015.04.033.
Gould, S. B., Sommer, M. S., Hadfi, K., Zauner, S., Kroth, P. G., & Maier, U. (2006). Protein targeting into the complex plastid of cryptophytes. *Journal of Molecular Evolution*, *62*, 674–681. http://dx.doi.org/10.1007/s00239-005-0099-y.
Gould, S. B., Waller, R. F., & McFadden, G. I. (2008). Plastid evolution. *Annual Review of Plant Biology*, *59*, 491–517. http://dx.doi.org/10.1146/annurev.arplant.59.032607.092915.
Green, B. R. (2004). The chloroplast genome of dinoflagellates—A reduced instruction set? *Protist*, *155*, 23–31. http://dx.doi.org/10.1078/1434461000161.
Gruber, A., Vugrinec, S., Hempel, F., Gould, S. B., Maier, U. G., & Kroth, P. G. (2007). Protein targeting into complex diatom plastids: Functional characterisation of a specific targeting motif. *Plant Molecular Biology*, *64*(5), 519–530. http://dx.doi.org/10.1007/s11103-007-9171-x.
Hackett, J. D., Yoon, H. S., Soares, M. B., Bonaldo, M. F., Casavant, T. L., Scheetz, T. E., et al. (2004). Migration of the plastid genome to the nucleus in a peridinin dinoflagellate. *Current Biology*, *14*(3), 213–218. http://dx.doi.org/10.1016/S0960-9822(04)00042-9.
Hansen, G., Botes, L., & De Salas, M. (2007). Ultrastructure and large subunit rDNA sequences of *Lepidodinium viride* reveal a close relationship to *Lepidodinium chlorophorum* comb. nov. (= *Gymnodinium chlorophorum*). *Phycological Research*, *55*(1), 25–41. http://dx.doi.org/10.1111/j.1440-1835.2006.00442.x.

He, D., Sierra, R., Pawlowski, J., & Baldauf, S. L. (2016). Reducing long-branch effects in multi-protein data uncovers a close relationship between Alveolata and Rhizaria. *Molecular Phylogenetics and Evolution*, *101*, 1–7. http://dx.doi.org/10.1016/j.ympev.2016.04.033.

Hehenberger, E., Imanian, B., Burki, F., & Keeling, P. J. (2014). Evidence for the retention of two evolutionary distinct plastids in dinoflagellates with diatom endosymbionts. *Genome Biology and Evolution*, *6*(9), 2321–2334. http://dx.doi.org/10.1093/gbe/evu182.

Hopkins, J., Fowler, R., Krishna, S., Wilson, I., Mitchell, G., & Bannister, L. (1999). The plastid in *Plasmodium falciparum* asexual blood stages: A three-dimensional ultrastructural analysis. *Protist*, *150*(3), 283–295. http://dx.doi.org/10.1016/S1434-4610(99)70030-1.

Howe, C. J., Nisbet, R. E. R., & Barbrook, A. C. (2008). The remarkable chloroplast genome of dinoflagellates. *Journal of Experimental Botany*, *59*(5), 1035–1045. http://dx.doi.org/10.1093/jxb/erm292.

Imanian, B., & Keeling, P. J. (2007). The dinoflagellates *Durinskia baltica* and *Kryptoperidinium foliaceum* retain functionally overlapping mitochondria from two evolutionarily distinct lineages. *BMC Evolutionary Biology*, 7, 172. http://dx.doi.org/10.1186/1471-2148-7-172.

Janouškovec, J., Gavelis, G. S., Burki, F., Dinh, D., Bachvaroff, T. R., Gornik, S. G., et al. (2017). Major transitions in dinoflagellate evolution unveiled by phylotranscriptomics. *Proceedings of the National Academy of Sciences of the United States of America*, *114*(2), E171–E180. http://dx.doi.org/10.1073/pnas.1614842114.

Janouškovec, J., Horák, A., Barott, K. L., Rohwer, F. L., & Keeling, P. J. (2012). Global analysis of plastid diversity reveals apicomplexan-related lineages in coral reefs. *Current Biology*, *22*(13), R518–R519. http://dx.doi.org/10.1016/j.cub.2012.04.047.

Janouškovec, J., Horák, A., Oborník, M., Lukeš, J., & Keeling, P. J. (2010). A common red algal origin of the apicomplexan, dinoflagellate, and heterokont plastids. *Proceedings of the National Academy of Sciences of the United States of America*, *107*(24), 10949–10954. http://dx.doi.org/10.1073/pnas.1003335107.

Janouškovec, J., Sobotka, R., Lai, D. H., Flegontov, P., Koník, P., Komenda, J., et al. (2013). Split photosystem protein, linear-mapping topology, and growth of structural complexity in the plastid genome of *Chromera velia*. *Molecular Biology and Evolution*, *30*(11), 2447–2462. http://dx.doi.org/10.1093/molbev/mst144.

Janouškovec, J., Tikhonenkov, D. V., Burki, F., Howe, A. T., Kolísko, M., Mylnikov, A. P., et al. (2015). Factors mediating plastid dependency and the origins of parasitism in apicomplexans and their close relatives. *Proceedings of the National Academy of Sciences of the United States of America*, *112*(33), 10200–10207. http://dx.doi.org/10.1073/pnas.1423790112.

Keeling, P. J. (2013). The number, speed, and impact of plastid endosymbioses in eukaryotic evolution. *Annual Review of Plant Biology*, *64*, 583–607. http://dx.doi.org/10.1146/annurev-arplant-050312-120144.

Kilian, O., & Kroth, P. G. (2005). Identification and characterization of a new conserved motif within the presequence of proteins targeted into complex diatom plastids. *Plant Journal*, *41*(2), 175–183. http://dx.doi.org/10.1111/j.1365-313X.2004.02294.x.

Kissinger, J. C., & DeBarry, J. (2011). Genome cartography: Charting the apicomplexan genome. *Trends in Parasitology*, *27*(8), 345–354. http://dx.doi.org/10.1016/j.pestbp.2011.02.012.Investigations.

Kite, G. C., & Dodge, J. D. (1988). Cell and chloroplast ultrastructure in *Gyrodinium aureolum* and *Gymnodinium galatheanum*. Two marine dinoflagellates containing an unusual carotenoid. *Sarsia*, *73*, 131–138. http://dx.doi.org/10.1080/00364827.1988.10420679.

Köhler, S. (2005). Multi-membrane-bound structures of Apicomplexa: I. The architecture of the *Toxoplasma gondii* apicoplast. *Parasitology Research*, *96*(4), 258–272. http://dx.doi.org/10.1007/s00436-005-1338-2.

Kořený, L., Lukeš, J., & Oborník, M. (2010). Evolution of the haem synthetic pathway in kinetoplastid flagellates: An essential pathway that is not essential after all? *International Journal for Parasitology*, *40*(2), 149–156. http://dx.doi.org/10.1016/j.ijpara.2009.11.007.

Kořený, L., & Oborník, M. (2011). Sequence evidence for the presence of two tetrapyrrole pathways in *Euglena gracilis*. *Genome Biology and Evolution*, *3*(1), 359–364. http://dx.doi.org/10.1093/gbe/evr029.

Kořený, L., Oborník, M., & Lukeš, J. (2013). Make it, take it, or leave it: Heme metabolism of parasites. *PLoS Pathogens*, *9*(1) . e1003088 http://dx.doi.org/10.1371/journal.ppat.1003088.

Kořený, L., Sobotka, R., Janouškovec, J., Keeling, P. J., & Oborník, M. (2011). Tetrapyrrole synthesis of photosynthetic chromerids is likely homologous to the unusual pathway of apicomplexan parasites. *The Plant Cell*, *23*(9), 3454–3462. http://dx.doi.org/10.1105/tpc.111.089102.

Kořený, L., Sobotka, R., Kovářová, J., Gnipová, A., Flegontov, P., Horváth, A., et al. (2012). Aerobic kinetoplastid flagellate *Phytomonas* does not require heme for viability. *Proceedings of the National Academy of Sciences of the United States of America*, *109*(10), 3808–3813. http://dx.doi.org/10.1073/pnas.1201089109.

Kotabová, E., Kaňa, R., Jarešová, J., & Prášil, O. (2011). Non-photochemical fluorescence quenching in Chromera velia is enabled by fast violaxanthin de-epoxidation. *FEBS Letters*, *585*(12), 1941–1945. http://dx.doi.org/10.1016/j.febslet.2011.05.015.

LaGier, M. J., Tachezy, J., Stejskal, F., Kutisova, K., & Keithly, J. S. (2003). Mitochondrial-type iron-sulfur cluster biosynthesis genes (*IscS* and *IscU*) in the apicomplexan *Cryptosporidium parvum*. *Microbiology*, *149*(12), 3519–3530. http://dx.doi.org/10.1099/mic.0.26365-0.

Lane, C. E., van den Heuvel, K., Kozera, C., Curtis, B. A., Parsons, B. J., Bowman, S., et al. (2007). Nucleomorph genome of *Hemiselmis andersenii* reveals complete intron loss and compaction as a driver of protein structure and function. *Proceedings of the National Academy of Sciences of the United States of America*, *104*(50), 19908–19913. http://dx.doi.org/10.1073/pnas.0707419104.

Lang-Unnasch, N., Reith, M. E., Munholland, J., & Barta, J. R. (1998). Plastids are widespread and ancient in parasites of the phylum Apicomplexa. *International Journal for Parasitology*, *28*(11), 1743–1754. http://dx.doi.org/10.1016/S0020-7519(98)00136-2.

Lau, A. O. T., McElwain, T. F., Brayton, K. A., Knowles, D. P., & Roalson, E. H. (2009). Babesia bovis: A comprehensive phylogenetic analysis of plastid-encoded genes supports green algal origin of apicoplasts. *Experimental Parasitology*, *123*(3), 236–243. http://dx.doi.org/10.1016/j.exppara.2009.07.007.

Lemgruber, L., Kudryashev, M., Dekiwadia, C., Riglar, D. T., Baum, J., Stahlberg, H., et al. (2013). Cryo-electron tomography reveals four-membrane architecture of the *Plasmodium* apicoplast. *Malaria Journal*, *12*(1), 25. http://dx.doi.org/10.1186/1475-2875-12-25.

Leung, S. K., & Wong, J. T. Y. (2009). The replication of plastid minicircles involves rolling circle intermediates. *Nucleic Acids Research*, *37*(6), 1991–2002. http://dx.doi.org/10.1093/nar/gkp063.

Lim, L., & McFadden, G. I. (2010). The evolution, metabolism and functions of the apicoplast. *Philosophical Transactions of the Royal Society of London. Series B, Biological Sciences*, *365*(1541), 749–763. http://dx.doi.org/10.1098/rstb.2009.0273.

Lubián, L. M., Montero, O., Moreno-Garrido, I., Huertas, I. E., Sobrino, C., González-del Valle, M., et al. (2000). Nannochloropsis (Eustigmatophyceae) as source of commercially valuable pigments. *Journal of Applied Phycology*, *12*, 249–255. http://dx.doi.org/10.1023/A:1008170915932.

Lucas, I. A. N., & Vesk, M. (1990). The fine structure of two photosynthetic species of Dinophysis (Dinophysiales, Dinophyceae). *Journal of Phycology*, *26*(2), 345–357. http://dx.doi.org/10.1111/j.0022-3646.1990.00345.x.

Lukeš, J., Leander, B. S., & Keeling, P. J. (2009). Cascades of convergent evolution: The corresponding evolutionary histories of euglenozoans and dinoflagellates. *Proceedings of the National Academy of Sciences of the United States of America*, *106*(Suppl. 1), 9963–9970. http://dx.doi.org/10.1073/pnas.0901004106.

Matsumoto, T., Shinozaki, F., Chikuni, T., Yabuki, A., Takishita, K., Kawachi, M., et al. (2011). Green-colored plastids in the dinoflagellate genus *Lepidodinium* are of core chlorophyte origin. *Protist*, *162*(2), 268–276. http://dx.doi.org/10.1016/j.protis.2010.07.001.

Matsuzaki, M., Kuroiwa, H., Kuroiwa, T., Kita, K., & Nozaki, H. (2008). A cryptic algal group unveiled: A plastid biosynthesis pathway in the oyster parasite *Perkinsus marinus*. *Molecular Biology and Evolution*, *25*(6), 1167–1179. http://dx.doi.org/10.1093/molbev/msn064.

McFadden, G. I., & Roos, D. S. (1999). Apicomplexan plastids as drug targets. *Trends in Microbiology*, 7(8), 328–333. http://dx.doi.org/10.1016/S0966-842X(99)01547-4.

McFadden, G. I., & Yeh, E. (2017). The apicoplast: Now you see it, now you don't. *International Journal for Parasitology*, *47*(2–3), 137–144. http://dx.doi.org/10.1016/j.ijpara.2016.08.005.

Minge, M. A., Shalchian-Tabrizi, K., Torresen, O. K., Takishita, K., Probert, I., Inagaki, Y., et al. (2010). A phylogenetic mosaic plastid proteome and unusual plastid-targeting signals in the green-colored dinoflagellate *Lepidodinium chlorophorum*. *BMC Evolutionary Biology*, *10*, 191. http://dx.doi.org/10.1186/1471-2148-10-191.

Moore, R. B., Oborník, M., Janouškovec, J., Chrudimský, T., Vancová, M., Green, D. H., et al. (2008). A photosynthetic alveolate closely related to apicomplexan parasites. *Nature*, *451*(7181), 959–963. http://dx.doi.org/10.1038/nature06635.

Morrissette, N. S., & Sibley, L. D. (2002). Cytoskeleton of apicomplexan parasites. *Microbiology and Molecular Biology Reviews*, *66*(1), 21–38. http://dx.doi.org/10.1128/MMBR.66.1.21.

Mungpakdee, S., Shinzato, C., Takeuchi, T., Kawashima, T., Koyanagi, R., Hisata, K., et al. (2014). Massive gene transfer and extensive RNA editing of a symbiotic dinoflagellate plastid genome. *Genome Biology and Evolution*, *6*(6), 1408–1422. http://dx.doi.org/10.1093/gbe/evu109.

Nash, E. A., Nisbet, R. E. R., Barbrook, A. C., & Howe, C. J. (2008). Dinoflagellates: A mitochondrial genome all at sea. *Trends in Genetics*, *24*(7), 328–335. http://dx.doi.org/10.1016/j.tig.2008.04.001.

Nassoury, N., Cappadocia, M., & Morse, D. (2003). Plastid ultrastructure defines the protein import pathway in dinoflagellates. *Journal of Cell Science*, *116*(Pt. 14), 2867–2874. http://dx.doi.org/10.1242/jcs.00517.

Neuhaus, H. E., & Emes, M. J. (2000). Nonphotosynthetic metabolism in plastid. *Annual Review Plant Biology*, *51*, 111–140.

Nuismer, S. L., & Otto, S. P. (2004). Host-parasite interactions and the evolution of ploidy. *Proceedings of the National Academy of Sciences of the United States of America*, *101*(30), 11036–11039. http://dx.doi.org/10.1073/pnas.0403151101.

Oborník, M., & Green, B. R. (2005). Mosaic origin of the heme biosynthesis pathway in photosynthetic eukaryotes. *Molecular Biology and Evolution*, *22*(12), 2343–2353. http://dx.doi.org/10.1093/molbev/msi230.

Oborník, M., Janouškovec, J., Chrudimský, T., & Lukeš, J. (2009). Evolution of the apicoplast and its hosts: From heterotrophy to autotrophy and back again. *International Journal for Parasitology*, *39*(1), 1–12. http://dx.doi.org/10.1016/j.ijpara.2008.07.010.

Oborník, M., Jirků, M., Šlapeta, J. R., Modrý, D., Koudela, B., & Lukeš, J. (2002). Notes on coccidian phylogeny, based on the apicoplast small subunit ribosomal DNA. *Parasitology Research, 88*(4), 360–363. http://dx.doi.org/10.1007/s00436-001-0557-4.

Oborník, M., Kručinská, J., & Esson, H. (2016). Life cycles of chromerids resemble those of colpodellids and apicomplexan parasites. *Perspectives in Phycology, 3*(1), 21–27. http://dx.doi.org/10.1127/pip/2016/0038.

Oborník, M., & Lukeš, J. (2013). Cell biology of chromerids: Autotrophic relatives to apicomplexan parasites. *International Review of Cell and Molecular Biology, 306*, 333–369. http://dx.doi.org/10.1016/B978-0-12-407694-5.00008-0.

Oborník, M., & Lukeš, J. (2015). The organellar genomes of *Chromera* and *Vitrella*, the phototrophic relatives of apicomplexan parasites. *Annual Review of Microbiology, 69*(1), 129–144. http://dx.doi.org/10.1146/annurev-micro-091014-104449.

Oborník, M., Modrý, D., Lukeš, M., Černotíková-Stříbrná, E., Cihlář, J., Tesařová, M., et al. (2012). Morphology, ultrastructure and life cycle of *Vitrella brassicaformis* n. sp., n. gen., a novel chromerid from the Great Barrier Reef. *Protist, 163*(2), 306–323. http://dx.doi.org/10.1016/j.protis.2011.09.001.

Oborník, M., Vancová, M., Lai, D. H., Janouškovec, J., Keeling, P. J., & Lukeš, J. (2011). Morphology and ultrastructure of multiple life cycle stages of the photosynthetic relative of apicomplexa, *Chromera velia. Protist, 162*(1), 115–130. http://dx.doi.org/10.1016/j.protis.2010.02.004.

Okamoto, N., & Keeling, P. J. (2014). The 3D structure of the apical complex and association with the flagellar apparatus revealed by serial TEM tomography in *Psammosa pacifica*, a distant relative of the apicomplexa. *PLoS One, 9*(1), e84653. http://dx.doi.org/10.1371/journal.pone.0084653.

Owens, T. G., Gallagher, J. C., & Alberte, R. S. (1987). Photosynthetic light-harvesting function of violaxanthin in *Nannochloropsis* spp. (Eustigmatophyceae). *Journal of Phycology, 23*(1), 79–85. http://dx.doi.org/10.1111/j.0022-3646.1987.00079.x.

Park, M. G., Kim, M., & Kim, S. (2014). The acquisition of plastids/phototrophy in heterotrophic dinoflagellates. *Acta Protozoologica, 53*(1), 39–50. http://dx.doi.org/10.4467/16890027AP.14.005.1442.

Patron, N. J., & Waller, R. F. (2007). Transit peptide diversity and divergence: A global analysis of plastid targeting signals. *BioEssays, 29*(10), 1048–1058. http://dx.doi.org/10.1002/bies.20638.

Patron, N. J., Waller, R. F., Archibald, J. M., & Keeling, P. J. (2005). Complex protein targeting to dinoflagellate plastids. *Journal of Molecular Biology, 348*, 1015–1024. http://dx.doi.org/10.1016/j.jmb.2005.03.030.

Patron, N. J., Waller, R. F., & Keeling, P. J. (2006). A tertiary plastid uses genes from two endosymbionts. *Journal of Molecular Biology, 357*(5), 1373–1382. http://dx.doi.org/10.1016/j.jmb.2006.01.084.

Petersen, J., Ludewig, A. K., Michael, V., Bunk, B., Jarek, M., Baurain, D., et al. (2014). *Chromera velia*, endosymbioses and the rhodoplex hypothesis—Plastid evolution in cryptophytes, alveolates, stramenopiles, and haptophytes (CASH lineages). *Genome Biology and Evolution, 6*(3), 666–684. http://dx.doi.org/10.1093/gbe/evu043.

Pienaar, R. N., Sakai, H., & Horiguchi, T. (2007). Description of a new dinoflagellate with a diatom endosymbiont, *Durinskia capensis* sp. nov. (Peridiniales, Dinophyceae) from South Africa. *Journal of Plant Research, 120*(2), 247–258. http://dx.doi.org/10.1007/s10265-006-0047-y.

Pradel, G., & Schlitzer, M. (2010). Antibiotics in malaria therapy and their effect on the parasite apicoplast. *Current Molecular Medicine, 10*(3), 335–349. http://dx.doi.org/10.2174/156652410791065273.

Ralph, S. A., Foth, B. J., Hall, N., & McFadden, G. I. (2004a). Evolutionary pressures on apicoplast transit peptides. *Molecular Biology and Evolution, 21*(12), 2183–2194. http://dx.doi.org/10.1093/molbev/msh233.

Ralph, S. A., van Dooren, G. G., Waller, R. F., Crawford, M. J., Fraunholz, M. J., Foth, B. J., et al. (2004b). Metabolic maps and functions of the *Plasmodium falciparum* apicoplast. *Nature Reviews Microbiology, 2*(3), 203–216. http://dx.doi.org/10.1038/nrmicro843.

Roos, D. S., Crawford, M. J., Donald, R. G., Kissinger, J. C., Klimczak, L. J., & Striepen, B. (1999). Origin, targeting, and function of the apicomplexan plastid. *Current Opinion in Microbiology, 2*(4), 426–432. http://dx.doi.org/10.1016/S1369-5274(99)80075-7.

Schnepf, E., & Deichgraber, G. (1984). "Myzocytosis," a kind of endocytosis with implications to compartmentalization in endosymbiosis. *Naturwissenschaften, 71*, 218–219.

Ševčíková, T., Horák, A., Klimeš, V., Zbránková, V., Demir-Hilton, E., Sudek, S., et al. (2015). Updating algal evolutionary relationships through plastid genome sequencing: Did alveolate plastids emerge through endosymbiosis of an ochrophyte? *Scientific Reports, 5*(1), 10134. http://dx.doi.org/10.1038/srep10134.

Sheiner, L., & Striepen, B. (2013). Protein sorting in complex plastids. *Biochimica et Biophysica Acta, 1833*(2), 352–359. http://dx.doi.org/10.1016/j.bbamcr.2012.05.030.

Simpson, A. G. B., & Patterson, D. J. (1996). Ultrastructure and identification of the predatory flagellate *Colpodella pugnax* Cienkowski (Apicomplexa) with a description of *Colpodella turpis* n. sp. and a review of the genus. *Systematic Parasitology, 33*(3), 187–198. http://dx.doi.org/10.1007/BF01531200.

Slamovits, C. H., & Keeling, P. J. (2008). Plastid-derived genes in the nonphotosynthetic alveolate *Oxyrrhis marina*. *Molecular Biology and Evolution, 25*(7), 1297–1306. http://dx.doi.org/10.1093/molbev/msn075.

Smith, D. R., & Keeling, P. J. (2015). Mitochondrial and plastid genome architecture: Reoccurring themes, but significant differences at the extremes. *Proceedings of the National Academy of Sciences of the United States of America, 112*(33), 10177–10184. http://dx.doi.org/10.1073/pnas.1422049112.

Sommer, M. S., Gould, S. B., Lehmann, P., Gruber, A., Przyborski, J. M., & Maier, U. (2007). Der1-mediated preprotein import into the periplastid compartment of chromalveolates? *Molecular Biology and Evolution, 24*(4), 918–928. http://dx.doi.org/10.1093/molbev/msm008.

Steidinger, K. A., Truby, E. W., & Dawes, C. J. (1978). Ultrastructure of the red tide dinoflagellate Gymnodinium breve. I. General description. *Journal of Phycology, 14*, 72–79.

Stiller, J. W., Schreiber, J., Yue, J., Guo, H., Ding, Q., & Huang, J. (2014). The evolution of photosynthesis in chromist algae through serial endosymbioses. *Nature Communications, 5*, 5764. http://dx.doi.org/10.1038/ncomms6764.

Stork, S., Moog, D., Przyborski, J. M., Wilhelmi, I., Zauner, S., & Maier, U. G. (2012). Distribution of the SELMA translocon in secondary plastids of red algal origin and predicted uncoupling of ubiquitin-dependent translocation from degradation. *Eukaryotic Cell, 11*(12), 1472–1481. http://dx.doi.org/10.1128/EC.00183-12.

Taylor, F. (1987). The biology of dinoflagellates. Botanical monographs. vol. 21. Hoboken, NJ: Wiley-Blackwell.

Toso, M. A., & Omoto, C. K. (2007). Gregarina niphandrodes may lack both a plastid genome and organelle. *Journal of Eukaryotic Microbiology, 54*(1), 66–72. http://dx.doi.org/10.1111/j.1550-7408.2006.00229.x.

Waller, R. F., Gornik, S. G., Kořený, L., & Pain, A. (2016). Metabolic pathway redundancy within the apicomplexan-dinoflagellate radiation argues against an ancient chromalveolate plastid. *Communicative & Integrative Biology, 9*(1), e1116653. http://dx.doi.org/10.1080/19420889.2015.1116653.

Waller, R. F., Keeling, P. J., Donald, R. G., Striepen, B., Handman, E., Lang-Unnasch, N., et al. (1998). Nuclear-encoded proteins target to the plastid in *Toxoplasma gondii* and *Plasmodium falciparum*. *Proceedings of the National Academy of Sciences of the United States of America*, *95*(21), 12352–12357. http://dx.doi.org/10.1073/pnas.95.21.12352.

Wilson, R. J. M. (2002). Progress with parasite plastids. *Journal of Molecular Biology*, *319*(2), 257–274. http://dx.doi.org/10.1016/S0022-2836(02)00303-0.

Wilson, R. J., Denny, P. W., Preiser, P. R., Rangachari, K., Roberts, K., Roy, A., et al. (1996). Complete gene map of the plastid-like DNA of the malaria parasite *Plasmodium falciparum*. *Journal of Molecular Biology*, *261*(2), 155–172. http://dx.doi.org/10.1006/jmbi.1996.0449.

Woehle, C., Dagan, T., Martin, W. F., & Gould, S. B. (2011). Red and problematic green phylogenetic signals among thousands of nuclear genes from the photosynthetic and apicomplexa-related *Chromera velia*. *Genome Biology and Evolution*, *3*(1), 1220–1230. http://dx.doi.org/10.1093/gbe/evr100.

Woo, Y. H., Ansari, H., Otto, T. D., Klinger, C. M., Kolísko, M., Michálek, J., et al. (2015). Chromerid genomes reveal the evolutionary path from photosynthetic algae to obligate intracellular parasites. *eLife*, *4*, 1–41. http://dx.doi.org/10.7554/eLife.06974.

Zhang, Z., Green, B. R., & Cavalier-Smith, T. (1999). Single gene circles in dinoflagellate chloroplast genomes. *Nature*, *400*(6740), 155–159. http://dx.doi.org/10.1038/22099.

Zhu, G., Marchewka, M. J., & Keithly, J. S. (2000). *Cryptosporidium parvum* appears to lack a plastid genome. *Microbiology*, *146*(2), 315–321. http://dx.doi.org/10.1055/s-2001-18357.

CHAPTER SEVEN

Biology of Haptophytes: Complicated Cellular Processes Driving the Global Carbon Cycle

Yoshinori Tsuji*[,1], Masaki Yoshida†
*Kwansei Gakuin University, Sanda, Hyogo, Japan
†University of Tsukuba, Tsukuba, Ibaraki, Japan
[1]Corresponding author: e-mail address: tsuji.yoshinori.fu@kwansei.ac.jp

Contents

Advances in Botanical Research, Volume 84
ISSN 0065-2296
http://dx.doi.org/10.1016/bs.abr.2017.07.002

Abstract

Haptophytes are one of the photosynthetic microalgae with red-algal-derived chloroplasts, and thought to be among the most important primary producers in oceans. Calcifying haptophytes (coccolithophores) have a significant effect on the global carbon cycle, as they fix dissolved inorganic carbon through photosynthesis and calcification. Given that these two processes occur in a single cell, intracellular ion and metabolite traffic is expected to be more complex than in other algae. Haptophytes synthesise and accumulate various compounds during photosynthesis, such as long chain unsaturated ketones (alkenones), β-glucan, mannitol, acid polysaccharides (APs), and dimethylsulphoniopropionate. Some compounds are not equally produced in all haptophytes but are produced by specific species or groups. For example, APs are produced only in coccolithophores and suggested to support morphogenesis of calcite scales. The structure of APs is highly species-specific. Alkenones are produced in only five haptophyte species regardless of their calcification ability. The occurrence of such variations in carbon metabolism across species is also a marked characteristic of haptophytes. This chapter introduces basic cellular features of haptophytes and focuses on features of their carbon metabolism and calcification.

1. INTRODUCTION

Haptophyte algae are a dominant marine primary producer in modern oceans, as they occupy 30–50% of total chlorophyll *a* biomass in oceans (Liu et al., 2009). Calcifying haptophyte algae, called coccolithophores, play a central role in the global carbon cycle by fixing dissolved inorganic carbon (DIC) through photosynthesis and calcification (Holligan, Viollier, Harbour, Camus, & Champagne-Philippe, 1983; Poulton, Adey, Balch, & Holligan, 2007). A massive quantity of calcified cells has been sedimented throughout geological time, as seen in the White Cliffs of Dover; thus, coccolithophores contribute to sequester atmospheric CO_2 as limestone. A representative coccolithophore, *Emiliania huxleyi*, is thought to be most important species, as this alga is distributed worldwide and frequently forms blooms that cover $>100,000\,km^2$ of the ocean's surface (Tyrrell & Merico, 2004). The importance of haptophytes is well recognised in the field of biogeochemistry due to their effects. However, research on cell biology of haptophytes is insufficient, particularly compared with diatoms in which molecular biological approaches are highly advanced.

Haptophytes have unique cellular processes, such as intracellular calcification, and production of unique carbon compounds, such as alkenones and acid polysaccharides (APs). Haptophytes vary greatly in terms of appearance and carbon metabolism. A morphological and genomic characterisation of *E. huxleyi* suggested the occurrence of extensive diversity even within this single species (Hagino et al., 2011; Paasche, 2001; Read et al., 2013). Given the genetic diversity of haptophytes, it is difficult to establish a single model to explain the cellular processes of all haptophytes. Currently, the cosmopolitan coccolithophore *E. huxleyi* is considered a representative haptophyte, so considerable research has focused on this species. This chapter summarises the general features of haptophytes and describes their calcification and carbon metabolism with a focus on *E. huxleyi*.

2. MORPHOLOGY OF HAPTOPHYTES

Haptophyte algae are unicellular, free-living microalgae found in aquatic environments. The most characteristic morphology of haptophytes is calcified scales known as coccoliths. Coccolith-bearing haptophytes are called coccolithophores. Haptophytes also bear nonmineralised organic scales. The morphology of these cell coverings is a diagnostic taxonomic character for the inter/intraphylum classification. The diagram of a representative haptophyte cell is shown in Fig. 1. Haptophyte cells are surrounded

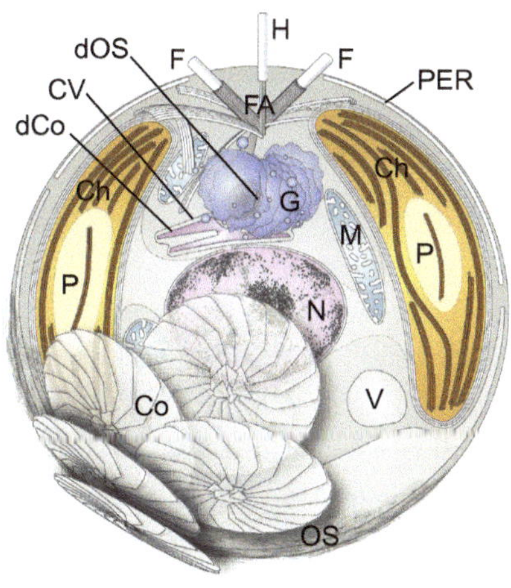

Fig. 1 A diagram of a representative cell of haptophyte algae, based on Inouye (1997) and Billard and Inouye (2004). Ch, chloroplast; Co, coccolith; CV, coccolith forming vesicle; dCo, developing heterococcolith; dOS, developing organic scale; F, flagellum; FA, flagellar apparatus; G, Golgi apparatus; H, haptonema; M, mitochondrion; N, nucleus; OS, organic scale; P, pyrenoid; PER, peripheral endoplasmic reticulum; V, vacuole.

by a cytoplasmic membrane and subjacent peripheral endoplasmic reticulum (ER). Most haptophytes have one or two yellow to golden-brown chloroplasts containing pyrenoids within a cell. The chloroplasts are surrounded by four envelope membranes. The outermost membrane is continuous with the nuclear envelope, which is known as the chloroplast ER (CER). The chloroplasts contain chlorophyll *a* and *c* and some carotenoids. Among carotenoids, 19′-hexanoyloxyfucoxantin is specific to haptophytes, and thus, this pigment is considered to be a chemical marker of haptophytes and some dinoflagellates possessing haptophyte-derived chloroplasts. Limited species of haptophytes, *Papposphaera*, *Trigonaspis*, and some other genera are reported to be heterotrophic; no chloroplast has been observed (Thomsen & Østergaard, 2015). The mitochondria of haptophytes bear tubular cristae. Carbohydrates are stored as chrysolaminarin, a water-soluble β-1,3 glucan, and paramylon-like storage carbohydrates are found in some species of Pavlovophyceae (Bendif et al., 2011). Haptophyte cells typically have two equal or unequal flagella. Although some coccolithophores have extremely short flagella, the genera *Prymnesium* and *Chrysochromulina* have two long flagella, and exceptionally *C. quadrikonta* is known to be a quadriflagellate species (Kawachi & Inouye, 1993). Haptophytes have a thin filamentous organelle, termed a haptonema, which is similar to a flagellum but differs in inner structure, especially in the arrangement of microtubules and its behaviour. The haptonema is a distinctive and unique structure of haptophytes; the name Haptophyta is derived from this organelle.

2.1 Haptonema

Haptonema is named from the Greek word *hapsis* (=touch) based on one of its unique functions. A haptonema emerges from the anterior tip of haptophyte cells (Fig. 1). Because the haptonema resembles flagella under light microscope, it had been mistaken as a third flagellum until 1955 (Parke, Manton, & Clarke, 1955). A haptonematal basal body forms a complex with flagellar basal bodies and related microtubular roots, which together form a flagellar apparatus. The behaviour of haptonema is also coordinated with flagella. A haptonema comprises six to seven microtubules in the free part of the organelle, and eight to nine microtubules at the base. Haptonematal microtubules are surrounded by the ER, which is thought to be related to a calcium-induced response of the haptonema. Haptonema is occasionally much longer than the flagella and plays several roles, such as inducing an avoidance response, capturing and gathering food particles, and adhering to the substratum (Kawachi & Inouye, 1994; Kawachi, Inouye, Maeda, & Chihara, 1991). Haptonema is highly developed in the Prymnesiales, particularly in the genus *Chrysochromulina*. The haptonema cannot beat like a flagellum does but can bend or coil. The haptonema coiling is thought to be a function of obstacle avoidance, and it occurs within extremely short time, 1/60 to 1/100 s (Kawachi et al., 1991). Simultaneously, the beating of flagella changes to the opposite direction, and the cell escapes from the obstacle. Haptonema is also used for capturing food particles. In *C. spinifera*, a functional role of the haptonema in collaboration with the organic spine scales in the feeding process has been reported (Kawachi & Inouye, 1995). Food particles, e.g., bacteria, adhere to the organic scales of *C. spinifera* and then gather together at the tip of the haptonema. Finally, the particles are engulfed by phagocytosis into a food vacuole. Except for the genus *Chrysochromulina*, haptonema is generally short and its function is limited. In some Isochrysidales (e.g. *Imantonia rotunda*, *Dicrateria inornata*, and *Emiliania huxleyi*), their haptonemas are degraded and haptonematal microtubules are also lost. *Pleurochrysis* and *Hymenomonas* in Coccolithales and members of Pavlovophyceae also have a shrunken haptonema, but the microtubules remain present.

2.2 Coccoliths

Haptophyte algae form a huge variety of mineralised scales (Fig. 2). The most common and distinctive scale among haptophytes is a finely ornamented calcified scale, called a coccolith. Coccolith includes two kinds of calcified scales, heterococcolith and holococcolith, distinguished by their arrangement of calcium carbonate crystals. Coccoliths are distributed in haptophytes exclusively, especially in Isochrysidales and Coccolithales. They were named after the Greek words, *kokkos* (=berry, grain) and *lithos* (=stone), by Thomas Henry Huxley (Huxleyi, 1858). He observed coccoliths in ocean bed sediment. Historically, coccoliths had been first observed and described by Christian Gottfried Ehrenberg (Ehrenberg, 1836); however, he thought they were mineral structures generated by abiotic factors. Coccoliths were also thought to be nonbiological matter by later researchers at the beginning of coccolith research. In 1860, George Charles Wallich found coccoliths joined to form complete a sphere (coccosphere) and proposed an organic origin of coccoliths (Wallich, 1860). Following a publication by Henry Clifton Sorby (Sorby, 1861), the organic origin

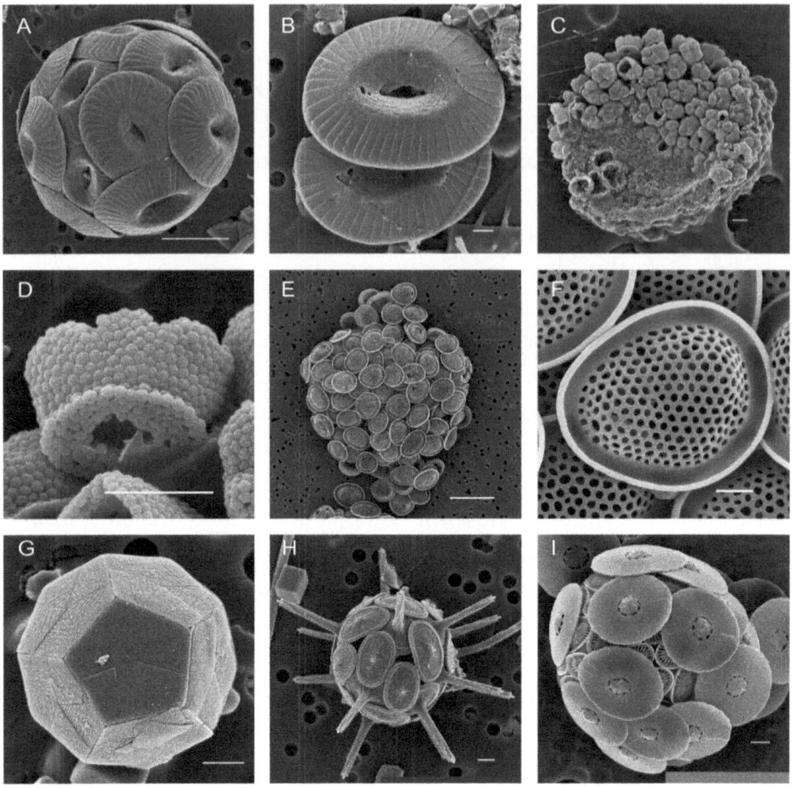

Fig. 2 Scanning electron microscope (SEM) micrographs of mineralised scales in haptophytes. (A–D and G–I) calcified scales, (E and F) silicified scales. (A and B) Heterococcolith (discolith) of *Coccolithus pelagicus*. (C and D) Holococcolith of *Calyptrosphaera* sp. (E and F) Perforated siliceous scales of *Hyalolithus neolepis*. (G and I) Characteristic heterococcoliths. (G) Polygonal pentalith of *Braarudosphaera bigelowii* (H) Rhabdolith of *Rhabdosphaera clavigera*. Prominent rod-shaped part is pentamerous radially symmetrical. (I) *Syracosphaera anthos*, having distinctive double-layered heterococcoliths. Scale bars: A: 5 μm; B–D, F–I: 1 μm, E: 10 μm.

of coccoliths was generally accepted. The study of coccoliths and coccolithophores has flourished since the 1960s, as electron microscopes became popular in the field of phycology and palaeontology. The structure of coccoliths has been well researched, and descriptive terminology of coccolith morphology was organised comprehensively, mainly based on electron microscopy (Young et al., 1997).

Heterococcoliths consist of interlocked, variously formed calcite crystal elements, produced by diploid haptophyte cells. Heterococcoliths are generated in a coccolith-forming vesicle (CV) in the cytoplasm and secreted to the cell surface (Fig. 1). The CV is a large, flat compartment attached to the nuclear envelope, derived from Golgi vesicles. Initially, a thin cellulosic template is formed within the CV and then crystal deposition occurs around the rim of the template. A finished heterococcolith is ejected to the cell surface of the anterior end, near the flagellar apparatus. The heterococcoliths of *Emiliania* and *Gephyrocapsa* consist of distal and proximal shields, which interlock to form a coccosphere. Some other species also have one or two shields, which occasionally overlap but do not interlock (Winter & Siesser, 1994). Heterococcoliths have wide variation in shape and structure (Fig. 2). The discolith of *Coccolithus pelagicus* (Fig. 2A and B) is one of the typical heterococcoliths having two shields and a concave central area. The pentalith of *Braarudosphaera bigelowii* (Fig. 2G) is a pentagonal heterococcolith consisting of five parts of laminating segments. *Rhabdosphaera clavigera* possesses a heterococcolith with a prominent central area, called the rhabdolith (Fig. 2H). A peculiar example, *Syracosphaera anthos*, has distinctive double-layered heterococcoliths (Fig. 2I).

In contrast, holococcoliths consist of many small homomorphic (typically rhomboidal or hexagonal) calcite crystals, and are formed extracellularly by haploid cells. Representative holococcoliths of *Calyptrosphaera* are shown in Fig. 2C and D. Unlike heterococcoliths, calcite crystals of holococcoliths are not interlocked but bonded to each other by organic material. As a first step of holococcolith formation, an organic scale is secreted on the cell surface and then crystals are deposited on the scale within a vacuolar region separated by an organic envelope (de Vrind-de Jong, Emburg, & de Vrind, 1994).

As mentioned above, haploid and diploid haptophyte cells show different coccolith morphologies, which has occurred dual nomenclature problem that a single species was occasionally described as separate species. For example, the diploid *Coccolithus pelagicus* with heterococcoliths and the haploid holococcolith-bearing *Crystalolithus hyalinus* had been thought to be separate species, although they are an identical species (Parke & Adams, 1960). An increasing number of such species have been discovered through flow cytometry analysis measuring cellular DNA content and molecular phylogenetic analysis based on 18S ribosomal RNA (rRNA) genes. To reorganise described species, a comprehensive list of extant haptophytes, including extensive documentation of life cycle combinations, has been periodically reported (Jordan & Green, 1994; Jordan, Cross, & Young, 2004).

2.3 Organic Scales

Both coccolithophores and noncoccolith-bearing flagellate species have several types of organic scales, with a few exceptions such as *Emiliania huxleyi* or colonial forms of *Phaeocystis*. An organic scale is one of the most plesiomorphic morphological characters along with the haptonema. Organic scales of haptophytes are composed mainly of cellulose, synthesised in the Golgi body near the flagellar apparatus and aligned on the cell surface of the anterior area (Leadbeater, 1970). Two major types of surface prints on the organic scales are known, the quadrant radial pattern (Fig. 3A) and the concentric pattern (Fig. 3B), and this difference would reflect the nuclear phase (haploid or diploid) of the cell (Billard & Inouye, 2004). Furthermore, the basket-shaped scales (Fig. 3C) and net-like scales (Fig. 3D) have been reported in *Prymnesium* species. It has been known that *Chrysoculter rhomboideus* has particularly large organic rhomboidal scales that possess two tubular projections at the longitudinal ends, nearly touching each other (Nakayama, Yoshida, Noel, Kawachi, & Inouye, 2005).

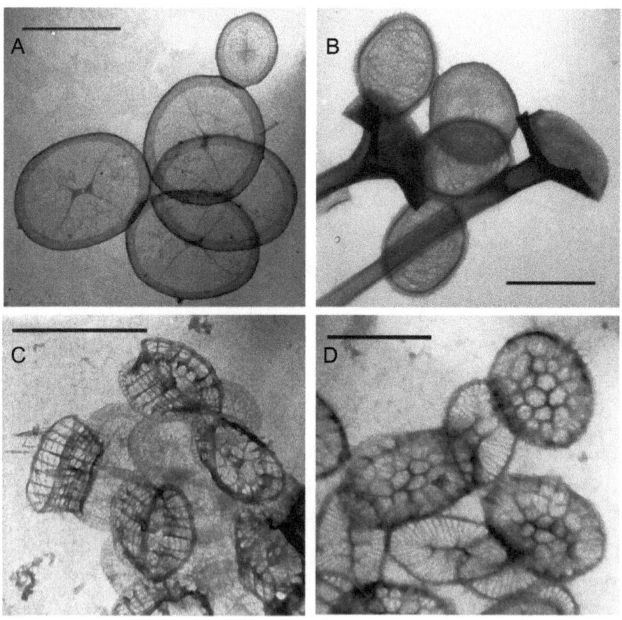

Fig. 3 Organic body scales of Prymnesiales. (A) *Chrysochromulina* sp., (B) *C. spinifera*, (C) *Prymnesium calathiferum*, and (D) *P. nemamethecum*. Scale bars: A, B: 1 μm; C, D: 500 nm.

2.4 Siliceous Scales

Although calcification is well developed in haptophytes, very few species in Prymnesiales harness silica to mineralise their scales. *Hyalolithus neolepis* has large siliceous scales around its spherical cell (Yoshida, Noel, Nakayama, Naganuma, & Inouye, 2006). The scales are hat-shaped, irregular ellipsoidal in top view, perforated by several small pores, and are deposited on cells in several layers (Fig. 2E and F). The siliceous scales are produced in silica deposition vesicles in the posterior region of the cell, and this manner is completely different from coccolith formation. Silicification is very rare in haptophytes, and thus far, it is only known to occur in *Hyalolithus neolepis* and *Prymnesium parvum* (Hibberd, 1980; Pienaar, 1980). An enigmatic siliceous scale-bearing organism, *Petasaria heterolepis*, was reported by Moestrup (1979), and some morphological similarities between *Hyalolithus* and *Petasaria* have been suggested (Jordan et al., 2015; Yoshida et al., 2006). However, it remains unclear whether *Petasaria* belongs to haptophytes or not, due to lack of ultrastructural intracellular observations and molecular information for *Petasaria.*

3. LIFE CYCLE AND ECOLOGY OF HAPTOPHYTES

Haptophytes grow by means of mitotic cell division. Prior to mitosis, the flagellar apparatus containing a haptonematal basal body is duplicated so that a cell with four flagella and two haptonemas occasionally occurs. The initiation of chloroplast division is elongation of the pyrenoid, and then the CER and inner two chloroplast membranes invaginate at the midpoints to finish the division of the chloroplast. Division of the Golgi apparatus also occurs prior to nuclear division. During mitosis, the nuclear envelope disappears in most haptophytes (Graham, Graham, & Wilcox, 2008), whereas the Pavlovophyceae is not in this case. After completion of nuclear division and reconstruction of the nuclear envelope, cytokinesis occurs by invagination of the cell membrane (Hibberd, 1980; Hori & Green, 1994; Manton, 1964). Cell coverings are equally shared by daughter cells, and new scales are generated to supply a deficiency after finishing the cell division (Fresnel & Probert, 2005). Cell division of haptophytes usually occurs in dark conditions, when photosynthetic activity decreases (Graham et al., 2008). Sexual reproduction was first confirmed in *Pleurochrysis pseudoroscoffensis* (Gayral & Fresnel, 1983), and has been widely reported in both coccolith and non-coccolith species (Cros, Kleijne, Zeltner, Billard, & Young, 2000;

Fresnel, 1994; Malinverno, Triantaphyllou, Dimiza, & Young, 2008; Manton & Oates, 1980; Noél, Kawachi, & Inouye, 2004).

Limited species of haptophytes, e.g. *Pavlova granifera*, *C. parva*, and coccolithophore *Hymenomonas roseola*, have been exclusively reported from freshwater environments (Green, 1973; Heimdal, 1996; Lackey, 1939). *Chrysotila lamellosa* and some other species are known to live in both freshwater and brackish environments. Most haptophytes are marine and are distributed from polar to tropical regions. With a very few exceptions, haptophytes are generally autotrophic or mixotrophic, free-living organisms; thus, haptophyte are one of the major primary producers in marine ecosystems. Coccolithophore *Papposphaera* and related species are notably heterotrophic as described above. *Reticulofenestra sessilis*, a coccolithophore in Isochrysidales, occurs almost exclusively with a centric diatom of the genus *Thalassiosira* in presumably symbiotic associations (Young et al., 2003).

4. PHYLOGENY AND CLASSIFICATION OF HAPTOPHYTES

About 280 species of haptophytes have been described, mainly based on the morphology of cell coverings (Graham et al., 2008). Based on morphological and molecular characters, haptophytes are thought to be a monophyletic group. They can be divided into two classes, Pavlovophyceae and Prymnesiophyceae (Edvardsen et al., 2000). A phylogenetic tree of haptophytes is shown in Fig. 4. The class Pavlovophyceae includes only four genera, *Pavlova*, *Diacronema*, *Exanthemachrysis*, and *Rebecca*. Pavlovophyceae is considered to be the most plesiomorphic group among haptophytes. They have two unequal flagella, small knobby organic scales, and distinctive sterols and pigments (Van Lenning et al., 2003). In contrast to Pavlovophyceae, the class Prymnesiophyceae (Coccolithophyceae) consists of four well recognised orders (Phaeocystales, Prymnesiales, Isochrysidales, and Coccolithales) (Adl et al., 2012), including over 70 contemporary genera (e.g., *Chrysochromulina*, *Prymnesium*, *Phaeocystis*, *Emiliania*, *Isochrysis*, *Coccolithus*, and *Pleurochrysis*). The genus *Phaeocystis* is a colonial haptophyte and typically occurs embedded in spherical gelatinous colonies (Parke & Manton, 1971). *Phaeocystis* cells have characteristic chitin fibrils, occasionally considered as a kind of trichocyst. These fibrils form a pentagram orientation when uncoiled, and the function of these fibrils is still unknown (Chrétiennot-Dinet et al., 1997). Prymnesiales and *Chrysochromulina* are motile, noncalcified groups of haptophytes. Unlike Pavlovophyceae, flagella

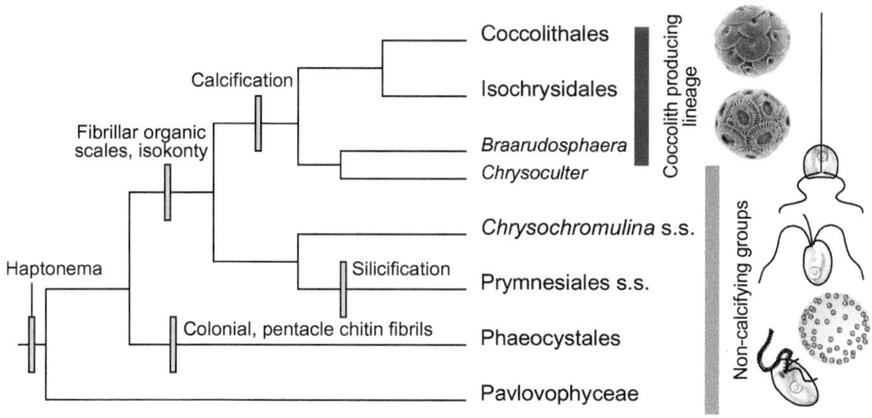

Fig. 4 A phylogenetic tree of haptophyte algae showing major groups, based on Edvardsen et al. (2000), Nakayama et al. (2005), Takano, Hagino, Tanaka, Horiguchi, and Okada (2006), and Yoshida et al. (2006). Minor groups exclusively consisting of environmental sequences are omitted.

of these groups are equal or almost equal and have no ornament on their surface. Species distinctions in these groups are based on morphology of organic scales and molecular data, e.g., nuclear 18S rRNA or chloroplast *rbcL* gene sequences. Some species have been found to produce ichthyotoxins, which are available as a chemical taxonomic character in Prymnesiales and *Chrysochromulina*. Coccolithales and Isochrysidales are coccolith-producing lineages that are representative of haptophytes. Approximately 200 living species are known as coccolithophores (Winter & Siesser, 1994), and many more fossil species have been described. The fossil record of coccolithophores begins over 200 million years ago (Bown, Lees, & Young, 2004), and flourished in the Late Cretaceous when massive limestone had been deposited (Young, Bown, & Burnett, 1994). Fossil records of coccolithophores indicate that 93% of existing coccolithophore species became extinct at the K/T boundary. The genus *Braarudosphaera* is one of few extant lineages of coccolithophores which survived beyond the K/T boundary. *Braarudosphaera bigelowii* is a very unique organism in terms of the morphology and arrangement of pentaliths and the presence of endosymbiotic nitrogen-fixing cyanobacteria (Fig. 2G; Hagino, Onuma, Kawachi, & Horiguchi, 2013). The phylogenetic position of *Chrysoculter rhomboideus*, which has unique organic scales but lacks coccoliths, is enigmatic (Takano et al., 2006); *C. rhomboideus* forms a clade with unidentified haptophytes recognised exclusively based on 18S rRNA gene sequence (Moon-van der Staay, De Wachter, & Vaulot, 2001) and this clade is sister to the coccolithophorids (Nakayama et al., 2005; Yoshida et al., 2006).

5. PHOTOSYNTHESIS AND CALCIFICATION OF HAPTOPHYTES

5.1 General Features of the CO_2-Concentrating Mechanism in Marine Phytoplankton

Most dissolved inorganic carbon (DIC) exists as HCO_3^- (ca. 2 mM) in seawater equilibrated with the current atmospheric CO_2 level, whereas the concentration of CO_2, which is a Rubisco substrate, is low (<15 μM at >15°C) (Badger et al., 1998; Raven, 1994). This concentration of CO_2 is lower than the Rubisco $K_m(CO_2)$ of most marine phytoplankton (Badger et al., 1998). Additionally, the slow diffusion rate of dissolved CO_2 and slow spontaneous equilibrium between HCO_3^- and CO_2 could also be constraints for CO_2 acquisition in the aquatic environment. To overcome the CO_2 limitation in photosynthesis, marine phytoplankton have developed a CO_2-concentrating mechanism (CCM), by which cells actively uptake DIC from external media and supply CO_2 efficiently to Rubisco, resulting in high affinity photosynthesis for DIC (Giordano, Beardall, & Raven, 2005; Reinfelder, 2011). Eukaryotic CCM is comprised of active transporters of DIC across plasma and/or chloroplast membranes and efficient generation of CO_2 from HCO_3^- in the pyrenoid where most of the Rubisco resides. Carbonic anhydrases (CAs) located in various cellular compartments are also important for acquisition and delivery of DIC (Aizawa & Miyachi, 1986; Giordano et al., 2005; Hopkinson, Dupont, & Matsuda, 2016; Samukawa, Shen & Hopkinson, 2014). For instance, periplasmic (external) CA has been hypothesised to facilitate the conversion of HCO_3^- into CO_2, by which depletion of CO_2 at the cell surface is prevented and a diffusive CO_2 gradient across the plasma membrane is maintained (Aizawa & Miyachi, 1986). In the green alga *Chlamydomonas reinhardtii*, an α-type carbonic anhydrase (CAH3) localised in the thylakoid lumen facilitates CO_2 supply from HCO_3^- in the pyrenoid (Karlsson et al., 1998). In the marine diatom *Phaeodactylum tricornutum*, a θ-type of CA (θ-CA) is located in the lumen of pyrenoid-penetrating thylakoids and is involved in CO_2 supply, akin to CAH3 in *C. reinhardtii* (Kikutani et al., 2016).

5.2 CCM as Excess Energy Dissipation System in Coccolithophores

The most dominant coccolithophore *E. huxleyi* has a significant effect on the global carbon cycle, and numerous physiological studies have been conducted on the CCM in this species (see Paasche, 2001 for review). The preferred form of DIC (i.e. CO_2 or HCO_3^-) for photosynthesis depends on the strain. Some *E. huxleyi* strains use CO_2 for photosynthesis (Sekino & Shiraiwa, 1994; Sikes & Wheeler, 1982), while other strains apparently utilise HCO_3^- as the major photosynthesis substrate (Table 1; Herfort et al., 2002; Rost et al., 2003). A comparative study using three different *E. huxleyi* strains revealed the occurrence of interstrain variations in DIC utilisation (Stojkovic et al., 2013). In strain B11, affinity for DIC is reduced by treatment with 4,4′-diisothiocyano-2,2′-stilbenedisulfonic acid (DIDS) that is an inhibitor for solute carrier (SLC) family transporters (Herfort et al., 2002), suggesting that SLC of this strain would mediate direct HCO_3^- uptake as reported in the marine diatom *P. tricornutum* (Nakajima, Tanaka, & Matsuda, 2013). Alternatively, some strains of *E. huxleyi* appear to facilitate DIC uptake using external CA, which mediates the conversion of HCO_3^- to CO_2 at the cell surface, resulting in enhanced diffusive entry of CO_2 (Table 1).

A transcription analysis in *E. huxleyi* revealed that the expression of putative CCM components (including α- and γ-CAs) is induced when external DIC is <1 mM, which is lower than the current DIC level in seawater (ca. 2 mM) (Mackinder et al., 2011). Consistent with this report, the induction of external CA activity under conditions of <1 mM DIC has also been documented (Nimer, Iglesias-Rodriguez, & Merrett, 1997). These data suggest that CAs in *E. huxleyi* are not fully induced under the current atmospheric CO_2 level.

Active uptake and intracellular accumulation of DIC have been reported in many *E. huxleyi* strains (Table 1). Nevertheless, photosynthetic affinity for DIC in *E. huxleyi* is generally rather low compared to other algae such as diatoms and noncalcifying haptophytes. Values of the half-saturation DIC concentration ($K_{0.5}$) in *E. huxleyi* are typically 0.5–3 mM, while noncalcifying haptophytes have much higher affinity for DIC (Table 1). Given that seawater under the present atmospheric CO_2 level (0.04%) contains 2 mM DIC (Raven, 1994), photosynthesis by most *E. huxleyi* strains is unsaturated with respect to DIC (Table 1, Reinfelder, 2011).

The mismatch between active DIC uptake and low affinity for DIC observed in *E. huxleyi* can be explained by the inefficiency of the CCM

Table 1 Characteristics of DIC Uptake in Various Haptophytes

Species	Strain	DIC accumulation	$K_{0.5}$ for DIC	CA Activity		Reference[a]
				External	Internal	
Emiliania huxleyi	Bigelow 88E	2-fold[b]	0.4 mM[b]	Yes[b]	Yes[b]	Nimer, Merrett, and Brownlee (1996) Nimer, Guan, and Merrett (1994)
	B11	n.d.	3.2 mM	No[c]	n.d.	Herfort, Thake, and Roberts (2002)
	PML B92/11	n.d.	1.56 mM	Yes (very low)[d]	Yes	Rost et al. (2003)
	CS-363	n.d.	0.078 mM	Yes	n.d.	Stojkovic, Beardall, and Matear (2013)
	CS-369	n.d.	0.705 mM	Yes	n.d.	
	CS-812	n.d.	0.046 mM	Yes	n.d.	
	NIES 837	16-fold	5.5 mM	No	No	Sekino and Shiraiwa (1994)
Gephyrocapsa ocenica	PCC NZ90	n.d.	n.d.	No	No	Nimer, Ling, Brownlee, and Merrett (1999)
Isochrysis galbana	?	n.d.	n.d.	Yes	Yes	Bhatti, Huertas, and Colman (2002)
	?	8.3 fold	7.6 μM	Yes	Yes	Burns and Beardall (1987)
Phaeocystis globosa	?	n.d.	122 μM	Yes	Yes	Rost et al. (2003)

[a]The detailed experimental conditions (light intensity, pH of media, measurement methods) are described in the references.
[b]Cells at stationary growth phase.
[c]Cells were grown under 2 mM DIC, which was equal to air-equilibrated seawater.
[d]External CA activity was detected, but at very low levels. Membrane-impermeable CA inhibitor did not alter the photosynthetic parameter on this strain, suggesting a negligible role of external CA in this *E. huxleyi* strain.
n.d., No data.

possibly due to a high efflux of unfixed CO_2 from cells. Energy-dependent active uptake and simultaneous leakage of CO_2 is apparently futile but this phenomenon is observed in many phytoplanktons. Such energy-dependent DIC cycling during photosynthesis may be a mechanism to dissipate excess light energy (Huertas, Espie, Colman, & Lubian, 2000; Tchernov, Silverman, Luz, Reinhold, & Kaplan, 2003; Tchernov et al., 1997). The rate of CO_2 efflux in *E. huxleyi* is controlled by light, and this alga can switch the CCM mode from net CO_2 uptake to efflux when light intensity increases (Tchernov et al., 2003). The flexibility of the CCM modes between net CO_2 uptake and efflux is a potential advantage to acclimate to a broad range of light intensities. *E. huxleyi* tends to bloom in the stratified surface layer where nutrients are exhausted and cells are exposed to high light intensity (Tyrrell & Merico, 2004). In such conditions, energy consumption by nutrient assimilation is suppressed and efficient dissipation of excess light energy is required to survive. The CCM is expected to contribute to dissipate excess light energy under such conditions.

5.3 Functional Link Between Calcification and Photosynthesis

The substrate for calcification is HCO_3^- in most *E. huxleyi* strains (Paasche, 2001; Sekino & Shiraiwa, 1994; Sikes & Wheeler, 1982). Calcification proceeds intracellularly, and H^+ or CO_2 is released in this process according to the following equations.

$$Ca^{2+} + HCO_3^- \rightarrow CaCO_3 + H^+ \quad (1)$$

$$H^+ + HCO_3^- \rightarrow CO_2 + H_2O \quad (2)$$

To avoid acidification of the calcification site, protons must be removed from the cell. Alternatively, protons facilitate hydration of HCO_3^- to yield CO_2 (Eq. 2). If CO_2 generated by calcification is efficiently supplied to Rubisco, calcification can function as the CCM to facilitate CO_2 fixation. Although early studies suggested that calcification supports photosynthetic CO_2 fixation (Sikes, Row, & Wilbur, 1980), more recent studies support the independence of these two processes (Herfort, Loste, Meldrum, & Thake, 2004; Leonardos, Read, Thake, & Young, 2009). Photosynthetic activity is unaffected when calcification is suppressed by decreasing the concentration of Ca^{2+} ($[Ca^{2+}]$) in the growth medium, demonstrating the independence of calcification and photosynthesis (Herfort et al., 2004; Leonardos et al., 2009). Bach et al. reported that calcification rate strongly depends on $[HCO_3^-]$ in the medium, whereas growth and organic carbon production are primarily correlated with [CO_2] (Bach et al., 2013). The calcification rate in *E. huxleyi* is simply defined by the availability of HCO_3^-, and there seems to be no close relationship between organic carbon production and calcification. Calcification is performed in CV, which is localised in the cytosol, whereas CO_2 fixation proceeds in the pyrenoid located in the centre part of the chloroplast stroma. Such spatial separation of the two processes would disable efficient fixation of CO_2 generated by calcification.

5.4 Molecular Components Involved in Ion Transport

Coccolithophores have evolved through secondary endosymbiosis and therefore have complex cellular membrane systems, such as four-membrane-bound chloroplasts and specialised CVs (Marsh, 1999). To support $CaCO_3$ precipitation in the CV, high Ca^{2+} and HCO_3^- influx into the CV, and rapid removal of H^+ from the CV is necessary. Additionally, calcification and photosynthesis proceed in parallel, resulting in complex movement of metabolites and other ions across organelles. Therefore, inter-organelle ion traffic is highly complex in coccolithophores compared with that in other algae. However, most of the ion transport mechanisms remain unknown. Suffrian, Schulz, Gutowska, Riebesell, and Bleich (2011) found that the internal pH of *E. huxleyi* changes readily in response to a pH change in the external medium, indicating high H^+ permeability of the coccolithophore plasma membrane. H^+ transport is not mediated by CO_2 or HCO_3^- transport, as short-term changes in [CO_2] or [HCO_3^-] in fixed pH medium do not alter internal pH (Suffrian et al., 2011). Taylor, Chrachri, Wheeler, Goddard, and Brownlee (2011) identified voltage-gated H^+ channels (H_v1) in the coccolithophores of *E. huxleyi* and *Coccolithus pelagicus*, which are involved in high H^+ permeation. The heterologously expressed H_v1 from *E. huxleyi* (EhH_v1) and *C. pelagicus* (CpH_v1) has been characterised and its properties are similar to those of the metazoan H_v1 (Taylor et al., 2011). Measurements of H^+ conductance in *C. pelagicus* cells using the patch-clamp technique (Taylor et al., 2011) revealed that outward H^+ conductance is activated by depolarisation and a decrease of internal pH. These data suggest that H_v1 allows high H^+ permeability in the plasma membrane, by which H^+ produced by calcification is efficiently exported from the cytosol. Besides calcification, H^+ is derived from various cellular processes, and the rate of H^+ production in the cytosol is highly variable. Therefore, in addition to removing H^+ during calcification, H_v1 likely plays a much more general role as a pH stabiliser by opening/closing the H^+ gate depending on the level of cytosolic H^+.

Another candidate involved in H^+ transport is a vacuolar-type ATPase (V-type ATPase) that was found to be localised on the CV membrane of *Pleurochrysis* (Araki & Gonzalez, 1998; Corstjens, Araki, & Gonzalez, 2001). The V-type ATPase was predicted to direct H^+ into the CV, based on sequence similarity with the homologues of higher plants (Corstjens et al., 2001). Thus, this V-type ATPase is unlikely to participate in removing H^+ from CVs. As eukaryotic V-ATPases are generally localised on the ER and lysosome, the haptophyte V-type ATPase may be mainly located on the endomembrane system rather than the CVs (Taylor, Brownlee, & Wheeler, 2017). Currently, the exact role of V-type ATPase is unknown but an expression analysis strongly suggested involvement of this ATPase in calcification (Mackinder et al., 2011).

In contrast, transporters delivering HCO_3^- and Ca^{2+} into CVs are poorly characterised. Possible candidates involved in ion transport for calcification have been identified based on a transcriptome analysis. These include a SLC4-type HCO_3^- transporter, an ER-type Ca^{2+}-ATPase (ECA), and a Ca^{2+}/H^+ exchanger (CAX) (Mackinder et al., 2011). In the diatom *P. tricornutum*, plasma membrane SLC4 transporters mediate uptake of HCO_3^- into the cytosol (Nakajima et al., 2013). The expression level of the gene encoding a SLC4 homologue (also called anion exchanger-like 1, AEL1) in a calcifying *E. huxleyi* strain is much higher than that in a noncalcifying strain, suggesting that the SLC4 homologue supplies HCO_3^- required for calcification (Mackinder et al., 2011; von Dassow et al., 2009). Additionally, two putative Ca^{2+}/H^+ exchangers (CAX3 and CAX4) have been strongly suggested to play a role in Ca^{2+} delivery, because expression of these genes is upregulated in a calcifying *E. huxleyi* compared with that in a noncalcifying strain. Besides the CAX proteins, a putative Ca^{2+}-ATPase (ECA2) is also a potential candidate involved in Ca^{2+} delivery. Expression of the ECA2 gene is upregulated in calcifying conditions (>10 mM Ca^{2+} in the medium) compared with noncalcifying conditions (0 mM Ca^{2+}). Although the expression patterns of these candidate transporters strongly suggest that they play an important role in calcification, more research is needed to clarify their actual functioning.

A vacuole-like compartment that accumulates high concentrations of disordered Ca^{2+} and polyphosphate has been identified (Sviben et al., 2016). This compartment is proposed to supply Ca^{2+} to CVs (Sviben et al., 2016). Ion transporters are likely distributed in various membranes, such as the ER, CVs, and vacuole-like compartments, and then regulate ion traffic during calcification. Thus, a subcellular localisation analysis is essential to reveal the functions of these candidate ion transporters. Recently, a stable expression system has been established in the coccolithophore *Pleurochrysis carterae* and expression of the green fluorescence protein (GFP) has been successfully achieved (Endo et al., 2016). This transformation technique will facilitate molecular studies on calcification.

6. INITIAL PROCESS OF PHOTOSYNTHETIC CARBON METABOLISM

6.1 Active Anaplerotic β-Carboxylation

Some research suggests that diatoms use the C_4 photosynthetic pathway during photosynthesis, but there is no evidence of the occurrence of C_4 photosynthesis in haptophytes so far (Tsuji, Suzuki, & Shiraiwa, 2009). Direct CO_2 fixation by the Calvin–Benson cycle has been demonstrated by short-term ^{14}C tracer experiments in both noncalcifying haptophytes and a coccolithophore (Beardall, Mukerji, Glover, & Morris, 1976; Tsuji et al., 2009), suggesting that most haptophytes perform C_3 photosynthesis. In *E. huxleyi*, a significant portion of C_3 compounds derived from the Calvin–Benson cycle is converted into a C_4 compound (aspartate [Asp]) through β-carboxylation (Tsuji et al., 2009). The β-carboxylation reaction yields oxaloacetic acid (OAA) from C_3 compounds, such as phosphoenolpyruvate (PEP) or pyruvate, and the OAA is converted into stable C_4 compounds, such as malate or Asp. Generally, this reaction serves as an anaplerotic reaction to replenish intermediates of the tricarboxylic acid (TCA) cycle withdrawn for amino acid biosynthesis. Although anaplerotic β-carboxylation is required of all organisms to sustain the TCA cycle, *E. huxleyi* likely has a relatively high flux in this reaction compared with that of green algae and higher plants, as 10–15% of fixed ^{14}C is incorporated into Asp within 60 s (Tsuji et al., 2009). Such high flux of anaplerotic β-carboxylation has been reported in other algae such as diatoms and brown algae, suggesting that this process is a common feature of secondary algae that possess red algal-derived chloroplasts.

According to genomic information and an enzymatic assay with a cell-free extract, *E. huxleyi* has multiple OAA-producing enzymes presumably located in different compartments; a phosphoenolpyruvate carboxylase (EhPEPC) was predicted to be localised in mitochondria, two pyruvate carboxylases (EhPYC1 and EhPYC2) were predicted to function in chloroplasts and mitochondria, respectively (Tsuji & Shiraiwa, 2016; Tsuji, Suzuki, & Shiraiwa, 2012). Of these, localization of EhPYC1 was experimentally confirmed (Tsuji et al., 2012). Subcellular localisation of OAA-producing enzymes is different between *E. huxleyi* and higher plants. In plants, a cytosolic PEPC plays a major role in the anaplerotic reaction, PYC is absent, and no OAA-producing enzymes are present in the chloroplasts, except for rice that possesses a chloroplastic PEPC (Izui, Matsumura, Furumoto, & Kai, 2004; Masumoto et al., 2010).

Besides chloroplastic EhPYC1, the other OAA-producing enzymes (EhPEPC and EhPYC2) are predicted to be present in mitochondria (Tsuji et al., 2012). These enzymes are expected to supply OAA to the TCA cycle independently from EhPYC1 in chloroplasts. Multiple OAA-producing enzymes in different compartments would be responsible for the high Asp synthetic activity observed in the short-term *E. huxleyi* ^{14}C-labelling experiments. Curiously, malate also accumulates in various phototrophs but no accumulation of malate has been observed in *E. huxleyi* (Obata et al., 2013; Tsuji et al., 2009). The redundancy of OAA-producing enzymes suggests the occurrence of precise regulation of OAA production to meet different demands in different organelles of *E. huxleyi*.

6.2 Metabolite Transport Across Four Chloroplast Membranes

Haptophytes have complex chloroplasts, the so-celled secondary chloroplasts, surrounded by four membranes. Therefore, metabolites need to transport across the four membranes between the chloroplast stroma and cytosol in haptophytes, and this transport process would be more complex compared to that in organisms with primary chloroplasts. Metabolite transporters on chloroplasts have been studied mainly in higher plants (Facchinelli & Weber, 2011; Tsuji & Shiraiwa, 2016), and firstly we explain them briefly. Plant chloroplasts are surrounded by two envelope membranes. The outer membrane is highly permeable to metabolites due to the existence of pore-forming proteins (Flügge & Benz, 1984). In contrast, the inner membrane is a physical barrier to metabolite transport and requires transporters for permeation of metabolites. In higher plants, the family of plastidic phosphate translocators (pPTs) play a crucial role in metabolite transport across the inner envelope membrane (Fliege, Flügge, Werdan, & Heldt, 1978; Flügge, 1999). The representative pPT is a triose phosphate/phosphate translocator (TPT), which mediates the exchange of an inorganic phosphate for a triose phosphate, mainly as a form of dihydroxyacetone phosphate or

3-phosphoglycerate (3-PGA). The TPT plays a major role in exporting fixed carbon from the chloroplasts of plants and green algae (Facchinelli & Weber, 2011; Flügge, 1999). Another important pPT is the PEP/phosphate translocator (PPT), which mediates exchange of PEP and an inorganic phosphate (Fischer et al., 1997). PEP is utilised as a precursor in various biosynthetic pathways, such as the shikimate pathway, methylerythritol phosphate (MEP) pathway, and fatty acid biosynthesis, which play crucial roles in chloroplast development (Flügge, Häusler, Ludewig, & Gierth, 2011; Streatfield et al., 1999). PEP is not directly supplied from the Calvin cycle in the photosynthetic chloroplasts of higher plants, because the enzymes converting 3-PGA to PEP (i.e. enolase and/or 3-PGA mutase) are missing (Flügge et al., 2011 and references therein).

In contrast to higher plants, most features of metabolite transport in the four membrane-bound chloroplasts in secondary algae are speculative. According to genomic information, heterokontophytes and haptophytes have putative pPTs, which are closely related to red algal TPT (Moog, Rensing, Archibald, Maier, & Ullrich, 2015; Weber, Linka, & Bhattacharya, 2006). The diatom *P. tricornutum* has at least nine putative TPTs and the subcellular localisation of four of them has been determined (Moog et al., 2015). Four putative TPTs were localised in one of three chloroplast membranes, except for the third outermost membrane that corresponds to the outer membrane of primary chloroplast. These studies suggest that putative TPTs of secondary algae would be involved in the export of triose phosphates from the secondary chloroplasts whereas no experimental data are available on substrate spectrum. Notably, substrate preference of pPTs is generally unpredictable from their primary amino acid sequences (Lim, Linka, Mullin, Weber, & McFadden, 2010); thus, an experimental examination of the substrate spectrum of putative TPTs is necessary.

Although PPT is conserved in green and red algae (Flügge et al., 2011; Linka, Jamai, & Weber, 2008; Streatfield et al., 1999), no obvious PPTs have been identified in either haptophytes or diatoms (Weber et al., 2006). Therefore, the pathway to supply PEP into the secondary chloroplasts is unclear. It is possible that PEP is directly supplied from Calvin cycle intermediates. This is supported by the fact that *E. huxleyi* has putative chloroplastic isoforms of enzymes converting 3-PGA to PEP (3-PGA mutase and enolase), although the localisation of these enzymes has not been experimentally examined (Tsuji & Shiraiwa, 2016; Tsuji et al., 2009). Alternatively, it is possible that as yet unidentified PEP transporters exist. Several transporters besides pPTs have been identified in plants: For example, dicarboxylate transporters (DiT), which are involved in 2-oxoglutarate import and the malate valve (Kinoshita et al., 2011), and a sodium-dependent pyruvate transporter (BASS2) (Furumoto et al., 2011). No homologues for DiT have been identified in red algae and secondary algae with red-algal-derived chloroplasts. Thus, further study is needed to understand central carbon metabolism in secondary algae including haptophytes.

7. COMPLEX FATE OF FIXED CARBON IN HAPTOPHYTES

7.1 Unique Photosynthetic Products

Unlike green algae and higher plants, haptophytes produce soluble β-glucans as a carbon/energy storage source instead of starch. In addition to β-glucans, haptophytes accumulate large quantities of lipid (Fernández, Marañón, & Balch, 1996; Pond & Harris, 1996; Tsuji, Yamazaki, Suzuki, & Shiraiwa, 2015; Tsuji et al., 2009). Triacylglycerol (TAG) is a typical storage lipid in most microalgae (Yoshida, Tanabe, Yonezawa, & Watanabe, 2012), whereas some specific haptophytes accumulate extremely long chain (C_{37}–C_{40}) unsaturated ketones called alkenones (Fig. 5; Pond & Harris, 1996; Volkman, Everitt, & Allen, 1986). Although the function of alkenones in haptophytes is controversial, recent researches strongly suggests that alkenones are storage compounds utilised as an energy/carbon source during the dark period (Eltgroth, Watwood, & Wolfe, 2005; Epstein, D'Hondt, & Hargraves, 2001; Pan & Sun, 2011; Tsuji et al., 2015). In *E. huxleyi*, alkenones are the major storage compounds and the contribution of β-glucans is less significant. We introduce several unique photosynthetic products of particular haptophytes and discuss whole carbon metabolism in this section.

7.2 β-Glucans

Glucans are glucose polymers produced as dominant photosynthetic products in diverse phototrophs (Suzuki & Suzuki, 2013). Green algae and higher plants produce insoluble α-glucans (starch) consisting of amylose (α-1,4-linked glucoses) and amylopectin (α-1,4- and α-1,6-linked glucoses). In contrast, secondary algae containing red algal-derived chloroplasts produce β-glucans instead of α-glucans (Suzuki & Suzuki, 2013). Most haptophytes produce soluble β-glucans containing β-(1→3)- and β-(1→6)-linkages with the exception of *Pavlova lutheri*, which produces crystallised glucan

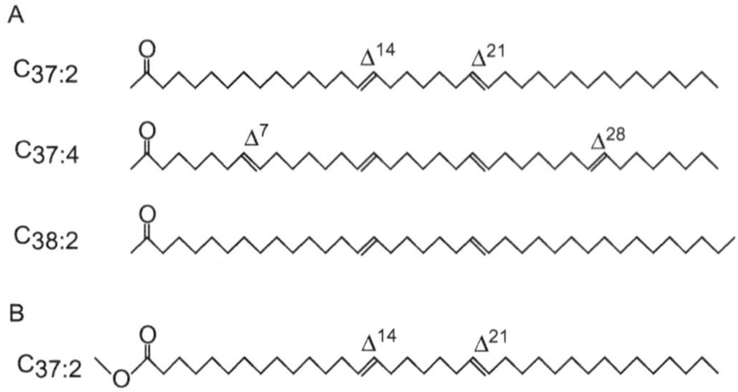

Fig. 5 Structures of alkenones and related compounds. (A) Structure of methyl alkenones. (B) Structure of $C_{37:2}$ alkenoate.

(Table 2; Hirokawa et al., 2008; Janse et al., 1996; Sadovskaya et al., 2014; Vårum et al., 1986). Most noncalcifying haptophytes produce β-glucans consisting of β-(1 → 3)-linked glucoses as the main chain with branched β-(1 → 6)-linkages (Table 2; Janse et al., 1996; Sadovskaya et al., 2014). This structure is the same as that of chrysolaminarin, which is a storage β-glucan in diatoms (Paulsen & Myklestad, 1978). On the other hand, β-glucans of coccolithophores contain the β-(1 → 6)-linkage as a backbone with β-(1 → 3) linkage branches (Hirokawa et al., 2008; Vårum et al., 1986). Additionally, the molecular weight of β-glucans in coccolithophores is much larger than that of a typical chrysolaminarin (Table 2).

Enzymes involved in β-glucan synthesis have not been identified in haptophytes, but an early research on diatoms and *Euglena* suggested that UDP-glucose is a precursor for β-1,3-glucan (Roessler, 1987). Furthermore, knock-down or knock-out of UDP-glucose pyrophosphorylase results in decreased accumulation of β-glucans in *P. tricornutum* (Zhu et al., 2016). No experimental data are available for haptophytes, but β-glucans might be synthesised from UDP-glucose as in diatoms. In several haptophytes, significant accumulation of β-glucans in the light and subsequent consumption in the dark have been reported, indicating that β-glucans serve as a major carbon/energy storage in the dark (Hirokawa et al., 2008; Janse et al., 1996).

7.3 Alkenones

Some haptophytes produce very long chain unsaturated ketones called alkenones (Conte & Eglinton, 1993; Rontani, Prahl, & Volkman, 2006). Alkenones have unique features in their structure, such as extremely long

Table 2 Structural Features of β-Glucan From Diatoms and Haptophytes

Group	Organisms	Main Chain	Molecular Weight	Degree of Polymerisation	Form in the Cell	References
Diatom	*Skeletonema costatum*	β-1,3-linked glucose	6200	n.d.	Water soluble	Paulsen and Myklestad (1978)
	Mixed culture of freshwater diatoms	β-1,3-linked glucose[a]	3500	21	Water soluble	Beattie, Hirst, and Percival (1961)
Noncalcifying haptophytes	*Isochrysis galbana*	β-1,6-linked glucose	n.d.	n.d.	Water soluble	Sadovskaya et al. (2014)
	Phaeocystis globosa	β-1,3-linked glucose	3500	20	Water soluble	Janse et al. (1996)
	Pavlova lutheri	n.d.	n.d.	n.d.	Crystalline	Kiss and Triemer (1988)
Coccolithophores	*Emiliania huxleyi*	mainly β-1,6-linked glucose[b]	n.d.	106	Water soluble	Vårum, Kvam, Myklestad, and Paulsen (1986)
	Pleurochrysis haptonemofera	n.d.[c]	34,000	203	Water soluble	Hirokawa et al. (2008)

[a]The main chain also contains a tiny amount of β-1,6-linkages.
[b]The main chain also contains a small amount of β-1,3-linkages.
[c]Details were not analysed but the ratio of β-1,6-linkages to β-1,3-linkages was 3:2.
n.d., No data.

chains (C_{37}–C_{40}) and the occurrence of *trans*-double bonds and keto groups (Fig. 5). Five species, including *E. huxleyi*, *Gephyrocapsa oceanica*, *Tisochrysis lutea* (formerly *Isochrysis galbana*, Tahiti strain), *Chrysotile lamellose,* and *Isochrysis galbana*, are known to produce alkenones and related compounds such as alkenoates. These alkenone producers accumulate no or negligible amounts of TAG, which is a typical storage lipid in various algae.

Several possible functions of alkenones have been proposed, such as buoyancy compounds to cancel the weight of coccoliths (Fernández et al., 1996; Sawada & Shiraiwa, 2004), membrane components (Dunstan, Volkman, Barrett, & Garland, 1993; Prahl, Muehlhausen, & Zahnle, 1988), and storage compounds to supply energy/carbon in the dark (Eltgroth et al., 2005; Epstein et al., 2001). Early studies assumed that alkenones were membrane lipids, because the number of double bonds increases when cells are grown under low temperatures. This response is similar to that observed in membrane lipids in which the number of *cis*-double bonds increases to sustain membrane fluidity. However, unlike typical membrane lipids containing *cis*-double bonds, alkenones contain *trans*-double bonds, which do not alter linear structure (Fig. 5). Subcellular fractionation has also demonstrated that alkenones are not in cellular membranes (Eltgroth et al., 2005; Sawada & Shiraiwa, 2004). Therefore, alkenones are unlikely to be membrane lipids.

More recent research involving subcellular fractionation and microscopic analyses strongly suggests that alkenones accumulate in lipid droplets in the cytosol (Eltgroth et al., 2005; Shi et al., 2015). Analyses of lipid composition in isolated lipid droplets revealed that alkenones account for >70% of lipids (Eltgroth et al., 2005; Shi et al., 2015), and thus the lipid droplets are called 'alkenone bodies' (Shi et al., 2015). A proteome analysis identified that alkenone bodies contain a V-type H^+-ATPase, suggesting that alkenone bodies are derived from the budding of the ER or CER membrane (Shi et al., 2015).

When cells are kept in the dark, the alkenone content decreases concomitantly with the size decrease of alkenone bodies (Eltgroth et al., 2005; Epstein et al., 2001; Pan & Sun, 2011; Tsuji et al., 2015). This fact strongly suggests that alkenones are a carbon/energy storage source to maintain cellular activities in the dark. Alkenones and β-glucans may play similar roles in carbon/energy storage. However, in *E. huxleyi*, the partitioning of photosynthetically fixed carbons into alkenones is much higher than that into β-glucans, suggesting that alkenones contribute significantly to supply energy and carbon skeletons (Tsuji et al., 2015).

7.4 Acid Polysaccharides

Coccolithophores produce APs, which are structurally diverse (Fichtinger-Schepman, Kamerling, Versluis, & Vliegenthart, 1981; Hirokawa, Fujiwara, & Tsuzuki, 2005; Marsh, 1999; Ozaki, Sakuda, & Nagasawa, 2007; Westbroek, de Jong, Dam, & Bosch, 1973). APs are involved in coccolith morphogenesis by controlling nucleation and growth of $CaCO_3$ crystals in the CV (Marsh, 1999). In vitro research has demonstrated that APs have a high capacity to bind Ca^{2+} and inhibit precipitation of $CaCO_3$ (Borman et al., 1982; Ozaki et al., 2007). These features suggest that APs play an important role in the coccolith morphogenesis. Different haptophyte species produce different types of APs, suggesting a contribution of APs to the species-specific shapes of coccoliths. *Pleurochrysis carterae* has four APs called PS-1, PS-2, PS-3, and the coccolith matrix AP (CMAP) (Fig. 6; Marsh, 1999; Ozaki et al., 2007). In *P. carterae*, PS-1 and PS-2 form a particulate complex with Ca^{2+} in Golgi-derived vesicles, which are fused to the CV where mineralisation occurs (Marsh, 1999). Both PS-1 and PS-2 are associated or embedded on the surface of $CaCO_3$ crystals and are ultimately displayed on the cell surface as a coccolith. Thus, APs are structural components and are unlikely to function as carbon/energy storage.

The structures of PS-2 and CMAP of *Pleurochrysis* have been determined (Fig. 6; Marsh et al., 1992; Ozaki et al., 2007). Interestingly, both APs are structurally very similar, but their inhibitory effect on in vitro $CaCO_3$ precipitation is opposite; CMAP significantly inhibits $CaCO_3$ precipitation at a low concentration (1 μg ml^{-1}), whereas PS2 does not even at a high concentration (100 μg ml^{-1}) (Ozaki et al., 2007). However, the PS-2 mutant of *P. carterae* shows a strong inhibition in coccolith formation, suggesting a significant contribution of PS-2 to in vivo mineralisation (Marsh & Dickinson, 1997).

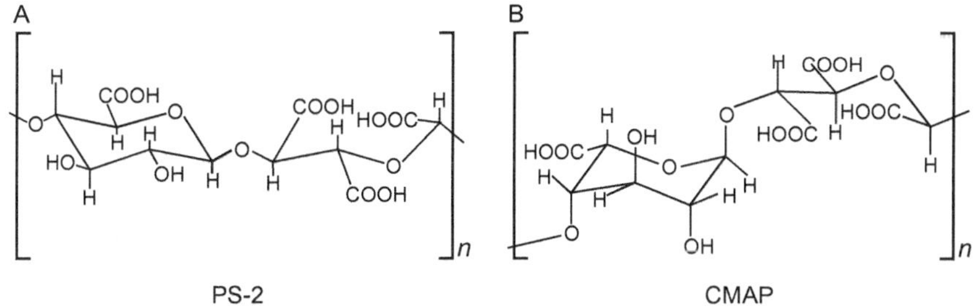

Fig. 6 Structure of acid polysaccharides (APs) from *Pleurochrysis*. (A) The structure of PS-2 from *P. carterae* (Marsh, Chang, & King, 1992). (B) The structure of coccolith matrix acidic polysaccharide (CMAP) from *P. haptonemofera (Ozaki et al., 2007)*.

In contrast to *Pleurochrysis*, *E. huxleyi* has only an AP, the so-called coccolith polysaccharide (CP), as a dominant organic component of its coccoliths (Fichtinger-Schepman et al., 1981; Kayano & Shiraiwa, 2009). The *E. huxleyi* CP also has the ability to bind Ca^{2+} and inhibit in vitro $CaCO_3$ crystallisation (Borman et al., 1982). The structure of the CP is highly complex and consists of a mannose chain containing acidic residues (sulphate esters) and side chains containing diverse sugars and acidic residues, such as xylose, rhamnose, and galacturonic acid. Similar to the *P. carterae* APs, the CP would be involved in coccolith morphogenesis by controlling $CaCO_3$ crystallisation in the CV.

7.5 Mannitol

In addition to high-molecular-weight compounds, substantial differences also exist in the profiles of low-molecular-weight compounds (LMCs) among organisms. Green algae and higher plants accumulate sucrose and fructose as major LMCs, whereas *E. huxleyi* predominantly accumulates mannitol and tiny amounts of sucrose and fructose (Obata et al., 2013). The cellular concentration of mannitol is estimated at 34.8 mM (Obata et al., 2013). Mannitol may be an energy/carbon storage source in *E. huxleyi*, based on its high accumulation level. Mannitol likely plays different physiological roles in different organisms (Iwamoto & Shiraiwa, 2005). In brown algae, mannitol functions as a compatible solute (Reed, Davison, Chudek, & Foster, 1985) or an energy/carbon storage source (Yamaguchi, Ikawa, & Nishizawa, 1966). The red alga *Caloglossa* lives in the estuarine zone and utilises mannitol as a compatible solute to acclimate to changes in the salinity of the environment (Iwamoto, Kawanobe, Ikawa, & Shiraiwa, 2003; Karsten et al., 1992).

A genomic search identified two possible pathways for mannitol synthesis in *E. huxleyi* (Obata et al., 2013). One is the higher plant-type pathway in which fructose 6-phosphate (F6P) is isomerised to mannose 6-phosphate, which in turn is reduced to mannitol 1-phosphate (M1P), and then M1P is dephosphorylated to produce mannitol. Another pathway is the brown algal type, in which F6P is directly reduced to M1P by mannitol 1-phosphate dehydrogenase (Iwamoto & Shiraiwa, 2005; Obata et al., 2013). Currently, it is unknown which is the major pathway in *E. huxleyi*.

7.6 Dimethylsulphoniopropionate

Another important LMC produced by haptophytes is dimethylsulphoniopropionate (DMSP), which functions as an osmolyte and antioxidant

(Gillian Malin & Kirst, 1997; Stefels, 2000; Sunda, Kieber, Kiene, & Huntsman, 2002). DMSP produced by phytoplankton is degraded into volatile dimethylsulphide (DMS), and then DMS oxidation products act as cloud condensation nuclei (CCN). Although marine phytoplankton have been proposed as major contributors to CCN production, accumulating studies suggested that phytoplankton-derived DMS is less significant in regulation of the climate change (Quinn & Bates, 2011). Rather, DMSP producers are important driving forces of the oceanic sulphur cycle. Among various phytoplankton, haptophytes tend to accumulate high concentrations of DMSP (Keller, Kiene, Matrai, & Bellows, 1999; Gill Malin & Steinke, 2004; Gillian Malin & Kirst, 1997).

Typically, cellular DMSP concentrations exceed 100 mM in haptophytes, dinoflagellates (Malin & Steinke, 2004), occupying 50–70% of total organic sulphur in the cell (Matrai & Keller, 1994). DMSP functions as a compatible solute. In noncalcifying haptophytes, such as *Phaeocystis* sp. and *Prymnesium parvum*, cellular DMSP concentration increases with the salinity of the growth medium (Dickson & Kirst, 1987; Stefels, 2000). More recent research has demonstrated that DMSP plays a role as an antioxidant, because DMSP and its derivatives (DMS, acrylate, dimethylsulphoxide, and methane sulfinic acid) effectively remove hydroxyl radicals and other reactive oxygen species (Sunda et al., 2002). Treatments that trigger oxidative stress in *E. huxleyi* significantly increase cellular DMSP concentration compared with that in untreated cells (Sunda et al., 2002). Additionally, cellular DMSP levels increase when growth temperature decreases (van Rijssel & Gieskes, 2002). The change in cellular DMSP concentration in response to various environmental factors suggests that DMSP plays multiple roles in haptophytes besides its function as a compatible solute.

In marine phytoplankton, including *E. huxleyi*, DMSP synthesis has been deduced from tracer experiments with ^{35}S-methionine (Gage et al., 1997). Methionine is deaminated into 4-methylthio-2-oxobutylate (MTOB), which is reduced to 4-methylthio-2-hydroxybutyrate and subsequent methylation yields 4-dimethylsulphonio-2-hydroxybutyrate (DMSHB). Finally, DMSHB is converted into dimethylsulphoniopropionate (DMSP) by oxidative decarboxylation (Gage et al., 1997). Synthesis of DMSP has been suggested to serve as remobilisation of cellular nitrogen under nitrogen-limited conditions, because NH_3 is released from methionine during the first step. Gage et al. (1997) found that nitrogen depletion stimulates DMSP accumulation, whereas another study found no clear relationship between nitrogen availability and DMSP accumulation (Keller et al.,

1999). Therefore, more research is required to explore the hypothesis that DMSP synthesis supports nitrogen remobilisation.

8. CARBON PARTITIONING INTO ALKENONES, β-GLUCANS, AND OTHER COMPOUNDS IN *E. huxleyi*

As mentioned above, *E. huxleyi* produces a variety of compounds by photosynthesis, such as alkenones, β-glucans, APs, mannitol, and DMSP. These compounds have been independently studied by researchers in different fields, and simultaneous measurements of carbon flux into each compound had not been conducted until recently. As a consequence, although alkenones, β-glucans, and mannitol have been suggested to be energy/carbon storage compounds, the capacity and contribution of each compound to storage have been elusive. Fernández et al. (1996) reported that lipids are the most dominant carbon pool, occupying 45–60% of total organic carbon in *E. huxleyi*. The polysaccharide fraction is also significant, occupying 20–26%. However, because neither lipid nor polysaccharide fractions have been analysed further, carbon partitioning into alkenones, β-glucans, and APs is unknown. Our recent study established fine fractionation of *E. huxleyi* cells, allowing quantitative measurements of carbon flux into alkenones, APs, β-glucans, and mannitol (Fig. 7; Tsuji et al., 2015). Consistent with the results of Fernández et al. (1996), lipids were the most dominant product in *E. huxleyi*, occupying 36–40% of total fixed carbon. Further separation of alkenones from the lipid fraction revealed that alkenones occupy 17%. Compared with alkenones, carbon flux into β-glucans is very low (<1%). Fernández et al. (1996) reported significant flux into the polysaccharide fraction, but most carbons in this fraction are likely to be APs involved in calcification rather than β-glucans. Indeed, our study revealed significant flux (15–20%) of fixed carbon into APs, and most existed as extracellular coccoliths (Fig. 7; Tsuji et al., 2015). Mannitol was the most dominant compound in the LMC fraction in our experiments, as reported by Obata et al. (2013). However, the mannitol composition (< 5%) is less significant compared with alkenones (17%) (Fig. 7). These data indicate that excess organic carbon is mainly stored as alkenones, rather than as β-glucans. Taken together with previous studies showing a decrease in alkenones under dark conditions, alkenones are considered to be a major energy/carbon storage source in *E. huxleyi*. Fernández et al. (1996) observed reallocation of carbon from lipids to proteins in the dark. Although the authors did not examine alkenones, it is possible that

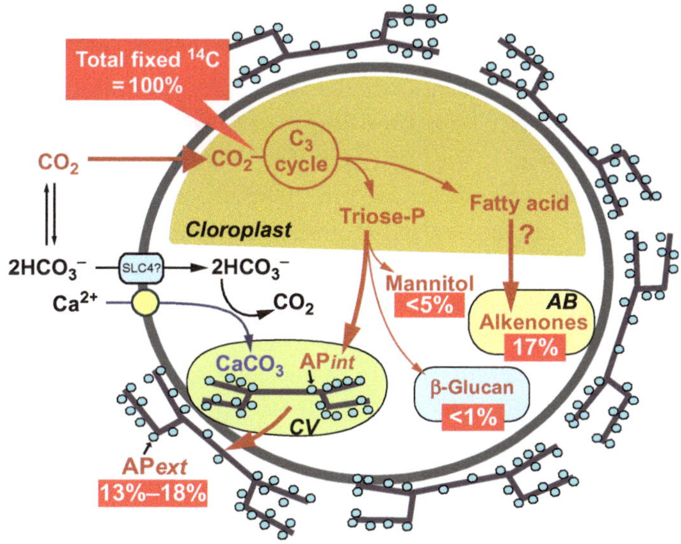

Fig. 7 Photosynthetic carbon partitioning in the coccolithophore *E. huxleyi* NIES837. Red allows indicate carbon flux from CO_2 to major carbon compounds (alenones, β-glucan, Aps, and mannitol). Each number represents the ratio of carbon compounds when total fixed carbon is 100%. AB, alkenone body; AP*int*, internal AP; AP*ext*, external AP; CV, coccolith vesicle.

alkenones are utilised as a carbon source for protein synthesis as well as a respiratory substrate because alkenones are a major constituent of the lipid fraction.

Carbon flux into alkenones does not change significantly in cells between the logarithmic growth and the stationary phase (Tsuji et al., 2015). A low carbon flux into β-glucans can also be observed regardless of the growth phase. These results suggest that high carbon flux into alkenones and low flux into β-glucans are intrinsic physiological properties of *E. huxleyi*, rather than a characteristic controlled by environmental factors. Consistent with our findings, Conte, Thompson, Lesley, and Harris (1998) also reported a constant ratio of alkenones to total organic carbon throughout the growth phase. Based on these data, alkenones are unlikely to be buoyancy compounds that cancel the heaviness of coccolith, because carbon flux into alkenones is fairly stable, even during the stationary phase where coccolith production is stimulated by nutrient limitations (Kayano & Shiraiwa, 2009; Paasche, 2001).

As described above, *E. huxleyi* carbon metabolism is characterised by high accumulation of alkenones and low accumulation of β-glucans

(Fig. 7). However, the reason why *E. huxleyi* accumulates alkenones instead of glucan is unknown. Furthermore, the reason why *E. huxleyi* and some haptophytes produce alkenones instead of TAG, a common storage lipid in other algae, is also unknown. One possible explanation is the photochemical stability of alkenones. Rontani et al. (2006) reported that *trans*-double bonds in alkenones are resistant to photochemical degradation compared with *cis*-double bonds. *E. huxleyi* frequently blooms in the stratified ocean surface in summer; thus, the photochemical stability of alkenones would be an advantage to survive the high irradiance conditions. Notably, the high accumulation of lipids and low accumulation of β-glucan observed in *E. huxleyi* are a not common feature of haptophytes. Significant accumulation of β-glucans (20–50 pg per cell) has been reported in the nonalkenone-producing coccolithophore *Pleurochrysis haptonemofera* (Hirokawa et al., 2008), which is comparable to that in the diatom *S. costatum*. More than 80% of β-glucans are degraded within 24 h when cells are transferred to the dark, suggesting that β-glucans are a primary energy/carbon storage source in *P. haptonemofera*. The difference between *E. huxleyi* and *P. haptonemofera* suggests the occurrence of substantial variations in carbon metabolism between alkenone-producing and nonalkenone-producing coccolithophores.

9. PHYSIOLOGICAL DIFFERENCES BETWEEN DIATOMS AND COCCOLITHOPHORES

Both diatoms and coccolithophores frequently bloom in the ocean. However, significant differences exist in the conditions triggering blooms between these two algal groups. Diatoms tends to bloom in well-mixed nutrient-rich conditions, whereas coccolithophores prefer the stratified surface layer where nutrients are depleted and cells are exposed to high irradiance (Tyrrell & Merico, 2004). Because each bloom type involves a variety of factors and conditions, it is difficult to explain all blooms using a single reason. However, diatoms and coccolithophores have different physiological properties of photosynthesis and nutrient assimilation, which might be related with the different conditions for blooms. Here, we discuss the physiological differences between diatoms and the representative coccolithophore *E. huxleyi*.

One important physiological property of diatoms is their large capacity for dark nitrate (NO_3^-) assimilation (Fig. 8). Because large amounts of reducing equivalents and ATP are required to convert NO_3^- into amino acids

(10 electrons and 1 ATP), NO_3^- assimilation is generally a light-dependent process. However, marine diatoms maintain significant NO_3^- assimilation activity in the dark (Clark, Flynn, & Owens, 2002; Needoba & Harrison, 2004). Based on measurements of cellular organic N content during growth under a light/dark cycle, dark NO_3^- assimilation rates are estimated at 30–50% of those in diatoms in light (Clark et al., 2002; Needoba & Harrison, 2004). In contrast to diatoms, *E. huxleyi* has up to a 10% dark assimilation rate compared with that in the light (Needoba & Harrison, 2004). The high rate of dark NO_3^- assimilation in diatoms supports their rapid growth under NO_3^- replete conditions (Clark et al., 2002). Diatoms have a large vacuole and can store inorganic nutrients (Fig. 8; Needoba & Harrison, 2004). The combination of NO_3^- stored in a vacuole and dark NO_3^- uptake serve as a source for dark NO_3^- assimilation (Needoba & Harrison, 2004). In contrast, *E. huxleyi* has no such vacuole and the intracellular NO_3^- pool is much smaller than that in diatoms (Fig. 8; Needoba & Harrison, 2004).

The reducing equivalents and ATP required for dark NO_3^- assimilation are supplied by the degradation of energy storage sources, most probably β-glucans (Handa, 1969; Vårum & Myklestad, 1984). β-glucans also supply 2-oxoglutarate, an NH_3 acceptor, through their degradation to C_3 compounds and the subsequent anaplerotic β-carboxylation reaction (Fig. 8; Granum & Myklestad, 1999). To supply 2-oxoglutarate, β-glucans are degraded into PEP or pyruvate via glycolysis, and then β-carboxylation of PEP/pyruvate (C_3) yields OAA (C_4), which replenishes the TCA cycle intermediates withdrawn for amino acid synthesis (Granum & Myklestad, 1999). To meet the carbon skeleton demand for N assimilation, diatoms have a significant CO_2 (or HCO_3^-) fixation rate via anaplerotic β-carboxylation in the dark, and this rate is up to 10% of photosynthetic CO_2 fixation (Morris, Yentsch, & Yentsch, 1971; Mortain-Bertrand, Descolas-Gros, & Jupin, 1988). This feature also differs from *E. huxleyi*, which has a 2% dark CO_2 fixation rate compared to the photosynthetic CO_2 fixation, indicating a low NO_3^- assimilation rate in the dark (Tsuji et al., 2009).

Diatoms accumulate excess carbon as a form of TAG in addition to β-glucans. However, the primary carbon/energy source of dark NO_3^- assimilation is likely to be β-glucan, because diatoms use β-glucans as their primary respiratory substrate and TAG is degraded when β-glucans are exhausted (Handa, 1969). Efficient CCM probably allows diatoms to store sufficient β-glucans during the day. Then, the accumulated β-glucans fuel carbon and energy for dark NO_3^- assimilation (Fig. 8). Again, this feature

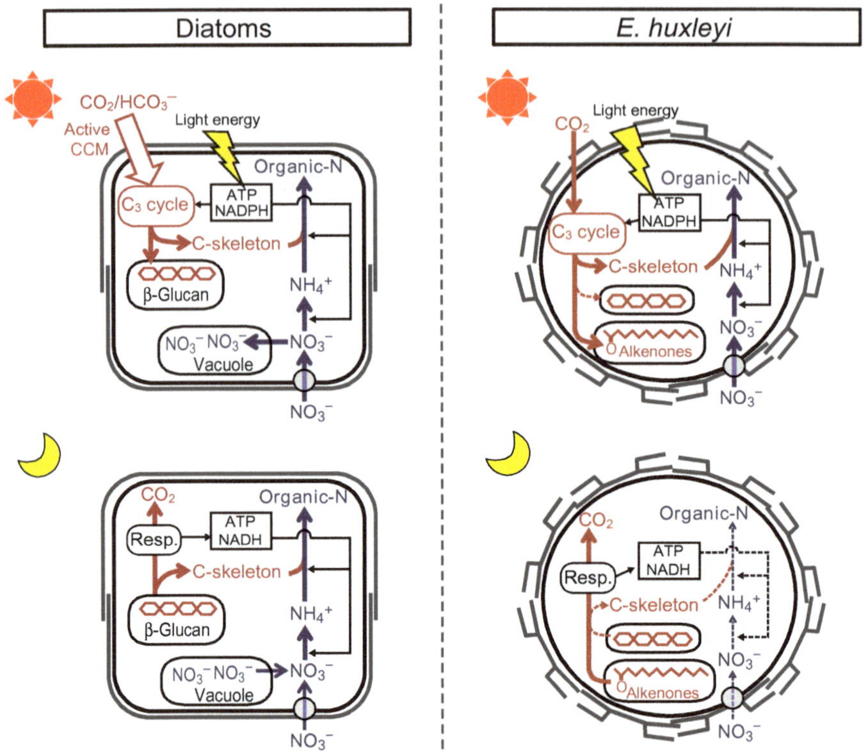

Fig. 8 Physiological differences of diatoms and the representative coccolithophore *E. huxleyi* with respect to light/dark metabolism. During the day, diatoms actively fix carbon with support of CCM and then accumulate excess carbon as β-glucan. A portion of the fixed carbon is converted into organic acids and then supplied as carbon skeletons (C-skeleton) for N assimilation. Light energy drives uptake and reduction of NO_3^-, and excess NO_3^- is accumulated in vacuoles. Unlike diatoms, *E. huxleyi* accumulates a low amount of β-glucan and instead accumulates alkenones as the major storage compound. *E. huxleyi* cannot store excess NO_3^- since this alga does not have vacuoles. At night, diatoms can continue NO_3^- assimilation using energy and reducing power supplied from the respiration of β-glucans. C-skeletons are also supplied from β-glucan via the anaplerotic β-carboxylation reaction. In contrast, *E. huxleyi* shows low or negligible flux of NO_3^- assimilation at night (shown as dotted lines).

differs from *E. huxleyi*, which has low CCM efficiency and accumulates alkenones rather than a β-glucan as a main storage (Tsuji et al., 2015).

E. huxleyi is superior at tolerating environmental stress compared with diatoms. Such high stress tolerance can be an advantage in the nutrient-limited surface layer, where cells cannot dissipate light energy by nutrient assimilation. *E. huxleyi* seems to have some mechanism to protect the photosynthetic machinery under stressed conditions, such as nutrient limitation.

When the maximum quantum yield of photosystem II (F_v/F_m) was monitored after transfer to an N-depleted condition, F_v/F_m in *E. huxleyi* was maintained at significantly high levels (50% of initial value) after 38 days of N-depletion (Loebl, Cockshutt, Campbell, & Finkel, 2010). In contrast, the F_v/F_m value of the diatom *T. pseudonana* rapidly dropped to 25% of the initial value within 10 days (Loebl et al., 2010). Different patterns of F_v/F_m decline during senescence of *E. huxleyi* and *T. pseudonana* have also been demonstrated (Franklin et al., 2012). Additionally, *E. huxleyi* has remarkable tolerance against high light intensity (Paasche, 2001; Ragni, Airs, Leonardos, & Geider, 2008; Tyrrell & Merico, 2004). *E. huxleyi* does not exhibit photoinhibition even under irradiance of 2000 μmol photons $m^{-2} s^{-1}$, which corresponds to direct sunlight in summer (Ragni et al., 2008). These data clearly indicate that *E. huxleyi* can sustain photosynthetic machinery under stressed conditions, whereas the detailed mechanism has not been revealed. This feature would be one reason explaining the dominance of *E. huxleyi* over diatoms under a stratified surface layer.

10. CONCLUDING REMARKS

Cellular processes involved in calcification and photosynthesis in calcifying haptophytes are complex. Despite their ecological importance, molecular studies involving haptophytes are highly limited, in contrast to research focusing on diatoms. Although important insights have been elucidated from the genomic sequence of *E. huxleyi*, careful inspection using genetic transformants is necessary to demonstrate the physiological significance of the specific genes or metabolic pathways that have been suggested by sequence-based analysis. Studies of haptophytes are limited by this step, because genetic transformation techniques have not yet been established. Recently, the coccolithophore *P. carterae* has been stably transformed, which will aid molecular studies of haptophytes as in diatoms. Additionally, considering the genetic diversity of haptophytes, comparative studies with different species, including both noncalcifying species and coccolithophores, will be required to reveal universal features conserved in all haptophytes and characteristics in specific species such as *E. huxleyi*.

ACKNOWLEDGEMENTS

Our work was supported by JSPS KAKENHI grant numbers 16H06557, 17K15326 and 15K16156 (to Y.T.), by the Science Research Promotion Fund of the Promotion and Mutual Aid Corporation for Private Schools of Japan (to Y.T.), by the Individual Special Research Fund of Kwansei Gakuin University. Special thanks are extended to the members of "GELPOD Cruise Shinkai-2000/Natsushima" for collecting the materials used in Fig. 2 of this chapter. This cruise was sponsored by the Japan Agency for Marine-Earth Science and Technology.

REFERENCES

Adl, S. M., Simpson, A. G. B., Lane, C. E., Lukeš, J., Bass, D., Bowser, S. S., et al. (2012). The revised classification of eukaryotes. *Journal of Eukaryotic Microbiology, 59*, 429–514.

Aizawa, K., & Miyachi, S. (1986). Carbonic anhydrase and CO_2 concentrating mechanisms in microalgae and cyanobacteria. *FEMS Microbiology Letters, 39*, 215–233.

Araki, Y., & Gonzalez, E. L. (1998). V- and P-type Ca^{2+}-stimulated atpases in a calcifying strain of *Pleurochrysis* sp. (haptophyceae). *Journal of Phycology, 34*, 79–88.

Bach, L. T., Mackinder, L. C. M., Schulz, K. G., Wheeler, G., Schroeder, D. C., Brownlee, C., et al. (2013). Dissecting the impact of CO_2 and pH on the mechanisms of photosynthesis and calcification in the coccolithophore *Emiliania huxleyi*. *The New Phytologist, 199*, 121–134.

Badger, M. R., Andrews, T. J., Whitney, S. M., Ludwig, M., Yellowlees, D. C., Leggat, W., et al. (1998). The diversity and coevolution of rubisco, plastids, pyrenoids, and chloroplast-based CO_2-concentrating mechanisms in algae. *Canadian Journal of Botany, 76*, 1052–1071.

Beardall, J., Mukerji, D., Glover, H. E., & Morris, I. (1976). The path of carbon in photosynthesis by marine phytoplankton. *Journal of Phycology, 12*, 409–417.

Beattie, A., Hirst, E. L., & Percival, E. (1961). Studies on the metabolism of the Chrysophyceae. Comparative structural investigations on leucosin (chrysolaminarin) separated from diatoms and laminarin from the brown algae. *Biochemical Journal, 79*, 531–537.

Bendif, E. M., Probert, I., Hervé, A., Billard, C., Goux, D., Lelong, C., et al. (2011). Integrative taxonomy of the Pavlovophyceae (Haptophyta): A reassessment. *Protist, 162*, 738–761.

Bhatti, S., Huertas, I. E., & Colman, B. (2002). Acquisition of inorganic carbon by the marine haptophyte *Isochrysis galbana* (Prymnesiophyceae). *Journal of Phycology, 38*, 914–921.

Billard, C., & Inouye, I. (2004). What is new in coccolithophore biology? In H. R. Thierstein & J. R. Young (Eds.), *Coccolithophores from molecular processes to global impact* (pp. 1–30). Berlin: Springer-Verlag.

Borman, A. H., Jong, E. W., Huizinga, M., Kok, D. J., Westbroek, P., & Bosh, L. (1982). The role in CaCO3 crystallization of an acid Ca2+-binding polysaccharide associated with coccoliths of *Emiliania huxleyi*. *European Journal of Biochemistry, 129*, 179–183.

Bown, P. R., Lees, J. A., & Young, J. R. (2004). Calcareous nannoplankton evolution and diversity through time. In *Coccolithophores* (pp. 481–508). Berlin, Heidelberg: Springer.

Burns, B. D., & Beardall, J. (1987). Utilization of inorganic carbon by marine microalgae. *Journal of Experimental Marine Biology and Ecology, 107*, 75–86.

Chrétiennot-Dinet, M. J., Giraud-Guille, M. M., Vaulot, D., Putaux, J. L., Saito, Y., & Chanzy, H. (1997). The chitinous nature of filaments ejected by *Phaeocystis* (Prymnesiophyceae). *Journal of Phycology, 33*, 666–672.

Clark, D. R., Flynn, K. J., & Owens, N. J. P. (2002). The large capacity for dark nitrate-assimilation in diatoms may overcome nitrate limitation of growth. *New Phytologist, 155*, 101–108.

Conte, M. H., & Eglinton, G. (1993). Alkenone and alkenoate distributions within the euphotic zone of the eastern North Atlantic: Correlation with production temperature. *Deep Sea Research Part I: Oceanographic Research Papers, 40*, 1935–1961.

Conte, H. M., Thompson, A., Lesley, D., & Harris, R. P. (1998). Genetic and physiological influences on the alkenone/alkenoate versus growth temperature relationship in *Emiliania huxleyi* and *Gephyrocapsa Oceanica*. *Geochimica et Cosmochimica Acta*, *62*, 51–68.

Corstjens, P. L. A. M., Araki, Y., & Gonzalez, E. L. (2001). A coccolithophorid calcifying vesicle with a vacuolar-type ATPase proton pump: Cloning and immunolocalization of the V_o subunit *c*. *Journal of Phycology*, *37*, 71–78.

Cros, L., Kleijne, A., Zeltner, A., Billard, C., & Young, J. R. (2000). New examples of holococcolith–heterococcolith combination coccospheres and their implications for coccolithophorid biology. *Marine Micropaleontology*, *39*, 1–34.

de Vrind-de Jong, E. W., Emburg, P. R., & de Vrind, J. P. M. (1994). Mechanisms of calcification: *Emiliania huxleyi* as a model system. In J. C. Green & B. S. C. Leadbeater (Eds.), *The haptophyte algae* (pp. 149–166). Oxford: Oxford Univ. Press.

Dickson, D. M. J., & Kirst, G. O. (1987). Osmotic adjustment in marine eukaryotic algae: The role of inorganic ions, quaternary ammonium, tertiary sulphonium and carbohydrate solutes. II. Prasinophytes and haptophytes. *New Phytologist*, *106*, 657–666.

Dunstan, G. A., Volkman, J. K., Barrett, S. M., & Garland, C. D. (1993). Changes in the lipid composition and maximisation of the polyunsaturated fatty acid content of three microalgae grown in mass culture. *Journal of Applied Phycology*, *5*, 71–83.

Edvardsen, B., Eikrem, W., Green, J. C., Andersen, R. A., Moon-van der Staay, S. Y., & Medlin, L. K. (2000). Phylogenetic reconstructions of the Haptophyta inferred from 18S ribosomal DNA sequences and available morphological data. *Phycologia*, *39*, 19–35.

Ehrenberg, C. G. (1836). *Bemerkungen uber feste mikroscopische anorganische Formen in den erdigen und derben Mineralien* (pp. 84–85). Berlin: Bericht. Verh. K. Preuss. Akad. Wiss.

Eltgroth, M. L., Watwood, R. L., & Wolfe, G. V. (2005). Production and cellular localization of neutral long-chain lipids in the haptophyte algae *Isochrysis galbana* and *Emiliania huxleyi*. *Journal of Phycology*, *41*, 1000–1009.

Endo, H., Yoshida, M., Uji, T., Saga, N., Inoue, K., & Nagasawa, H. (2016). Stable nuclear transformation system for the coccolithophorid alga *Pleurochrysis carterae*. *Scientific Reports*, *6*, 22252.

Epstein, B. L., D'Hondt, S., & Hargraves, P. E. (2001). The possible metabolic role of C_{37} alkenones in *Emiliania huxleyi*. *Organic Geochemistry*, *32*(6), 867–875.

Facchinelli, F., & Weber, A. P. M. (2011). The metabolite transporters of the plastid envelope: An update. *Frontiers in Plant Science*, *2*, 1–18.

Fernández, E., Marañón, E., & Balch, W. M. (1996). Intracellular carbon partitioning in the coccolithophorid *Emiliania huxleyi*. *Journal of Marine Systems*, *9*, 57–66.

Fichtinger-Schepman, A. M. J., Kamerling, J. P., Versluis, C., & Vliegenthart, J. F. G. (1981). Structural studies of the methylated, acidic polysaccharide associated with coccoliths of *Emiliania huxleyi* (lohmann) kamptner. *Carbohydrate Research*, *93*, 105–123.

Fischer, K., Kammerer, B., Gutensohn, M., Arbinger, B., Weber, A., Häusler, R. E., et al. (1997). A new class of plastidic phosphate translocators: A putative link between primary and secondary metabolism by the phosphoenolpyruvate/phosphate antiporter. *The Plant Cell*, *9*, 453–462.

Fliege, R., Flügge, U.-I., Werdan, K., & Heldt, H. W. (1978). Specific transport of inorganic phosphate, 3-phosphoglycerate and triosephosphates across the inner membrane of the envelope in spinach chloroplasts. *Biochimica et Biophysica Acta (BBA)*, *502*, 232–247.

Flügge, U. I. (1999). Phosphate translocators in plastids. *Annual Review of Plant Physiology and Plant Molecular Biology*, *50*, 27–45.

Flügge, U. I., & Benz, R. (1984). Pore-forming activity in the outer membrane of the chloroplast envelope. *FEBS Letters*, *169*(1), 85–89.

Flügge, U. I., Häusler, R. E., Ludewig, F., & Gierth, M. (2011). The role of transporters in supplying energy to plant plastids. *Journal of Experimental Botany*, *62*, 2381–2392.

Franklin, D. J., Airs, R. L., Fernandes, M., Bell, T. G., Bongaerts, R. J., Berges, J. A., et al. (2012). Identification of senescence and death in *Emiliania huxleyi* and *Thalassiosira pseudonana* : Cell staining, chlorophyll alterations, and dimethylsulfoniopropionate (DMSP) metabolism. *Limnology and Oceanography, 57*, 305–317.
Fresnel, J. (1994). A heteromorphic life cycle in two coastal coccolithophorids, *Hymenomonas lacuna* and *Hymenomonas coronata* (Prymnesiophyceae). *Canadian Journal of Botany, 72*, 1455–1462.
Fresnel, J., & Probert, I. (2005). The ultrastructure and life cycle of the coastal coccolithophorid *Ochrosphaera neapolitana* (Prymnesiophyceae). *European Journal of Phycology, 40*, 105–122.
Furumoto, T., Yamaguchi, T., Ohshima-Ichie, Y., Nakamura, M., Tsuchida-Iwata, Y., Shimamura, M., et al. (2011). A plastidial sodium-dependent pyruvate transporter. *Nature, 476*, 472–475.
Gage, D. A., Rhodes, D., Nolte, K. D., Hicks, W. A., Leustek, T., Cooper, A. J., et al. (1997). A new route for synthesis of dimethylsulphoniopropionate in marine algae. *Nature, 387*, 891–894.
Gayral, P., & Fresnel, J. (1983). Description, sexualite et cycle de developpement d'une nouvelle Coccolithophoracee (Prymnesiophyceae): *Pleurochrysis pseudoroscoffensis* sp. nov. *Protistologica, 19*, 245–261.
Giordano, M., Beardall, J., & Raven, J. A. (2005). CO_2 concentrating mechanisms in algae: Mechanisms, environmental modulation, and evolution. *Annual Review of Plant Biology, 56*, 99–131.
Graham, J. E., Graham, L. E., & Wilcox, L. W. (2008). Algae. In *Haptophytes* (2nd ed., pp. 170–185). San Francisco: Benjamin Cummings. Chapter 10.
Granum, E., & Myklestad, S. M. (1999). Effects of NH_4^+ assimilation on dark carbon fixation and β-1,3-glucan metabolism in the marine diatom *Skeletonema costatum* (bacillariophyceae). *Journal of Phycology, 35*, 1191–1199.
Green, J. C. (1973). Studies in the fine structure and taxonomy of flagellates in the genus *Pavlova*. II. A freshwater representative, *Pavlova granifera* (Mack) comb. nov. *British Phycological Journal, 8*, 1–12.
Hagino, K., Bendif, E. M., Young, J. R., Kogame, K., Probert, I., Takano, Y., et al. (2011). New evidence for morphological and genetic variation in the cosmopolitan coccolithophore *Emiliania huxleyi* (Prymnesiophyceae) from the *Cox1b-ATP4* genes. *Journal of Phycology, 47*, 1164–1176.
Hagino, K., Onuma, R., Kawachi, M., & Horiguchi, T. (2013). Discovery of an endosymbiotic nitrogen-fixing cyanobacterium UCYN-A in *Braarudosphaera bigelowii* (Prymnesiophyceae). *PLoS One, 8*, e81749.
Handa, N. (1969). Carbohydrate metabolism in the marine diatom *Skeletonema costatum*. *Marine Biology, 4*, 208–214.
Heimdal, B. R. (1996). Modern coccolithophorids. In C. R. Tomas (Ed.), *Identifying marine phytoplankton* (pp. 731–834). San Diego: Academic Press.
Herfort, L., Loste, E., Meldrum, F., & Thake, B. (2004). Structural and physiological effects of calcium and magnesium in *Emiliania huxleyi* (Lohmann) hay and Mohler. *Journal of Structural Biology, 148*, 307–314.
Herfort, L., Thake, B., & Roberts, J. (2002). Acquisition and use of bicarbonate by *Emiliania huxleyi*. *New Phytologist*, 427–436.
Hibberd, D. J. (1980). Prymnesiophytes (=Haptophytes). In E. R. Cox (Ed.), *Phytoflagellates* (pp. 273–317). Amsterdam, North Holland: Elsevier Science.
Hirokawa, Y., Fujiwara, S., Suzuki, M., Akiyama, T., Sakamoto, M., Kobayashi, S., et al. (2008). Structural and physiological studies on the storage beta-polyglucan of haptophyte *Pleurochrysis haptonemofera*. *Planta, 227*, 589–599.

Hirokawa, Y., Fujiwara, S., & Tsuzuki, M. (2005). Three types of acidic polysaccharides associated with coccolith of *Pleurochrysis haptonemofera*: Comparison with *Pleurochrysis carterae* and analysis using fluorescein-isothiocyanate-labeled lectins. *Marine Biotechnology*, *7*, 634–644.

Holligan, P. M., Viollier, M., Harbour, D. S., Camus, P., & Champagne-Philippe, M. (1983). Satellite and ship studies of coccolithophore production along a continental shelf edge. *Nature*, *304*, 339–342.

Hopkinson, B. M., Dupont, C. L., & Matsuda, Y. (2016). The physiology and genetics of CO2 concentrating mechanisms in model diatoms. *Current Opinion in Plant Biology*, *31*, 51–57.

Hori, T., & Green, J. C. (1994). Mitosis and cell division. In J. C. Green & B. S. C. Leadbeater (Eds.), *The haptophyte algae* (pp. 91–110). New York, NY: Cambridge University Press.

Huertas, I. E., Espie, G. S., Colman, B., & Lubian, L. M. (2000). Light-dependent bicarbonate uptake and CO_2 efflux in the marine microalga *Nannochloropsis gaditana*. *Planta*, *211*, 43–49.

Huxleyi, T. H. (1858). Appendix A. In J. Dayman (Ed.), *Deep sea soundings in the North Atlantic Ocean between Ireland and Newfoundland made in H.M.S. Cyclops, Commander Joseph Dayman in June and July 1857* (pp. 63–68). London: George E. Eyre and William Spottiswoode for Her Majesty's Stationery Office.

Inouye, I. (1997). Systematics of haptophyte algae in Asia-Pacific waters. *The Korean Journal of Phycology*, *12*, 247–261.

Iwamoto, K., Kawanobe, H., Ikawa, T., & Shiraiwa, Y. (2003). Characterization of salt-regulated mannitol-1-phosphate dehydrogenase in the red alga *Caloglossa continua*. *Plant Physiology*, *133*, 893–900.

Iwamoto, K., & Shiraiwa, Y. (2005). Salt-regulated mannitol metabolism in algae. *Marine Biotechnology*, *7*, 407–415.

Izui, K., Matsumura, H., Furumoto, T., & Kai, Y. (2004). Phosphoenolpyruvate carboxylase: A new era of structural biology. *Annual Review of Plant Biology*, *55*, 69–84.

Janse, I., Rijssel, M., Hall, P.-J., Gerwig, G. J., Gottschal, J. C., & Prins, R. A. (1996). The storage glucan of *Phaeocystis globosa* (Prymnesiophyceae) cells. *Journal of Phycology*, *32*, 382–387.

Jordan, R. W., Abe, K., Cruz, J., Eriksen, R., Guerreiro, C., Hagino, K., et al. (2015). Observations on the morphological diversity and distribution of two siliceous nannoplankton genera, *Hyalolithus* and *Petasaria*. *Micropaleontology*, *61*, 439–455.

Jordan, R. W., Cros, L., & Young, J. R. (2004). A revised classification scheme for living haptophytes. *Micropaleontology*, *50*, 55–79.

Jordan, R. W., & Green, J. C. (1994). A check-list of the extant haptophyta of the world. *Journal of the Marine Biological Association of the United Kingdom*, 7, 149–174.

Karlsson, J., Clarke, A. K., Chen, Z. Y., Hugghins, S. Y., Park, Y. I., Husic, H. D., et al. (1998). A novel alpha-type carbonic anhydrase associated with the thylakoid membrane in *Chlamydomonas reinhardtii* is required for growth at ambient CO2. *The EMBO Journal*, *17*, 1208–1216.

Karsten, U., West, J. A., Mostaert, A. S., King, R. J., Barrow, K. D., & Kirst, G. O. (1992). Mannitol in the red algal genus *Caloglossa* (Harvey) J. Agardh. *Journal of Plant Physiology*, *140*, 292–297.

Kawachi, M., & Inouye, I. (1993). *Chrysochromulina quadrikonta* sp. nov., a quadriflagellate member of the genus *Chrysochromulina* (Prymnesiophyceae = Haptophyceae). *Japanese. Journal of Phycology*, *41*, 221–230.

Kawachi, M., & Inouye, I. (1994). Observations on the flagellar apparatus of a Coccolithophorid, *Cruciplacolithus neohelis* (Prymnesiophyceae). *Journal of Plant Research*, *107*, 53–62.

Kawachi, M., & Inouye, I. (1995). Functional roles of the haptonema and the spine scales in the feeding process of *Chrysochromulina spinifera* (Fournier) Pienaar et Norris (Haptophyta = Prymnesiophyta). *Phycologia, 34,* 215–227.

Kawachi, M., Inouye, I., Maeda, M., & Chihara, M. (1991). The haptonema as a food-capturing device: Study on *Chrysochromulina hirta* (Prymnesiophyceae). *Phycologia, 30,* 563–573.

Kayano, K., & Shiraiwa, Y. (2009). Physiological regulation of coccolith polysaccharide production by phosphate availability in the coccolithophorid *Emiliania huxleyi. Plant & Cell Physiology, 50,* 1522–1531.

Keller, M. D., Kiene, R. P., Matrai, P. A., & Bellows, W. K. (1999). Production of glycine betaine and dimethylsulfoniopropionate in marine phytoplankton. I. Batch cultures. *Marine Biology, 135,* 237–248.

Kikutani, S., Nakajima, K., Nagasato, C., Tsuji, Y., Miyatake, A., & Matsuda, Y. (2016). Thylakoid luminal θ-carbonic anhydrase critical for growth and photosynthesis in the marine diatom *Phaeodactylum tricornutum. Proceedings of the National Academy of Sciences of the United States of America, 113,* 9828–9833.

Kinoshita, H., Nagasaki, J., Yoshikawa, N., Yamamoto, A., Takito, S., Kawasaki, M., et al. (2011). The chloroplastic 2-oxoglutarate/malate transporter has dual function as the malate valve and in carbon/nitrogen metabolism. *The Plant Journal, 65,* 15–26.

Kiss, J. Z., & Triemer, R. E. (1988). A comparative study of the storage carbohydrate granules from *Euglena* (Euglenida) and *Pavlova* (Prymnesiida). *The Journal of Protozoology, 35,* 237–341.

Lackey, J. B. (1939). Notes on plankton flagellates from the Scioto River (with descriptions of new forms). *Lloydia Cincinnati, 2,* 128–143.

Leadbeater, B. S. C. (1970). Preliminary observations on differences of scale morphology at various stages in the life cycle of '*Apistonema-Syracosphaera*' sensu von Stosch. *British Phycological Journal, 5,* 57–69.

Leonardos, N., Read, B., Thake, B., & Young, J. R. (2009). No mechanistic dependence of photosynthesis on calcification in the coccolithophorid *Emiliania huxleyi* (Haptophyta). *Journal of Phycology, 45,* 1046–1051.

Lim, L., Linka, M., Mullin, K. A., Weber, A. P. M., & McFadden, G. I. (2010). The carbon and energy sources of the non-photosynthetic plastid in the malaria parasite. *FEBS Letters, 584*(3), 549–554.

Linka, M., Jamai, A., & Weber, A. P. M. (2008). Functional characterization of the plastidic phosphate translocator gene family from the thermo-acidophilic red alga *Galdieria sulphuraria* reveals specific adaptations of primary carbon partitioning in green plants and red algae. *Plant Physiology, 148,* 1487–1496.

Liu, H., Probert, I., Uitz, J., Claustre, H., Aris-Brosou, S., Frada, M., et al. (2009). Extreme diversity in noncalcifying haptophytes explains a major pigment paradox in open oceans. *Proceedings of the National Academy of Sciences of the United States of America, 106,* 12803–12808.

Loebl, M., Cockshutt, A. M., Campbell, D. A., & Finkel, Z. V. (2010). Physiological basis for high resistance to photoinhibition under nitrogen depletion in *Emiliania huxleyi. Limnology and Oceanography, 55*(5), 2150–2160.

Mackinder, L., Wheeler, G., Schroeder, D., von Dassow, P., Riebesell, U., & Brownlee, C. (2011). Expression of biomineralization-related ion transport genes in *Emiliania huxleyi. Environmental Microbiology, 13,* 3250–3265.

Malin, G., & Kirst, G. O. (1997). Algal production of dimethyl sulfide and its atmospheric role. *Journal of Phycology, 33,* 889–896.

Malin, G., & Steinke, M. (2004). Dimethyl sulfide production: What is the contribution of the coccolithophores? In H. R. Thierstein & J. R. Young (Eds.), *Coccolithophores from molecular processes to global impact* (pp. 481–508). Berlin: Springer-Verlag.

Malinverno, E., Triantaphyllou, M., Dimiza, M. D., & Young, J. R. (2008). New possible hetero-holococcolithophore combinations within the genus *Syracosphaera*. *Journal of Nannoplankton Research, 30*, 1–8.

Manton, I. (1964). Observations with the electron microscope on the division cycle in the flagellate *Prymnesium parvum* Carter. *Journal of the Royal Microscopical Society, 83*, 317–325.

Manton, I., & Oates, K. (1980). *Polycrater galapagensis* gen. et sp. nov., a putative coccolithophorid from the Galapagos Islands with an unusual aragonitic periplast. *British Phycological Journal, 15*, 95–103.

Marsh, M. E. (1999). Biomineralization in coccolithophores. *Gravitational and Space Biology Bulletin, 12*, 5–14.

Marsh, M. E., Chang, D. K., & King, G. C. (1992). Isolation and characterization of a novel acidic polysaccharide containing tartrate and glyoxylate residues from the mineralized scales of a unicellular coccolithophorid alga *Pleurochrysis carterae*. *The Journal of Biological Chemistry, 267*, 20507–20512.

Marsh, M. E., & Dickinson, D. P. (1997). Polyanion-mediated mineralization—Mineralization in coccolithophore (*Pleurochrysis carterae*) variants which do not express PS2, the most abundant and acidic mineral-associated polyanion in wild-type cells. *Protoplasma, 199*, 9–17.

Masumoto, C., Miyazawa, S. I., Ohkawa, H., Fukuda, T., Taniguchi, Y., Murayama, S., et al. (2010). Phospho*enol*pyruvate carboxylase intrinsically located in the chloroplast of rice plays a crucial role in ammonium assimilation. *Proceedings of the National Academy of Sciences of the United States of America, 107*, 5226–5231.

Matrai, P. A., & Keller, M. D. (1994). Total organic sulfur and dimethylsulfoniopropionate in marine phytoplankton: Intracellular variations. *Marine Biology, 119*, 61–68.

Moestrup, Ø. (1979). Identification by electron microscopy of marine nanoplankton from New Zealand, including the description of four new species. *New Zealand Journal of Botany, 17*, 61–95.

Moog, D., Rensing, S. A., Archibald, J. M., Maier, U. G., & Ullrich, K. K. (2015). Localization and evolution of putative triose phosphate translocators in the diatom *Phaeodactylum tricornutum*. *Genome Biology and Evolution*, 7, 2955–2969.

Moon-van der Staay, S. Y., De Wachter, R., & Vaulot, D. (2001). Oceanic 18S rDNA sequences from picoplankton reveal unsuspected eukaryotic diversity. *Nature, 409*, 607–610.

Morris, I., Yentsch, C. M., & Yentsch, C. S. (1971). Relationship between light carbon dioxide fixation and dark carbon dioxide fixation by marine algae. *Limnology and Oceanography, 16*, 854–858.

Mortain-Bertrand, A., Descolas-Gros, C., & Jupin, H. (1988). Pathway of dark inorganic carbon fixation in two species of diatoms: Influence of light regime and regulator factors on diel variations. *Journal of Plankton Research, 10*, 199–217.

Nakajima, K., Tanaka, A., & Matsuda, Y. (2013). SLC4 family transporters in a marine diatom directly pump bicarbonate from seawater. *Proceedings of the National Academy of Sciences of the United States of America, 110*, 1767–1772.

Nakayama, T., Yoshida, M., Noel, M. H., Kawachi, M., & Inouye, I. (2005). Ultrastructure and phylogenetic position of *Chrysoculter rhomboideus* gen. et sp. nov. (Prymnesiophyceae), a new flagellate haptophyte from Japanese coastal waters. *Phycologia, 44*, 369–383.

Needoba, J. A., & Harrison, P. J. (2004). Influence of low light and a light: Dark cycle on NO_3^- uptake, intracellular NO_3^-, and nitrogen isotope fractionation by marine phytoplankton. *Journal of Phycology, 40*, 505–516.

Nimer, N. A., Guan, Q., & Merrett, M. J. (1994). Extra- and intra-cellular carbonic anhydrase in relation to culture age in a high-calcifying strain of *Emiliania huxleyi* Lohmann. *New Phytologist, 126*, 601–607.

Nimer, N. A., Iglesias-Rodriguez, M. D., & Merrett, M. J. (1997). Bicarbonate utilization by marine phytoplankton species. *Journal of Phycology, 33*, 625–631.

Nimer, N. A., Ling, M. X., Brownlee, C., & Merrett, M. J. (1999). Inorganic carbon limitation, exofacial carbonic anhydrase activity, and plasma membrane redox activity in marine phytoplankton species. *Journal of Phycology, 35*, 1200–1205.

Nimer, N. A., Merrett, M. J., & Brownlee, C. (1996). Inorganic carbon transport in relation to culture age and inorganic carbon concentration in a high-calcifying strain of *Emiliania huxleyi* (Prymnesiophyceae). *Journal of Phycology, 32*, 813–818.

Noél, M. H., Kawachi, M., & Inouye, I. (2004). Induced dimórphic life cycle of a coccolithophorid, *Calyptrosphaera sphaeroidea* (Prymnesiophyceae, Haptophyta). *Journal of Phycology, 40*, 112–129.

Obata, T., Schoenefeld, S., Krahnert, I., Bergmann, S., Scheffel, A., & Fernie, A. (2013). Gas-chromatography mass-spectrometry (GC-MS) based metabolite profiling reveals mannitol as a major storage carbohydrate in the coccolithophorid alga *Emiliania huxleyi. Metabolites, 3*, 168–184.

Ozaki, N., Sakuda, S., & Nagasawa, H. (2007). A novel highly acidic polysaccharide with inhibitory activity on calcification from the calcified scale "coccolith" of a coccolithophorid alga, *Pleurochrysis haptonemofera. Biochemical and Biophysical Research Communications, 357*(4), 1172–1176.

Paasche, E. (2001). A review of the coccolithophorid *Emiliania huxleyi* (Prymnesiophyceae), with particular reference to growth, coccolith formation, and calcification-photosynthesis interactions. *Phycologia, 40*, 503–529.

Pan, H., & Sun, M.-Y. (2011). Variations of alkenone based paleotemperature index ($U_{37}^{K'}$) during *Emiliania huxleyi* cell growth, respiration (auto-metabolism) and microbial degradation. *Organic Geochemistry, 42*, 678–687.

Parke, M., & Adams, J. (1960). The motile (*Crystalolithus hyalinus* Gaarder and Markali) and nonmotile stages in the life history of *Coccolithus pelagicus* (Wallich) Schiller. *Journal of the Marine Biological Association of the United Kingdom, 39*, 263–274.

Parke, M., & Manton, I. (1971). Observations on the fine structure of zoids of the genus Phaeocystis (Haptophyceae). *Journal of the Marine Biological Association of the United Kingdom, 51*, 927–941.

Parke, M., Manton, I., & Clarke, I. (1955). Studies on marine flagellate II. Three new species of *Chrysochromulina. Journal of the Marine Biological Association of the United Kingdom, 34*, 579–609.

Paulsen, B. S., & Myklestad, S. (1978). Structural studies of the reserve glucan produced by the marine diatom *Skeletonema costatum* (grev.) Cleve. *Carbohydrate Research, 62*, 386–388.

Pienaar, R. N. (1980). Observations on the structure and composition of the cyst of *Prymnesium* (Prymnesiophyceae). *Annual Conference Proceedings of Electron Microscopy Society of South Africa, 10*, 73–74.

Pond, D. W., & Harris, R. P. (1996). The lipid composition of the coccolithophore *Emiliania huxleyi* and its possible ecophysiological significance. *Journal of the Marine Biological Association of the United Kingdom, 76*, 579–594.

Poulton, A. J., Adey, T. R., Balch, W. M., & Holligan, P. M. (2007). Relating coccolithophore calcification rates to phytoplankton community dynamics: Regional differences and implications for carbon export. *Deep Sea Research Part II: Topical Studies in Oceanography, 54*, 538–557.

Prahl, F. G., Muehlhausen, L. A., & Zahnle, D. L. (1988). Further evaluation of long-chain alkenones as indicators of paleoceanographic conditions. *Geochimica et Cosmochimica Acta, 52*, 2303–2310.

Quinn, P. K., & Bates, T. S. (2011). The case against climate regulation via oceanic phytoplankton sulphur emissions. *Nature, 480*, 51–56.

Ragni, M., Airs, R. L., Leonardos, N., & Geider, R. J. (2008). Photoinhibition of psii in *Emiliania huxleyi* (haptophyta) under high light stress: The roles of photoacclimation, photoprotection, and photorepair. *Journal of Phycology*, *44*, 670–683.

Raven, J. A. (1994). Carbon fixation and carbon availability in marine phytoplankton. *Photosynthesis Research*, *39*, 259–273.

Read, B. A., Kegel, J., Klute, M. J., Kuo, A., Lefebvre, S. C., Maumus, F., et al. (2013). Pan genome of the phytoplankton *Emiliania* underpins its global distribution. *Nature*, *499*, 209–213.

Reed, R. H., Davison, I. R., Chudek, J. A., & Foster, R. (1985). The osmotic role of mannitol in the Phaeophyta: An appraisal. *Phycologia*, *24*, 35–47.

Reinfelder, J. R. (2011). Carbon concentrating mechanisms in eukaryotic marine phytoplankton. *Annual Review of Marine Science*, *3*, 291–315.

Roessler, P. G. (1987). UDP glucose pyrophosphorylase activity in the diatom *Cyclotella cryptica*. Pathway of chrysolaminarin biosynthesis. *Journal of Phycology*, *23*, 494–498.

Rontani, J.-F., Prahl, F. G., & Volkman, J. K. (2006). Re-examination of the double bond positions in alkenones and derivatives: Biosynthetic implications. *Journal of Phycology*, *42*, 800–813.

Rost, B., Burkhardt, S., Sültemeyer, D., Riebesell, U., Burkhardt, S., & Sültemeyer, D. (2003). Carbon acquisition of bloom-forming marine phytoplankton. *Limnology and Oceanography*, *48*, 55–67.

Sadovskaya, I., Souissi, A., Souissi, S., Grard, T., Lencel, P., Greene, C. M., et al. (2014). Chemical structure and biological activity of a highly branched (1→3,1→6)-β-D-glucan from *Isochrysis galbana*. *Carbohydrate Polymers*, *111*, 139–148.

Samukawa, M., Shen, C., Hopkinson, B. M., & Matsuda, Y. (2014). Localization of putative carbonic anhydrases in the marine diatom, *Thalassiosira pseudonana*. *Photosynthesis Research*, *121*, 235–249.

Sawada, K., & Shiraiwa, Y. (2004). Alkenone and alkenoic acid compositions of the membrane fractions of *Emiliania huxleyi*. *Phytochemistry*, *65*, 1299–1307.

Sekino, K., & Shiraiwa, Y. (1994). Accumulation and utilization of dissolved inorganic carbon by a marine unicellular coccolithophorid, *Emiliania huxleyi*. *Plant Cell Physiology*, *35*, 353–361.

Shi, Q., Araie, H., Bakku, R. K., Fukao, Y., Rakwal, R., Suzuki, I., et al. (2015). Proteomic analysis of lipid body from the alkenone-producing marine haptophyte alga *Tisochrysis lutea*. *Proteomics*, *15*, 4145–4158.

Sikes, C. S., Row, R. D., & Wilbur, K. M. (1980). Photosynthesis and coccolith formation: Inorganic carbon sources and net inorganic reaction of deposition. *Limnology and Oceanography*, *25*, 248–261.

Sikes, C. S., & Wheeler, A. P. (1982). Carbonic anhydrase and carbon fixation in coccolithophorids. *Journal of Phycology*, *18*, 423–426.

Sorby, H. C. (1861). On the organic origin of the so-called 'crystalloids' of the chalk. *Annals and Magazine of Natural History*, *8*, 193–200. Series 3.

Stefels, J. (2000). Physiological aspects of the production and conversion of DMSP in marine algae and higher plants. *Journal of Sea Research*, *43*, 183–197.

Stojkovic, S., Beardall, J., & Matear, R. (2013). CO_2-concentrating mechanisms in three southern hemisphere strains of *Emiliania huxleyi*. *Journal of Phycology*, *49*, 670–679.

Streatfield, S. J., Weber, A., Kinsman, E. A., Häusler, R. E., Li, J., Post-Beittenmiller, D., et al. (1999). The phospho*enol*pyruvate/phosphate translocator is required for phenolic metabolism, palisade cell development, and plastid-dependent nuclear gene expression. *The Plant Cell*, *11*, 1609–1622.

Suffrian, K., Schulz, K. G., Gutowska, M. A., Riebesell, U., & Bleich, M. (2011). Cellular pH measurements in Emiliania huxleyi reveal pronounced membrane proton permeability. *New Phytologist, 190*, 595–608.

Sunda, W., Kieber, D. J., Kiene, R. P., & Huntsman, S. (2002). An antioxidant function for DMSP and DMS in marine algae. *Nature, 418*, 317–320.

Suzuki, E., & Suzuki, R. (2013). Variation of storage polysaccharides in phototrophic microorganisms. *Journal of Applied Glycoscience, 60*, 21–27.

Sviben, S., Gal, A., Hood, M. A., Bertinetti, L., Politi, Y., Bennet, M., et al. (2016). A vacuole-like compartment concentrates a disordered calcium phase in a key coccolithophorid alga. *Nature Communications*, 7, 11228.

Takano, Y., Hagino, K., Tanaka, Y., Horiguchi, T., & Okada, H. (2006). Phylogenetic affinities of an enigmatic nannoplankton, *Braarudosphaera bigelowii* based on the SSU rDNA sequences. *Marine Micropaleontology, 60*, 145–156.

Taylor, A. R., Brownlee, C., & Wheeler, G. (2017). Coccolithophore cell biology: Chalking up progress. *Annual Review of Marine Science, 9*, 283–310.

Taylor, A. R., Chrachri, A., Wheeler, G., Goddard, H., & Brownlee, C. (2011). A voltage-gated H^+ channel underlying pH homeostasis in calcifying coccolithophores. *PLoS Biology, 9*, e1001085.

Tchernov, D., Hassidim, M., Luz, B., Sukenik, A., Reinhold, L., & Kaplan, A. (1997). Sustained net CO_2 evolution during photosynthesis by marine microorganism. *Current Biology*, 7, 723–728.

Tchernov, D., Silverman, J., Luz, B., Reinhold, L., & Kaplan, A. (2003). Massive light-dependent cycling of inorganic carbon between oxygenic photosynthetic microorganisms and their surroundings. *Photosynthesis Research*, 77, 95–103.

Thomsen, H. A., & Østergaard, J. B. (2015). Coccolithophorids in polar waters: *Trigonaspis* spp. revisited. *Acta Protozoologica, 54*, 85–96.

Tsuji, Y., & Shiraiwa, Y. (2016). Distinctive features of photosynthetic carbon metabolism in secondary endosymbiotic algae. *Perspectives in Phycology, 3*, 3–9.

Tsuji, Y., Suzuki, I., & Shiraiwa, Y. (2009). Photosynthetic carbon assimilation in the coccolithophorid *Emiliania huxleyi* (Haptophyta): Evidence for the predominant operation of the C_3 cycle and the contribution of β-carboxylases to the active anaplerotic reaction. *Plant & Cell Physiology, 50*, 318–329.

Tsuji, Y., Suzuki, I., & Shiraiwa, Y. (2012). Enzymological evidence for the function of a plastid-located pyruvate carboxylase in the haptophyte alga *Emiliania huxleyi*. A novel pathway for the production of C_4 compounds. *Plant & Cell Physiology, 53*, 1043–1052.

Tsuji, Y., Yamazaki, M., Suzuki, I., & Shiraiwa, Y. (2015). Quantitative analysis of carbon flow into photosynthetic products functioning as carbon storage in the marine coccolithophore, *Emiliania huxleyi*. *Marine Biotechnology*, 17, 428–440.

Tyrrell, T., & Merico, A. (2004). *Emiliania huxleyi*: Bloom observations and the conditions that induce them. In H. R. Thierstein & J. R. Young (Eds.), *Coccolithophores from molecular processes to global impact* (pp. 481–508). Berlin: Springer-Verlag.

Van Lenning, K., Latasa, M., Estrada, M., Sáez, A. G., Medlin, L., Probert, I., et al. (2003). Pigment signatures and phylogenetic relationship of the Pavlovophyceae (Haptophyta). *Journal of Phycology, 39*, 379–389.

van Rijssel, M., & Gieskes, W. W. C. (2002). Temperature, light, and the dimethylsulfoniopropionate (DMSP) content of *Emiliania huxleyi* (Prymnesiophyceae). *Journal of Sea Research, 48*, 17–27.

Vårum, K. M., Kvam, B. J., Myklestad, S., & Paulsen, B. S. (1986). Structure of a food-reserve β-d-glucan produced by the haptophyte alga *Emiliania huxleyi* (lohmann) hay and mohler. *Carbohydrate Research, 152*, 243–248.

Vårum, K. M., & Myklestad, S. (1984). Effects of light, salinity and nutrient limitation on the production of β-1,3-d-glucan and exo-d-glucanase activity in *Skeletonema costatum* (Grev.) Cleve. *Journal of Experimental Marine Biology and Ecology, 83*, 13–25.
Volkman, J. K., Everitt, D. A., & Allen, D. I. (1986). Some analyses of lipid classes in marine organisms, sediments and seawater using thin-layer chromatography—Flame ionisation detection. *Journal of Chromatography A, 356*, 147–162.
von Dassow, P., Ogata, H., Probert, I., Wincker, P., Da Silva, C., Audic, S., et al. (2009). Transcriptome analysis of functional differentiation between haploid and diploid cells of *Emiliania huxleyi*, a globally significant photosynthetic calcifying cell. *Genome Biology, 10*, R114.
Wallich, G. C. (1860). Results of soundings in the North Atlantic. *Annals and Magazine of Natural History, 6*, 457–458. Series 3.
Weber, A. P. M., Linka, M., & Bhattacharya, D. (2006). Single, ancient origin of a plastid metabolite translocator family in plantae from an endomembrane-derived ancestor. *Eukaryotic Cell, 5*, 609–612.
Westbroek, P., de Jong, E. W., Dam, W., & Bosch, L. (1973). Soluble intracrystalline polysaccharides from coccoliths of *Coccolithus huxleyi* (Lohmann) Kamptner (I). *Calcified Tissue Research, 12*, 227–238.
Winter, A., & Siesser, W. G. (1994). *Coccolithophores*. New York, NY: Cambridge University Press.
Yamaguchi, T., Ikawa, T., & Nishizawa, K. (1966). Incorporation of radioactive carbon from H14CO3 – into sugar constituents by a brown alga, *Eisenia bicyclis*, during photosynthesis and its fate in the dark. *Plant & Cell Physiology, 7*, 217–229.
Yoshida, M., Noel, M. H., Nakayama, T., Naganuma, T., & Inouye, I. (2006). A haptophyte bearing siliceous scales: Ultrastructure and phylogenetic position of *Hyalolithus neolepis* gen. et sp. nov. (Prymnesiophyceae, Haptophyta). *Protist, 157*, 213–234.
Yoshida, M., Tanabe, Y., Yonezawa, N., & Watanabe, M. M. (2012). Energy innovation potential of oleaginous microalgae. *Biofuels, 3*, 761–781.
Young, J. R., Bergen, J. A., Bown, P. R., Burnett, J. A., Fiorentino, A., Jordan, R. W., et al. (1997). Guidelines for coccolith and calcareous nannofossil terminology. *Palaeontology, 40*, 875–912.
Young, J. R., Bown, P. R., & Burnett, J. A. (1994). Paleontological perspectives. In J. C. Green & B. S. C. Leadbeater (Eds.), *The haptophyte algae* (pp. 379–392). Oxford: Oxford Univ. Press.
Young, J. R., Geisen, M., Cros, L., Kleijne, A., Sprengel, C., Probert, I., et al. (2003). A guide to extant coccolithophore taxonomy. *Journal of Nannoplankton Research*, 1–125. Special Issue 1.
Zhu, B. H., Shi, H. P., Yang, G. P., Lv, N. N., Yang, M., & Pan, K. H. (2016). Silencing UDP-glucose pyrophosphorylase gene in *Phaeodactylum tricornutum* affects carbon allocation. *New Biotechnology, 33*, 237–244.

CHAPTER EIGHT

Cryptomonads: A Model Organism Sheds Light on the Evolutionary History of Genome Reorganization in Secondary Endosymbioses

Goro Tanifuji*,[1], Naoko T. Onodera†
*National Museum of Nature and Science, Tsukuba, Ibaraki, Japan
†Dalhousie University, Halifax, NS, Canada
[1]Corresponding author: e-mail address: gorot@kahaku.go.jp

Contents

Abstract

The cryptomonads are ubiquitous in the earth's hydrosphere. Most members of this unicellular group are photosynthetic and retain red alga-derived plastids. The significant feature of cryptomonads from an evolutionary and biological point of view is that they contain the residual nucleus of a eukaryotic endosymbiont, the so-called nucleomorph, which is direct evidence of eukaryote–eukaryote endosymbiosis. Besides cryptomonads,

Advances in Botanical Research, Volume 84
ISSN 0065-2296
http://dx.doi.org/10.1016/bs.abr.2017.06.005

this unusual organelle has been found only in chlorarachniophytes so far. In the first half of this chapter, we briefly describe cryptomonad morphology, classification, and phylogeny. The evolutionary history of auto- or heterotrophic lifestyle transitions in cryptomonads is discussed. In the latter part, we focus on the recent outcomes of comparative genomics and review perspectives on the genome reorganization process that occurs during the integration of two eukaryotes into one organism.

1. INTRODUCTION: GENERAL CHARACTERISTICS OF CRYPTOMONAD PROTISTS

Cryptomonads are small (5–50 μm) unicellular biflagellates found widely in aquatic settings including marine, brackish, and freshwater environments (Clay, Kugrens, & Lee, 1999; Hoef-Emden & Archibald, 2017; Kugrens & Lee, 1991; Kugrens, Lee, & Hill, 2002). Many of them are photosynthetic, but also some are found to be nonphotosynthetic (Deane et al., 2002; Hoef-Emden, Marin, & Melkonian, 2002; Hoef-Emden & Melkonian, 2003; Kim & Archibald, 2013; Martin-Cereceda et al., 2010; McFadden, Gilson, & Hill, 1994). In this chapter, cryptomonads are designated as an organismal group, which consists of plastid-containing (Cryptophyceae) and plastid-lacking species (Goniomonadea including the genus *Goniomonas*) (Cavalier-Smith, 2013; Ruggiero et al., 2015a, 2015b). Cryptomonad cell shapes are flattened and asymmetric and have been likened to a grain of rice (Fig. 1A–C; Clay et al., 1999; Hoef-Emden & Archibald, 2017; Klaveness, 1985; Kugrens & Lee, 1991; Kugrens et al., 2002). Two flagella protrude from the subapical region of the cell with a slight shift to one side. At the base of the flagella, a furrow (ventral groove)–gullet (tubular invagination) system is present and that penetrates into the cell (Clay et al., 1999; Hoef-Emden & Archibald, 2017; Kugrens et al., 2002). Large ejectosomes (extrusive organelles) (e.g. Anderson, 1962) can be characteristically observed under light microscopes and these coiled ribbon-like structures form a line along the furrow–gullet system within the cell (Clay et al., 1999; Hoef-Emden & Archibald, 2017; Kugrens et al., 2002; Morrall & Greenwood, 1980). Smaller ejectosomes also exist beneath the cell surface. The two flagella are slightly unequal in length. Flagellar hairs consisting of a basal shaft and a thin terminal filament are attached to the main flagellar axes (Hibberd, Greenwood, & Griffiths, 1971; Hoef-Emden & Archibald, 2017; Kugrens et al., 2002). Some strains have two rows of flagellar hairs on longer flagellum and one row of hairs on shorter flagellum. The different patterns of one/two rows of flagellar hairs

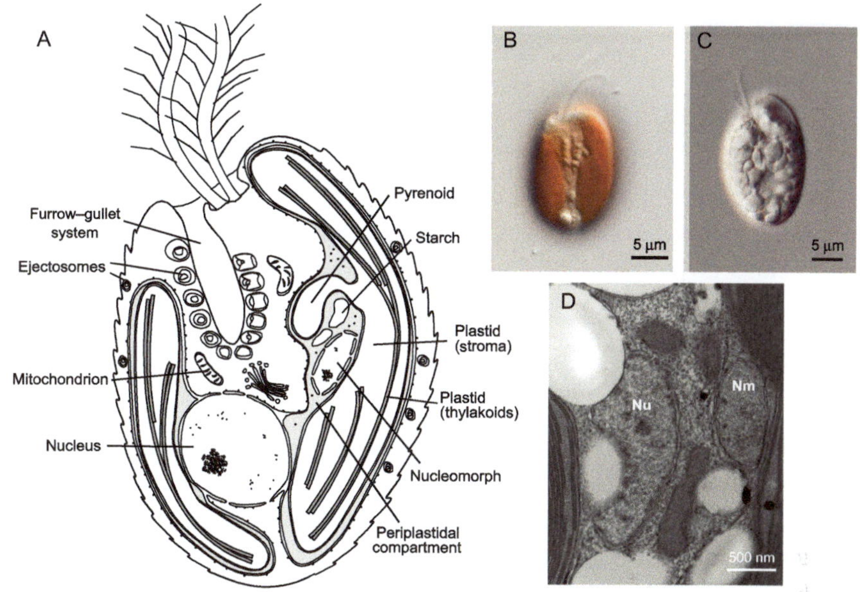

Fig. 1 Basic features of cryptomonads. (A) Drawing of a cryptomonad cell. The outermost membrane of the plastid is contiguous with the nuclear envelope. Nucleomorphs reside in the periplastidal compartment (PPC) between the second and third plastid membranes. Ejectosomes line the furrow–gullet system (large type) or are located beneath the cell surface (small type). (B) Light micrograph of the photosynthetic *Cryptomonas curvata* CCAP 979/52. (C) Light micrograph of the nonphotosynthetic *Cryptomonas paramecium* CCAP 977/2A. (D) Transmission electron micrograph of *Guillardia theta* CCMP 327. The nucleus (Nu) and nucleomorph (Nm) are labelled.

are observed depending on the species (Clay et al., 1999; Hoef-Emden & Archibald, 2017; Kugrens, Lee, & Andersen, 1987; Kugrens et al., 2002). The plasma membrane of the cell is sandwiched between the inner and the surface periplasts composed of proteinaceous substances. Scanning electron microscopy shows the unique patterns of the inner periplast component (IPC) (e.g. hexagonal, polygonal, or sheet-like) and also the surface periplast component that often appears to be regularly formed plates, rosulate scales, and/or a fibrous coat (Brett & Wetherbee, 1985; Clay et al., 1999; Gillott & Gibbs, 1983; Hibberd et al., 1971; Hoef-Emden & Archibald, 2017; Kugrens & Lee, 1987; Kugrens et al., 1987, 2002; Martin-Cereceda et al., 2010; Whetherbee, Hill, & McFadden, 1985).

The origin of the plastids in cryptomonads was found to be a red algal endosymbiont (Cavalier-Smith et al., 1996; Douglas, Durnford, & Morden, 1990; Douglas & Penny, 1999). The plastid-containing cryptomonads possess a single H-shaped plastid with four membrane envelopes (Hoef-Emden & Archibald, 2017; Kugrens et al., 2002). The outermost plastid membrane is contiguous with the outer nuclear membrane of the host. The photosynthetic pigments are chlorophylls $a+c$ (c_2, strictly speaking), carotenoids, and biliproteins (Hill & Rowan, 1989; Wedemayer, Kidd, & Glazer, 1996). Although red algae (and cyanobacteria) contain two types of light-harvesting biliproteins, phycocyanin (blue) and phycoerythrin (red), the extant cryptomonad species possess either phycocyanin or phycoerythrin (Apt, Collier, & Grossman, 1995; Wedemayer et al., 1996). Due to the differences in pigment composition, cryptomonad plastids exhibit diverse colourations such as olive-green, brown, red, and blue. However, the cell's colour can change depending on the cell's condition (Clay et al., 1999). Since biliproteins are embedded in the thylakoid lumen, the thylakoids of cryptomonad plastids are generally thicker compared with plastids of other photosynthetic eukaryotes (Broughton, Howe, & Hiller, 2006; Gantt, Edwards, & Provasoli, 1971; Hill & Rowan, 1989).

An intriguing morphological feature, the nucleomorph, the relic nucleus derived from the eukaryotic endosymbiont, is retained in the space between the second and third plastid membranes (Greenwood, 1974; Moore & Archibald, 2009; Tanifuji & Archibald, 2014). This space is called the periplastidal compartment (PPC) and corresponds to the ancient cytosolic space of the eukaryotic endosymbiont (Fig. 1D; Gillott & Gibbs, 1980). Nucleomorphs contain their own DNA (Ludwig & Gibbs, 1985), and the transcripts from the nucleomorph genome are translated using red alga-derived ribosomes in the PPC. Cryptomonads can be recognized by their characteristic cell shapes and their swimming motion under light microscopy. However, electron microscopy is necessary for cryptomonad identification at the genus/species level (Hoef-Emden & Archibald, 2017). Cell shapes, periplast patterns, the positional relation of the host nucleus and/or pyrenoid with the nucleomorph, the morphology of the furrow–gullet system, and other detailed morphologies are available as taxonomic characters to describe the cryptomonad genera/species (Hibberd et al., 1971; Hill, 1991; Hoef-Emden & Archibald, 2017; Klaveness, 1985; Kugrens & Lee, 1987, 1991; Kugrens et al., 1987, 2002; Santore, 1984). In addition, a spectrogram analysis of pigment composition is helpful to recognize genera, since the different types of biliproteins in cryptomonads have diverged (Hoef-Emden & Archibald, 2017; Wedemayer et al., 1996). On the other hand, while these characters mentioned above provide a useful overview of cryptomonad systematics, the classification based on traditional studies has been found to be inconsistent with the recent molecular phylogenies (Deane et al., 2002; Hoef-Emden, 2007; Hoef-Emden et al., 2002; Hoef-Emden & Melkonian, 2003; Shalchian-Tabrizi et al., 2008) and revision or integration of some genera is necessary for a rigorous classification of cryptomonads.

Cryptomonads are distributed across diverse aquatic environments such as offshore, coastal and tide-pool marine waters, freshwater lakes and ponds, and they are present throughout the year (Alcocer, Lugo, Sánchez, & Escobar, 1998; Hill, 1991; Klaveness, 1988; Phlips, Havens, & Lopes, 2008). Cold and tropical areas are also habitats of cryptomonad species. In addition, cryptomonads are found to adapt themselves to low-light conditions (Gervais, 1997; Hammer, Schumann, & Schubert, 2002). Their ubiquitous presence and abundance in various environments suggest potential importance as a primary producer globally and as prey in aquatic food chains for planktonic predators (Pedrós-Alió, Massana, Latasa, Garcí-Cantizano, & Gasol, 1995; Roberts & Laybourn-Parry, 1999; Tirok & Gaedke, 2007; Weisse & Kirchhoff, 1997).

2. SECONDARY ORGANISMS: CRYPTOMONADS

2.1 Challenges for Cryptomonad Classification

Currently, more than 100 species of cryptomonad, with approximately 20 genera, have been described (Hoef-Emden & Archibald, 2017). Since they exhibit distinct cell colours, cryptomonad genera were divided by cell colours as an early taxonomic approach prior to electron microscopy becoming available (*Cryptomonas* = brown, *Rhodomonas* = red, and *Chroomonas* = blue) (Pringsheim, 1944). Since the cell colours of cryptomonads vary depending on the culture conditions and starving states, the systematic treatments based on their colourations might have led to confusion in cryptomonad literature. Nevertheless, the spectrum analyses revealed that at least eight types of distinct biliproteins are produced across cryptomonads, and that each clonal strain stably contains only one type (Glazer & Wedemayer, 1995; Hill & Rowan, 1989; Hoef-Emden, 2008). The cryptomonad biliproteins were modified in the course of evolution and are named according to the different wavelength of their respective absorption maxima (e.g. phycoerythrin 545—PE545 and phycocyanin 630—PC630). This feature of listing the biliprotein type is helpful for cryptomonad classification, even now. Nevertheless, some different genera that are well supported by both morphology and recent molecular data share the same type of biliprotein (e.g. PE545 in *Rhodomonas* and *Guillardia*) (Fig. 2; Deane et al., 2002; Hoef-Emden et al., 2002; Marin, Klingberg, & Melkonian, 1998). Conversely, the case that close relatives contain different biliproteins is known as well (e.g. *Chroomonas/Hemiselmis/Komma* clade) (Hoef-Emden, 2008; Hoef-Emden & Archibald, 2017). Thus, using only pigments is not sufficient for the identification of cryptomonad genera.

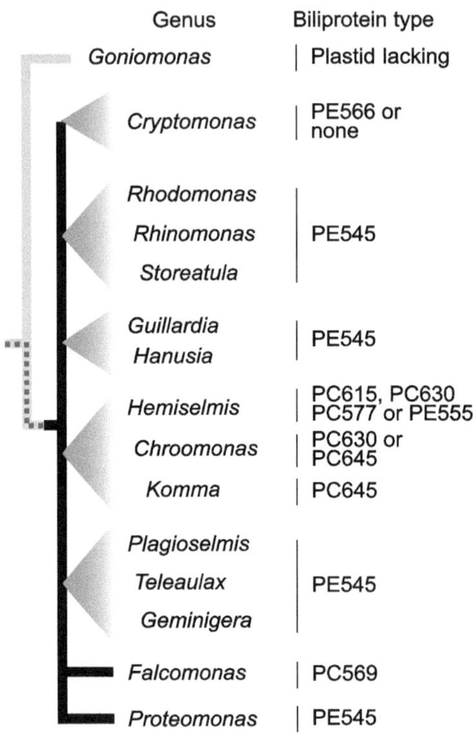

Fig. 2 Schematic phylogeny of the cryptomonads with pigment information. *Phylogenetic backbone was adapted from Deane, J. A., Isabbelle, M., Strachan, I. M., Saunders, G. W., Hill, D. R. A., & McFadden, G. I. (2002). Cryptomonad evolution: Nuclear 18S rDNA phycologeny versus cell morphology and pigmentation.* Journal of Phycology, *38, 1236–1244; Hoef-Emden, K., Marin, B., & Melkonian, M. (2002). Nuclear and nucleomorph SSU rDNA phylogeny in the cryptophyta and the evolution of cryptophyte diversity.* Journal of Molecular Evolution, *55, 161–179; Tanifuji, G., Onodera, N. T., & Hara, Y. (2010). Nucleomorph genome diversity and its phylogenetic implications in cryptomonad algae.* Phycological Research, *58, 230–237. Data of biliprotein was taken from Hoef-Emden, K., & Archibald, J. M. (2017). Cryptophyta (cryptomonads). In J. M. Archibald, A. G. B. Simpson, C. Slamovits, L. Margulis, M. Melkonian, D. J. Chapman, & J. O. Corliss (Eds.),* Handbook of the protists. *Springer-Verlag.*

The search for a single trait diagnostic for each lineage of genus/species in cryptomonads is challenging. In reality, a combination of multiple morphological characters is necessary to describe cryptomonads at the genus/species level. Pigment composition, the periplast patterns (a regular protein structure associated with the plasma membrane), the morphology of the rhizostyle (a microtubular structure directly connected to the basal body of the flagellum), the flagellar hair patterns, the morphotype of the furrow–gullet system, and the position of the nucleomorph (and pyrenoid) in the cells could be used as morphological characters defining cryptomonad genera/species (see review by Hoef-Emden & Archibald, 2017). However, traditional classifications based on a combination of these morphological characters are often inconsistent with ones done by a molecular phylogenetic approach (Deane et al., 2002; Hoef-Emden, 2005, 2007, 2008; Hoef-Emden & Archibald, 2017; Hoef-Emden et al., 2002; Marin et al., 1998). It is clear that revisiting the classification of cryptomonads is necessary.

Although the knowledge of sexual reproduction in cryptomonads is still limited, some studies suggested that the broad range of cryptomonad species is capable of sexual reproduction (Hill & Wetherbee, 1986; Hoef-Emden & Melkonian, 2003; Kugrens & Lee, 1988). *Proteomonas sulcata*, for example, is known to possess two morphotypes with relatively large and small cells in a monoclonal culture (Hill & Wetherbee, 1986). While the larger cells exhibited keeled rhizostyles and sheet-like IPC, the smaller one possessed nonkeeled rhizostyles and the IPC was composed of hexagonal plates. Also, Hill and Wetherbee (1986) measured the amount of DNA in the two different cell types of *P. sulcata* and showed that the smaller cells contained half of the DNA compared with larger cells (Hill & Wetherbee, 1986). This strongly suggested that the smaller cells are haploid and the larger cells are diploid derived from fused haploid cells through sexual reproduction. Hoef-Emden and Melkonian (2003) pointed out that hidden dimorphism diversity might have caused confusion in cryptomonad classification schemes because the different morphotypes of close relatives were possibly described as different genera/species. They showed that a dimorphic *Cryptomonas* strain exhibits both characters previously used to distinguish between the genus *Cryptomonas* (e.g. with polygonal plates) and the genus *Campylomonas* (e.g. with sheet-like). Based on molecular phylogeny, it was suggested that *Campylomonas* represents the alternate morph of *Cryptomonas*. The same situation was seen between other phylogenetically close genera in cryptomonads; *Storeatula* with sheet-like and *Rhodomonas*+

Rhinomonas with rectangular/hexagonal plates were close relatives. *Plagioselmis* and *Germinigera* + *Teleaulax* might also be in the same situation since their periplast patterns are hexagonal plates and sheet-like, respectively (Deane et al., 2002; Hoef-Emden & Archibald, 2017; Hoef-Emden et al., 2002; Majaneva et al., 2014; Shalchian-Tabrizi et al., 2008). Thus, to uncover the true diversity and nature of cryptomonads, more detailed analyses with consensus between the morphological and the molecular data are needed (Hoef-Emden, 2007; Hoef-Emden & Melkonian, 2003; Lane & Archibald, 2008).

In the taxonomic levels above genera, cryptomonads could be divided into at least two groups, Cryptophyceae and Goniomonadea (Cavalier-Smith, 2013; Ruggiero et al., 2015a, 2015b). The most significant difference between these two groups is probably the presence/absence of plastids. Goniomonadea consists of the single phagotrophic genus *Goniomonas* and lacks any structures corresponding to a plastid (Kim & Archibald, 2013; Martin-Cereceda et al., 2010; McFadden, Gilson, & Hill, 1994; Von der Heyden, Chao, & Cavalier-Smith, 2004). All species of Cryptophyceae contain a plastid and most of them are photosynthetic (Clay et al., 1999; Deane et al., 2002; Hoef-Emden & Archibald, 2017; Hoef-Emden et al., 2002; Kugrens et al., 2002). *Cryptomonas*, *Rhodomonas*, *Hemiselmis*, *Chroomonas*, and *Guillardia* are representative genera of Cryptophyceae. Although some *Cryptomonas* were found to have lost photosynthetic ability (e.g. *Cryptomonas paramecium*) (Fig. 1C), they still retain nonphotosynthetic plastids with a reduced plastid genome (Donaher et al., 2009; Hoef-Emden & Melkonian, 2003; Hoef-Emden, Tran, & Melkonian, 2005). In addition, *Goniomonas* displays different characters from the plastid-containing cryptomonads such as an extremely flattened cell body and large rectangular periplast plates (Kim & Archibald, 2013; Kugrens & Lee, 1991; Martin-Cereceda et al., 2010). Molecular phylogeny supports the sister relationship between Cryptophyceae and Goniomonadea (Kim & Archibald, 2013; Martin-Cereceda et al., 2010; McFadden, Gilson, & Hill, 1994; Von der Heyden et al., 2004). In addition to the cultured strains, molecular data, based on environmental DNAs, suggest the presence of hidden cryptomonad lineages distinct from Cryptophyceae and Goniomonadae (Kim & Archibald, 2013; Shalchian-Tabrizi et al., 2008). Recently, a novel plastid-lacking strain, *Hemiarma marina*, previously assigned to the cryptomonad clade based only on environmental DNA, was described (Shiratori & Ishida, 2016). Since *H. marina* and cryptomonads share some morphological features including the presence of a periplast and ejectosomes, this species was described as a novel cryptomonad genus. Thus, cryptomonads consist of Cryptophyceae, Goniomonadea, and the currently cryptic group containing *H. marina.*

Various molecular data suggested that the kathablepharids (i.e. class Leucocryptea) is a distant sister group of cryptomonads (Burki, Okamoto, Pombert, & Keeling, 2012; Okamoto, Chantangsi, Horák, Leander, & Keeling, 2009; Okamoto & Inouye, 2005; Yabuki et al., 2014). They share similar morphological features such as the presence of bipartite ejectosomes (Clay & Kugrens, 1999; Lee & Kugrens, 1991; Okamoto & Inouye, 2005). In addition, *Palpitomonas bilix*, a marine heterotrophic biflagellate, was found to be basal to the phylogenetic clade consisting of cryptomonads + kathablepharids (Yabuki, Inagaki, & Ishida, 2010; Yabuki et al., 2014), suggesting that the eukaryotic assemblage, "Cryptista", is comprised of cryptomonads, kathablepharids (class Leucocryptea), and *P. bilix* (class Palpitea) (Fig. 3; Cavalier-Smith, 2013; Ruggiero et al., 2015a, 2015b; Yabuki et al., 2014). On the other hand, ejectosomes, characteristic structures in cryptomonads and kathablepharids, were not observed in *P. bilix.* Yabuki et al. (2014) found transcripts coding for major proteins of

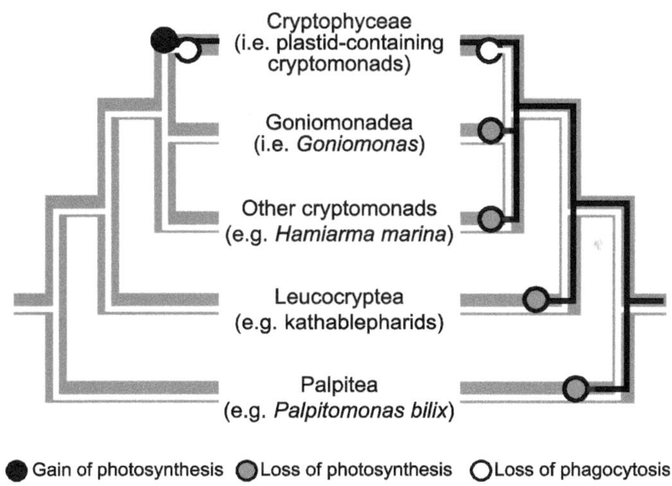

Fig. 3 Two possible scenarios of the evolution of plastid acquisition or loss in Cryptista organisms. Topology on the *left* shows the scenario that the secondary endosymbiosis between the eukaryotic host and red alga occurred in the common ancestor of extant plastid-containing cryptomonads (i.e. Cryptophyceae). The topology on the *right* follows the "chromalveolate hypothesis", in which the loss of the plastid in each ancestor of extant plastid-lacking Cryptista organisms occurred independently. *Data taken from Yabuki, A., Kamikawa, R., Ishikawa, S. A., Kolisko, M., Kim, E., Tanabe, A. S., et al. (2014).* Palpitomonas bilix *represents a basal cryptist lineage: Insight into the character evolution in Cryptista.* Scientific Reports, 4, *4641.*

ejectosomes from the cryptomonads and kathablepharids, but detected no homologues in *P. bilix* (Yabuki et al., 2014; Yamagishi, Kai, & Kawai, 2012). This is consistent with the absence of the ejectosome in *P. bilix*, suggesting that the ejectosome was established in the common ancestor of cryptomonads and kathablepharids.

2.2 The Secondary Plastid of Cryptomonads

Ultrastructural observations using transmission electron microscopy revealed a significant amount of diversity in plastid morphology (Gibbs, 1978; Whatley & Whatley, 1981). The plastids of plants, green algae, red algae, and glaucophytes were found to be surrounded by double membranes, while other photosynthetic eukaryotes such as euglenoids, heterokonts, and cryptomonads possess four (or three) membrane-bound plastids. Prompted by consideration of plastid pigment compositions, researchers begun to speculate that independent endosymbioses in multiple lineages gave rise to the plastid diversity in eukaryotes (e.g. Gray & Doolittle, 1982). The plastids with double membrane envelopes are now believed to be the result of an endosymbiosis between a eukaryote and a cyanobacterium (i.e. primary endosymbiosis) (Gould, Waller, & McFadden, 2008; Reyes-Prieto, Weber, & Bhattacharya, 2007). Subsequently, organisms with primary plastids were engulfed by other eukaryotes, giving rise to secondary plastids with four or three envelopes (i.e. secondary endosymbiosis).

According to endosymbiosis theory, the inner pair of four plastid membranes in secondary endosymbiotic organisms is considered to be homologous to the primary plastid membranes (Gould et al., 2008; Reyes-Prieto et al., 2007). The outermost and second outermost membranes are supposed to be derived from the host's phagosomal and the endosymbiont's plasma membrane in secondary endosymbiosis, respectively (Cavalier-Smith, 2000). Cryptomonads are a good example, with compelling evidence, of the past secondary endosymbiotic event because they contain the nucleomorphs (residual nuclei derived from the eukaryotic endosymbiont) and the ribosomes in the space, which corresponds to the endosymbiont's cytosol (PPC) between the outer and the inner pairs of plastid membranes (Archibald, 2007; Cavalier-Smith, 2002; Gillott & Gibbs, 1980; Moore & Archibald, 2009; Tanifuji & Archibald, 2014).

The nucleomorph was first reported by Greenwood (1974). Ludwig and Gibbs (1985) showed by 4′-6-diamidino-2-phenylindole-staining and transmission microscopy that the nucleomorphs contain DNAs. Subsequently, McFadden (1990) showed that eukaryotic ribosomal RNAs (rRNAs) exist in the PPC and the nucleolus-like structure of nucleomorphs (McFadden, 1990). These studies strongly suggested that the nucleomorph was derived from the nucleus of eukaryotic endosymbiont. Douglas and her colleagues suggested that cryptomonad plastids are derived from a red alga based on molecular phylogenetic studies inferred from the large subunit of Rubisco (Douglas et al., 1990). Also, Douglas, Murphy, Spencer, and Gray (1991) isolated two distinct eukaryotic rRNA sequences representing nuclear and nucleomorph genes and showed that the cryptomonads with nucleomorphs are evolutionary chimeras between two eukaryotes (Douglas et al., 1991). These results were supported by subsequent molecular phylogenetic analyses with taxon- and gene-rich data. It is currently accepted that the plastid-containing cryptomonads acquired their plastids from a red alga (Cavalier-Smith et al., 1996; Douglas & Penny, 1999). Nevertheless, the precise origin of the ancestral endosymbiont of cryptomonads in the red algal lineage is still unclear.

One of the intriguing aspects of cryptomonad plastid evolution is changes of biliprotein compositions and their physical distribution. The biliproteins of cryptomonads evolved from red algal phycobilisomes. A red algal phycobilisome consists of three types of biliproteins (phycoerythrin, phycocyanin, and allophycocyanin) and linker proteins (Adir, 2005; Arteni et al., 2008), which form several rods and the core subunits. Three allophycocyanins along with linker proteins are the central core of phycobilisomes. Tri- or hexameric heterodimers of phycocyanin and phycoerythrin form the rods. This high-molecular light-harvesting protein complex on the thylakoids transfers the energy. On the other hand, cryptomonads do not have phycobilisomes, presumably due to loss of allophycocyanin. Since biliproteins in cryptomonads are embedded into the thylakoid lumen, unlike red algae (Ludwig & Gibbs, 1989b), the thylakoids of cryptomonads appear to be relatively thicker than that of other photosynthetic eukaryotes. Furthermore, while red algae contain both phycocyanin and phycoerythrin, a cryptomonad species contains only either phycocyanin or phycoerythrin (Glazer & Wedemayer, 1995; Hill & Rowan, 1989; Hoef-Emden, 2008). Apt et al. (1995) suggested that the ancestor of cryptomonads retained a phycoerythrin. The different types of cryptomonad biliproteins have evolved from phycoerythrin with modifications of the linear tetrapyrrole chromophores. Therefore, the "phycocyanins" of the extant cryptomonads are modified versions of the former phycoerythrin from the ancestor (Glazer & Wedemayer, 1995). At least eight types of biliproteins have been identified with different wavelengths of respective absorption maxima (three types of phycoerythrin and five types of phycocyanin).

In general, the plastid genomes contain incomplete sets of genes for all the plastid functions, and many essential proteins are supplied by the nuclear genome (Bock & Timmis, 2008; Gould, Maier, & Martin, 2015; Martin & Herrmann, 1998; van Dooren, Schwartzbach, Osafune, & McFadden, 2001). As with other photosynthetic organisms, many nuclear-encoded proteins (approximately 750) were predicted as plastid-targeted proteins in the model cryptomonad, *Guillardia theta* (Curtis et al., 2012). For primary plastids, transit peptides upstream of mature proteins act as a "tag" to allow passage across the double plastid membranes. Since the secondary plastids are surrounded by two (or one) more membranes, in addition to ones derived from the primary plastids, plastid-targeted proteins contain bipartite-targeting leaders consisting of signal and transit peptides at the N-terminus (van Dooren et al., 2001). Gould and colleagues have investigated the plastid-targeting mechanism of several nuclear-encoded proteins in *G. theta* using a heterologous import experimental system of diatom transformants (Gould, Fan, Hempel, Maier, & Kloesgen, 2007; Gould, Sommer, Hadfi, et al., 2006; Gould, Sommer, Kroth, et al., 2006). Based on these experimental data and preexisting knowledge regarding the plastid import machinery, they proposed a protein import schema for cryptomonads. Under this schema, the signal peptide is responsible for passage through the outermost plastid membrane that is decorated with 80S ribosomes like the rough endoplasmic reticulum (ER). Targeted proteins cross the outermost membrane in a cotranslational manner. After cleavage at the AXAF motif between the signal and the transit peptide, the newly exposed transit peptide mediates transport across the remaining three membranes. The SEC61 complex and the ER-associated degradation-like (ERAD-like) machinery are considered to be the translocators of the outmost and the second outermost plastid membranes, respectively. Also, as in other primary and secondary plastids, TIC/TOC (translocons at the inner/outer envelope membrane of chloroplasts) complexes are probably involved in protein import across the inner pair of plastid membranes (Gould et al., 2015; Gould, Sommer, Kroth, et al., 2006). In addition, proteins targeted to the thylakoid lumen such as biliproteins need to cross a fifth membrane (thylakoid membrane). In those cases, a twin arginine motif following the transit peptide is responsible for passage through the thylakoid membrane. Thus, some proteins targeted to the thylakoid lumen such as a subunit of biliprotein (cpeA) contain a tripartite topogenic signal at the N-terminus (Gould et al., 2007).

2.3 Insight Into Plastid Evolution in Secondary Endosymbiosis—Various States of Cryptomonad Plastids

Cryptomonad plastids might be key to understanding the detailed process of plastid acquisition in a broader range of eukaryotes because they possess the different evolutionary states of plastids.

The timing of plastid acquisition in a cryptomonad ancestor has been a controversial issue. As mentioned earlier, cryptomonads are divided into at least two groups consisting of the plastid-containing group (Cryptophyceae) and the plastid-lacking phagotrophic group *Goniomonas* (Goniomonadea). The kathablepharids (class Leucocryptea) and *P. bilix* (class Palpitea) are the closest sister groups to cryptomonads in Cryptista (Burki, Okamoto, et al., 2012; Cavalier-Smith, 2013; Kim & Archibald, 2013; Martin-Cereceda et al., 2010; Okamoto et al., 2009; Okamoto & Inouye, 2005; Von der Heyden et al., 2004; Yabuki et al., 2014). While *Goniomonas*, kathablepharids, and *P. bilix* are phagotrophic, the organisms of Cryptophyceae do not exhibit phagotrophy even in the nonphotosynthetic *Cryptomonas* species (Hoef-Emden & Melkonian, 2003; Hoef-Emden, Shrestha, et al., 2005). Thus, one possible scenario of the plastid acquisition process in cryptomonads is that plastid acquisition and the loss of phagotrophy occurred at some points leading to Cryptophyceae after divergence from the Goniomonadea lineage (Fig. 3, left; Yabuki et al., 2014). Under this scenario, Goniomonadea, kathablepharids, and *P. bilix* were ancestrally nonphotosynthetic (i.e. never possessed a plastid). On the other hand, Cavalier-Smith (1999) proposed the "Chromalveolate hypothesis", which posited that the red alga-derived plastids of cryptomonads, ochrophytes (photosynthetic stramenopiles), haptophytes, and myzozoans (alveolates with plastids: i.e. dinoflagellates, apicomplexans, and chrompodellids) share a common origin (Cavalier-Smith, 1999). Under this hypothesis, another option to consider is that Cryptista organisms independently lost their plastids at different times in the ancestors of Goniomonadea, kathablepharids, and *P. bilix* (Fig. 3, right). Both scenarios are possible, and a definitive answer has not yet been found. In any case, elucidating the evolutionary history of Cryptista plastids would be helpful in delineating the true nature of the evolution of red alga-derived plastids in eukaryotes. The specific relationship between the phylogeny of cryptomonads and the chromalveolate hypothesis is also discussed later.

Nonphotosynthetic plastids are found in freshwater *Cryptomonas* species (e.g. *C. paramecium*). Since they are easily distinguishable from other coloured strains, noncoloured *Cryptomonas* species had been classified as *Chilomonas* previously (Clay et al., 1999; Deane et al., 2002; Hoef-Emden et al., 2002). However, since the molecular phylogenetic analyses revealed that "*C. paramecium*" is nested within the *Cryptomonas* clade, *Chilomonas* was subsequently integrated into the genus *Cryptomonas* (Hoef-Emden & Melkonian, 2003). In addition to *C. paramecium*, two more noncoloured polyphyletic groups appeared in the *Cryptomonas* clade, suggesting that the ancestor of *Cryptomonas* species lost the photosynthetic ability on at least three different occasions (Hoef-Emden, 2005; Hoef-Emden & Melkonian, 2003). Hoef-Emden (2005) pointed out that the potential mixotrophy of the ancestral *Cryptomonas* might be a prerequisite for a change to a heterotrophic lifestyle. As described earlier, if Goniomonadea have never possessed plastids, extant cryptomonads would represent three different evolutionary states, (1) before plastid acquisition (*Goniomonas*), (2) after the acquisition of photosynthetic plastids (the coloured Cryptophyceae species), and (3) after the loss of photosynthesis (the noncoloured *Cryptomonas*). Therefore, cryptomonads would be a good model to gain insight into a series of evolutionary steps of pre- and postplastid acquisition and after the secondary loss of photosynthesis.

Moreover, it is known that cryptomonads themselves are engulfed by other phagotrophic protists and behave as temporary plastids. This phenomenon, referred to as "kleptoplasty", is seen in some harmful dinoflagellates and a ciliate (e.g. *Dinophysis* spp., *Nusuttodinium acidotum*, *Pfiesteria piscicida*, and *Mesodinium rubrum*) (Fields & Rhodes, 1991; Gustafson, Stoecker, Johnson, Van Heukelem, & Sneider, 2000; Minnhagen & Janson, 2006; Onuma & Horiguchi, 2015; Schnepf, Winter, & Mollenhauer, 1989; Takishita, Koike, Maruyama, & Ogata, 2002). Host cells are believed to take advantage of the photosynthesis before completely digesting the cryptomonad plastids (i.e. kleptochloroplast). It is noteworthy that a certain degree of diversity in host-cryptomonad specificity and of control capacity over the endosymbionts is seen in kleptoplasty involving dinoflagellates. For example, *Nusuttodinium poecilochroum* and *Nusuttodinium latum* seem to engulf any cryptomonad (Horiguchi & Pienaar, 1992). Their kleptochloroplasts exhibit different colours such as red, blue, and yellow depending on the ingested cryptomonad species, and sometimes they even display two colours in the same cell due to possessing two different cryptomonad plastids (Horiguchi & Pienaar, 1992). On the other hand, *Nusuttodinium aeruginosum* only engulfs a particular cryptomonad, *Chroomonas* (Fields & Rhodes, 1991; Schnepf et al., 1989).

While *N. poecilochroum* completely digests the cryptomonad plastid in a few days, *N. aeruginosum* is capable of distributing the plastid to daughter cells. Furthermore, *N. aeruginosum* can maintain the cryptomonad nucleus in addition to the plastid, and they distribute the cryptomonad nucleus to either of two daughter cells (Onuma & Horiguchi, 2015). The plastid without the cryptomonad nucleus in the same cell shrinks faster than that with the cryptomonad nucleus. Although *N. aeruginosum* is not able to control the cryptomonad nuclear division, it appears to take advantage of the cryptomonad plastid for a longer period than *N. poecilochroum and N. latum* (Onuma & Horiguchi, 2015). Thus, some researchers suggest that the kleptochloroplasts may represent primitive stages in the process of the engulfed endosymbionts transforming into permanent plastids.

A similar story of kleptoplasty is seen in Cryptista organism other than cryptomonads. Although all known kathablepharids lack permanent plastids, *Hatena arenicola* is known to temporarily retain an engulfed green algal cell (i.e. *Nephroselmis*) (Okamoto & Inouye, 2006). The engulfed green algal cell is distributed to either of two daughter cells, resulting in green and non-coloured cells. Like the story of dinoflagellate/cryptomonad, the unequal distribution of the green algal cell in *Hatena* could be the primitive stage of establishing a permanent plastid.

In summary, as a lineage, cryptomonads display several aspects that would be helpful in understanding the processes by which secondary endosymbioses occur such as (i) having red alga-derived plastids, (ii) having different plastid states such as photosynthetic or nonphotosynthetic, and (iii) used as a temporary plastid. The future analyses of molecular, biochemical, and morphological studies of cryptomonad plastids can shed light on various issues of secondary endosymbioses.

2.4 The Phylogenetic Position of Cryptomonads in the Eukaryotic Tree of Life

Eukaryotes are known to be divided into at least six supergroups including archaeplastids, opisthokonts, excavates, SAR (stramenopiles, alveolates, and Rhizaria), Amoebozoa, and others (Adl et al., 2005, 2012; Burki, 2014). Mainly by use of genome wide data, researchers have attempted to uncover the true nature of the eukaryotic phylogeny (Brown et al., 2013; Brown, Spiegel, & Silberman, 2009; Burki, 2014; Burki, Flegontov, et al., 2012; Burki et al., 2009; Burki, Okamoto, et al., 2012; Derelle & Lang, 2012; Hampl et al., 2009; Janouškovec, Horák, Oborník, Lukeš, & Keeling,

2010; Moore et al., 2008; Williams, Foster, Cox, & Embley, 2013; Zaremba-Niedzwiedzka et al., 2017). Even within these studies, the phylogenetic position of cryptomonads is still unclear.

Cavalier-Smith (1986) placed cryptomonads in the "Chromista" together with ochrophytes (photosynthetic stramenopiles) and haptophytes (Cavalier-Smith, 1986). This proposal was based on the pigment compositions of the plastids and ultrastructural features. Cryptomonads, haptophytes, and ochrophytes commonly contain chlorophyll *c* in those with photosynthetic plastids, and the outermost membrane of the four membrane-bound plastids is contiguous with the nuclear envelop. Subsequently, Cavalier-Smith (1999) proposed the integration of Chromista and alveolates into Chromalveolata and posited that their plastids were derived from one endosymbiotic event in the common ancestor of chromalveolates (Cavalier-Smith, 1999). The nonphotosynthetic alveolates (e.g. ciliates) and stramenopiles (e.g. oomycetes) were explained by the multiple loss of plastids under the chromalveolate hypothesis. Later studies of molecular phylogeny supported the monophyly of the alveolates including dinoflagellates, apicomplexan parasites with nonphotosynthetic plastids (i.e. apicoplast), and ciliates without any plastid-like structure (Fast, Xue, Bingham, & Keeling, 2002; Harper, Waanders, & Keeling, 2005). Stramenopiles also show a phylogenetic affinity for alveolates (Burki et al., 2007; Harper et al., 2005; Van de Peer, Van der Auwera, & De Wachter, 1996).

The phylogenies inferred from the plastid genes of chromalveolates generally are consistent in their topologies. Several phylogenetic trees suggested that red algal plastids in secondary organisms are the product of a single secondary endosymbiosis (Janouškovec et al., 2010; Yoon, Hackett, Pinto, & Bhattacharya, 2002). Other evidence supporting the single origin of the red alga-derived plastids is seen in the plastid protein transport machinery. Because the plastids are surrounded by three or four membranes, a novel transport system is necessary to cross extra boundaries, in addition to the TIC/TOC translocons of the primary plastids (Bolte et al., 2009; Gould et al., 2015, 2008; Jarvis & Soll, 2002). The multiprotein complex known as SELMA [symbiont-specific ER-associated degradation-like machinery] is believed to be involved in traversing the second outermost membrane in four-membrane-bound red alga-derived plastids (Felsner et al., 2011; Gould et al., 2015; Hempel, Bullmann, Lau, Zauner, & Maier, 2009; Sommer et al., 2007; Stork et al., 2012). The SELMA protein complex consists of a dozen or more proteins originally involved in an

ERAD system and appears to have a single origin linked to red algal plastids (Archibald, 2015).

The observations above do not conflict with the chromalveolate hypothesis. However, the issue is that the phylogenetic positions of cryptomonad and haptophyte hosts remain challenging to place in the eukaryotic tree. Some phylogenetic analyses with multiple genes suggested a monophyletic origin of Cryptista and haptophytes (Burki et al., 2009; Hackett et al., 2007; Patron, Inagaki, & Keeling, 2007), resulting in the proposed Hacrobia clade (Okamoto et al., 2009). However, separate origins of Cryptista and haptophytes were also shown in several other phylogenetic analyses based on genome wide data (Brown et al., 2013; Burki, Flegontov, et al., 2012; Burki et al., 2016; Yabuki et al., 2014). And in many recent global phylogenies, cryptomonads and haptophytes do not show a significant affinity with either alveolates or stramenopiles (Brown et al., 2013; Burki, Flegontov, et al., 2012; Burki et al., 2016; Yabuki et al., 2014). Overall, whereas the plastids of chromalveolates appear to derive from a single endosymbiotic event, phylogenetic studies of the hosts are contradictory about supporting the single ancestry of the members of Chromalveolata. Thus, it is unclear which plastids in cryptomonads, haptophytes, ochrophytes, dinoflagellates, and apicomplexans share a common endosymbiotic origin.

Although the chromalveolate hypothesis tries to explain the diversity of secondary red algal plastids by a one-time endosymbiotic event (Fig. 4A), independent, multiple endosymbioses are another option to consider (Fig. 4B). In the case of secondary green algal plastids, the plastids of euglenoids and chlorarachniophytes showed phylogenetic affinities to Prasinophyceae and Ulvophyceae, respectively, suggesting that those plastids were derived from separate endosymbiotic events (Kamikawa et al., 2015; Rogers, Gilson, Su, McFadden, & Keeling, 2007; Suzuki, Hirakawa, Kofuji, Sugita, & Ishida, 2016; Turmel, Gagnon, O'Kelly, Otis, & Lemieux, 2009).

Some recent studies suggested that the plastids of cryptomonads, heterokonts, and haptophytes are a product of serial endosymbiosis. By using regression analyses based on a similarity search among genomic data of cryptomonad, heterokonts, and haptophyte, Stiller and colleagues have reported that the haptophyte has significant affinity to heterokonts, the cryptophyte appears a potentially significant level of similarity to heterokonts, and haptophyte and cryptomonad show less similarity compared with the

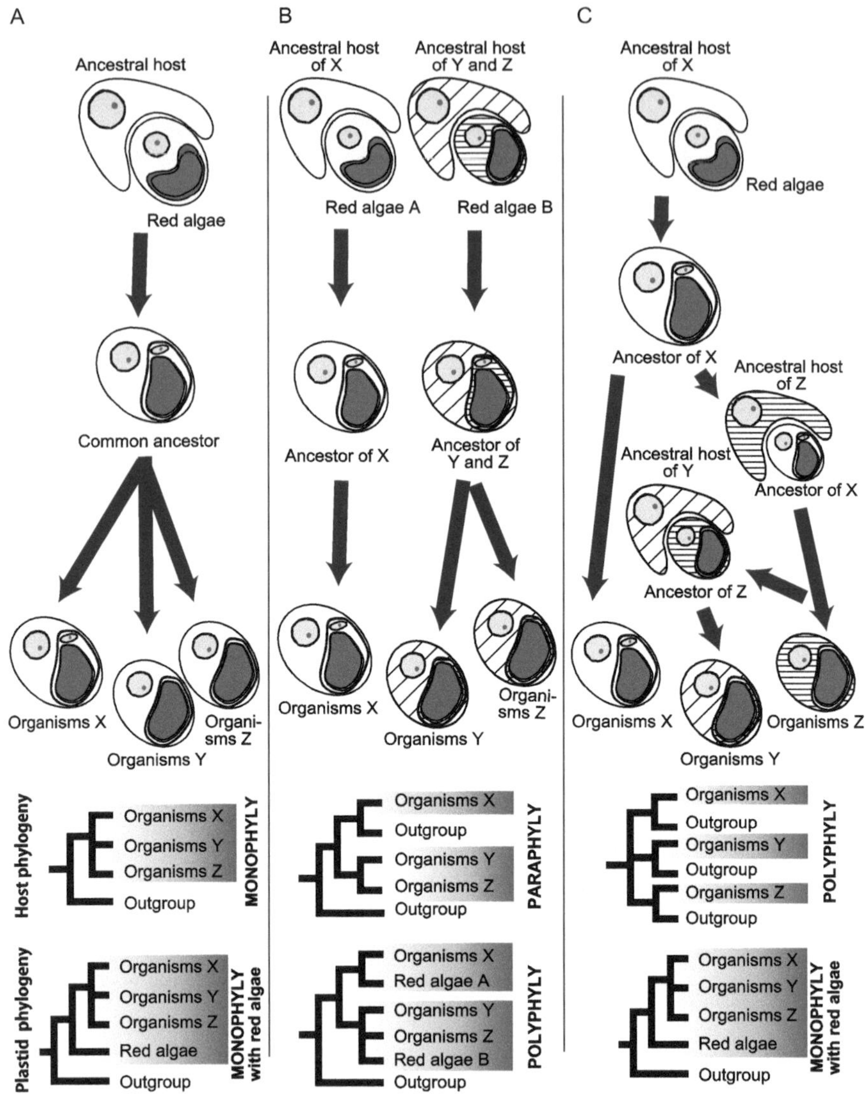

Fig. 4 Schematic diagram of three possible scenarios for the evolution of red alga-derived plastids (*above*) and the corresponding phylogenetic topologies of hosts and plastids (*below*). (A) Single origin of red alga-derived plastid. Descendants of the common ancestor evolved into the various extant eukaryotes with red alga-derived plastids. While plastid topology is consistent with recent phylogenetic analyses, host topology is poorly supported. (B) Independent endosymbioses between hosts and red algae. Although the host phylogeny is compatible with recent phylogenetic analyses, the plastid topology is inconsistent with recent analyses. (C) Cryptic serial endosymbioses. A single endosymbiosis between a eukaryotic host and a red alga took place. Subsequently, the red alga-derived plastid was engulfed by other eukaryotes. The tree

former two relations (Stiller et al., 2014). In this context, they proposed an intriguing hypothesis that, (i) the ancestor of cryptomonads acquired a plastid from red algae, (ii) subsequently, the heterokont ancestor engulfed a cryptomonad and retained its plastid, and (iii) finally haptophytes obtained a quaternary plastid from the heterokonts (Fig. 4C).

Furthermore, by use of comprehensive molecular data, Burki et al. (2016) showed a surprising tree topology that has Cryptista positioned within Archaeplastida (i.e. plants, green algae, red algae, and glaucophytes) (Burki et al., 2016). According to this topology, cryptomonads acquired their plastid through a different endosymbiotic event from other members of chromalveolates. In addition, the tree topology presented by Burki et al. (2016) does not conflict with multiple endosymbioses between red algae and the individual ancestors of cryptomonads, heterokonts, and haptophytes (Fig. 4B). Nevertheless, the authors discussed that the cryptic serial endosymbioses proposed by Stiller et al. (2014) are a likely scenario (Fig. 4C). If the red alga-derived plastids originated from different red algae, the phylogeny inferred from plastid genes would be polyphyletic (Fig. 4B; Janouškovec et al., 2010; Yoon et al., 2002). Therefore, the idea that a stable cryptomonad plastid was subsequently engulfed by other eukaryotes in a serial fashion is compatible with tree topologies of hosts and plastids in the chromalveolates (Fig. 4C).

The phylogenetic position of the cryptomonads will need to be tested from various perspectives such as biochemical assays in addition to the phylogenetic analyses. The serial endosymbiosis scenario is attractive in explaining the current inconsistency of the tree topologies of the hosts and the red alga-derived plastids. Although serial endosymbioses have been described in only a few dinoflagellates such as *Dinophysis* with a haptophyte-derived plastid and *Lepidodinium* with a green alga-derived plastid (Hackett, Anderson, Erdner, & Bhattacharya, 2004; Koike et al., 2005; Watanabe, Suda, Inouye, Sawaguchi, & Chihara, 1990), hidden endosymbiotic events might have had a greater impact on plastid diversity than we previously thought.

topologies are compatible with recent phylogeny. Organisms X are equivalent to the extant cryptomonads under each scenario. *Adapted from Burki, F., Kaplan, M., Tikhonenkov, D. V., Zlatogursky, V., Minh, B. Q., Radaykina, L. V., et al. (2016). Untangling the early diversification of eukaryotes: A phylogenomic study of the evolutionary origins of Centrohelida, Haptophyta and Cryptista.* Proceedings of the Royal Society B: Biological Sciences, 283, *20152802, with some modifications.*

3. NUCLEOMORPH GENOMICS: A MODEL FOR INVESTIGATION OF GENOME REORGANIZATION BETWEEN EUKARYOTES

3.1 Reductive Genome Evolution of Endosymbionts

When one organism is incorporated into another through endosymbiosis, one can expect that there will be numerous modifications in the structures and metabolic capacities of the two organisms as they are establishing their intimate relationship. The genes/genomes of both host and endosymbiont also need to be reorganized as two biologically different organisms are integrated into one organism. From a host's point of view, there is the need to internalize and control the organism that is being taken up. Conversely, the endosymbiont is necessary to adapt to an intracellular lifestyle having previously been free-living.

Genome reduction of the endosymbiont is the most striking aspect that can be observed as a result of genome reorganization. For example, plastids, well-known organelles in which photosynthesis takes place, and are derived from a cyanobacterium engulfed by a heterotrophic unicellular eukaryote (primary endosymbiosis), now retain only ~5% of the original genome (~100–150 kb) compared with the median size of modern-day cyanobacteria (3.5 Mb) (Beck, Knoop, Axmann, & Steuer, 2012; Martin & Herrmann, 1998; Martin et al., 2002). Mitochondria, which also originated through endosymbiosis and trace their origin back to alpha-proteobacteria, possess ~30 kb genomes (median), while the genome size of present alpha-proteobacteria is ~3 Mb (Gray, Lang, & Burger, 2004).

Such size reduction in endosymbiont genomes is explained by the elimination of endosymbiont genes that are not essential any longer as the adaptation proceeds (e.g. extracellular proteins and motile mechanisms). In addition to the simple loss of genes, the reduction is accomplished by massive transfer of endosymbiont genes to the host nuclear genome. Indeed, plastid and mitochondrial-derived genes that encode organelle-targeted proteins are scattered across host chromosomes (Bock & Timmis, 2008; Matsuo, Ito, Yamauchi, & Obokata, 2005; van Dooren et al., 2001). Since only a portion of the necessary protein-coding genes are retained in the organelle genome, it is impossible to perform all the required plastid/mitochondrial functions without the nuclear-encoded proteins. The nuclear-encoded proteins possess "entry pass tickets" (i.e. targeting peptides) at the N-terminus for correct transport to the plastid/mitochondrial compartments

(Bolte et al., 2009; Gould et al., 2008; Jarvis & Soll, 2002). The vast majority of indispensable proteins that function in plastids and mitochondria are under the control of the host (which is one possible reason for referring to an enslaved endosymbiont as a permanent "organelle" (Keeling, 2010; Keeling & Archibald, 2008; Keeling, McCutcheon, & Doolittle, 2015; McFadden, 2014; Nakayama & Archibald, 2012), and due to this host-dependent regulation, endosymbionts/organelles are no longer able to return to an autonomous lifestyle. Endosymbiotic gene reorganizations involving changes in host chromosomes and the reduction of the endosymbiont genome as outlined above are cases of eukaryotic-prokaryotic associations, with particular focus on ones now known as organelles. What, then, is eukaryote–eukaryote endosymbiosis like, with respect to gene reorganization?

Secondary endosymbiotic organisms arose from colourless eukaryotes engulfing primary endosymbiotic organisms, and the plastid acquisitions are believed to have happened multiple times (Gould et al., 2008; Reyes-Prieto et al., 2007). Most of integrated eukaryotic endosymbionts have lost their mitochondria except in a few cases (e.g. dinoflagellate with diatom (Carty & Cox, 1986; Dodge & Crawford, 1969) and marine amoeba with kinetoplastid endosymbiont (Tanifuji, Kim, et al., 2011)), and it is conceivable that functional overlaps in host and endosymbiont mitochondria might have led to the loss of the endosymbiont mitochondria. It is clear that genome reduction of the endosymbiont nucleus has occurred in general, since most secondary organisms are completely devoid of the endosymbiont nucleus. However, during this process of integration, how were the endosymbiont genomes that encoded large numbers of plastid-targeted proteins reorganized? The gene transfers and genome reorganization of the endosymbiont nuclear genomes might be more complex than the reorganizations that occurred in the prokaryotic endosymbiont.

Here we introduce nucleomorph genomes as an example of reduction in eukaryotic endosymbiont genomes. Nucleomorphs are the remnant nuclei of endosymbionts found in two different lineages, cryptomonads, and chlorarachniophytes (Fig. 5; Archibald, 2007; Archibald & Lane, 2009; Cavalier-Smith, 2002; Gilson & McFadden, 2002; Moore & Archibald, 2009; Tanifuji & Archibald, 2014). As of March 2017, they are the only lineages that still maintain endosymbiont nuclei. They differ with regard to their origins; the former arose from a colourless biflagellated host by engulfing a red alga (Cavalier-Smith et al., 1996; Douglas et al., 1990; Douglas & Penny, 1999), and the latter, derived from a endosymbiotic event

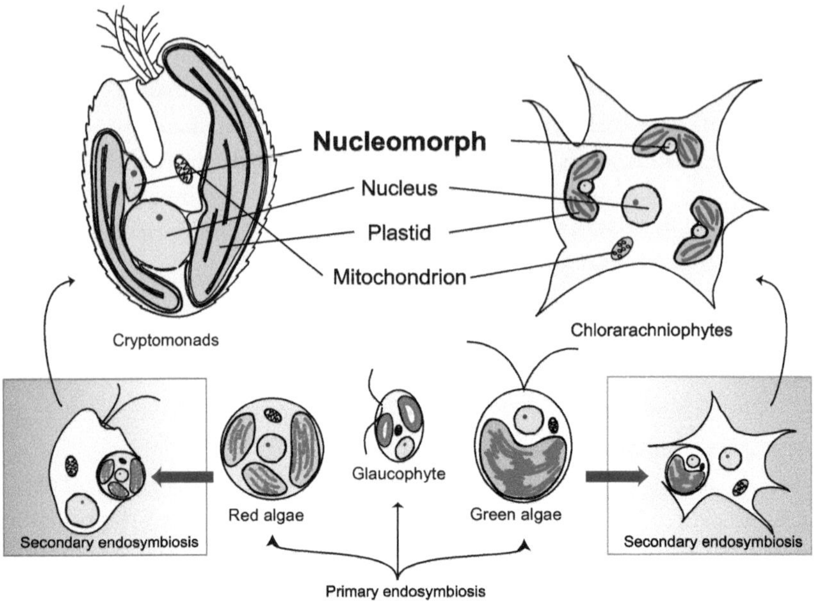

Fig. 5 A schematic diagram of the secondary endosymbioses giving rise to cryptomonads and chlorarachniophytes. The four genome-containing compartments are indicated.

between a colourless cercozoan protist and a green alga (Fig. 5; Ishida, Cao, Hasegawa, Okada, & Hara, 1997; Ishida, Green, & Cavalier-Smith, 1999; McFadden, Gilson, Hofmann, Adcock, & Maier, 1994; McFadden, Gilson, & Waller, 1995). As mentioned earlier, since most of the secondary endosymbiotic organisms have lost their endosymbiont nucleus, nucleomorphs were seen as an intermediate stage of genome reduction through endosymbiosis (Archibald & Lane, 2009; Gilson et al., 2006; Tanifuji, Onodera, & Hara, 2010), although the fate of nucleomorphs (intermediate or end-point) is an as yet unresolved question (Curtis et al., 2012). Also, the current state of comparative analyses of nucleomorph genomes has revealed significant similarities of their genome architectures, suggesting convergent evolution because of their distinct origins (Suzuki, Shirato, Hirakawa, & Ishida, 2015; Tanifuji & Archibald, 2014; Tanifuji, Onodera, et al., 2011).

So far, four nucleomorph genomes have been completely sequenced from both cryptomonads and chlorarachniophytes (Douglas et al., 2001; Gilson et al., 2006; Lane et al., 2007; Moore, Curtis, Mills, Tanifuji, & Archibald, 2012; Suzuki et al., 2015; Tanifuji, Onodera, Brown, et al.,

2014; Tanifuji, Onodera, et al., 2011). Also, the nuclear genome sequences of the model organisms, *G. theta* (cryptomonad) and *Bigelowiella natans* (chlorarachniophyte), are available (Curtis et al., 2012). Three distinct types of organisms, including photosynthetic nucleomorph-containing species, nonphotosynthetic nucleomorph-containing species, and nonphotosynthetic nucleomorph-lacking species, exist in cryptomonads. Here we summarize and discuss recent progress in understanding genome evolution in secondary endosymbiosis based on the comparative genomics of these nucleomorph-bearing organisms.

3.2 The Discovery of Nucleomorphs

By the 1970s, electron microscope technologies had developed enough to enable researchers to explore cells at the ultrastructural level, and the new techniques facilitated the observation of cells of various organisms. Electron microscopy of photosynthetic organisms across various lineages revealed that the structure of plastid membrane envelopes varies with respect to the numbers of layers surrounding plastids (Gibbs, 1978; Whatley & Whatley, 1981). Plants, green and red algae, and glaucophytes have double-membrane-bound plastids, while in other plastid-bearing organisms plastids are enclosed by three- or four-layered membranes (e.g. euglenoids, dinoflagellates, cryptophytes) (Gray & Doolittle, 1982). This multiplicity of layers of plastid membrane envelopes is one of the key pieces of evidence, together with biochemical assays of photosynthetic pigments such as chlorophylls, for supporting the theory of endosymbiosis that explains the origin of eukaryotic cells (Gray & Doolittle, 1982; Rowan, 1984). It is now widely accepted that the groups that harbour double-membrane-bound plastids are ones that originated via the endosymbiosis between a colourless eukaryote and a photosynthetic prokaryote (cyanobacterium) (i.e. primary endosymbiosis), and groups with one or two additional membrane arose by further engulfment of primary endosymbiotic organisms by other unrelated eukaryotes (i.e. secondary endosymbiosis) (Gould et al., 2008; Reyes-Prieto et al., 2007).

In 1974, the nucleomorph was discovered for the first time from one of the cryptomonad species, and recognized as an organelle whose structure was similar to a nucleus, being surrounded by two membranes and retaining nucleolus-like structures (Greenwood, 1974). The most striking point was that this nucleus-like structure resided in a space between the second and third layers of plastid membranes, that is, a space which is equivalent to the cytosol of a secondary endosymbiont (PPC). Subsequently, Ludwig and Gibbs (1985) found that this organelle contains DNA, suggesting a secondary endosymbiotic origin for nucleomorphs (Ludwig & Gibbs, 1985). Chlorarachniophyte's nucleomorph, on the other hand, was found by Hibberd and Norris (1984), and it also contained DNA (Ludwig & Gibbs, 1989a). Phylogenetic analyses using nucleomorph genes uncovered the sources of these two independently acquired organelles and it turned out to be a red algal origin for cryptomonads (Cavalier-Smith et al., 1996; Douglas et al., 1990; Douglas & Penny, 1999), and a green algal origin for chlorarachniophytes (Ishida et al., 1997, 1999; McFadden et al., 1995). Their host origins have also been studied and for cryptomonads, the host lineage is still under consideration (see above), and for chlorarachniophytes, its hosts are closely related to organisms that belong to Cercozoa (Burki et al., 2009; Hampl et al., 2009). These two nucleomorph-bearing lineages originated from different hosts and different endosymbionts, and accordingly, they have been hailed as a "smoking gun" that indicates that secondary endosymbioses have occurred independently multiple times (Archibald, 2007; Archibald & Lane, 2009; Moore & Archibald, 2009; Tanifuji & Archibald, 2014).

3.3 The Three Chromosomes and Genome Size Diversity

Initial studies on nucleomorph genomes were mostly karyotypic analyses by pulsed-field gel electrophoreses and Southern hybridizations using ribosomal DNA probes (Eschbach, Hofmann, Maier, Sitte, & Hansmann, 1991; Gilson & McFadden, 1996; Ishida, Endo, & Koike, 2011; Lane & Archibald, 2006; Lane et al., 2006; Phipps, Donaher, Lane, & Archibald, 2008; Rensing, Goddemeier, Hofmann, & Maier, 1994; Silver et al., 2007; Tanifuji et al., 2010). To date, karyotypic analyses of nucleomorph genomes have been carried out in ~50 cryptomonad and ~20 chlorarachniophyte strains (Fig. 6). It has been shown that their chromosome sizes range from ~100 to 400 kb, and all the nucleomorph genomes from both cryptomonads and chlorarachniophytes possess three linear chromosomes. Furthermore, both lineages share similarities with regard to their genome architectures in that regions of rRNA operons are found at subtelomeric regions of the chromosomes (Lane & Archibald, 2006, 2008; Silver, Moore, & Archibald, 2010). However, the rRNA operons differ in their composition (i.e. sometime only 5S rRNAs are found) and direction. The common traits shared by nucleomorphs in cryptomonads and

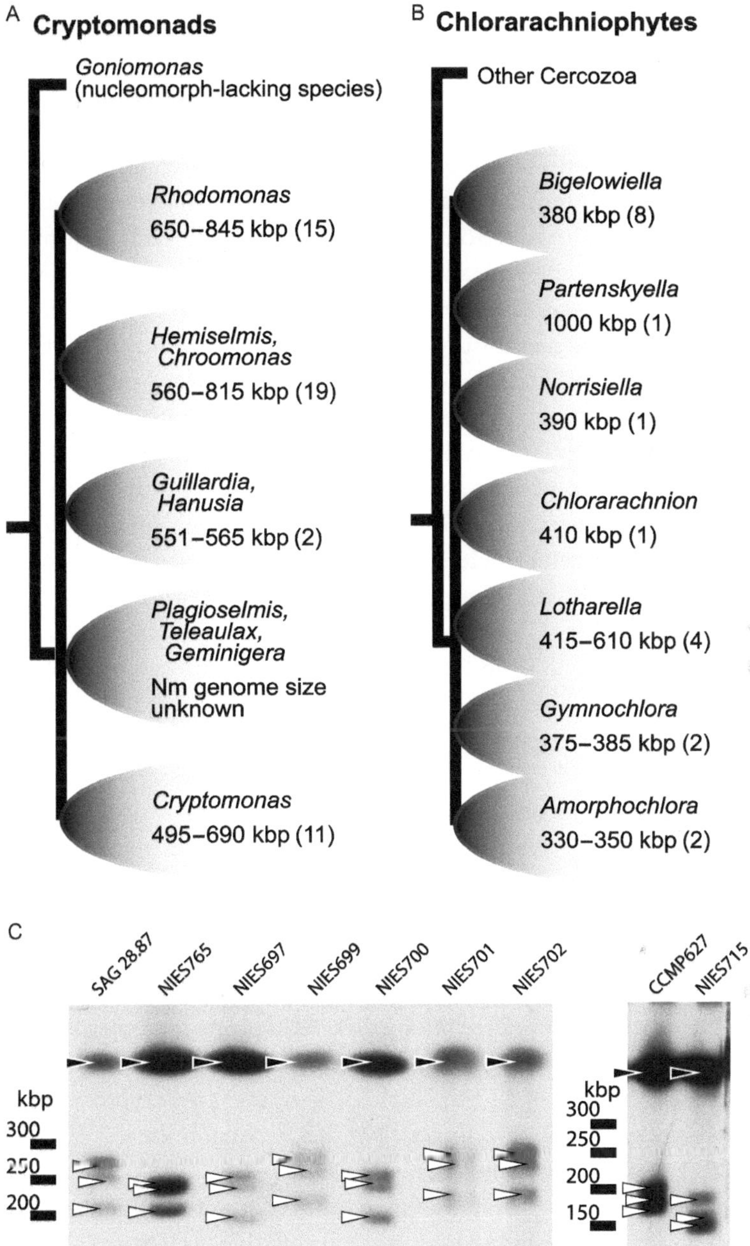

Fig. 6 Nucleomorph genome size variation in (A) cryptomonads and (B) chlorarachniophytes, and (C) images of nucleomorph karyotypic analyses by a combination of pulsed-field gel electrophoresis (PFGE) and Southern blot analyses using SSU rDNA probes. *White* and *black arrowheads* indicate the nucleomorph and the nuclear chromosomes, respectively. The schematic trees and genome size estimates were taken from Tanifuji and Archibald (2014). Strains shown in PFGE images are as follows: SAG 28.87—*Pyrenomonas helgolandii*; NIES765—*Rhodomonas duplex*; NIES697—*R. acuta*; NIES699—*R. atrorosea*; NIES700—*R. baltica*; NIES701—*R. chrysoidea*; NIES702—*R. falcata*; CCMP327—*Guillardia theta*; NIES715—*Chilomonas paramecium* (i.e. *Cryptomonas paramecium*).

chlorarachniophytes are considered to be a result of convergent evolution, indicating that nucleomorph genomes from both lineages were exposed to similar selective pressures during the process of reductive genome evolution. Cavalier-Smith pointed out that nucleomorph chromosomes might have been shaped in this way due to physical shrinkage of the nucleomorph itself (Cavalier-Smith, 2002). As is the case with one of the microsporidian parasites, *Encephalitozoon cuniculi*, which harbours a 2.9 Mb genome comprised of 11 chromosomes, ribosomal DNA operons are found in subtelomeric regions (Katinka et al., 2001). This suggests that there is a link between reductive genome evolution and ribosomal DNA operons at the ends of chromosomes. However, there has been no further data supporting this idea. As well, why nucleomorphs always retain three chromosomes remains a mystery.

The copy number of nucleomorph genomes has also been investigated and it appears to be four copies of the nucleomorph genome in the cryptomonad species, *G. theta*, and two copies in the chlorarachniophyte *B. natans* (Hirakawa & Ishida, 2014). Thus, the number of chromosomes in nucleomorphs will be 12 (3 chromosomes × 4 copies) in *G. theta* and 6 chromosomes in *B. natans* (3 chromosomes × 2 copies). Hirakawa and Ishida (2014) pointed out that an increase in the copy number is seen, not only in nucleomorphs but also in other reductive genomes of plastids, in mitochondria, and even in some symbiotic bacteria (e.g. Buchnera) of Arthropoda (Hirakawa & Ishida, 2014; Komaki & Ishikawa, 1999). These seemingly conflicting phenomena of genome reductions vs increases in copy number may generally be seen in the genomes that have undergone the process of reductive genome evolution.

Nucleomorph genomes are the smallest eukaryotic genomes known, and the sizes in both nucleomorph-bearing lineages are distributed across a similar range, with 0.45–0.87 Mb/haploid genomes in cryptomonads and 0.33–1 Mb in chlorarachniophytes (Fig. 6; Eschbach et al., 1991; Ishida et al., 2011; Lane & Archibald, 2006; Lane et al., 2006; Phipps et al., 2008; Rensing et al., 1994; Silver et al., 2007; Tanifuji et al., 2010). Genomes in free-living eukaryotes of close relatives (ones with a relatively small genome size) to the endosymbionts range from 10–16 Mb in Cyanidiales (red algae) (Matsuzaki et al., 2004; Moreira, Lopezarchilla, Amils, & Marin, 1994; Muravenko, Selyakh, Kononenko, & Stadnichuk, 2001), ~20 Mb in *Micromonas* (green algae) (Worden et al., 2009), and ~12.6 Mb in *Ostreococcus* (green algae) (Derelle et al., 2006), suggesting that the nucleomorph genomes are degraded by ~90% or more compared with their ancestral relatives. When taking a closer look, the nucleomorph genome sizes are divergent within each clade, and the genome size ranges of the major clades overlap (Fig. 6). For example, in the cryptomonad clade that consists of *Rhodomonas* species, the nucleomorph genome sizes are 650–845 kb, and in the clade of *Hemiselmis* and *Chroomonas*, sizes range between 560 and 815 kb. This overlap in genome sizes between clades suggests that size transitions occurred even after the divergence of the major clades.

3.4 Compaction of Nucleomorph Genomes

Nucleomorph genomes have been completely sequenced in four cryptomonad species including *G. theta* (Douglas et al., 2001), *Hemiselmis andersenii* (Lane et al., 2007), *C. paramecium* (Tanifuji, Onodera, et al., 2011), and *Chroomonas mesostigmatica* (Moore et al., 2012), and in four chlorarachniophyte species—*B. natans* (Gilson et al., 2006), *Lotharella oceanica* (Tanifuji, Onodera, Brown, et al., 2014), *Amorphochlora amoebiformis* (Suzuki et al., 2015), and *Lotharella vacuolata* (Suzuki et al., 2015). A summary of each nucleomorph genome is shown in Table 1. In addition to the nucleomorph genomes, the complete sequencing of host nuclear genomes has been carried out in *G. theta* and *B. natans* (Curtis et al., 2012). The G + C content of the nuclear genomes is found to be ~53% and ~45% in *G. theta* and *B. natans*, respectively, while the nucleomorph genomes show extreme A + T richness when compared to host genomes, with ~25–35% G + C contents (Table 1). As well as nucleomorph genomes, a bias towards A + T bases has been generally seen in other reduced genomes (e.g. plastid, mitochondria, and symbionts of Arthropoda) (Moran, 1996; Nakabachi et al., 2006; Smith, 2009), and A + T richness may be one of the key features of reduced genomes.

The average intergenic spacer size in nucleomorph genomes is 90–200 bp and gene densities are 0.9–1.1 kb/gene, which indicates highly compacted genomes compared with the host nuclear genomes (0.284 kb/gene in cryptomonads and 0.229 kb/gene in chlorarachniophytes) (Table 1; Curtis et al., 2012; Douglas et al., 2001; Gilson et al., 2006; Lane et al., 2007; Moore et al., 2012; Suzuki et al., 2015; Tanifuji, Onodera, Brown, et al., 2014; Tanifuji, Onodera, et al., 2011). Lane et al. (2007) demonstrated, with a comparative analysis of nucleomorph genomes from *G. theta* and *H. andersenii*, that the average protein length and the average size of intergenic spacers in nucleomorph genomes are significantly shorter than ones in genomes of

Table 1 Overview of Published Nucleomorph Genome Sequences

Species	***Guillardia theta***	***Hemiselmis andersenii***	***Cryptomonas paramecium***	***Chroomonas mesostigmatica***
Genome size (kb)	Total 550.5	Total 571.4	Total 485.9	Total 702.9
	chr.1 195.9	chr.1 207.3	chr.1 177.0	chr.1 243.9
	chr.2 180.6	chr.2 184.6	chr.2 159.7	chr.2 121.6
	chr.3 173.9	chr.3 179.4	chr.3 149.1	chr.3 226.1
G + C content (%)	26.4	25.2	26.05	26.0
# of protein-coding genes of genes	487	472	466	505
Mean intergenic distance (base pairs)	93	132	102	200
Gene density (genes/kilobase pairs)	0.977	1.09	1.07	0.83
# of introns	17	0	2	24
# of plastid-associated genes	31	31	18	31
Species	***Bigelowiella natans***	***Lotharella oceanica***	***Lotharella vacuolata***	***Amorphochlora amoebiformis***
Genome size (kb)	Total 372.9	Total ~611.6	Total 431.8	Total 373.9
	chr.1 140.6	chr.1 ~210	chr.1 166.1	chr.1 131.9
	chr.2 134.1	chr.2 207.5	chr.2 141.6	chr.2 124.0
	chr.3 98.1	chr.3 194.1	chr.3 124.0	chr.3 118.0

Table 1 Overview of Published Nucleomorph Genome Sequences—cont'd

Species	*Bigelowiella natans*	*Lotharella oceanica*	*Lotharella vacuolata*	*Amorphochlora amoebiformis*
G + C content (%)	28.5	33.0	24.7	30.0
# of protein-coding genes (total/nonredundant)	288 (288)	596 (338)	319 (294)	295 (300)
Mean intergenic distance (base pairs)	112	163	134	86
Gene density (genes/kilobase pairs)	0.89	1.04	0.83	0.91
# of introns	865	1021	1028	793
# of plastid-associated genes	17	17	17	17

Numbers may slightly vary from the original articles, depending on updated analyses. The structural RNA genes are not counted here.

Data taken from Douglas, S. E., Zauner, S., Fraunholz, M., Beaton, M., Penny, S., Deng, L. T., et al. (2001). The highly reduced genome of an enslaved algal nucleus. *Nature, 410*, 1091–1096; Gilson, P. R., Su, V., Slamovits, C. H., Reith, M. E., Keeling, P. J., & McFadden, G. I. (2006). Complete nucleotide sequence of the chlorarachniophyte nucleomorph: Nature's smallest nucleus. *Proceedings of the National Academy of Sciences of the United States of America, 103*, 9566–9571; Lane, C. E., van den Heuvel, K., Kozera, C., Curtis, B. A. Parsons, B. J., Bowman, S., & Archibald, J. M. (2007). Nucleomorph genome of *Hemiselmis andersenii* reveals complete intron loss and compaction as a driver of protein structure and function. *Proceedings of the National Academy of Sciences of the United States of America, 104*, 19908–19913; Moore, C. E., Curtis, B. A., Mills, T., Tanifuji, G., & Archibald, J. M. (2012). Nucleomorph genome sequence of the cryptophyte alga *Chroomonas mesostigmatica* CCMP1168 reveals lineage-specific gene loss and genome complexity. *Genome Biology and Evolution, 4*, 1162–1175; Suzuki, S., Shirato, S., Hirakawa, Y., & Ishida, K. (2015). Nucleomorph genome sequences of two chlorarachniophytes, *Amorphochlora amoebiformis* and *Lotharella vacuolata*. *Genome Biology and Evolution, 7*, 1533–1545; Tanifuji, G., Kim, E., Onodera, N. T., Gibeault, R., Dlutek, M., Cawthorn, R. J., et al. (2011). Genomic characterization of *Neoparamoeba pemaquidensis* (Amoebozoa) and its kinetoplastid endosymbiont. *Eukaryotic Cell, 10*(8), 1143–1146; Tanifuji, G., Onodera, N. T., Wheeler, T. J., Dlutek, M., Donaher, N., & Archibald, J. M. (2011). Complete nucleomorph genome sequence of the nonphotosynthetic alga *Cryptomonas paramecium* reveals a core nucleomorph gene set. *Genome Biology and Evolution, 3*, 44–54; Tanifuji, G., Onodera, N. T., Brown, M. W., Curtis, B. A., Roger, A. J., Wong, G. K. S., et al. (2014). Nucleomorph and plastid genome sequences of the chlorarachniophyte *Lotharella oceanica*: Convergent reductive evolution and frequent recombination in nucleomorph-bearing algae. *BMC Genomics, 15*, 374; Tanifuji, G., Onodera, N. T., Moore, C. E., & Archibald, J. M. (2014). Reduced nuclear genomes maintain high gene transcription levels. *Molecular Biology and Evolution, 31*, 625–635.

free-living organisms including red algae (Lane et al., 2007). This suggests that during the process of genome reductive evolution, nucleomorph genomes have been reduced much more intensively in a streamlined manner, not only by simple losses of genes and endosymbiotic gene transfers (EGTs) but also by shortening intergenic spacer sizes and protein lengths.

The most notable difference in genome structure between cryptomonad and chlorarachniophyte nucleomorphs is the number of introns. In cryptomonad nucleomorphs, there are 17 introns in *G. theta*, 24 in *Ch. mesostigmatica*, and 2 in *C. paramecium*. In *H. andersenii*, no introns were found, possibly reflecting the number of spliceosome-related genes in nucleomorph genomes (Douglas et al., 2001; Lane et al., 2007; Moore et al., 2012; Tanifuji, Onodera, et al., 2011). In fact, no spliceosomal genes were found in *H. andersenii*, whereas up to 15 spliceosomal genes are found in other cryptomonad species (Douglas et al., 2001; Lane et al., 2007; Moore et al., 2012; Tanifuji, Onodera, et al., 2011). In contrast, in chlorarachniophytes, numerous tiny introns (~20 nt in size, 800–1000 introns) exist in nucleomorph genomes and 40 proteins for spliceosomes are encoded (Gilson et al., 2006; Suzuki et al., 2015; Tanifuji, Onodera, Brown, et al., 2014). Conserved shared intron positions among the *B. natans* nucleomorph, and the *Arabidopsis* and *Chlamydomonas* nuclear genomes suggest that the introns in nucleomorph genomes are inherited from a green algal ancestor (Gilson et al., 2006; Roy & Penny, 2007). The smallness of the intron sizes in chlorarachniophytes may be explained by reductive pressure that shortens the introns because intron sizes in free-living organisms are usually larger in the range of 150–300 nt (Bowler et al., 2008; Curtis et al., 2012; Derelle et al., 2006; Kaul et al., 2000; Merchant et al., 2007; Worden et al., 2009). Nevertheless, whether this difference in the number of introns in cryptomonads is the result of a different evolutionary trajectory or is one of the characters that were inherited from their ancestors is not yet resolved. Regardless, it is clear that the nucleomorph genomes are markedly compact compared to the nuclear genomes of free-living protists.

3.5 Are Nucleomorphs in an Intermediate State or End-Point of Reductive Genome Evolution?

The existence of nucleomorphs in cryptomonads and chlorarachniophytes is an enigma. There are no simple and compelling answers to the question why these organelles still persist only in cryptomonads and chlorarachniophytes in spite of being completely lost in other secondary endosymbiotic organisms. In the case of plastid and mitochondrial genomes, the persistence of these two reduced, organellar genomes has been reasonably explained by redox potential regulation of genes: the regulation of expressed genes that reside in the organellar genomes is strictly controlled by oxidation–reduction reactions that take place within the organelle (Allen, 1993, 2003). Thus, genes in organelle genomes are constrained, so the complete elimination of the organelle genome leads to the collapse of the organelle itself. However, for the nucleomorph genomes, the redox regulation hypothesis cannot be applied because they are in entirely different situations from plastids and mitochondria.

Then, are the nucleomorphs in an intermediate state of the process of genome degradation, or have they gone through different evolutionary processes from other secondary endosymbiotic organisms where there are no traces of an endosymbiont-derived nucleus? As mentioned earlier, comparisons of the number of chromosomes, genome sizes, and architectures (plus gene content described later), in cryptomonad and chlorarachniophyte nucleomorphs, display high similarities and these clearly indicate that all nucleomorph genomes have undergone similar evolutionary pressures. To get insights into the evolutionary processes of nucleomorphs, more detailed comparisons of nucleomorph genomes either among cryptomonad/chlorarachniophyte species or between these two lineages have been carried out.

3.5.1 Limited Number of Plastid-Related Genes in Nucleomorph Genomes

It seems that, prior to the complete genome sequencing of nucleomorphs, many researchers thought that the nucleomorphs played an important role for plastid functions because the photosynthetic ability is one of the main functions that is acquired via secondary endosymbiosis and nuclear genomes generally encode numerous plastid-associated proteins (Gilson, Maier, & McFadden, 1997; Maier, Fraunholz, Zauner, Penny, & Douglas, 2000; McFadden et al., 1997; Palmer & Delwiche, 1996). And accordingly, it might be expected that many genes for plastid-targeted proteins still remain in nucleomorph genomes. However, the first complete sequence of a nucleomorph genome, from the cryptomonad *G. theta*, by Douglas et al. (2001) revealed that there was only a small set of plastid-associated proteins (Douglas et al., 2001). There are 548 genes in the *G. theta* nucleomorph genome, of which 487 encode proteins (Table 1). Surprisingly, only 31 plastid-associated genes were found and most of the other functionally known protein genes were housekeeping genes that are responsible for gene transcription and translation such as genes for RNA polymerases, ribosomal proteins, and elongation factors (another major category, proteins of unknown function are discussed in a later section). Complete sequencing of nucleomorph genomes has been done for other cryptomonad species, *H. andersenii* and *Ch. mesostigmatica*, and of 472 and 505 protein genes in the respective genomes, the exact same set of 31 plastid-associated genes was found in both lineages (Lane et al., 2007; Moore et al., 2012). *C. paramecium*, also a cryptomonad, lost its photosynthetic ability after the acquisition of a plastid. Its complete nucleomorph genome was found to contain 466 protein genes, and only 18 genes encode plastid-associated proteins. These 18 genes are a subset of the 31 plastid-associated proteins found in the other 3 photosynthetic cryptomonads (Tanifuji, Onodera, et al., 2011). In chlorarachniophytes, complete sequences have been done for 4 species, and, of ~300 unique protein genes after removing multicopy genes from the 300–600 protein genes that are encoded in their nucleomorph genomes, only 17 are annotated as plastid-associated genes, with the exact same set of genes among species (Gilson et al., 2006; Suzuki et al., 2015; Tanifuji, Onodera, Brown, et al., 2014).

Thus, in the nucleomorph genomes examined so far, only a small proportion (3–6%) of encoded proteins are associated with plastid functions, and the majority of function-predicted proteins of nucleomorph genomes are for housekeeping in both cryptomonads and chlorarachniophytes. The presence of a large number of housekeeping genes in nucleomorph genomes cannot explain the persistence of nucleomorphs in these lineages because these genes are highly conserved across all eukaryotic groups. Gilson et al. (2006) hypothesized that these housekeeping genes still remain to enable the expression of the small set of plastid-associated genes, and once all these plastid-associated genes in nucleomorph genomes are successfully transferred to the host nuclear genomes, the nucleomorph may disappear (Gilson et al., 2006). Moreover, Tanifuji, Onodera, et al. (2011) suggested that in the nucleomorph genome of the nonphotosynthetic cryptomonad, *C. paramecium*, not only photosynthetic-related genes but also genes that are not directly responsible for photosynthesis are missing (e.g. gyrA and gyrB), so there are no strong reasons for these plastid-associated genes to stay in the nucleomorph genomes (Tanifuji, Onodera, et al., 2011). How about the other protein genes? Are there any reasons for them to be encoded in nucleomorph genomes?

3.5.2 Lineage Specific and Convergent Loss of Genes

A comparative analysis of the nucleomorph genomes of cryptomonad and chlorarachniophytes suggests that the ones in chlorarachniophytes might be further along in terms of genome reduction. 90% of housekeeping genes are conserved in cryptomonads and chlorarachniophytes. Tanifuji, Onodera, et al. (2011) referred to 217 shared housekeeping genes among *G. theta*, *H. andersenii*, and *C. paramecium* as the "core set" of nucleomorph genes and compared it with 127 housekeeping genes in the chlorarachniophyte, *B. natans* (spliceosomal genes were excluded from the comparison) (Tanifuji, Onodera, et al., 2011). As a result, 81% of housekeeping genes in the *B. natans* nucleomorph genome (98/120) are shared with ones in cryptomonads. The majority of genes in the *B. natans* nucleomorph genome fall into the subset of the core genes in cryptomonads, suggesting that the nucleomorph genomes in chlorarachniophytes are in a state of advanced endosymbiotic genome reduction compared with ones in cryptomonads. Additional sequences of chlorarachniophyte nucleomorph genomes and comparisons among them provided further support that the majority of chlorarachniophyte housekeeping genes match those in the cryptomonad core subset (Suzuki et al., 2015; Tanifuji, Onodera, Brown, et al., 2014). It is also in agreement with the fact that chlorarachniophyte nucleomorph genomes accumulate more mutations than ones in cryptomonads, when compared with ones from their ancestral lineage (Patron, Rogers, & Keeling, 2006). Therefore, it is likely that, if the fundamental sets of genes are similar between cryptomonads and chlorarachniophytes, the genes in cryptomonad nucleomorphs that have disappeared from the chlorarachniophyte nucleomorph genomes will also disappear in the future. In other words, their presence in the cryptomonad nucleomorph genomes is no guarantee that they will persist in the nucleomorph. In fact, several essential genes detected in one cryptomonad species (e.g. genes for ribosomal proteins) were not found in other cryptomonad species in a comparison of nucleomorph genomes. It appeared that some lineage-specific loss of genes had occurred in the *H. andersenii* nucleomorph, in which the spliceosomal introns and key components of the spliceosome (e.g. prp8) are missing (Lane et al., 2007). Moreover, Moore et al. (2012) found a lineage specific loss of proteasome-related genes in the *Ch. mesostigmatica* nucleomorph (Moore et al., 2012). Given that a lack of proteasome-related genes is also seen in chlorarachniophyte nucleomorphs (Gilson et al., 2006; Suzuki et al., 2015; Tanifuji, Onodera, Brown, et al., 2014), and that these genes with signal peptides for targeting to the PPC have not been detected in the host nuclear genome in *B. natans* (Curtis et al., 2012), the removal of proteasome-related genes/proteins from the compartment of the endosymbiont at some time during secondary-endosymbiosis could be the result of convergent evolution.

Is it a plausible idea that the nucleomorphs are going to vanish? As mentioned earlier, at least from the perspective of comparative nucleomorph genomics, there has been no strong evidence for nucleomorphs being permanent residents. And if the existence of nucleomorphs is due to the maintenance of plastid functions, nucleomorphs might disappear after all plastid-associated genes in the nucleomorphs are moved to the host genome. Still, there remains the possibility that other reasons for the persistency of nucleomorphs exist. Perhaps with additional sequences of host nuclear genomes, we may find answers to this matter.

3.5.3 Nucleomorphs Are "Frozen"?

What is the process by which genes are lost from nucleomorph genomes? For the disappearance of an "essential" nucleomorph gene, the host nuclear genome must acquire a suitable homologue capable of being targeted back to the endosymbiont compartment. At least four scenarios are possible: (i) the nucleomorph gene could be transferred to the nuclear genome via EGT and acquire the necessary regulatory elements to allow expression as well as targeting peptides; (ii) homologues could be transferred laterally from outside of the cell (LGT) and subsequently obtain the necessary elements for expression and import; (iii) gene duplication of host-derived homologues could produce a new version of the protein targeted to the endosymbiont; or (iv) a single host-derived gene could have a role in both host and endosymbiont through dual targeting. Although the degree to which some or all of the above processes have contributed to the process of gene loss in the nucleomorph is unclear, the scenario iii is observed for the nuclear-encoded, PPC-targeted phosphoglucomutase, starch synthase class I–II, and cyclin-dependent kinase regulatory subunit proteins, which are not of nucleomorph origin but rather have evolved from duplicated host genes (Curtis et al., 2012). There is also an example of the scenario iv, in which different isoforms of aminoacyl-tRNA synthetases generated though alternative transcription/translation were found to be dually targeted to two subcellular compartments (Gile, Moog, Slamovits, Maier, & Archibald, 2015; Hirakawa, Burki, & Keeling, 2013). However, although the scenarios iii and iv could explain the loss of nucleomorph genes for eukaryotic core functions, which are common to both host and endosymbiont genomes (e.g. ribosomal subunits, nucleotide polymerase and elongation factors), it is not the case for the loss of unique genes such as plastid-targeted genes.

How and what steps are required for successful EGT? First, the endosymbiont DNA fragments containing particular genes need to be transferred to the host nucleus (Fig. 7A). In the second step, the genes that are transferred to the host genome need to acquire regulatory factors to be translated into proteins (Fig. 7B). Third, the acquisition of targeting signals is also needed for transport of proteins into the PPC/plastid (Fig. 7C). After these steps, genes can eventually be eliminated from the endosymbiont genome (Fig. 7D). How did genomes in many organisms succeed in the transfer of thousands of genes to the host nucleus from their endosymbiont genome? The simplest way could be transferring DNA fragments over and over again until the requirements are met and this is considered the means by which

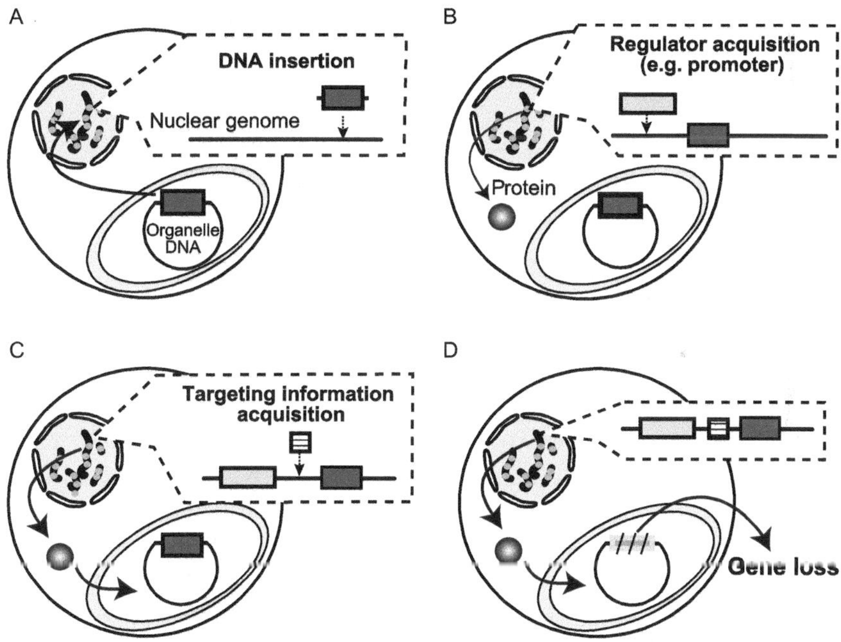

Fig. 7 Schematic diagrams of steps necessary for endosymbiotic gene transfer. (A) DNA fragments containing endosymbiont gene are transferred and inserted into the nuclear genome. (B) Gene regulator elements such as a promoter and a terminator become associated with the transferred endosymbiont-derived DNA. Regulators could be acquired after DNA transfer or transferred DNAs might co-opt existing host regulators. (C) Targeting signals are necessary for transport to the endosymbiont compartment. (D) The endosymbiont-encoded gene could be lost after the nuclear-encoded replacement becomes capable of working in the endosymbiont compartment.

transfers of plastid and mitochondria genes are accomplished. In plant nuclear genomes, it is known that DNA sequences that are nearly identical to ones in plastid genomes (nuclear plastid DNAs: NUPTs) have been dispersed across hundreds of host chromosomal sites and in the case of *Oryza sativa*, more than 800 kb of plastid DNA has accumulated in its nuclear genome (Matsuo et al., 2005; Smith, Crosby, & Lee, 2011). Furthermore, in tobacco plants, the frequency of DNA transfer was calculated by detecting markers for the plastid DNA fragment in the nuclear genome of every pollen, and the frequency of DNA transfer from plastid genome to host nucleus was estimated to be one transposition event for every 16,000 pollen grains (Huang, Ayliffe, & Timmis, 2003). Likewise, it has been reported that in yeast, DNA transfer has occurred from the mitochondrial genome to the host nucleus at high frequencies ($\sim 2 \times 10^{-5}$ per cell per generation) (Thorsness & Fox, 1990). Massive transfers of DNA fragments might explain how EGT has occurred successfully (Fig. 7).

Under this scenario, if many DNA fragments of the nucleomorph genomes were detected in the host nuclear genomes of cryptomonads and chlorarachniophytes, then this increases the possibility that the nucleomorphs will disappear in the future. To test this hypothesis, Curtis et al. (2012) looked for DNA pieces from the organelles in the nuclear genomes of *G. theta* and *B. natans* (Curtis et al., 2012). Unexpectedly, a few DNA fragments from the mitochondrial genome were detected (Curtis & Archibald, 2010; Curtis et al., 2012). However, there were no detectable fragments of plastid or nucleomorph DNAs (NUNMs) (Curtis et al., 2012). This indicates that there are few "recent" DNA transfers from the nucleomorph genome to the nuclear genome. If there were no further EGTs from the nucleomorph to the host genome in the future, to complete the loss of the nucleomorph genome or the loss of plastid-associated genes now encoded in nucleomorph genomes, the genes would have to be supplied to the nuclear genome via lateral gene transfers. Therefore, even if the nucleomorph genome seems to be in the middle of the process of disappearing, once the nuclear genome is factored into the scenario it appears that the nucleomorph is at the endpoint of genome reduction. Curtis et al. (2012) referred to it as a "frozen" state (Curtis et al., 2012).

Why are there no DNA transfers from nucleomorphs/plastids to the host nucleus? The "limited transfer window hypothesis" proposed by Barbrook, Howe, and Purton (2006) is potentially important in understanding the current situation with regard to the different degrees of DNA transfer from the nucleomorph and the other organelles (Barbrook et al., 2006). In this hypothesis, the frequency of EGT is correlated with the number of organelles per cell. A diagram of this hypothesis is shown in Fig. 8. If one of multiple organelles is lysed in a cell, this would not be lethal to the cell because the other organelles could compensate for its loss and some of the DNA from the degraded organelle may have a chance to move to the host nucleus (Fig. 8A). On the other hand, if there is only one organelle in the cell, then its lysis would be lethal and there would be little chance of transfer of organellar DNA to the nucleus (Fig. 8B). This hypothesis fits well when considering NUPTs in rice and *Chlamydomonas reinhardtii*, both of which are organisms that belong to the green plant (or green algal) lineage. Rice has multiple plastids in a cell and, as mentioned earlier, numerous NUPTs (~782 kb in total) have been identified in the nuclear genome. On the other hand, in *C. reinhardtii* (a green alga), only one plastid is retained in the cell, with hundreds of genome copies per the single organelle, and NUPTs total only ~1.9 kb in length (Barbrook et al., 2006; Smith et al., 2011). The correlation between the number of organelles and the total length of NUPTs as

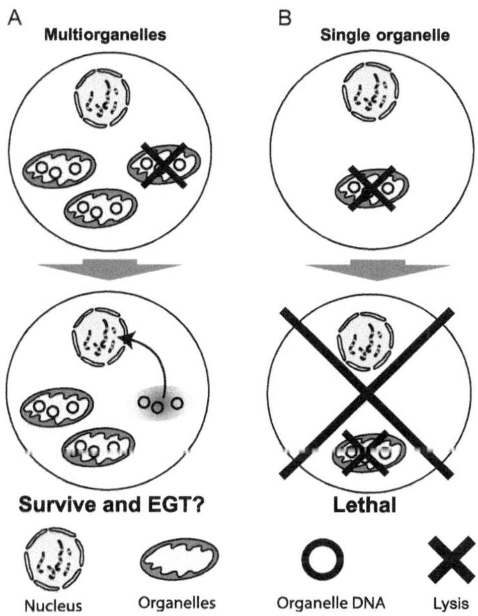

Fig. 8 Schematic diagram of the "limited transfer window" hypothesis proposed by Barbrook et al. (2006). (A) The multiple organelles scenario. Even after lysis of some of the organelles, the cell would survive. The floating DNAs of lysed organelles in the cytosol may be transferred to the nuclear genome. (B) The single organelle scenario. The lysis of the one and only organelle should cause cell death.

well as NUMTs in nuclear genomes is seen widely across eukaryotic lineages to some extent (Smith et al., 2011).

The cryptomonad *G. theta* and the chlorarachniophyte *B. natans* both harbour one plastid and one nucleomorph, and multiple mitochondria in a cell (Curtis et al., 2012). As mentioned earlier, NUPTs and NUNMs were not detected in the nuclear genomes of both organisms, but NUMTs were found, with 17 in *G. theta* and 13 in *B. natans*, showing a consistency with the limited transfer window hypothesis (Curtis & Archibald, 2010; Curtis et al., 2012). If the frequency of gene transfer is strongly affected by the number of organelles in a cell, then there may have been a stage when multiple plastids and multiple endosymbiont nuclei were present in a cell because PPC/plastid-targeted proteins are retained in nuclear genomes. Barbrook et al. (2006) also posit that endosymbiont division was probably not coordinated completely with that of the host in the initial stages of endosymbiosis, resulting in many endosymbionts per cell (Barbrook et al., 2006). In this context, at some point in the evolutionary path, only single plastids and endosymbiont nuclei were maintained (at least in *G. theta* and *B. natans*), resulting in the situation that gene transfers rarely occur (Curtis et al., 2012). Some species in cryptomonads and chlorarachniophytes have multiple organelles, so there is the expectation that NUNMs and NUPTs might be found in their nuclear genomes.

Still, the question of whether nucleomorphs will persist or disappear remains. Some "essential genes" in nucleomorph genomes appear to have disappeared without any replacements. For example, 17 protein genes including RNA polymerase II subunits, SMC proteins, and a replication initiation factor are missing in the nucleomorph genome of *G. theta* despite their presence in other cryptomonad nucleomorph genomes. Where have they gone? Of these 17 missing genes, it has been determined that 12 genes have been lost without any replacement in the host genome (Curtis et al., 2012). Also, 17 of 79 protein subunits usually used for making ribosomes are missing from the PPC of *B. natans* (Curtis et al., 2012). Although any compensations for missing proteins are not yet known, in my colleague's study, we noted that these missing proteins are part of a subunit that forms a protein complex (Tanifuji, Onodera, Moore, & Archibald, 2014). And we posited that other subunits might compensate for the missing subunit by an increase in transcription levels of nucleomorph genes since the transcription of nucleomorph genes tends to be higher than for homologous genes in the host nuclear genome (Tanifuji, Onodera, Moore, et al., 2014). Hirakawa, Suzuki, Archibald, Keeling, and Ishida (2014) demonstrated that protein-folding chaperons, hsp70 and hsp90, are transcribed at a much higher level than other nucleomorph genes, suggesting that these proteins may buffer the deleterious effects of mutated nucleomorph-encoded proteins (Hirakawa et al., 2014). In any case, it is clear that losses of conserved and essential genes have occurred and this should be taken into consideration when discussing the reductive genome evolution that occurs during endosymbiosis.

3.6 Genome Comparisons Between Prokaryotic and Eukaryotic Reduced Genomes

The nucleomorph-bearing organisms provide a rare opportunity for a direct comparison between eukaryotic and prokaryotic reduced genomes in the same cell. In addition, noncoloured *Cryptomonas* is the sole case of an organism retaining the nucleomorph and plastid genomes despite the loss of photosynthesis. In this section, we summarize recent reductive genome evolution from the aspect of organelle genome dynamics.

3.6.1 Comparison Between Plastid and Mitochondrial Genome Evolution

In cryptomonads, so far, three photosynthetic plastid genomes from *G. theta*, *Rhodomonas salina*, and *Teleaulax amphioxeia*, and one nonphotosynthetic plastid genome of *C. paramecium* have been completely sequenced (Donaher et al., 2009; Douglas & Penny, 1999; Khan et al., 2007; Kim et al., 2015). In the three photosynthetic cryptomonad plastids, the genome sizes range from 120 to 135 kb, and the nearly identical 145 protein genes occur in the same order and direction in the genomes. Thus, no genome rearrangements were observed. The genome size of *C. paramecium* is ~80 kb, considerably smaller than those of its photosynthetic relatives. Not surprisingly, the proteins directly related to the photosystem (e.g. *psa*, *psb*, and *pet* gene families) disappeared from the *C. paramecium* plastid genome, while 82 other protein genes for functions such as translation (e.g. ribosomal proteins), transcription, secretion, and biosynthesis are retained (Donaher et al., 2009). The coding capacity, including structural RNA genes, is 84.5%, which is not significantly different from that of genomes from photosynthetic cryptomonads (80%–88%). In Chlorarachniophytes, five complete plastid genomes have been sequenced—*B. natans*, *L. oceanica*, *Gymnochlora stellata*, *L. vacuolata*, and *Partenskyella glossopodia*. They are similar in size (70 kb) and contain nearly the same 60 protein genes except for a few unique ORFs (Rogers et al., 2007;

Suzuki et al., 2016; Tanifuji, Onodera, Brown, et al., 2014). No genome rearrangements were observed. Overall, the plastid genomes of cryptomonads (with the exception of nonphotosynthetic species) and chlorarachniophytes are very conserved.

Two mitochondrial genomes from both cryptomonads and chlorarachniophytes have been completely sequenced (Curtis et al., 2012; Hauth, Maier, Lang, & Burger, 2005; Kim et al., 2008; Tanifuji, Archibald, & Hashimoto, 2016). In cryptomonads, the mitochondrial genomes of *H. andersenii* and *R. salina* are 62 and 41 kb in size, and 36 and 40 protein genes were found, respectively (Hauth et al., 2005; Kim et al., 2008). Four protein genes including *rps1*, *sdh4*, *atp4*, and *tatA* are missing from the *H. andersenii* mitochondrial genome despite their presence in *R. salina*. The size difference between the two cryptomonad mitochondrial genomes is mainly due to the presence of a 20-kb large tandem repeat region only in the *H. andersenii* mitochondrial genome (Kim et al., 2008). In chlorarachniophytes, it was shown by pulsed-field gel electrophoresis analyses that mitochondrial genome sizes range from 34 to 180 kb (Gilson & McFadden, 1999; Gilson, Waller, & Mcfadden, 1995). Two completely sequenced mitochondrial genomes of *B. natans* and *L. oceanica* in chlorarachniophytes are approximately 37 kb in size and contain 27 and 24 protein genes, respectively (Curtis et al., 2012; Tanifuji et al., 2016). At least one protein gene, *rpl16*, present in the *B. natans* mitochondrial genome was transferred to the nuclear genome in *L. oceanica* (Tanifuji et al., 2016). While cryptomonads possess a circular mitochondrial genome as seen in the vast majority of eukaryotes, the mitochondrial genomes of chlorarachniophytes were found to have a linear structure (Gilson & McFadden, 1999; Gilson et al., 1995; Tanifuji et al., 2016). Frequent genome rearrangements were observed in both lineages of mitochondrial genomes (Kim et al., 2008; Tanifuji et al., 2016), whereas few rearrangements have been observed in the photosynthetic plastid genomes in cryptomonads and chlorarachniophytes (Suzuki et al., 2016; Tanifuji, Onodera, Brown, et al., 2014). In summary, it appears that the mitochondrial genomes have evolved more dynamically than the plastid genomes in both cryptomonads and chlorarachniophytes (discussed in more detail later).

3.6.2 The Hypothetical Proteins With No Similarity to Known Proteins

One of the significant differences between eukaryotic (nucleomorphs) and prokaryotic (plastids and mitochondria) organellar genomes is the number of protein genes with unknown functions. As mentioned in a previous section, most of the functionally known protein genes in nucleomorph genomes are "housekeeping genes" for transcription and translation, and ~90% of them are conserved among the different nucleomorph genomes (Douglas et al., 2001; Gilson et al., 2006; Lane et al., 2007; Moore et al., 2012; Suzuki et al., 2015; Tanifuji & Archibald, 2014; Tanifuji, Onodera, Brown, et al., 2014; Tanifuji, Onodera, et al., 2011). Nevertheless, nucleomorph genomes also contain an abundance of functionally unknown protein genes. In *G. theta*, for example, 234 of 487 nucleomorph protein genes have unknown functions. Tanifuji et al. (2011) designated these unknown proteins as nucleomorph orfans (nORFans) and examined how many nORFans are shared among cryptomonads (Tanifuji, Onodera, et al., 2011). Only ~12% (23 genes in total) of the nORFans are shared in cryptomonads, while ~90% (217 genes) of housekeeping genes are shared. Furthermore, 70%–81% (127–160 genes) of the nORFans showed no similarity to any known protein genes (i.e. true orfans) (Tanifuji, Onodera, et al., 2011). The same trend was observed in chlorarachniophytes in that only 10%–18% (18 of 100–150 genes) of nORFans are shared in four-way comparisons of chlorarachniophyte nucleomorph genomes (Suzuki et al., 2015; Tanifuji, Onodera, Brown, et al., 2014).

Thus, a clear inconsistency of protein set stability can be seen between the housekeeping genes and nORFans in nucleomorphs. Generally speaking, the orfans are rare in highly reduced genomes, presumably because only essential and highly conserved protein genes for core cellular processes tend to be retained, while specific or variable genes disappear. For reference, only seven and three orfans are found in cryptomonad plastid and mitochondrial genomes, respectively (the numbers are updated from the original publications due to later analyses) (Curtis et al., 2012; Donaher et al., 2009; Douglas & Penny, 1999; Hauth et al., 2005; Khan et al., 2007; Kim et al., 2008, 2015; Rogers et al., 2007; Suzuki et al., 2016; Tanifuji, Onodera, Brown, et al., 2014; Tanifuji et al., 2016).

It is noteworthy that many nORFans appear to have counterparts in syntenic regions in other nucleomorph genomes despite no detectable sequence similarity (Fig. 9; Douglas et al., 2001; Gilson et al., 2006; Lane et al., 2007; Moore et al., 2012; Suzuki et al., 2015; Tanifuji, Onodera, Brown, et al., 2014; Tanifuji, Onodera, et al., 2011). These syntenic nORFans are found at the same position within the syntenic block with the same functionally known genes in the same order and direction, suggesting that these

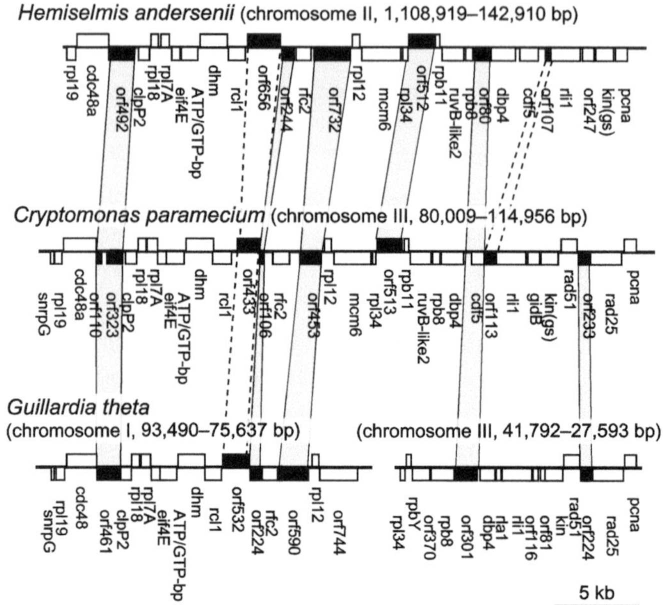

Fig. 9 Illustration of nORFans embedded in syntenic regions of the nucleomorph genomes of the cryptomonads *Hemiselmis andersenii*, *Cryptomonas paramecium*, and *Guillardia theta*. *Chroomonas mesostigmatica* is not included because of poor conservation of synteny. *Black boxes* indicate syntenic nORFans and *grey shadows* emphasize their conserved position within the same syntenic block. *Dotted lines* show syntenic nORFans with detectable amino acid sequence similarity. Genes indicated by *boxes above the lines* are transcribed *left* to *right*, those below the lines *right* to *left*. *Data taken from Douglas, S. E., Zauner, S., Fraunholz, M., Beaton, M., Penny, S., Deng, L. T., et al. (2001). The highly reduced genome of an enslaved algal nucleus.* Nature, 410, *1091–1096; Lane, C. E., van den Heuvel, K., Kozera, C., Curtis, B. A., Parsons, B. J., Bowman, S., & Archibald, J. M. (2007). Nucleomorph genome of* Hemiselmis andersenii *reveals complete intron loss and compaction as a driver of protein structure and function.* Proceedings of the National Academy of Sciences of the United States of America, 104, *19908–19913; Tanifuji, G., Kim, E., Onodera, N. T., Gibeault, R., Dlutek, M., Cawthorn, R. J., et al. (2011). Genomic characterization of* Neoparamoeba pemaquidensis *(Amoebozoa) and its kinetoplastid endosymbiont.* Eukaryotic Cell, 10*(8), 1143–1146; Tanifuji, G., Onodera, N. T., Wheeler, T. J., Dlutek, M., Donaher, N., & Archibald, J. M. (2011). Complete nucleomorph genome sequence of the nonphotosynthetic alga* Cryptomonas paramecium *reveals a core nucleomorph gene set.* Genome Biology and Evolution, 3, *44–54; Tanifuji, G., & Archibald, J. M. (2014). Nucleomorph comparative genomics. In L. Wolfgang (Ed.),* Endosymbiosis *(pp. 197–214). Vienna: Springer.*

genes are homologues. Expression analyses showed that all nORFans were transcribed. However, the significance of the abundant nORFans is currently a mystery (Hirakawa et al., 2014; Tanifuji, Onodera, Moore, et al., 2014).

3.6.3 Genome Rearrangement Frequency

It is known that the genome rearrangement frequencies are different among the three organellar genomes in nucleomorph-bearing organisms (Tanifuji et al., 2016). The protein genes of completely sequenced genomes in photosynthetic plastids are found to be in mostly the same order and direction within each cryptomonad and chlorarachniophyte (Douglas & Penny, 1999; Khan et al., 2007; Kim et al., 2015; Rogers et al., 2007; Suzuki et al., 2016; Tanifuji, Onodera, Brown, et al., 2014). Thus, few genome rearrangements were observed in the photosynthetic plastid genomes. On the other hand, syntenic blocks found in the nucleomorph and mitochondrial genomes suggest relatively high degrees of genome rearrangements (Fig. 10; Douglas et al., 2001; Gilson et al., 2006; Kim et al., 2015; Lane et al., 2007; Moore et al., 2012; Suzuki et al., 2015; Tanifuji et al., 2016; Tanifuji, Onodera, Brown, et al., 2014; Tanifuji, Onodera, et al., 2011).

Why are these rearrangement frequencies different? One of the significant differences between these organelle genomes is their origin; the nucleomorph genome is eukaryotic, while mitochondria and plastid genomes are derived from prokaryotic genomes. In general, eukaryotic and prokaryotic genomes are linear and circular, respectively, although many exceptions are known. However, this structural difference is not applicable to nucleomorph-bearing organisms because mitochondrial genomes of cryptomonads are circular, while those of chlorarachniophytes are linearized. Mitochondrial genomes from both lineages also show genome rearrangements. Although this idea is still controversial, one explanation for how genome stability is achieved is through intermolecular homologous recombination via inverted repeats of rRNA in plastid genomes. Strauss, Palmer, Howe, and Doerksen (1988) pointed out the possibility that large inverted repeats serve to stabilize genomes, buffering them against genome rearrangement via intermolecular recombination. This possibility has been discussed elsewhere (e.g. Vieira Ldo et al., 2014; Wei et al., 2013). Indeed, a large inverted repeat of the rRNA operon is characteristic of photosynthetic plastid genomes. Furthermore, the nonphotosynthetic cryptomonad *C. paramecium* has lost one side of its rRNA operon of the inverted repeat, and rearrangements are apparent in the region once sandwiched by inverted repeats (Fig. 10C; Donaher et al., 2009; Tanifuji et al., 2016). These observations are consistent with the hypothesis of Strauss et al. (1988), and it should be tested using more examples in the future.

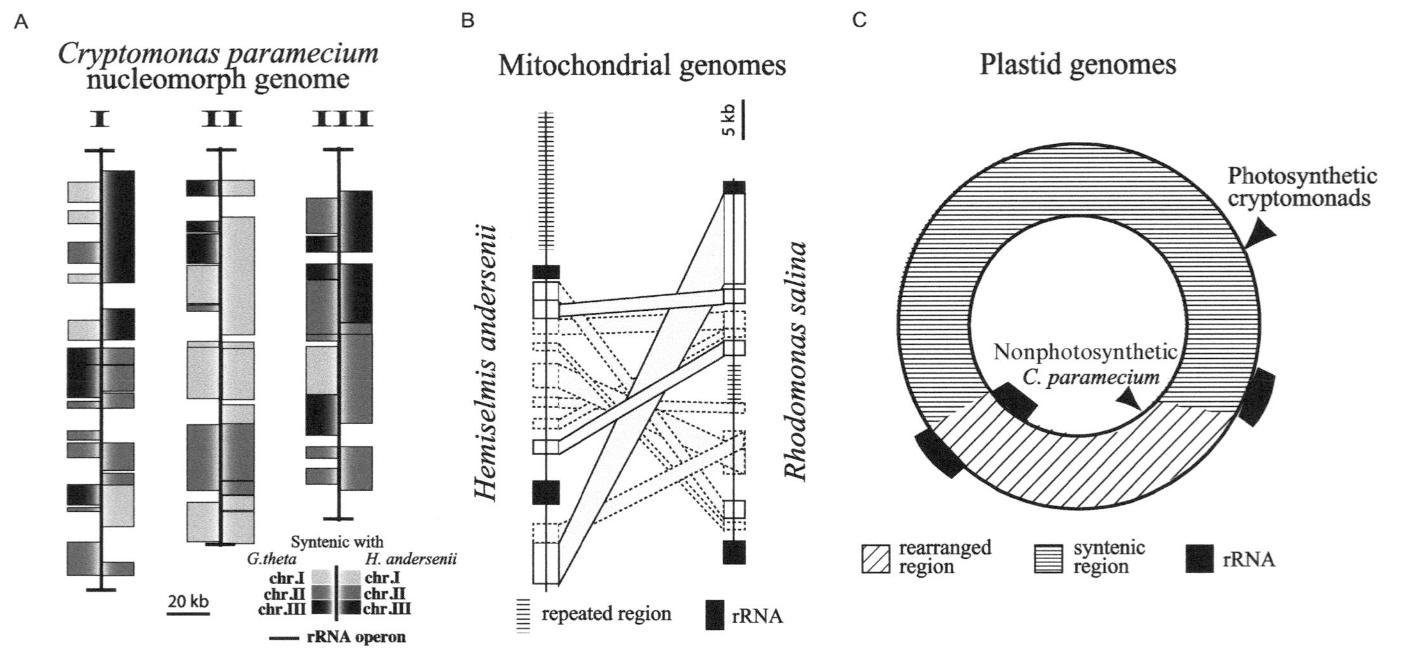

Fig. 10 Physical maps of cryptomonad organelle genomes with genome rearrangement information. (A) The physical map of the *Cryptomonas paramecium* nucleomorph genome. *Boxes* indicate syntenic regions with *Guillardia theta* (*left*) and *Hemiselmis andersenii* (*right*). Regions composed of four or more functionally known homologues with the same orientation in *G. theta* and *H. andersenii* were designated as "syntenic". *Chroomonas mesostigmatica* is not considered because of poor synteny. (B) Mitochondrial genome map. Circular DNAs of mitochondrial genomes are shown in the artificially linearized form. *Solid line boxes* indicate syntenic regions composed of three or more homologues in the same direction. The *dashed line boxes* show regions with two homologues. *Grey shadows* emphasize the syntenic region between the two mitochondrial genomes. (C) Physical map of plastid genomes with rearrangement information for photosynthetic and

nonphotosynthetic strains. The photosynthetic plastid genomes sequenced so far (*outside circle*) are fully syntenic with the exception of a few simple gene losses. Genome rearrangement compared with the photosynthetic plastid genomes is seen in the nonphotosynthetic *Cryptomonas paramecium* (the *inner circle*). The simple losses of photosynthetic genes (e.g. *psa* and *psb* gene families) are ignored. *Data taken from Donaher, N., Tanifuji, G., Onodera, N. T., Malfatti, S. A., Chain, P. S. G., Hara, Y., & Archibald, J. M. (2009). The complete plastid genome sequence of the secondarily nonphotosynthetic alga* Cryptomonas paramecium*: Reduction, compaction, and accelerated evolutionary rate.* Genome Biology and Evolution, 1, *439–448; Douglas, S. E., Murphy, C. A., Spencer, D. F., & Gray, M. W. (1991). Cryptomonad algae are evolutionary chimeras of two phylogenetically distinct unicellular eukaryotes.* Nature, 350, *148–151; Douglas, S. E., Zauner, S., Fraunholz, M., Beaton, M., Penny, S., Deng, L. T., et al. (2001). The highly reduced genome of an enslaved algal nucleus.* Nature, 410, *1091–1096; Hauth, A. M., Maier, U. G., Lang, B. F., & Burger, G. (2005). The* Rhodomonas salina *mitochondrial genome: Bacteria-like operons, compact gene arrangement and complex repeat region.* Nucleic Acids Research, 33, *4433–4442; Khan, H., Parks, N., Kozera, C., Curtis, B. A., Parsons, B. J., Bowman, S., & Archibald, J. M. (2007). Plastid genome sequence of the cryptophyte alga* Rhodomonas salina *CCMP1319: Lateral transfer of putative DNA replication machinery and a test of chromist plastid phylogeny.* Molecular Biology and Evolution, 24, *1832–1842; Kim, E., Lane, C. E., Curtis, B. A., Kozera, C., Bowman, S., & Archibald, J. M. (2008). Complete sequence and analysis of the mitochondrial genome of* Hemiselmis andersenii *CCMP644 (Cryptophyceae).* BMC Genomics, 9, *215; Kim, J. I., Yoon, H. S., Yi, G., Kim, H. S., Yih, W., & Shin, W. (2015). The plastid genome of the cryptomonad* Teleaulax amphioxeia. PLoS One, 10, *e0129284; Lane, C. E., van den Heuvel, K., Kozera, C., Curtis, B. A., Parsons, B. J., Bowman, S., & Archibald, J. M. (2007). Nucleomorph genome of* Hemiselmis andersenii *reveals complete intron loss and compaction as a driver of protein structure and function.* Proceedings of the National Academy of Sciences of the United States of America, 104, *19908–19913; Moore, C. E., Curtis, B. A., Mills, T., Tanifuji, G., & Archibald, J. M. (2012). Nucleomorph genome sequence of the cryptophyte alga* Chroomonas mesostigmatica *CCMP1168 reveals lineage-specific gene loss and genome complexity.* Genome Biology and Evolution, 4, *1162–1175; Tanifuji, G., Kim, E., Onodera, N. T., Gibeault, R., Dlutek, M., Cawthorn, R. J., et al. (2011). Genomic characterization of* Neoparamoeba pemaquidensis *(Amoebozoa) and its kinetoplastid endosymbiont.* Eukaryotic Cell, 10*(8), 1143–1146; Tanifuji, G., Onodera, N. T., Wheeler, T. J., Dlutek, M., Donaher, N., & Archibald, J. M. (2011). Complete nucleomorph genome sequence of the nonphotosynthetic alga* Cryptomonas paramecium *reveals a core nucleomorph gene set.* Genome Biology and Evolution, 3, *44–54.*

4. CONCLUDING REMARKS

Cryptomonads are useful organisms to gain insight into the evolutionary biology of secondary endosymbiosis. Studies on cryptomonads have yielded important insights into fundamental questions like the phylogeny of eukaryotes with red alga-derived plastids and processes that drive organelle genome evolution. Especially important has been the study of nucleomorphs, residual nuclei of secondary endosymbionts, because they represent a midpoint in the transition from a eukaryotic endosymbiont to a nuclear-lacking endosymbiotic compartment as seen in all other organisms with secondary plastids. The comparative genomics of the nucleomorph-bearing organisms, including chlorarachniophytes, provide rare opportunities to investigate the process of massive genome reorganization between two different eukaryotes during secondary endosymbioses. Nevertheless, many questions are still unresolved. Why do nucleomorphs persist in cryptomonads and chlorarachniophytes despite their absence in other secondary organisms? Why are some nucleomorph genes lost without any replacement? What are the functions of nORFans? Why is the number of nucleomorph chromosomes always three? Furthermore, the basic question of the phylogenetic relationship of cryptomonads to other eukaryotes persists and needs to be tested further using more datasets. As well, an exhaustive monograph on cryptomonad systematics is not yet available. Although studies using cryptomonads have provided much knowledge, it seems that we are still standing at the beginning of a great journey to unveil the evolution of complex life. Future studies exploring the various aspects of cryptomonads may shed light on the whole picture of the process of secondary endosymbiosis.

ACKNOWLEDGEMENTS

The authors thank Dr Bruce A. Curtis for critical advice and proofreading of an early draft of this manuscript, Dr Eunsoo Kim for providing the TEM picture of *Guillardia theta*, and Dr Takashi Shiratori for providing the light micrographs.

REFERENCES

Adir, N. (2005). Elucidation of the molecular structures of components of the phycobilisome: Reconstructing a giant. *Photosynthesis Research*, *85*, 15–32.

Adl, S. M., Simpson, A. G. B., Farmer, M. A., Andersen, R. A., Anderson, O. R., Barta, J. R., et al. (2005). The new higher level classification of eukaryotes with emphasis on the taxonomy of protists. *Journal of Eukaryotic Microbiology*, *52*, 399–451.

Adl, S. M., Simpson, A. G. B., Lane, C. E., Lukeš, J., Bass, D., Bowser, S. S., et al. (2012). The revised classification of eukaryotes. *Journal of Eukaryotic Microbiology, 59*, 429–493.
Alcocer, J., Lugo, A., Sánchez, M. d. R., & Escobar, E. (1998). Isabela Crater-Lake: A Mexican insular saline lake. *Hydrobiologia, 381*, 1–7.
Allen, J. F. (1993). Control of gene-expression by redox potential and the requirement for chloroplast and mitochondrial genomes. *Journal of Theoretical Biology, 165*, 609–631.
Allen, J. F. (2003). The function of genomes in bioenergetic organelles. *Philosophical Transactions of the Royal Society B: Biological Science, 358*, 19–37.
Anderson, E. (1962). A cytological study of *Chilomonas paramecium* with particular reference to the so-called trichocysts. *Journal of Protozoology, 9*, 380–395.
Apt, K. E., Collier, J. L., & Grossman, A. R. (1995). Evolution of the phycobiliproteins. *Journal of Molecular Biology, 248*, 79–96.
Archibald, J. M. (2007). Nucleomorph genomes: Structure, function, origin and evolution. *BioEssays, 29*, 392–402.
Archibald, J. M. (2015). Genomic perspectives on the birth and spread of plastids. *Proceedings of the National Academy of Sciences of the United States of America, 112*, 10147–10153.
Archibald, J. M., & Lane, C. E. (2009). Going, going, not quite gone: Nucleomorphs as a case study in nuclear genome reduction. *Journal of Heredity, 100*, 582–590.
Arteni, A. A., Liu, L. N., Aartsma, T. J., Zhang, Y. Z., Zhou, B. C., & Boekema, E. J. (2008). Structure and organization of phycobilisomes on membranes of the red alga *Porphyridium cruentum*. *Photosynthesis Research, 95*, 169–174.
Barbrook, A. C., Howe, C. J., & Purton, S. (2006). Why are plastid genomes retained in non-photosynthetic organisms? *Trends in Plant Science, 11*, 101–108.
Beck, C., Knoop, H., Axmann, I. M., & Steuer, R. (2012). The diversity of cyanobacterial metabolism: Genome analysis of multiple phototrophic microorganisms. *BMC Genomics, 13*, 56.
Bock, R., & Timmis, J. N. (2008). Reconstructing evolution: Gene transfer from plastids to the nucleus. *BioEssays, 30*(6), 556–566.
Bolte, K., Bullmann, L., Hempel, F., Bozarth, A., Zauner, S., & Maier, U. G. (2009). Protein targeting into secondary plastids. *Journal of Eukaryotic Microbiology, 56*, 9–15.
Bowler, C., Allen, A. E., Badger, J. H., Grimwood, J., Jabbari, K., Kuo, A., et al. (2008). The *Phaeodactylum* genome reveals the evolutionary history of diatom genomes. *Nature, 456*, 239–244.
Brett, S. J., & Wetherbee, R. (1985). A comparative study of periplast structure in *Cryptomonas cryophila* and *C. ovata* (Cryptophyceae). *Protoplasma, 131*, 23–31.
Broughton, M. J., Howe, C. J., & Hiller, R. G. (2006). Distinctive organization of genes for light-harvesting proteins in the cryptophyte alga *Rhodomonas*. *Gene, 369*, 72–79.
Brown, M. W., Sharpe, S. C., Silberman, J. D., Heiss, A. A., Lang, B. F., Simpson, A. G. B., et al. (2013). Phylogenomics demonstrates that breviate flagellates are related to opisthokonts and apusomonads. *Proceedings. Biological Sciences., 280*, 20131755.
Brown, M. W., Spiegel, F. W., & Silberman, J. D. (2009). Phylogeny of the "forgotten" cellular slime mold, *Fonticula alba*, reveals a key evolutionary branch within Opisthokonta. *Molecular Biology and Evolution, 26*, 2699–2709.
Burki, F. (2014). The eukaryotic tree of life from a global phylogenomic perspective. *Cold Spring Harbor Perspectives in Biology, 6*, a016147.
Burki, F., Flegontov, P., Oborník, M., Cihlar, J., Pain, A., Lukeš, J., et al. (2012). Re-evaluating the green versus red signal in eukaryotes with secondary plastid of red algal origin. *Genome Biology and Evolution, 4*, 738–747.
Burki, F., Inagaki, Y., Brate, J., Archibald, J. M., Keeling, P. J., Cavalier-Smith, T., et al. (2009). Large-scale phylogenomic analyses reveal that two enigmatic protist lineages, Telonemia and Centroheliozoa, are related to photosynthetic chromalveolates. *Genome Biology and Evolution, 1*, 231–238.

Burki, F., Kaplan, M., Tikhonenkov, D. V., Zlatogursky, V., Minh, B. Q., Radaykina, L. V., et al. (2016). Untangling the early diversification of eukaryotes: A phylogenomic study of the evolutionary origins of Centrohelida, Haptophyta and Cryptista. *Proceedings of the Royal Society B: Biological Sciences*, *283*, 20152802.

Burki, F., Okamoto, N., Pombert, J. F., & Keeling, P. J. (2012). The evolutionary history of haptophytes and cryptophytes: Phylogenomic evidence for separate origins. *Proceedings of the Royal Society B: Biological Sciences*, *279*, 2246–2254.

Burki, F., Shalchian-Tabrizi, K., Minge, M., Skjaeveland, A., Nikolaev, S. I., Jakobsen, K. S., et al. (2007). Phylogenomics reshuffles the eukaryotic supergroups. *PLoS One*, *2*, e790.

Carty, S., & Cox, E. R. (1986). *Kansodinium* gen. nov. and *Durinskia* gen. nov.: Two genera of freshwater dinoflagellates (Pyrrhophyta). *Phycologia*, *25*, 197–204.

Cavalier-Smith, T. (1986). The kingdom Chromista: Origin and systematics. In F. E. Round & D. J. Chapman (Eds.), *Vol. 4. Progress in phycological research* (pp. 309–347). Bristol: Biopress.

Cavalier-Smith, T. (1999). Principles of protein and lipid targeting in secondary symbiogenesis: Euglenoid, dinoflagellate, and sporozoan plastid origins and the eukaryote family tree. *Journal of Eukaryotic Microbiology*, *46*, 347–366.

Cavalier-Smith, T. (2000). Membrane heredity and early chloroplast evolution. *Trends in Plant Science*, *5*, 174–182.

Cavalier-Smith, T. (2002). Nucleomorphs: Enslaved algal nuclei. *Current Opinion in Microbiology*, *5*, 612–619.

Cavalier-Smith, T. (2013). Symbiogenesis: Mechanisms, evolutionary consequences, and systematic implications. *Annual Review of Ecology, Evolution and Systematics*, *44*, 145–172.

Cavalier-Smith, T., Couch, J. A., Thorsteinsen, K. E., Gilson, P. R., Deane, J. A., Hill, D. R. A., et al. (1996). Cryptomonad nuclear and nucleomorph 18S rRNA phylogeny. *European Journal of Phycology*, *31*, 315–328.

Clay, B., & Kugrens, P. (1999). Systematics of the enigmatic kathablepharids, including EM characterization of the type species, *Kathablepharis phoenikoston*, and new observations on *K. remigera* comb. nov. *Protist*, *150*, 43–59.

Clay, B. L., Kugrens, P., & Lee, R. E. (1999). A revised classification of Cryptophyta. *Botanical Journal of the Linnean Society*, *131*, 131–151.

Curtis, B. A., & Archibald, J. M. (2010). A spliceosomal intron of mitochondrial DNA origin. *Current Biology*, *20*, R919–R920.

Curtis, B. A., Tanifuji, G., Burki, F., Gruber, A., Irimia, M., Maruyama, S., et al. (2012). Algal genomes reveal evolutionary mosaicism and the fate of nucleomorphs. *Nature*, *492*, 59–65.

Deane, J. A., Isabbelle, M., Strachan, I. M., Saunders, G. W., Hill, D. R. A., & McFadden, G. I. (2002). Cryptomonad evolution: Nuclear 18S rDNA phycologeny versus cell morphology and pigmentation. *Journal of Phycology*, *38*, 1236–1244.

Derelle, E., Ferraz, C., Rombauts, S., Rouze, P., Worden, A. Z., Robbens, S., et al. (2006). Genome analysis of the smallest free-living eukaryote *Ostreococcus tauri* unveils many unique features. *Proceedings of the National Academy of Sciences of the United States of America*, *103*, 11647–11652.

Derelle, R., & Lang, B. F. (2012). Rooting the eukaryotic tree with mitochondrial and bacterial proteins. *Molecular Biology and Evolution*, *29*, 1277–1289.

Dodge, J. D., & Crawford, R. M. (1969). Observations on the fine structure of the eyespot and associated organelles in the dinoflagellate *Glenodinium foliaceum*. *Journal of Cell Science*, *5*, 479–493.

Donaher, N., Tanifuji, G., Onodera, N. T., Malfatti, S. A., Chain, P. S. G., Hara, Y., et al. (2009). The complete plastid genome sequence of the secondarily nonphotosynthetic

alga *Cryptomonas paramecium*: Reduction, compaction, and accelerated evolutionary rate. *Genome Biology and Evolution*, *1*, 439–448.
Douglas, S. E., Durnford, D. G., & Morden, C. W. (1990). Nucleotide sequence of the gene for the large subunit of ribulose-1,5-bisphosphate carboxylase/oxygenase from Cryptomonas phi: Evidence supporting the polyphyletic origin of plastids. *Journal of Phycology*, *26*, 500–508.
Douglas, S. E., Murphy, C. A., Spencer, D. F., & Gray, M. W. (1991). Cryptomonad algae are evolutionary chimeras of two phylogenetically distinct unicellular eukaryotes. *Nature*, *350*, 148–151.
Douglas, S. E., & Penny, S. L. (1999). The plastid genome of the cryptophyte alga, *Guillardia theta*: Complete sequence and conserved synteny groups confirm its common ancestry with red algae. *Journal of Molecular Evolution*, *48*, 236–244.
Douglas, S. E., Zauner, S., Fraunholz, M., Beaton, M., Penny, S., Deng, L. T., et al. (2001). The highly reduced genome of an enslaved algal nucleus. *Nature*, *410*, 1091–1096.
Eschbach, S., Hofmann, C. J., Maier, U. G., Sitte, P., & Hansmann, P. (1991). A eukaryotic genome of 660 kb: Electrophoretic karyotype of nucleomorph and cell nucleus of the cryptomonad alga, *Pyrenomonas salina*. *Nucleic Acids Research*, *19*, 1779–1781.
Fast, N. M., Xue, L., Bingham, S., & Keeling, P. J. (2002). Re-examining alveolate evolution using multiple protein molecular phylogenies. *Journal of Eukaryotic Microbiology*, *49*, 30–37.
Felsner, G., Sommer, M. S., Gruenheit, N., Hempel, F., Moog, D., Zauner, S., et al. (2011). ERAD components in organisms with complex red plastids suggest recruitment of a preexisting protein transport pathway for the periplastid membrane. *Genome Biology and Evolution*, *3*, 140–150.
Fields, S. D., & Rhodes, R. G. (1991). Ingestion and retention of *Chroomonas* spp. (Cryptophyceae) by *Gymnodinium acidotum* (Dinophyceae)1. *Journal of Phycology*, *27*, 525–529.
Gantt, E., Edwards, M. R., & Provasoli, L. (1971). Chloroplast structure of the Cryptophyceae. *Journal of Cell Biology*, *48*, 280–290.
Gervais, F. (1997). Diel vertical migration of *Cryptomonas* and *Chromatium* in the deep chlorophyll maximum of a eutrophic lake. *Journal of Plankton Research*, *19*, 533–550.
Gibbs, S. P. (1978). Chloroplasts of euglena may have evolved from symbiotic green-algae. *Canadian Journal of Botany-Revue Canadienne De Botanique*, *56*, 2883–2889.
Gile, G. H., Moog, D., Slamovits, C. H., Maier, U. G., & Archibald, J. M. (2015). Dual organellar targeting of aminoacyl-tRNA synthetases in diatoms and Cryptophytes. *Genome Biology and Evolution*, *7*, 1728–1742.
Gillott, M. A., & Gibbs, R. (1980). The cryptomonad nucleomorph: Its ultrastructure and evolutionary significance. *Journal of Phycology*, *16*, 558–568.
Gillott, M. A., & Gibbs, R. S. (1983). Comparison of the flagellar rootlets and periplast in two marine cryptomonads. *Canadian Journal of Botany*, *61*, 1964–1978.
Gilson, P. R., Maier, U. G., & McFadden, G. I. (1997). Size isn't everything: Lessons in genetic miniaturisation from nucleomorphs. *Current Opinion in Genetics & Development*, *7*, 800–806.
Gilson, P. R., & McFadden, G. I. (1996). The miniaturized nuclear genome of eukaryotic endosymbiont contains genes that overlap, genes that are cotranscribed, and the smallest known spliceosomal introns. *Proceedings of the National Academy of Sciences of the United States of America*, *93*, 7737–7742.
Gilson, P. R., & McFadden, G. I. (1999). Molecular, morphological and phylogenetic characterization of six chlorarachniophyte strains. *Phycological Research*, *47*, 7–19.
Gilson, P. R., & McFadden, G. I. (2002). Jam packed genomes—A preliminary, comparative analysis of nucleomorphs. *Genetica*, *115*, 13–28.

Gilson, P. R., Su, V., Slamovits, C. H., Reith, M. E., Keeling, P. J., & McFadden, G. I. (2006). Complete nucleotide sequence of the chlorarachniophyte nucleomorph: Nature's smallest nucleus. *Proceedings of the National Academy of Sciences of the United States of America, 103*, 9566–9571.

Gilson, P. R., Waller, R., & Mcfadden, G. (1995). Preliminary characterization of Chlorarachniophyte mitochondrial-DNA. *Journal of Eukaryotic Microbiology, 42*, 696–701.

Glazer, A. N., & Wedemayer, G. J. (1995). Cryptomonad biliproteins—An evolutionary perspective. *Photosynthesis Research, 46*, 93–105.

Gould, S. B., Fan, E., Hempel, F., Maier, U. G., & Kloesgen, R. B. (2007). Translocation of a phycoerythrin alpha subunit across five biological membranes. *Journal of Biological Chemistry, 282*, 30295–30302.

Gould, S. B., Maier, U. G., & Martin, W. F. (2015). Protein import and the origin of red complex plastids. *Current Biology, 25*, R515–R521.

Gould, S. B., Sommer, M. S., Hadfi, K., Zauner, S., Kroth, P. G., & Maier, U. G. (2006). Protein targeting into the complex plastid of cryptophytes. *Journal of Molecular Evolution, 62*, 674–681.

Gould, S. B., Sommer, M. S., Kroth, P. G., Gile, G. H., Keeling, P. J., & Maier, U. G. (2006). Nucleus-to-nucleus gene transfer and protein retargeting into a remnant cytoplasm of cryptophytes and diatoms. *Molecular Biology and Evolution, 23*, 2413–2422.

Gould, S. B., Waller, R. R., & McFadden, G. I. (2008). Plastid evolution. *Annual Review of Plant Biology, 59*, 491–517.

Gray, M. W., & Doolittle, W. F. (1982). Has the endosymbiont hypothesis been proven? *Microbiological Reviews, 46*, 1–42.

Gray, M. W., Lang, B. F., & Burger, G. (2004). Mitochondria of protists. *Annual Review of Genetics, 38*, 477–524.

Greenwood, A. (1974). The Cryptophyta in relation to phylogeny and photosynthesis. In J. Sanders & D. Goodchild (Eds.), *8th international congress of electron microscopy* (pp. 566–567). Canberra: Australian Academy of Sciences.

Gustafson, D. E., Jr., Stoecker, D. K., Johnson, M. D., Van Heukelem, W. F., & Sneider, K. (2000). Cryptophyte algae are robbed of their organelles by the marine ciliate *Mesodinium rubrum*. *Nature, 405*, 1049–1052.

Hackett, J. D., Anderson, D. M., Erdner, D. L., & Bhattacharya, D. (2004). Dinoflagellates: A remarkable evolutionary experiment. *American Journal of Botany, 91*, 1523–1534.

Hackett, J. D., Yoon, H. S., Li, S., Reyes-Prieto, A., Rummele, S. E., & Bhattacharya, D. (2007). Phylogenomic analysis supports the monophyly of cryptophytes and haptophytes and the association of rhizaria with chromalveolates. *Molecular Biology and Evolution, 24*, 1702–1713.

Hammer, A., Schumann, R., & Schubert, H. (2002). Light and temperature acclimation of *Rhodomonas salina* (Cryptophyceae): Photosynthetic performance. *Aquatic Microbial Ecology, 29*, 287–296.

Hampl, V., Hug, L., Leigh, J. W., Dacks, J. B., Lang, B. F., Simpson, A. G. B., et al. (2009). Phylogenomic analyses support the monophyly of Excavata and resolve relationships among eukaryotic "supergroups". *Proceedings of the National Academy of Sciences of the United States of America, 106*, 3859–3864.

Harper, J. T., Waanders, E., & Keeling, P. J. (2005). On the monophyly of chromalveolates using a six-protein phylogeny of eukaryotes. *International Journal of Systematic and Evolutionary Microbiology, 55*, 487–496.

Hauth, A. M., Maier, U. G., Lang, B. F., & Burger, G. (2005). The *Rhodomonas salina* mitochondrial genome: Bacteria-like operons, compact gene arrangement and complex repeat region. *Nucleic Acids Research, 33*, 4433–4442.

Hempel, F., Bullmann, L., Lau, J., Zauner, S., & Maier, U. G. (2009). ERAD-derived preprotein transport across the second outermost plastid membrane of diatoms. *Molecular Biology and Evolution, 26*(8), 1781–1790.

Hibberd, D. J., Greenwood, A. D., & Griffiths, H. B. (1971). Observations on the ultrastructure of the flagella and periplast in the Cryptophyceae. *British Phycological Journal, 6*, 61–72.

Hibberd, D. J., & Norris, R. E. (1984). Cytology and ultrastructure of *Chlorarachnion reptans* (Chlorarachniophyta Divisio Nova, Chlorarachniophyceae Classis Nova). *Journal of Phycology, 20*, 310–330.

Hill, D. R. A. (1991). A revised circumscription of *Cryptomonas* (Cryptophyceae) based on examination of Australian strains. *Phycologia, 30*, 170–188.

Hill, D. R. A., & Rowan, K. S. (1989). The biliproteins of the Cryptophyceae. *Phycologia, 28*, 455–463.

Hill, D. R. A., & Wetherbee, R. (1986). *Proteomonas sulcata* gen. et sp. nov. (Cryptophyceae), a cryptomonad with two morphologically distinct and alternating forms. *Phycologia, 25*, 521–543.

Hirakawa, Y., Burki, F., & Keeling, P. J. (2013). Dual targeting of aminoacyl-tRNA synthetases to the mitochondrion and complex plastid in chlorarachniophytes. *Journal of Cell Science, 125*, 6176.

Hirakawa, Y., & Ishida, K. I. (2014). Polyploidy of endosymbiotically derived genomes in complex algae. *Genome Biology and Evolution, 6*, 974–980.

Hirakawa, Y., Suzuki, S., Archibald, J. M., Keeling, P. J., & Ishida, K. (2014). Overexpression of molecular chaperone genes in nucleomorph genomes. *Molecular Biology and Evolution, 31*(6), 1437–1443.

Hoef-Emden, K. (2005). Multiple independent losses of photosynthesis and differing evolutionary rates in the genus *Cryptomonas* (Cryptophyceae): Combined phylogenetic analyses of DNA sequences of the nuclear and the nucleomorph ribosomal operons. *Journal of Molecular Evolution, 60*, 183–195.

Hoef-Emden, K. (2007). Revision of the genus *Cryptomonas* (Cryptophyceae) II: Incongruences between the classical morphospecies concept and molecular phylogeny in smaller pyrenoid-less cells. *Phycologia, 46*, 402–428.

Hoef-Emden, K. (2008). Molecular phylogeny of phycocyanin-containing cryptophytes: Evolution of biliproteins and geographical distribution. *Journal of Phycology, 44*, 985–993.

Hoef-Emden, K., & Archibald, J. M. (2017). Cryptophyta (cryptomonads). In J. M. Archibald, A. G. B. Simpson, & C. Slamovits (Eds.), *Handbook of the protists*. Cham: Springer International Publishing.

Hoef-Emden, K., Marin, B., & Melkonian, M. (2002). Nuclear and nucleomorph SSU rDNA phylogeny in the cryptophyta and the evolution of cryptophyte diversity. *Journal of Molecular Evolution, 55*, 161–179.

Hoef-Emden, K., & Melkonian, M. (2003). Revision of the genus *Cryptomonas* (Cryptophyceae): A combination of molecular phylogeny and morphology provides insights into a long-hidden dimorphism. *Protist, 154*, 371–409.

Hoef-Emden, K., Shrestha, R. P., Lapidot, M., Weinstein, Y., Melkonian, M., & Arad, S. (2005). Actin phylogeny and intron distribution in bangiophyte red algae (rhodoplantae). *Journal of Molecular Evolution, 61*, 360–371.

Hoef-Emden, K., Tran, H. D., & Melkonian, M. (2005). Lineage-specific variations of congruent evolution among DNA sequences from three genomes, and relaxed selective constraints on rbcL in *Cryptomonas* (Cryptophyceae). *BMC Evolutionary Biology, 5*, 56.

Horiguchi, T., & Pienaar, R. N. (1992). *Amphidinium latum* (Dinophyceae), a sand-dwelling dinoflagellate feeding on cryptomonads. *Japanese Journal Phycology (Sorui), 40*, 353–363.

Huang, C. Y., Ayliffe, M. A., & Timmis, J. N. (2003). Direct measurement of the transfer rate of chloroplast DNA into the nucleus. *Nature, 422*, 72–76.

Ishida, K., Cao, Y., Hasegawa, M., Okada, N., & Hara, Y. (1997). The origin of chlorarachniophyte plastids, as inferred from phylogenetic comparisons of amino acid sequences of EF-Tu. *Journal of Molecular Evolution*, *45*, 682–687.

Ishida, K., Endo, H., & Koike, S. (2011). *Partenskyella glossopodia* (Chlorarachniophyceae) possesses a nucleomorph genome of approximately 1 Mbp. *Phycological Research*, *59*, 120–122.

Ishida, K., Green, B. R., & Cavalier-Smith, T. (1999). Diversification of a chimaeric algal group, the chlorarachniophytes: Phylogeny of nuclear and nucleomorph small-subunit rRNA genes. *Molecular Biology and Evolution*, *16*, 321–331.

Janouškovec, J., Horák, A., Oborník, M., Lukeš, J., & Keeling, P. J. (2010). A common red algal origin of the apicomplexan, dinoflagellate, and heterokont plastids. *Proceedings of the National Academy of Sciences of the United States of America*, *107*(24), 10949–10954.

Jarvis, P., & Soll, J. (2002). Toc, tic, and chloroplast protein import. *Biochimica et Biophysica Acta*, *1590*, 177–189.

Kamikawa, R., Tanifuji, G., Kawachi, M., Miyashita, H., Hashimoto, T., & Inagaki, Y. (2015). Plastid genome-based phylogeny pinpointed the origin of the green-colored plastid in the dinoflagellate *Lepidodinium chlorophorum*. *Genome Biology and Evolution*, *7*, 1133–1140.

Katinka, M. D., Duprat, S., Cornillot, E., Metenier, G., Thomarat, F., Prensier, G., et al. (2001). Genome sequence and gene compaction of the eukaryote parasite *Encephalitozoon cuniculi*. *Nature*, *414*, 450–453.

Kaul, S., Koo, H. L., Jenkins, J., Rizzo, M., Rooney, T., Tallon, L. J., et al. (2000). Analysis of the genome sequence of the flowering plant *Arabidopsis thaliana*. *Nature*, *408*, 796–815.

Keeling, P. J. (2010). The endosymbiotic origin, diversification and fate of plastids. *Philosophical Transactions of the Royal Society B: Biological Sciences*, *365*, 729–748.

Keeling, P. J., & Archibald, J. M. (2008). Organelle evolution: What's in a name? *Current Biology*, *18*, R345–R347.

Keeling, P. J., McCutcheon, J. P., & Doolittle, W. F. (2015). Symbiosis becoming permanent: Survival of the luckiest. *Proceedings of the National Academy of Sciences of the United States of America*, *112*, 10101–10103. http://dx.doi.org/10.1073/pnas.1513346112.

Khan, H., Parks, N., Kozera, C., Curtis, B. A., Parsons, B. J., Bowman, S., et al. (2007). Plastid genome sequence of the cryptophyte alga *Rhodomonas salina* CCMP1319: Lateral transfer of putative DNA replication machinery and a test of chromist plastid phylogeny. *Molecular Biology and Evolution*, *24*, 1832–1842.

Kim, E., & Archibald, J. M. (2013). Ultrastructure and molecular phylogeny of the cryptomonad *Goniomonas avonlea* sp. nov. *Protist*, *164*, 160–182.

Kim, E., Lane, C. E., Curtis, B. A., Kozera, C., Bowman, S., & Archibald, J. M. (2008). Complete sequence and analysis of the mitochondrial genome of *Hemiselmis andersenii* CCMP644 (Cryptophyceae). *BMC Genomics*, *9*, 215.

Kim, J. I., Yoon, H. S., Yi, G., Kim, H. S., Yih, W., & Shin, W. (2015). The plastid genome of the cryptomonad *Teleaulax amphioxeia*. *PLoS One*, *10*, e0129284.

Klaveness, D. (1985). Classical and modern criteria for determining species of the Cryptophyceae. *Bulletin of the Plankton Society of Japan*, *32*, 111–123.

Klaveness, D. (1988). Ecology of the Cryptomonadida: A first review. In Sandgren (Ed.), *Growth and reproductive strategies of freshwater phytoplankton* (pp. 105–133). Cambridge: Cambridge University Press.

Koike, K., Sekiguchi, H., Kobiyama, A., Takishita, K., Kawachi, M., Koike, K., et al. (2005). A novel type of kleptoplastidy in *Dinophysis* (Dinophyceae): Presence of haptophyte-type plastid in *Dinophysis mitra*. *Protist*, *156*, 225–237.

Komaki, K., & Ishikawa, H. (1999). Intracellular bacterial symbionts of aphids possess many genomic copies per bacterium. *Journal of Molecular Evolution*, *48*, 717–722.

Kugrens, P., & Lee, R. E. (1987). An ultrastructural survey of cryptomonad periplasts using quick freezing freeze fracture techniques. *Journal of Phycology, 23*, 365–376.
Kugrens, P., & Lee, R. E. (1988). Ultrastructure of fertilization in a cryptomonad. *Journal of Phycology, 24*, 385–393.
Kugrens, P., & Lee, R. E. (1991). Organization of cryptomonads. In D. J. Patterson & J. Larsen (Eds.), *The biology of free-living heterotrophic flagellates* (pp. 2195–2233). Oxford: Clarendon.
Kugrens, P., Lee, R. E., & Andersen, R. A. (1987). Ultra structural variations in cryptomonad flagella. *Journal of Phycology, 23*, 511–518.
Kugrens, P., Lee, R. E., & Hill, D. R. A. (2002). Order Cryptomonadida. In J. J. Lee, G. F. Leedale, & P. Bradbury (Eds.), *An illustrated guide to the protozoa* (pp. 1111–1125). Lawrence: Society of Protozoologists.
Lane, C. E., & Archibald, J. M. (2006). Novel nucleomorph genome architecture in the cryptomonad genus *Hemiselmis*. *Journal of Eukaryotic Microbiology, 53*, 515–521.
Lane, C. E., & Archibald, J. M. (2008). New marine members of the genus *Hemiselmis* (Cryptomonadales, Cryptophyceae). *Journal of Phycology, 44*, 439–450.
Lane, C. E., Khan, H., MacKinnon, M., Fong, A., Theophilou, S., & Archibald, J. M. (2006). Insight into the diversity and evolution of the cryptomonad nucleomorph genome. *Molecular Biology and Evolution, 23*, 856–865.
Lane, C. E., van den Heuvel, K., Kozera, C., Curtis, B. A., Parsons, B. J., Bowman, S., et al. (2007). Nucleomorph genome of *Hemiselmis andersenii* reveals complete intron loss and compaction as a driver of protein structure and function. *Proceedings of the National Academy of Sciences of the United States of America, 104*, 19908–19913.
Lee, R. E., & Kugrens, P. (1991). *Katablepharis ovalis*, a colorless flagellate with interesting cytological characteristics. *Journal of Phycology, 27*, 505–513.
Ludwig, M., & Gibbs, S. P. (1985). DNA is present in the nucleomorph of cryptomonads—Further evidence that the chloroplast evolved from a eukaryotic endosymbiont. *Protoplasma, 127*, 9–20.
Ludwig, M., & Gibbs, S. P. (1989a). Evidence that the nucleomorphs of *Chlorarachnion reptans* (Chlorarachniophyceae) are vestigial nuclei: Morphology, division and DNA-DAPI fluorescence. *Journal of Phycology, 25*, 385–394.
Ludwig, M., & Gibbs, S. P. (1989b). Localization of phycoerythrin at the lumenal surface of the thylakoid membrane in *Rhodomonas lens*. *The Journal of Cell Biology, 108*, 875.
Maier, U. G., Fraunholz, M., Zauner, S., Penny, S., & Douglas, S. E. (2000). A nucleomorph-encoded CbbX and the phylogeny of RuBisCo regulators. *Molecular Biology and Evolution, 17*, 576–583.
Majaneva, M., Remonen, I., Rintala, J. M., Belevich, I., Kremp, A., Setala, O., et al. (2014). *Rhinomonas nottbecki* n. sp. (cryptomonadales) and molecular phylogeny of the family Pyrenomonadaceae. *Journal of Eukaryotic Microbiology, 61*(5), 480–492.
Marin, B., Klingberg, M., & Melkonian, M. (1998). Phylogenetic relationships among the Cryptophyta: Analyses of nuclear-encoded SSU rRNA sequences support the monophyly of extant plastid-containing lineages. *Protist, 149*, 265–276.
Martin, W., & Herrmann, R. G. (1998). Gene transfer from organelles to the nucleus: How much, what happens, and why? *Plant Physiology, 118*, 9–17.
Martin, W., Rujan, T., Richly, E., Hansen, A., Cornelsen, S., Lins, T., et al. (2002). Evolutionary analysis of *Arabidopsis*, cyanobacterial, and chloroplast genomes reveals plastid phylogeny and thousands of cyanobacterial genes in the nucleus. *Proceedings of the National Academy of Sciences of the United States of America, 99*, 12246–12251.
Martin-Cereceda, M., Roberts, E. C., Wootton, E. C., Bonaccorso, E., Dyal, P., Guinea, A., et al. (2010). Morphology, ultrastructure, and small subunit rDNA phylogeny of the marine heterotrophic flagellate *Goniomonas* aff. *amphinema*. *Journal of Eukaryotic Microbiology, 57*, 159–170.

Matsuo, M., Ito, Y., Yamauchi, R., & Obokata, J. (2005). The rice nuclear genome continuously integrates, shuffles, and eliminates the chloroplast genome to cause chloroplast-nuclear DNA flux. *Plant Cell, 17*, 665–675.

Matsuzaki, M., Misumi, O., Shin-I, T., Maruyama, S., Takahara, M., Miyagishima, S. Y., et al. (2004). Genome sequence of the ultrasmall unicellular red alga *Cyanidioschyzon merolae* 10D. *Nature, 428*, 653–657.

McFadden, G. I. (1990). Evidence that cryptomonad chloroplasts evolved from photosynthetic eukaryotic endosymbionts. *Journal of Cell Science, 95*, 303–308.

McFadden, G. I. (2014). Origin and evolution of plastids and photosynthesis in eukaryotes. *Cold Spring Harbor Perspectives in Biology, 6*, a016105.

McFadden, G. I., Gilson, P. R., Douglas, S. E., Cavalier-Smith, T., Hofmann, C. J., & Maier, U. G. (1997). Bonsai genomics: Sequencing the smallest eukaryotic genomes. *Trends in Genetics, 13*, 46–49.

McFadden, G. I., Gilson, P. R., & Hill, D. R. A. (1994). Goniomonas—Ribosomal-RNA sequences indicate that this phagotrophic flagellate is a close relative of the host component of cryptomonads. *European Journal of Phycology, 29*, 29–32.

McFadden, G. I., Gilson, P. R., Hofmann, C. J., Adcock, G. J., & Maier, U. G. (1994). Evidence that an amoeba acquired a chloroplast by retaining part of an engulfed eukaryotic alga. *Proceedings of the National Academy of Sciences of the United States of America, 91*, 3690–3694.

McFadden, G. I., Gilson, P. R., & Waller, R. F. (1995). Molecular phylogeny of Chlorarachniophytes based on plastid ribosomal-RNA and rbcL sequences. *Archiv fur Protistenkunde, 145*, 231–239.

Merchant, S. S., Prochnik, S. E., Vallon, O., Harris, E. H., Karpowicz, S. J., Witman, G. B., et al. (2007). The *Chlamydomonas* genome reveals the evolution of key animal and plant functions. *Science, 318*, 245–251.

Minnhagen, S., & Janson, S. (2006). Genetic analyses of *Dinophysis* spp. support kleptoplastidy. *FEMS Microbiology Ecology, 57*, 47–54.

Moore, C. E., & Archibald, J. M. (2009). Nucleomorph genomes. *Annual Review of Genetics, 43*, 251–264.

Moore, C. E., Curtis, B. A., Mills, T., Tanifuji, G., & Archibald, J. M. (2012). Nucleomorph genome sequence of the cryptophyte alga *Chroomonas mesostigmatica* CCMP1168 reveals lineage-specific gene loss and genome complexity. *Genome Biology and Evolution, 4*, 1162–1175.

Moore, R. B., Oborník, M., Janouškovec, J., Chrudimský, T., Vancová, M., Green, D. H., et al. (2008). A photosynthetic alveolate closely related to apicomplexan parasites. *Nature, 451*, 959–963.

Moran, N. A. (1996). Accelerated evolution and Muller's ratchet in endosymbiotic bacteria. *Proceedings of the National Academy of Sciences of the United States of America, 93*, 2873–2878.

Moreira, D., Lopezarchilla, A. I., Amils, R., & Marin, I. (1994). Characterization of two new thermoacidophilic microalgae—Genome organization and comparison with *Galdieria sulphuraria. FEMS Microbiology Letters, 122*, 109–114.

Morrall, S., & Greenwood, A. D. (1980). A comparison of the periodic substructure of the trichocysts of the Cryptophyceae and Prasinophyceae. *Biosystems, 12*, 71–83.

Muravenko, O. V., Selyakh, I. O., Kononenko, N. V., & Stadnichuk, I. N. (2001). Chromosome numbers and nuclear DNA contents in the red microalgae *Cyanidium caldarium* and three *Galdieria* species. *European Journal of Phycology, 36*, 227–232.

Nakabachi, A., Yamashita, A., Toh, H., Ishikawa, H., Dunbar, H. E., Moran, N. A., et al. (2006). The 160-kilobase genome of the bacterial endosymbiont *Carsonella. Science, 314*, 267.

Nakayama, T., & Archibald, J. M. (2012). Evolving a photosynthetic organelle. *BMC Biology, 10*, 35.

Okamoto, N., Chantangsi, C., Horák, A., Leander, B. S., & Keeling, P. J. (2009). Molecular phylogeny and description of the novel katablepharid *Roombia truncata* gen. et sp. nov., and establishment of the Hacrobia taxon nov. *PLoS One*, *4*(9), e7080.
Okamoto, N., & Inouye, I. (2005). The katablepharids are a distant sister group of the Cryptophyta: A proposal for Katablepharidophyta divisio nova/Kathablepharida phylum novum based on SSU rDNA and beta-tubulin phylogeny. *Protist*, *156*, 163–179.
Okamoto, N., & Inouye, I. (2006). *Hatena arenicola* gen. et sp. nov., a katablepharid undergoing probable plastid acquisition. *Protist*, *157*, 401–419.
Onuma, R., & Horiguchi, T. (2015). Kleptochloroplast enlargement, karyoklepty and the distribution of the cryptomonad nucleus in *Nusuttodinium* (= *Gymnodinium*) *aeruginosum* (Dinophyceae). *Protist*, *166*, 177–195.
Palmer, J. D., & Delwiche, C. F. (1996). Second-hand chloroplasts and the case of the disappearing nucleus. *Proceedings of the National Academy of Sciences of the United States of America*, *93*, 7432–7435.
Patron, N. J., Inagaki, Y., & Keeling, P. J. (2007). Multiple gene phylogenies support the monophyly of cryptomonad and haptophyte host lineages. *Current Biology*, *17*, 887–891.
Patron, N. J., Rogers, M. B., & Keeling, P. J. (2006). Comparative rates of evolution in endosymbiotic nuclear genomes. *BMC Evolutionary Biology*, *6*, 46.
Pedrós-Alió, C., Massana, R., Latasa, M., Garcí-Cantizano, J., & Gasol, J. M. (1995). Predation by ciliates on a metalimnetic *Cryptomonas* population: Feeding rates, impact and effects of vertical migration. *Journal of Plankton Research*, *17*, 2131–2154.
Phipps, K. D., Donaher, N. A., Lane, C. E., & Archibald, J. M. (2008). Nucleomorph karyotype diversity in the freshwater cryptophyte genus *Cryptomonas*. *Journal of Phycology*, *44*, 11–14.
Phlips, E. J., Havens, K. E., & Lopes, M. R. M. (2008). Seasonal dynamics of phytoplankton in two Amazon flood plain lakes of varying hydrologic connectivity to the main river channel. *Fundamental and Applied Limnology*, *172*, 99–109.
Pringsheim, E. G. (1944). Some aspects of taxonomy in the Cryptophyceae. *The New Phytologist*, *43*, 143–150.
Rensing, S. A., Goddemeier, M., Hofmann, C. J. B., & Maier, U. G. (1994). The presence of a nucleomorph hsp70 gene is a common feature of cryptophyta and chlorarachniophyta. *Current Genetics*, *26*, 451–455.
Reyes-Prieto, A., Weber, A. P. M., & Bhattacharya, D. (2007). The origin and establishment of the plastid in algae and plants. *Annual Review of Genetics*, *41*, 147–168.
Roberts, E. C., & Laybourn-Parry, J. (1999). Mixotrophic cryptophytes and their predators in the Dry Valley lakes of Antarctica. *Freshwater Biology*, *41*, 737–746.
Rogers, M. B., Gilson, P. R., Su, V., McFadden, G. I., & Keeling, P. J. (2007). The complete chloroplast genome of the chlorarachniophyte *Bigelowiella natans*: Evidence for independent origins of chlorarachniophyte and euglenid secondary endosymbionts. *Molecular Biology and Evolution*, *24*, 54–62.
Rowan, K. S. (1984). *Photosynthetic pigments of algae*. Cambridge: Cambridge University Press.
Roy, S. W., & Penny, D. (2007). Patterns of intron loss and gain in plants: Intron loss-dominated evolution and genome-wide comparison of *O. sativa* and *A. thaliana*. *Molecular Biology and Evolution*, *24*, 171–181.
Ruggiero, M. A., Gordon, D. P., Orrell, T. M., Bailly, N., Bourgoin, T., Brusca, R. C., et al. (2015a). Correction: A higher level classification of all living organisms. *PLoS One*, *10*(6), e0130114.
Ruggiero, M. A., Gordon, D. P., Orrell, T. M., Bailly, N., Bourgoin, T., Brusca, R. C., et al. (2015b). A higher level classification of all living organisms. *PLoS One*, *10*, e0119248.
Santore, U. J. (1984). Some aspects of taxonomy in the Cryptophyceae. *The New Phytologist*, *98*, 627–646.

Schnepf, E., Winter, S., & Mollenhauer, D. (1989). *Gymnodinium aeruginosum* (Dinophyta): A blue-green dinoflagellate with a vestigial, anucleate, cryptophycean endosymbiont. *164*, 75–91.

Shalchian-Tabrizi, K., Brate, J., Logares, R., Klaveness, D., Berney, C., & Jakobsen, K. S. (2008). Diversification of unicellular eukaryotes: Cryptomonad colonizations of marine and fresh waters inferred from revised 18S rRNA phylogeny. *Environmental Microbiology*, *10*, 2635–2644.

Shiratori, T., & Ishida, K. I. (2016). A new heterotrophic cryptomonad: *Hemiarma marina* n. g., n. sp. *Journal of Eukaryotic Microbiology*, *63*, 804–812.

Silver, T. D., Koike, S., Yabuki, A., Kofuji, R., Archibald, J. M., & Ishida, K. I. (2007). Phylogeny and nucleomorph karyotype diversity of chlorarachniophyte algae. *Journal of Eukaryotic Microbiology*, *54*, 403–410.

Silver, T. D., Moore, C. E., & Archibald, J. M. (2010). Nucleomorph ribosomal DNA and telomere dynamics in chlorarachniophyte algae. *Journal of Eukaryotic Microbiology*, *57*, 453–459.

Smith, D. R. (2009). Unparalleled GC content in the plastid DNA of *Selaginella*. *Plant Molecular Biology*, *71*, 627–639.

Smith, D. R., Crosby, K., & Lee, R. W. (2011). Correlation between nuclear plastid DNA abundance and plastid number supports the limited transfer window hypothesis. *Genome Biology and Evolution*, *3*, 365–371.

Sommer, M. S., Gould, S. B., Lehmann, P., Gruber, A., Przyborski, J. M., & Maier, U. G. (2007). Der1-mediated preprotein import into the periplastid compartment of chromalveolates? *Molecular Biology and Evolution*, *24*, 918–928.

Stiller, J. W., Schreiber, J., Yue, J., Guo, H., Ding, Q., & Huang, J. (2014). The evolution of photosynthesis in chromist algae through serial endosymbioses. *Nature Communications*, *5*, 5764.

Stork, S., Moog, D., Przyborski, J. M., Wilhelmi, I., Zauner, S., & Maier, U. G. (2012). Distribution of the SELMA translocon in secondary plastids of red algal origin and predicted uncoupling of ubiquitin-dependent translocation from degradation. *Eukaryotic Cell*, *11*, 1472–1481.

Strauss, S. H., Palmer, J. D., Howe, G. T., & Doerksen, A. H. (1988). Chloroplast genomes of two conifers lack a large inverted repeat and are extensively rearranged. *Proceedings of the National Academy of Sciences of the United States of America*, *85*, 3898–3902.

Suzuki, S., Hirakawa, Y., Kofuji, R., Sugita, M., & Ishida, K. (2016). Plastid genome sequences of *Gymnochlora stellata*, *Lotharella vacuolata*, and *Partenskyella glossopodia* reveal remarkable structural conservation among chlorarachniophyte species. *Journal of Plant Research*, *129*, 581–590.

Suzuki, S., Shirato, S., Hirakawa, Y., & Ishida, K. (2015). Nucleomorph genome sequences of two chlorarachniophytes, *Amorphochlora amoebiformis* and *Lotharella vacuolata*. *Genome Biology and Evolution*, 7, 1533–1545.

Takishita, K., Koike, K., Maruyama, T., & Ogata, T. (2002). Molecular evidence for plastid robbery (Kleptoplastidy) in *Dinophysis*, a dinoflagellate causing diarrhetic shellfish poisoning. *Protist*, *153*, 293–302.

Tanifuji, G., & Archibald, J. M. (2014). Nucleomorph comparative genomics. In L. Wolfgang (Ed.), *Endosymbiosis* (pp. 197–214). Vienna: Springer.

Tanifuji, G., Archibald, J. M., & Hashimoto, T. (2016). Comparative genomics of mitochondria in chlorarachniophyte algae: Endosymbiotic gene transfer and organellar genome dynamics. *Scientific Reports*, *6*, 21016.

Tanifuji, G., Kim, E., Onodera, N. T., Gibeault, R., Dlutek, M., Cawthorn, R. J., et al. (2011). Genomic characterization of *Neoparamoeba pemaquidensis* (Amoebozoa) and its kinetoplastid endosymbiont. *Eukaryotic Cell*, *10*(8), 1143–1146.

Tanifuji, G., Onodera, N. T., Brown, M. W., Curtis, B. A., Roger, A. J., Wong, G. K. S., et al. (2014). Nucleomorph and plastid genome sequences of the chlorarachniophyte *Lotharella oceanica*: Convergent reductive evolution and frequent recombination in nucleomorph-bearing algae. *BMC Genomics, 15*, 374.

Tanifuji, G., Onodera, N. T., & Hara, Y. (2010). Nucleomorph genome diversity and its phylogenetic implications in cryptomonad algae. *Phycological Research, 58*, 230–237.

Tanifuji, G., Onodera, N. T., Moore, C. E., & Archibald, J. M. (2014). Reduced nuclear genomes maintain high gene transcription levels. *Molecular Biology and Evolution, 31*, 625–635.

Tanifuji, G., Onodera, N. T., Wheeler, T. J., Dlutek, M., Donaher, N., & Archibald, J. M. (2011). Complete nucleomorph genome sequence of the nonphotosynthetic alga *Cryptomonas paramecium* reveals a core nucleomorph gene set. *Genome Biology and Evolution, 3*, 44–54.

Thorsness, P. E., & Fox, T. D. (1990). Escape of DNA from mitochondria to the nucleus in *Saccharomyces cerevisiae*. *Nature, 346*, 376–379.

Tirok, K., & Gaedke, U. (2007). Regulation of planktonic ciliate dynamics and functional composition during spring in Lake Constance. *Aquatic Microbial Ecology, 49*, 87–100.

Turmel, M., Gagnon, M. C., O'Kelly, C. J., Otis, C., & Lemieux, C. (2009). The chloroplast genomes of the green algae *Pyramimonas*, *Monomastix*, and *Pycnococcus* shed new light on the evolutionary history of Prasinophytes and the origin of the secondary chloroplasts of Euglenids. *Molecular Biology and Evolution, 26*, 631–648.

Van de Peer, Y., Van der Auwera, G., & De Wachter, R. (1996). The evolution of stramenopiles and alveolates as derived by "substitution rate calibration" of small ribosomal subunit RNA. *Journal of Molecular Evolution, 42*, 201–210.

van Dooren, G. G., Schwartzbach, S. D., Osafune, T., & McFadden, G. I. (2001). Translocation of proteins across the multiple membranes of complex plastids. *Biochimica et Biophysica Acta, 1541*, 34–53.

Vieira Ldo, N., Faoro, H., Rogalski, M., Fraga, H. P., Cardoso, R. L., de Souza, E. M., et al. (2014). The complete chloroplast genome sequence of *Podocarpus lambertii*: Genome structure, evolutionary aspects, gene content and SSR detection. *PLoS One, 9*, e90618.

Von der Heyden, S., Chao, E. E., & Cavalier-Smith, T. (2004). Genetic diversity of goniomonads: An ancient divergence between marine and freshwater species. *European Journal of Phycology, 39*, 343–350.

Watanabe, M. M., Suda, S., Inouye, I., Sawaguchi, T., & Chihara, M. (1990). *Lepidodinium viride* gen. et sp. nov. (Gymnodiniales, Dinophyta), a green dinoflagellate with a chlorophyll a- and b-containing endosymbiont. *Journal of Phycology, 26*, 741–751.

Wedemayer, G. J., Kidd, D. G., & Glazer, A. N. (1996). Cryptomonad biliproteins: Bilin types and locations. *Photosynthesis Research, 48*, 163–170.

Wei, L., Xin, Y., Wang, D., Jing, X., Zhou, Q., Su, X., et al. (2013). *Nannochloropsis* plastid and mitochondrial phylogenomes reveal organelle diversification mechanism and intragenus phylotyping strategy in microalgae. *BMC Genomics, 14*, 534.

Weisse, T., & Kirchhoff, B. (1997). Feeding of the heterotrophic freshwater dinoflagellate *Peridiniopsis berolinense* on cryptophytes: Analysis by flow cytometry and electronic particle counting. *Aquatic Microbial Ecology, 12*, 153–164.

Whatley, J. M., & Whatley, F. R. (1981). Chloroplast evolution. *New Phytologist, 87*, 233–247.

Whetherbee, R., Hill, D. R. A., & McFadden, G. I. (1985). Periplast structure of the cryptomonad flagellate *Hemiselmis brunnescens*. *Protoplasma, 131*, 11–22.

Williams, T. A., Foster, P. G., Cox, C. J., & Embley, T. M. (2013). An archaeal origin of eukaryotes supports only two primary domains of life. *Nature, 504*, 231–236.

Worden, A. Z., Lee, J. H., Mock, T., Rouze, P., Simmons, M. P., Aerts, A. L., et al. (2009). Green evolution and dynamic adaptations revealed by genomes of the marine picoeukaryotes Micromonas. *Science*, *324*, 268–272.

Yabuki, A., Inagaki, Y., & Ishida, K. (2010). *Palpitomonas bilix* gen. Et sp. nov.: A novel deep-branching heterotroph possibly related to Archaeplastida or Hacrobia. *Protist*, *161*, 523–538.

Yabuki, A., Kamikawa, R., Ishikawa, S. A., Kolisko, M., Kim, E., Tanabe, A. S., et al. (2014). *Palpitomonas bilix* represents a basal cryptist lineage: Insight into the character evolution in Cryptista. *Scientific Reports*, *4*, 4641.

Yamagishi, T., Kai, A., & Kawai, H. (2012). Trichocyst ribbons of a cryptomonads are constituted of homologs of R-body proteins produced by the intracellular parasitic bacterium of *Paramecium*. *Journal of Molecular Evolution*, *74*, 147–157.

Yoon, H. S., Hackett, J. D., Pinto, G., & Bhattacharya, D. (2002). The single, ancient origin of chromist plastids. *Proceedings of the National Academy of Sciences of the United States of America*, *99*, 15507–15512.

Zaremba-Niedzwiedzka, K., Caceres, E. F., Saw, J. H., Backstrom, D., Juzokaite, L., Vancaester, E., et al. (2017). Asgard archaea illuminate the origin of eukaryotic cellular complexity. *Nature*, *541*, 353–358.

CHAPTER NINE

Secondary Plastids of Euglenophytes

Anna M.G. Vanclová[1], Lucia Hadariová, Štěpánka Hrdá, Vladimír Hampl[1]
Faculty of Science, Charles University in Prague, Prague, Czech Republic
[1]Corresponding authors: e-mail address: vanclova@gmail.com; vlada@natur.cuni.cz

Contents

Abstract

Euglenophytes obtained their plastids from a primary green alga related to extant genus *Pyramimonas*. The relatively recent establishment of this new organelle is an intriguing evolutionary phenomenon worth studying and comparing with other secondary plastids with a regard to their similarities and differences. A remarkably fast evolution driven by rapid intron gain and diversification is observed in euglenid plastid genomes which often tend to swell in size and rearrange while keeping the gene content stable. As a result of the secondary endosymbiosis, the plastid is wrapped in an additional membrane which makes any protein, metabolite, or ion transporting routes more complicated. In the case of protein import, secretory pathway-derived, signal peptide-dependent mechanism involving the endoplasmic reticulum, Golgi, and vesicular transport were recruited. The plastid endosymbiosis also served as a source of various oddities concerning metabolic pathways as the new organelle contained some of the enzymes and pathways already present in the host. Thus, several cases of division of labour and specialization can be observed, as well as simple redundancies which might be in fact just transitory and will eventually disappear in the future course of evolution. Endosymbiotic and lateral gene transfers were quite common in the

Advances in Botanical Research, Volume 84
ISSN 0065-2296
http://dx.doi.org/10.1016/bs.abr.2017.06.008

ancestors of euglenophytes, especially in the case of plastid proteins many of which were demonstrated to have originated not only from the green-algal endosymbiont but also from a spectrum of nongreen lineages. The circumstances of the nongreen-algal gene gains are unclear. Another evolutionary phenomenon occurring in euglenophytes is the secondary loss of plastid or its photosynthetic capacity. This process gave rise to a number of distinct species which no longer possess the ability to photosynthesize. Interestingly, this "bleaching" process can be induced in the laboratory, enabling to study the process of plastid loss in vitro.

1. INTRODUCTION: WHAT ARE THE EUGLENOPHYTES AND WHY TO CARE ABOUT THEM

Euglenids are a group of flagellate protists belonging to the phylum Euglenozoa (Excavata), along with kinetoplastids—a group made famous by its important pathogenic members such as trypanosomes—diplonemids, and symbiontids (also termed Postgaardea) (Adl et al., 2012; Cavalier-Smith, 2016). Euglenids are a highly diversified taxon in regard to their nutritional strategies. The ancestral mode of nutrition was probably bacteriovory which was replaced by eukaryovory in a large portion of the diversity; this shift is believed to have taken place once in the evolution. One lineage of eukaryovores then lost the phagotrophic ability and became osmotrophic. This lineage is currently termed primary osmotrophs (Leander, 2004; Leander, Esson, & Breglia, 2007; Leander, Triemer, & Farmer, 2001). However, some researchers propose an alternative hypothesis according to which eukaryovory was the ancestral nutritional strategy common not only to euglenids but also to kinetoplastids, diplonemids, and symbiontids, and all other feeding modes are derived from it (Cavalier-Smith, 2016). One lineage within eukaryovorous euglenids gained a green secondary plastid in presumably one endosymbiotic event. All these photoautotrophs form one robust phylogenetic clade termed the Euglenophyta. The term "euglenophytes" thus refers to phototrophic (or secondarily osmotrophic) euglenids that harbour (or once harboured) a plastid, while the term "euglenids" addresses the whole group of organisms of various nutritional strategies. The euglenophytes were one of the first protists to be discovered and described. The first documented observation of an euglenophyte, presumably the genus *Euglena*, and description of its typical slime-like mode of movement, metaboly, was carried out by John Harris at the end of the 17th century (Harris, 1695) and the first species of euglenophytes were described in the beginning of the 19th century by Ehrenberg (1830). Several isolated lineages within euglenophytes lost the photosynthetic pigments, switching to osmotrophic mode of nutrition independently of the earlier mentioned primary osmotrophs; these organisms are termed secondary osmotrophs and retain a colourless plastid despite not using it for photosynthesis. Additionally, one obligatory–mixotrophic species, *Rapaza viridis*, has a photosynthetically active green plastid but at the same time requires green-algal (*Tetraselmis*) prey to survive and it is the only euglenid which combines photoautotrophy with phagotrophy. It is suspected to represent a link between the eukaryovorous euglenids and the phototrophic euglenophytes. This is also suggested by its phylogenetic position as sister lineage to all other euglenophytes (Fig. 1; Yamaguchi, Yubuki, & Leander, 2012). However, it is still unclear whether the *R. viridis* plastid is stable or transient and if it is related to the plastids of other euglenophytes. The closest known non-photosynthetic sister of euglenophytes is the eukaryovorous genus *Teloprocta* (former *Heteronema*) *scaphurum* (Cavalier-Smith, 2016; Lax & Simpson, 2013).

Euglenophytes have been intensively studied for a long time thanks to their relatively large size, distinctive and aesthetically pleasing appearance, and easy collection and cultivation, and so approximately 950 species of euglenophytes have been described to this day (Guiry & Guiry, 2017). The inner taxonomy of euglenophytes is currently relatively clearly resolved (Fig. 1; Karnkowska et al., 2015; Kim, Linton, & Shin, 2015; Kim & Shin, 2008; Linton et al., 2010; Triemer et al., 2006; Zakryś, Milanowski, & Karnkowska, 2017). The group splits into two orders: Eutreptiales and Euglenales. The Eutreptiales exhibit a number of features which are considered ancestral, such as the adaptation to marine environment and presence of two emergent flagella, and contain two genera: *Eutreptia* and *Eutreptiella*. The Euglenales are more diverse, they have only one emergent flagellum, and they are thought to be almost exclusively freshwater dwelling. The Eutreptiales are probably paraphyletic, with the Euglenales being their inner group (Cavalier-Smith, 2016). The Euglenales are further split into two distinctive families, Euglenaceae and Phacaceae. The Euglenaceae include eight genera: *Euglena*, *Euglenaria*, *Euglenaformis*, *Cryptoglena*, *Monomorphina*, *Colacium*, *Trachelomonas*, and *Strombomonas*. This group is highly diverse in regard to the shape, number, and position of the plastids, and also overall morphology of the cell which may exhibit various shapes—some members, i.e., genera *Trachelomonas* and *Strombomonas*, are even surrounded

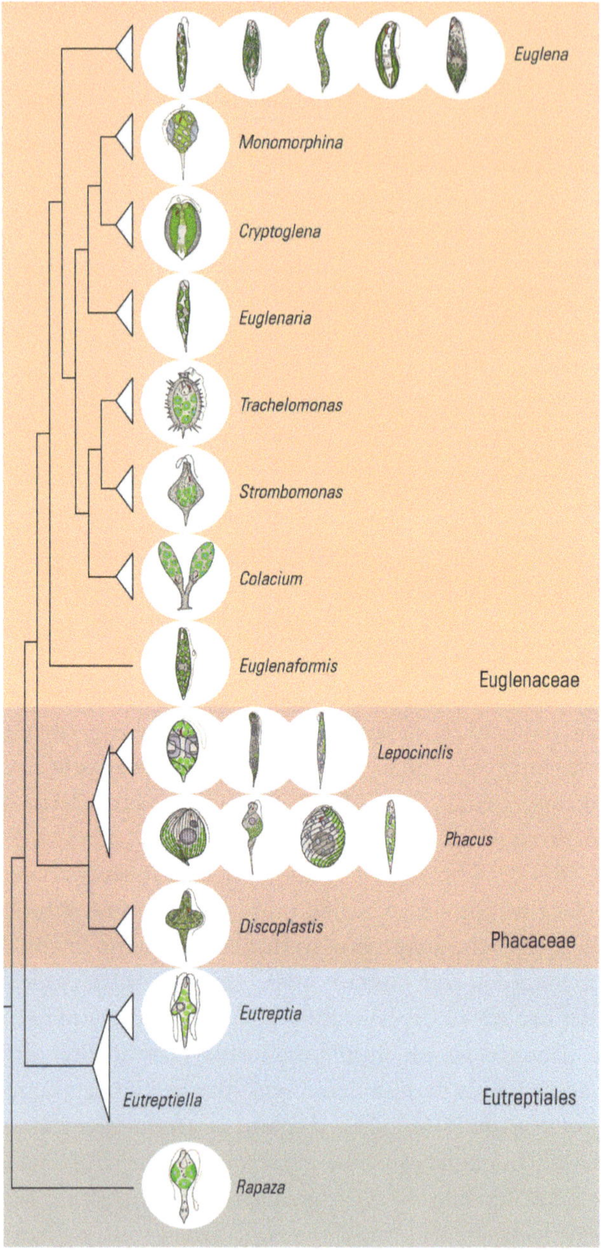

Fig. 1 Simplified phylogenetic tree of the 14 euglenophyte genera (including *Rapaza*) with examples of cell and plastid morphologies. One to five different illustrative species are shown for each genus, depending on its relative intrageneric plastid morphological diversity.

by mineralized extracellular lorica—and types and degrees of motility—some members have pellicles of low flexibility and prefer swimming using flagellum to metaboly, while some species in the genus *Colacium* are completely sessile. The Phacaceae include three genera: *Phacus*, *Lepocinclis*, and *Discoplastis*, the first two of which have a rigid pellicle, which is flattened or even helical in shape, and are not able to move using metaboly (Cavalier-Smith, 2016; de M. Bicudo & Menezes, 2016).

Euglenophytes are currently studied for their possible utilization in biotechnology and other applied sciences, especially as potential producers of biofuels and also nutritional supplements due to their capacity to synthesize various lipids and compounds with antioxidative properties such as tocopherols and carotenoids (Krajčovič, Vesteg, & Schwartzbach, 2015; Ogawa et al., 2015; Teerawanichpan & Qiu, 2010).

2. ORIGIN OF EUGLENOPHYTE PLASTIDS: EARLY OR LATE, GREEN OR RED?

Before the age of molecular phylogenetics, euglenophytes were traditionally classified as a subgroup or sister of green algae due to their plastid morphology and pigment content, especially the combination of chlorophylls *a* and *b*, which is otherwise unique to green plants and algae and chlorarachniophytes, and the chlorophylls *a*- and *b*-containing plastids were correctly assumed to have arisen only once in the evolution. However, it was suspected for a long time that something is amiss here because most of the nonplastid morphology, physiology, and biochemistry of these organisms were strikingly dissimilar to green algae. With the (re)invention of the endosymbiotic theory by Lynn Margulis in 1967 (Sagan, 1993) and its general acceptance, a completely new paradigm opened for evolutionary biology and taxonomy trying to resolve unclear phenomena and phylogenetic relationships such as this one. A decade later, the secondary endosymbiotic origin of the euglenophyte plastid was proposed by Gibbs (1978, 1981) and euglenids were placed along with kinetoplastids into a new phylum Euglenozoa by Cavalier-Smith (1981). The first sequenced genes from the model euglenid *Euglena gracilis* (Douglas & Turner, 1991; Morden & Golden, 1991), followed shortly by the complete sequencing of its plastid genome (Hallick et al., 1993), brought a molecular evidence for this claim. In 2009, a wider analysis of phylogenetic relationship within green algae with a focus on prasinophytes and also including sequences from secondary plastids was performed by Turmel, Gagnon, O'Kelly, Otis, and Lemieux

(2009) which resulted in pinpointing the origin of the euglenid plastid to the Pyramimonadales clade in the prasinophytes with the genus *Pyramimonas* as the most suspect source. Following studies on other euglenid plastid genomes (Bennett & Triemer, 2015; Bennett, Wiegert, & Triemer, 2014; Dabbagh & Preisfeld, 2016; Hrdá, Fousek, Szabová, Hampl, & Vlček, 2012; Pombert, James, Janouškovec, & Keeling, 2012; Wiegert, Bennett, & Triemer, 2012, 2013) confirmed *Pyramimonas parkeae* as the closest living relative of euglenophyte plastids.

The relative position of the plastid acquisition in the euglenid lineage was disputed in the past (Bodył, Mackiewicz, & Milanowski, 2010; Hannaert et al., 2003) and even though it is currently considered resolved and placed at the root of the extant euglenophytes the evolutionary history of euglenids seems more complex and traces of an interesting story about lateral gains and secondary losses of genes—and possibly endosymbionts or even organelles—can be read from the available sequence data of euglenids and their not-so-distant relatives kinetoplastids.

Kinetoplastids, a group known mainly for their parasitic members of considerable epidemiological significance of the genera *Trypanosoma* and *Leishmania*, have been shown to contain genes related to cyanobacteria or green plastids. Moreover, some of these genes are homologues of enzymes of the Calvin cycle which have been supposedly recruited for glycolysis-related functions in the glycosome (Hannaert et al., 2003) (i.e. peroxisome-derived organelle invented specifically by kinetoplastids and diplonemids; Morales et al., 2016; Rybicka, 1996). In reaction to this fact plus other occurrences of cyanobacterial-like genes in heterotrophic protists, a plastid-early hypothesis was formulated. This hypothesis placed the primary plastid acquisition much earlier in the evolution and suggested that all bikonts except the Archaeplastida underwent its secondary loss (Andersson & Roger, 2002; Maruyama, Matsuzaki, Misawa, & Nozaki, 2009). The less controversial theories focused on Euglenozoa suggested that the current secondary plastid of euglenophytes could have originated before the split of kinetoplastids and other euglenozoans and become lost in all lineages but euglenophytes (Bodył et al., 2010; Hannaert et al., 2003) resulting in the presence of "green" genes dispersed throughout the whole group. The possibility of a cryptic, euglenophyte-unrelated plastid gain and loss within the kinetoplastid lineage, was also considered (Bodył et al., 2010; Leander, 2004; Martin & Borst, 2003). The alternative hypothesis to all the previously mentioned suggests a mere lateral gene transfer from cyanobacteria or primary or secondary plastid-containing eukaryote based solely on the "you are what you eat" notion (Doolittle, 1998) without a requirement for a previously existing stable relationship between the host and the endosymbiont/organelle (Bodył et al., 2010; Leander, 2004; Maruyama et al., 2009).

The mosaicism in the evolutionary origins of euglenophyte genes provokes similar questions and similar approaches to their resolving. Euglenophytes contain undisputable amount of laterally transferred genes originating from rhodophytes or secondary algae containing red-algal plastids (e.g. haptophytes or stramenopiles), i.e. genes gained from other sources than the ancestor of the current plastid related to *Pyramimonas*. Most of these genes play a role in the typical plastid metabolic pathways and processes. The examples of these "nongreen" genes include several enzymes of the Calvin cycle (fructose bisphosphatase, glyceraldehyde 3-phosphate dehydrogenase, phosphoribulokinase, ribose 5-phosphate isomerase, and triosephosphate isomerase), several enzymes of the tetrapyrrole biosynthesis (glutamate 1-semialdehyde 2,1-aminotransferase, uroporphyrinogen decarboxylase, coproporphyrinogen oxidase, and protoporphyrinogen oxidase), glycolytic enzyme glucokinase, and tocopherol biosynthesis enzyme homogentisate phytyltransferase (Lakey & Triemer, 2016; Markunas & Triemer, 2016; Maruyama, Suzaki, Weber, Archibald, & Nozaki, 2011). Whether similar plastid-related genes, green or nongreen, are present in heterotrophic euglenids, and to what extent, is currently unknown.

Where do these genes come from? The least controversial explanation is simple lateral gene transfer from eukaryotic prey, congruent with the "you are what you eat" hypothesis (Doolittle, 1998). Many eukaryovorous euglenids prey on algae, often in a rather generalist way. This nutritional strategy represents a clear prerequisite for the plastid acquisition as well as a logical source of the lateral gene transfer from various phototrophs. But how come that the transferred genes were retained by the euglenid predator despite having photosynthetic or other plastidal function and being virtually useless to a heterotrophic organism? The answer could be that the said euglenid predator was not a pure heterotroph but rather a mixotroph which already possessed the plastid or *Pyramimonas*-related endosymbiont not yet fully transformed into a true organelle, much like the contemporary species *R. viridis* (Yamaguchi et al., 2012). Genes obtained from the algal prey could have been immediately recruited for a function in a preexisting plastid-localized process, allowing the reductive evolution of the plastid/endosymbiont genome and its further integration into the host. This evolutionary process was recently suggested as a major driving force of the organellogenesis of the chromatophore of *Paulinella chromatophora* (Nowack et al., 2016) (a rhizarian amoeba with a plastid-like organelle, chromatophore, acquired via second independent primary endosymbiosis; Marin, Nowack, & Melkonian, 2005).

Alternative hypotheses count with the possibility of euglenophyte predecessors harbouring other endosymbiont(s) or transient or even stable plastid(s) in their evolutionary past which were lost and replaced with the plastid from *Pyramimonas*-related alga. This model is based on the "shopping bag" hypothesis and reflects the current state of knowledge regarding the transient relationships, repeated gains and losses, and overall plastid fluidity observed in dinoflagellates (Bodył et al., 2010; Howe, Barbrook, Nisbet, Lockhart, & Larkum, 2008; Larkum, Lockhart, & Howe, 2007; O'Neill, Trick, Henrissat, & Field, 2015). If this hypothesis was correct, at least one "red" endosymbiont or preplastid was present in the ancestor of euglenophytes at some point. It is even possible that some of these plastid-like symbioses or similar ecological relationships could have taken place long time before the acquisition of the extant green plastid, explaining the presence of genes of algal origin in heterotrophic euglenids (Bodył et al., 2010; Leander, 2004; Maruyama et al., 2011) and making some of the implications of the plastid-early hypothesis true. The plastid-related genes it had left in the nuclear genome of the euglenid could have facilitated the acquisition of a brand new plastid and its "enslavement" by reductive genome evolution, quite opposite to the *Paulinella*-like model proposed earlier. Another imaginable scenario is the coexistence of red and green endosymbionts/preplastids in a single cell for a certain period of time during which the host had time to gain genes from both before ultimately "deciding" to let the red one go and keep the green one.

These hypotheses are not mutually exclusive and the truth can well lie somewhere between these proposed models—or somewhere else entirely.

3. PLASTID MORPHOLOGY: DISPLAY OF DIVERSITY

Not much can be said about the morphology of euglenophyte plastids in general because their shapes, sizes, numbers, positions, and other characteristics vary greatly among different genera and even species or strains (Ciugulea & Triemer, 2010; Leedale, 1967).

The most basic universal characteristic is perhaps that the euglenophyte plastids are enveloped by three membranes which reflect its secondary endosymbiotic origin and evolutionary history: the inner two membranes are generally believed to be homologous with the two membranes of primary plastid (i.e. originating in the two envelope membranes of cyanobacteria), the additional one is of an eukaryotic ancestry, derived either from the euglenid endomembrane system or from the cytoplasmic membrane of the green-algal ancestor of the plastid (Gibbs, 1978; Lefort-Tran, 1981). The same number of plastid membranes is described in dinoflagellates with peridinin-containing plastid, while four membranes are generally more common occurrence in secondary plastids. The outermost membrane of euglenophyte plastid is not spatially continuous with the ER as in the case of some other organisms with secondary plastids such as cryptophytes, haptophytes, and heterokonts (Bolte et al., 2009; Maier, Zauner, & Hempel, 2015). However, substantial communication via vesicles was observed to take place between the endomembrane system and plastid, and it was proposed that its outermost membrane might in fact act as a part of the secretory system (Sulli & Schwartzbach, 1995, 1996). This would be crucial for the targeting of ER-synthesized proteins and other molecules into the plastid.

Euglenophyte plastids are very variable in terms of shape: they can be spherical, oval, partially flattened, disc-shaped, watch-glass-shaped, biconcave or biconvex, with various numbers of lobes and prominences, or even ribbon- or star-shaped (Leedale, 1967). A certain plastid shape is usually specific to a certain monophyletic or paraphyletic group of euglenophytes. The amount of plastids per cell ranges from several to several dozens and it is also usually species-specific but in this case a certain range is conserved rather than a concrete number: in the model species *E. gracilis*, for example, the number of plastids varies from 6 to 12 (Leedale, 1967). This variability might be linked to the fact that plastid replication can take place either synchronously with the cell division or independently of it.

Euglenophyte plastids have well-developed thylakoids (also termed *discs* in older literature) which are organized in elongated stacks (also termed *lamellae* or *bands*) of varying number of layers (usually three) instead of grana (i.e. relatively high cylindrical stacks of thylakoids present in plant plastids) (Ben-Shaul, Schiff, & Epstein, 1966; Gibbs, 1970; Gibor & Granick, 1962).

Many euglenophyte plastids contain visible pyrenoids (i.e. dense regions where most of the carbon fixation enzymes are localized): these may be either naked or covered by a watch glass-shaped paramylon cap, either from one side (haplopyrenoids) or from both sides (diplopyrenoids). Under transmission electron microscopy, the pyrenoids appear as conspicuously delimited regions of dense granulation (Gibbs, 1970). Pyrenoids are absent in plastids of the family Phacaceae and also in some members of the Euglenaceae (Gibbs, 1970; Leedale, 1967). The presence/absence and appearance of pyrenoids can also change during the life of a single cell as a result of changes in environmental factors—light intensity, length of the light/dark cycle, starvation, nutritional value of the medium, and growth phase of the culture.

Euglenophytes possess a single eyespot (*stigma*), light-sensing, granular, red/orange-coloured, and carotenoid-containing organelle, which enables positive or negative phototaxis. In contrast to green algae, where eyespots exist within the plastids, the euglenophyte eyespot is located in the cytoplasm near the base of the flagellum (Benedetti & Checcucci, 1975; Iseki et al., 2002; Osafune & Schiff, 1980b; Walne & Arnott, 1967). However, it was probably derived from plastid as well (Walne & Arnott, 1967).

4. PLASTID GENOMES: SPED-UP EVOLUTION AND INTRONS GONE HAYWIRE

To this date, 17 euglenophyte plastid genomes (cpDNAs) have been published. The sampling covers most of the diversity of the Euglenaceae family with 14 cpDNAs (6 from the genus *Euglena* and 8 from other genera), one cpDNA of the Phacaceae family and two cpDNAs from the two genera of the Eutreptiales (Bennet, Wiegert, & Triemer, 2012; Bennett & Triemer, 2015; Bennett et al., 2014; Dabbagh & Preisfeld, 2016; Gockel & Hachtel, 2000; Hallick et al., 1993; Hrdá et al., 2012; Kasiborski, Bennett, Linton, & Lane, 2016; Pombert et al., 2012; Wiegert et al., 2012, 2013). Euglenophyte cpDNAs generally take the form of a circular chromosome. Four genomes are not complete and have not been circularized due to unknown number of repetitive sequences (Kasiborski et al., 2016) or unknown number of the ribosomal operons (Wiegert et al., 2012, 2013). Their basic characteristics including the comparison with the cpDNA of *P. parkeae* (Turmel et al., 2009) are summarized in Table 1.

The cpDNA of *P. parkeae,* the closest known relative of the euglenophyte plastid, is 101,605 bp long and contains 110 genes (Turmel et al., 2009). These values were seemingly reduced during the secondary

Table 1 Characteristics of the Plastid Genomes of 17 Euglenophytes and *Pyramimonas parkeae* According to Sequences Deposited in GenBank

Taxonomy	Species/Strain	cp Genome Size (bp)	Number of Genes	Number of Introns	GC Content (%)
	Euglena gracilis Z	143,171	90	145	26.1
	E. gracilis var. *bacillaris*	132,034	90	134	25.8
	Euglena longa	73,345	57	60	22.4
	Euglena viridis epitype	91,616	92	77	26.4
	E. viridis SAG 1224-17d	76,156	92	77	26.2
	Euglena mutabilis	86,975	92	76	26.7
	Monomorphina aenigmatica	74,746	92	53	29.4
	Monomorphina parapyrum	80,147	93	80	28.0
	Cryptoglena skujai	106,843	92	84	26.3
	Euglenaria anabaena	88,487	93	82	28.0
	Trachelomonas volvocina	85,392*	93	94	27.3
	*Strombomonas acuminata**	144,166	93	110	26.6
	*Colacium vesiculosum**	128,892	92	130	26.1
	Euglenaformis proxima	94,185	91	113	26.9
	*Phacus orbicularis**	65,992	90	66	27.2
	*Eutreptia viridis**	65,523	86	27	28.6
	Eutreptiella gymnastica	67,623	87	8	34.3
	Pyramimonas parkeae	101,605	110	1	34.7

The "number of genes" indicates both protein and RNA coding but duplicates and ORFs were not included. The genome sizes marked by *asterisk* denote incomplete genomes. The first three columns marked taxonomy symbolize phylogenetic relationships between the organisms—merged rows represent defined clades; the Euglenales are coloured in *orange* and additionally divided into the Euglenaceae (*light orange*) and the Phacaceae (*light red*), the Eutreptiales are coloured in *yellow* and the prasinophytes are coloured in *purple*.

plastid establishment—22 genes (e.g. all genes of NADH-plastoquinone oxidoreductase of plastidal respiratory chain) were lost or transferred into the nucleus of the common ancestor of euglenophytes (Fig. 2). The highly conserved core content of euglenophyte cpDNAs consists of 89 genes (including *rrn5* not present in *P. parkeae*). These include 32 genes for photosynthetic proteins, 5 genes for transcription/translation proteins, 22 genes for ribosomal proteins, 3 rRNAs, and 27 tRNAs. One to four of these genes were not found in some lineages but their absence does not show any

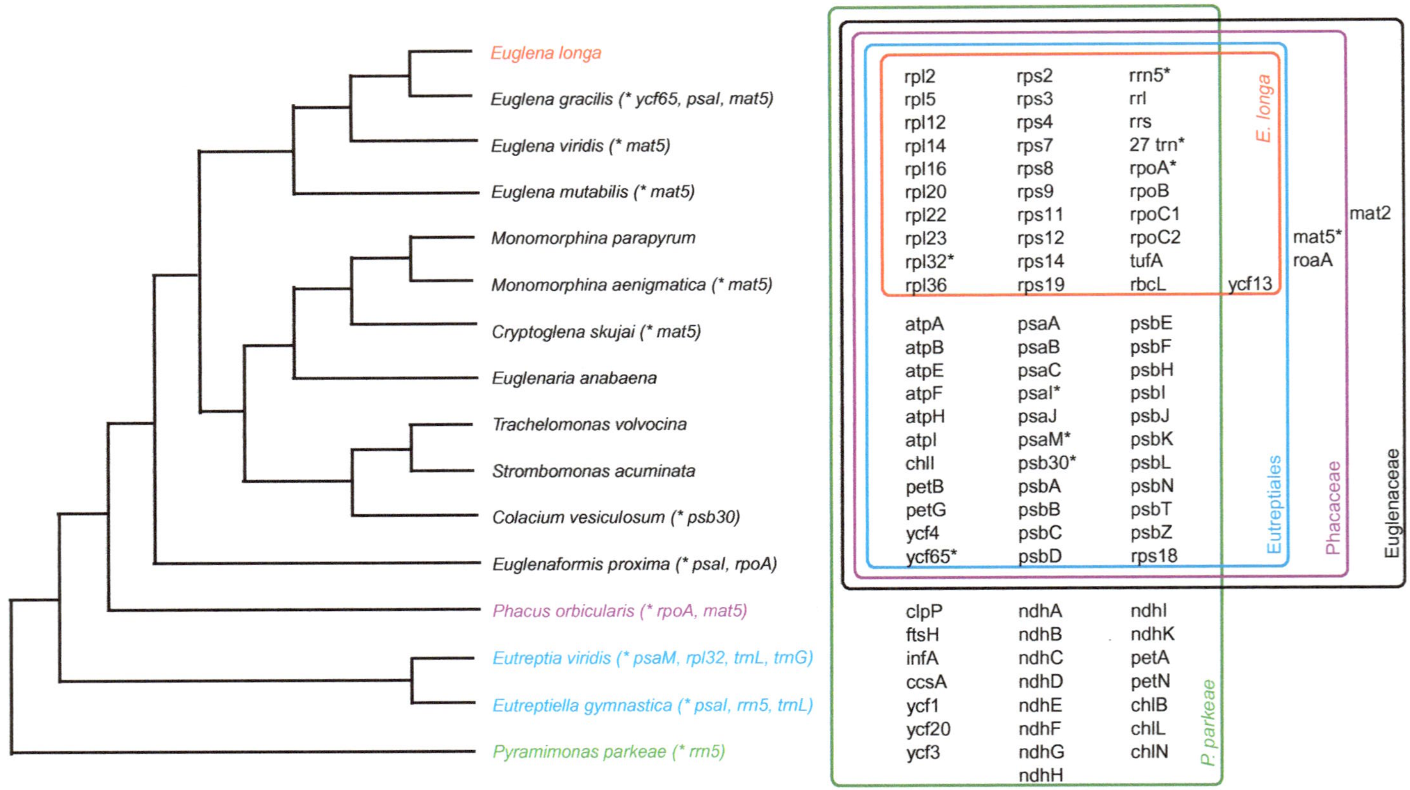

Fig. 2 Gene contents of the plastid genomes in 14 euglenophytes and *Pyraminonas parkeae*. *Asterisks* denote the genes missing in certain plastid genomes. Schematic phylogenetic tree is drawn according to Bennett and Triemer (2015).

phylogenetic pattern. Four genes were gained after the secondary endosymbiosis: the Eureptiales have obtained *ycf13* (synonymous to the intron-encoded maturase *mat1*), the Phacaceae possess three additional genes, *ycf13*, *mat5* (found within the *psbA* gene of *Lepocinclis spirogyroides*—GenBank record), and *roaA* (ribosomal operon-associated gene), and the Euglenaceae possess *mat2* in addition to the three aforementioned genes (Fig. 2; Bennett & Triemer, 2015). The *mat5* gene was lost in several independent instances within the Euglenales (Bennett & Triemer, 2015; Kasiborski et al., 2016). Unsurprisingly, genes for photosynthetic proteins have been lost in *Euglena longa* whose plastid has no photosynthetic activity. The gene contents are summarized in Fig. 2.

The plastid genome organization differs between *P. parkeae*, which contains two inverted repeats with ribosomal operon and large and small single copy region (Turmel et al., 2009), and most euglenophytes, which do not have this quadripartite arrangement. The cpDNA of *Eutreptiella gymnastica* represents a single exception as it contains two inverted repeats (one of them discontinuous), each with rRNA operon (Hrdá et al., 2012). The rRNA operon in *E. gracilis* and *E. longa* is organized in three tandemly repeated copies (Gockel & Hachtel, 2000; Hallick et al., 1993). The replication origins of euglenophyte cpDNAs are presumably located in the VNTR (variable number of tandem repeats) region (Koller & Delius, 1982; Ravel-Chapuis, Heizmann, & Nigon, 1982). Overall, the genes are arranged into 15 conserved gene clusters whose order and orientation have been excessively rearranged and it is hard to trace and reconstruct the course of these rearrangements (Dabbagh & Preisfeld, 2016).

The most noticeable trend in the euglenophyte plastid genome evolution is undoubtedly the rapid intron gain. The numbers of recognized introns in the genomes change over time as the annotations improve and so the numbers provided in Table 1 reflect the status of genome annotations published in the GenBank database as of July 2017. Remarkably, the cpDNA of *Pyramimonas* contains only a single intron like those of other prasinophyte algae such as *Ostreococcus tauri* and *Pycnococcus provasoli* (Turmel et al., 2009). The number of introns started to grow after the plastid acquisition by the ancestral euglenid before the split of Eutreptiales and Euglenales: cpDNAs of *Eutreptiella* and *Eutreptia* which represent the deepest splits from other euglenophytes already contain 8 and 27 introns, respectively (Hrdá et al., 2012; Wiegert et al., 2012). An intron of *psbC* gene, which carries an intron-encoded maturase *ycf13*, is considered the ancestral intron because it is the only homologous intron in euglenophyte cpDNAs (Bennett &

Triemer, 2015; Pombert et al., 2012). The major intron gain and amplification took place within the Euglenales lineage resulting in several species containing over 100 introns in their cpDNAs, the most extreme case being the model *E. gracilis* strain *Z* with at least 145 introns (Bennett & Triemer, 2015; Bennett et al., 2014; Hallick et al., 1993; Wiegert et al., 2013). The intron accumulation seems to be an ongoing process which is afoot at this very moment and it could be potentially very interesting to review some of the "old" sequences and see if they changed during the long-term cultivation by humans during a tiny part of the period of their sped-up evolution.

Not only are the euglenophyte plastid introns numerous, they are also unique in their structure. Majority of them are classified as group II introns, which are known from prokaryotes and mitochondria and plastids of eukaryotes. They function as self-splicing ribozymes and contain six stem-loop-forming domains and a conserved 5′-border motif (GUGYG). Group II introns are mobile elements and their mobility is mediated by maturases (Bonen & Vogel, 2001; Sheveleva & Hallick, 2004). The euglenophyte plastid group II introns are often significantly shorter (the average length of euglenophyte group II intron is 463 nt—approximately 100 nt shorter than the average group II intron of liverwort; Dabbagh & Preisfeld, 2016) with some of the conserved domains being missing and/or divergent beyond recognition. In addition to group II introns, euglenophyte plastid genomes contain group III introns which are exclusive to these organisms. They seem to be extremely derived form of group II introns which is much shorter (of average length around 100 nt) and lacks almost all core structures retaining only one modified conserved domain on 3′-end and a degenerate 5′-border motif (the consensus is NUNNG) (Bonen & Vogel, 2001; Doetsch, 2000; Doetsch, Favreau, Kuscuoglu, Thompson, & Hallick, 2001; Jenkins, Hong, & Hallick, 1995; Thompson, Copertino, Thompson, Favreau, & Hallick, 1995). To make matters even more complicated, euglenophyte plastid genomes also contain a number of twintrons—introns nested inside other introns (termed internal and external introns, respectively) which are spliced subsequently. Twintrons come in different types: group II, group III, and mixed or even complex ones where multiple internal introns (Copertino, Christopher, & Hallick, 1991; Copertino & Hallick, 1991, 1993) or even additional introns inside an internal intron (Drager & Hallick, 1993) can be observed in one external intron. In several cases, the recent conversion of a simple intron into a twintron is traceable on certain insertion sites. For example, six loci containing a twintron in the *E. gracilis* cpDNA contain only a single intron in *Monomorphina aenigmatica*.

In most of these cases, the introns of *M. aenigmatica* are orthologous to the external introns of their twintron counterparts in *E. gracilis* (Pombert et al., 2012). This further supports the notion that intron and twintron propagation in euglenophyte cpDNAs is a recent and probably still active process.

The relatively low GC content of the cpDNAs of euglenophytes is also linked to the intron accumulation since euglenid plastid introns are generally AT-rich and tend to bias the overall GC content of the whole genome. This is especially apparent in the case of *E. longa* whose cpDNA has the lowest number of genes while still being rich in introns, as a result its GC content is extremely low (22.4%) (Gockel & Hachtel, 2000).

5. PLASTID BIOGENESIS AND HOUSEKEEPING: HOW TO MAKE IT AND HOW TO CONTROL IT

The crucial part of organellogenesis is the ability of the host cell to take control over the replication and biogenesis of the formerly independent endosymbiotic hostage. The development of plastids of euglenophytes was studied in detail in the model species *E. gracilis* and the following subchapter will be based mainly on the findings regarding this particular organism.

Euglena plastids develop from proplastids which are nascent or dark-induced regressed stages with low or no photosynthetic activity, small number of thylakoids, and low amount of photosynthetic pigments stored mostly in their precursor forms (protochlorophyll and protochlorophyllide).

The plastid development is triggered by light and influenced by other environmental stimuli, especially availability of organic carbon sources which act as catabolic repressors that modulate the metabolic mode of the cell towards the more efficient one in the given situation. Glucose, ethanol, and acetate were shown to inhibit chlorophyll synthesis, while malate and succinate do not seem to function as catabolic repressors. Ethanol was also shown to inhibit synthesis or activity of various other proteins of carbon fixation pathway, light-harvesting antennae, and plastid housekeeping system (Horrum & Schwartzbach, 1980; Monroy & Schwartzbach, 1984; Schwartzbach, 2017). As a result, the cell performs the energetically expensive switch to the autotrophic nutritional mode and develops mature plastids only when carbon sources consumable by the relatively energetically cheap glyoxylate pathway become unavailable.

The DNA replication in plastids can take place independently of the nuclear DNA replication. However, the process of plastid greening and development is directed by the nucleus on the transcriptional and translational level. The proteosynthetic apparatus of the developing plastids receives an external boost as its nucleus-encoded components (ribosomal proteins and initiation and elongation factors) are upregulated by light, synthesized in the cytoplasm, and transported into proplastids as well as other nucleus-encoded plastid proteins (Bingham & Schiff, 1979; Bingham & Shiff, 1979; Egan & Carell, 1972; Fox et al., 1980; Kraus & Spremulli, 1988). The amounts of rRNA, tRNA, and other nucleic acids which are fairly low in inactive proplastids increase up to threefold after illumination (Egan & Carell, 1972). The plastid genes are expressed constitutively regardless of light or dark conditions and the light increases the transcriptional activity but does not change the composition and ratio of the expressed genes (Geimer et al., 2009).

Protochlorophyll and protochlorophyllide are phototransformable molecules which are believed to act as photoreceptors on the thylakoid membranes of the developing plastids (perhaps in cooperation with a yet unspecified photoreceptor specialized to blue light sensing). These precursors are converted to chlorophyll and play a role in the initiation of de novo chlorophyll synthesis (Egan, Dorsky, & Schiff, 1975; Kirk, 1970; Stern, Epstein, & Schiff, 1964; Stern, Schiff, & Epstein, 1964).

The proplastids start to grow in size and new thylakoids are formed by the invagination of the innermost membrane and eventually start fusing and stacking into lamellae (Ben-Shaul et al., 1966). Thylakoid formation and chlorophyll synthesis are mutually dependent so the time dynamics of these two processes are concurrent. This development starts soon after the induction by light but it progresses very slowly during the initial lag phase and sets off rapidly after approximately 6 h. The length of the lag can be, however, significantly modified by the adjustment of the light conditions, namely preillumination. While the concentration of chlorophyll grows steadily, the concentration of carotenoids remains more or less the same—the final ratio between these two types of pigments is ca. 2.5:1 in the mature plastids (Stern, Schiff, et al., 1964). The interconnected carbon fixation and oxygen production start several hours after the plastid differentiation induction and grow rapidly after approximately 10 h. Around this time, the plastids swell quickly gaining roughly three times the original mass and thylakoid amount. It was noticed that the number of initial proplastids in *E. gracilis* is usually around 30 per cell while the final number of mature plastids is around 10. The logical deduction is that the proplastids actually form aggregates of three and fuse around this time off the differentiation causing a significant leap in the plastids size and metabolic capacity. The process of the plastid maturation is completed in about 72 h (Ben-Shaul et al., 1966).

A vast majority of proteins which are synthesized and inserted into thylakoids during this process are coded in the nucleus and translated on cytoplasmic ribosomes—the gene expression is almost completely under the control of the host cell (Bingham & Shiff, 1979), only a small number of proteins are synthesized in the plastid (Schwartzbach & Schiff, 1974). These gene products are imported into plastids based on their rather complex-targeting signals.

The plastid protein import mechanism is much more complicated in euglenophytes in comparison to plants and is far from clearly resolved to this date. A schematic of the protein transport into euglenophyte plastid is shown in Fig. 3. Plastid-targeted precursor proteins (preproteins) coded in nucleus generally have an N-terminal signal peptide which is very similar to the classical signal peptide of proteins destined to the secretory pathway. The signal peptide is supposedly recognized by the signal recognition particle right after its translation and protrusion from the cytoplasmic ribosome and the rest of the translation takes place on the rough ER and the nascent preprotein is cotranslationally imported into the ER lumen. Subsequently, the signal peptide is cleaved by a signal peptidase in the ER revealing the following part of the signal whose amino acid composition is similar to that of a canonical plant chloroplast-targeting signal, the so-called transit peptide: it is rich in serine, threonine, and alanine; depleted in aspartic and glutamic acid; and has a slight positive charge as a result. The length of the transit peptides is quite variable and ranges from 36 to 135 amino acid residues in *E. gracilis* (Durnford & Gray, 2006). The origin of these transit peptides could be related to *cis*- and *trans*-splicing of short introns which were identified in some of these sequences (Vesteg et al., 2010). In euglenophytes, nucleus-encoded plastid preproteins can be divided into two classes based on the presence or absence of the third, highly hydrophobic domain which follows immediately after the transit peptide (Durnford & Gray, 2006) and acts as a stop-transfer transmembrane anchor. Preproteins are transported from the ER to the Golgi apparatus and they are either packed into a transport vesicle (class II) or anchored in its membrane (class I) with the transit peptide inside and the rest of the protein protruding into the cytoplasm (Durnford & Gray, 2006; Sulli, Fang, Muchhal, & Schwartzbach, 1999; Sulli & Schwartzbach, 1996). Vesicles from Golgi fuse with the outermost plastid membrane by a yet unknown mechanism. This process is resistant to *N*-ethylmaleimide that interacts with the *N*-ethylmaleimide-sensitive factor (NSF) and inhibits

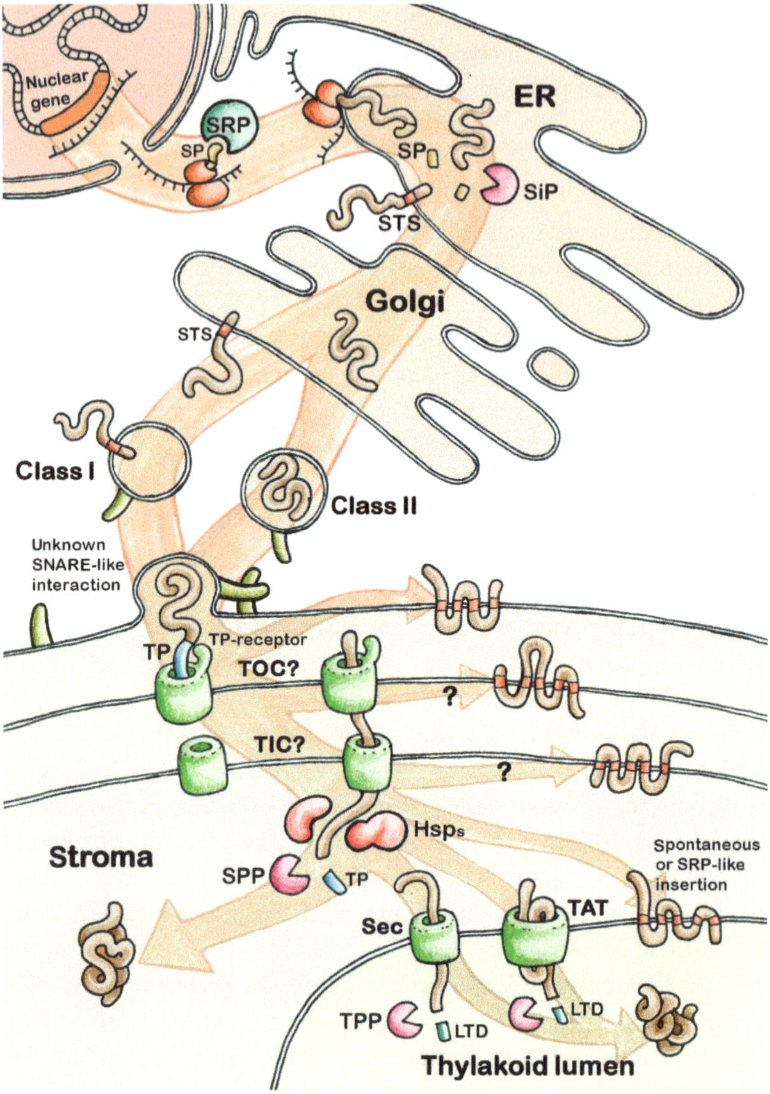

Fig. 3 Schematic of the plastid protein import pathway in euglenophytes. Nuclear-encoded plastid-targeted proteins are cotranslationally transported into the ER by the same mechanism as proteins destined into the secretory pathway based on their signal peptide (SP). They are either completely inserted into the ER lumen or remain anchored in its membrane by a hydrophobic stop-transfer signal (STS). Their SPs are cleaved by the signal peptidase (SiP) in the ER lumen. They pass through Golgi and are loaded onto vesicles which then fuse with the outermost plastid membrane via an unknown, SNARE-independent mechanism. Proteins destined to plastid stroma or thylakoids pass the remaining two membranes via transit peptide (TP)-dependent process which is probably facilitated by plant TIC- and TOC-like complexes. In plastid

soluble NSF attachment protein receptor (SNARE)-dependent membrane fusion. This suggests that SNAREs are not utilized in the vesicular transport of plastid preproteins (Slávíková et al., 2005). After reaching the inter-membrane space between the outermost and the middle plastid membrane, the transit peptide is apparently recognized by a yet unknown receptor and the preprotein is pulled through the inner two membranes, most probably by a system homologous to the TOC/TIC (translocons at the outer/inner envelope membrane of chloroplasts) machinery which is present in primary plastids of plants (Shi & Theg, 2013) and generally conserved (albeit reduced in the number of subunits) among other secondary plastids (Maier et al., 2015; Sheiner & Striepen, 2013). The transit peptides of euglenid plastid proteins were repeatedly demonstrated to function in heterologous system and successfully direct the protein into plant plastid both in vitro and in vivo (Inagaki, Fujita, Hase, & Yamamoto, 2000; Shashidhara, Limsli, Shackletonii, Robinsonii, & Smith, 1992; Slávíková et al., 2005). These results suggest the presence of a TOC/TIC-like machinery in euglenophyte plastids. However, no direct proof of existence of these protein complexes in euglenophytes was brought to this date.

The transit peptide is presumably cleaved by stromal processing peptidase and the protein assumes its native conformation with the help of putative translocase-associated and stromal chaperones (Maier et al., 2015; Shi & Theg, 2013). Proteins destined to the thylakoid membrane or lumen are supposedly further sorted based on their additional signals (N-terminal, C-terminal, or internal) and inserted into thylakoids by specialized import pathways known from primary plastids such as the Sec-dependent pathway or the twin-arginine translocation pathway (Jarvis & Robinson, 2004).

The transcription of genes encoded in the plastid genome was observed to take place constitutively with practically all genes including several unknown ORFs and pseudogenes being expressed and present as mRNAs in the plastid, albeit in very low concentration. The transcripts sometimes form very large and relatively stable polycistronic units which are subsequently subjected to extensive posttranscriptional modifications. The plastid

---

stroma, the TPs are cleaved by the stromal processing peptidase (SPP) and the proteins are sorted into their final localization (stroma, thylakoid membrane, or thylakoid lumen) and folded to their mature conformation. Proteins destined to thylakoid lumen are transported based on their lumen-targeting domains (LTD) which are then cleaved by the thylakoid-processing peptidase (TPP). Proteins of the plastid envelope are inserted into their target membrane via an unknown mechanism or spontaneously.

gene expression does not react to environmental stimuli in terms of quality but rather on quantitative level. Moreover, some genes seem to have very short turnover and to be expressed only to be degraded very quickly. In summary, regulation of the gene expression in euglenophyte plastid might be taking place predominantly on the level of posttranscriptional RNA processing and/or translation rather than transcription (Geimer et al., 2009; Thompson et al., 1995). A small portion of plastid transcripts was shown to have polyadenylated 3′-ends in *E. gracilis*. The low ratios of polyadenylated vs nonpolyadenylated RNAs (only 1/350 to 1/100,000) suggest that the polyadenylation serves as a signal for exonucleolytic degradation as it does in the case of primary plastids (Záhonová et al., 2014).

6. PLASTID METABOLISM: A FACTORY WITH REDUNDANT PRODUCTION LINES

In this chapter, findings on biosynthetic pathways in euglenophyte plastid will be briefly summarized with focus on their potentially unique characteristics and differences from other groups which are interesting from evolutionary point of view. Almost all studies cited in this section were performed on the model euglenophyte *E. gracilis*, and generalizations based on them should be taken with a grain of salt.

Photosynthetic apparatus of the *E. gracilis* plastid is comparable to that of green algae and plants in regard to the function and architecture. Photosynthesis is a subject of tight regulation by light (Yoshida et al., 2016) and it reacts negatively to both extremes in light conditions as the photosynthetic activity is inhibited or even completely compromised by both darkness and the excessive illumination by visible or UV light (Richter, Helbling, Streb, & Häder, 2007). Additionally, the photosynthetic capacity of *E. gracilis* is affected by salinity (Gonzalez-Moreno, Gomez-Barrera, Perales, & Moreno-Sanchez, 1997) and other ion concentrations (Ferroni, Baldisserotto, Fasulo, Pagnoni, & Pancaldi, 2004; Krajčovič et al., 2015).

E. gracilis synthesizes chlorophylls *a* and *b* via the standard plastid-localized pathway from aminolevulinic acid. There are several spectroscopically distinguishable precursor forms of chlorophylls (i.e. protochlorophyll, protochlorophyllide, and phytyl-protochlorophyllide), and developing plastids especially contain these precursors to a great extent, and their ratio, dynamics, and interconversions are believed to play a major role in regulation of biosynthesis of chlorophyll and possibly other compounds (Kirk, 1970).

For example, protochlorophyllide has been shown to inhibit further steps in chlorophyll biosynthesis until its light-induced conversion (Schwartzbach, Schiff, & Goldstein, 1975).

Aminolevulinic acid can be synthesized in two ways: through the C5 pathway from glutamate or through the C4 (Shemin) pathway from glycine and succinyl-CoA; most organisms utilize only one of these pathways. In *E. gracilis*, aminolevulinic acid was proven to be synthesized via the C5 pathway and for some time the organism was thought to lack the C4 pathway (Gomez-Silva, Timko, & Schiff, 1985). However, the situation was recently shown to be more complex as two pathways for the synthesis of aminolevulinic acid followed by the synthesis of haem have been predicted in *Euglena*. One pathway is predicted to be partially localized in either the cytosol or mitochondrion to produce haem for mitochondria. This pathway utilizes aminolevulinic acid synthesized in the C4 pathway. Another pathway is localized in the plastid to produce chlorophyll and it utilizes aminolevulinic acid synthesized in the C5 pathway (Kořený & Oborník, 2011).

Carotenoid synthesis seems to be linked to chlorophyll production and it was even hypothesized that one could be synthesized from another due to the observed reciprocal proportion of these two types of pigments (Wolken & Mellon, 1956). The regulation of biosynthesis of carotenoids and chlorophyll indeed seems to be tuned and subjected to reciprocal feedback but the exact mechanism of it remains unclear. Major *E. gracilis* carotenoids are antheraxanthin (more than 80%), β-carotene (11%), and neoxanthin (7%), which represent 99% of total carotenoids. Minor carotenoids include cryptoxanthin, γ-carotene, ζ-carotene, echinenone, hydroxyechinenone, and its derivate canthaxanthin (also termed euglenanone in older literature as it was believed to be unique to euglenids) (Krinsky & Goldsmith, 1960). The first metabolite in carotenoid synthesis pathway is geranylgeranyl diphosphate (GGPP) which is synthesized in putatively plastidal methylerythritol phosphate/deoxyxylulose phosphate (MEP/DOXP) pathway—as opposed to the GGPP synthesized in mitochondrial mevalonate pathway which is used for the synthesis of phytols and sterols in *E. gracilis*. Therefore, there are two distinct GGPP pools in *E. gracilis* and carotenoids and phytols are synthesized by independent pathways (Kim, Filtz, & Proteau, 2004).

A variety of terpenoids are synthesized in the *E. gracilis* plastid: plastoquinone, α-tocopherol, α-tocopherolquinone, phytylquinone, nonaprenyl, octaprenyl, nonaprenyl toluquinone, octaprenyl toluquinone, and phytyl pyrophosphate (Griffiths, Threlfall, & Goodwin, 1967; Thomas & Threlfall, 1974). Their production is not directly linked to photosynthesis but it is light dependent; terpenoid formation is mostly inhibited in dark-grown cultures as opposed to ubiquinone which is synthesized in mitochondrion and whose production level does not correlate with different light conditions (Griffiths et al., 1967). Polyprenylation/phytylation or nonoxidative decarboxylation of homogentisate were proposed as a mechanism of this reaction (Thomas & Threlfall, 1974). Tocopherols (compounds collectively termed as vitamin E) are present in all *E. gracilis* cell fractions including plastids where 97% of their bulk is represented by α-tocopherol which is thought to be synthesized in situ from other tocopherols which are transported from outside the plastid (Shigeoka, Onishi, Nakano, & Kitaoka, 1986). This compound functions as a protective agent against reactive oxygen species and its synthesis modulation by various substrates and conditions was studied in the view of potential biotechnological or pharmaceutical application (Fujita, Aoyagi, Ogbonna, & Tanaka, 2008; Fujita, Ogbonna, Tanaka, & Aoyagi, 2009). Ascorbic acid (vitamin C) is also an important antioxidative protectant. Remarkably, in *E. gracilis* it is synthesized by the alternative terminal enzyme, L-galactonolactone dehydrogenase, instead of L-gulonolactone oxidase which was lost multiple times and then replaced by an alternative enzyme in multiple phototrophic lineages (Wheeler, Ishikawa, Pornsaksit, & Smirnoff, 2015).

E. gracilis also synthesizes galactolipids (monogalactosyl and digalactosyl diglycerides) and sulpholipids (mainly sulphoquinovosyl diglyceride) in its plastid (Davies, Mercer, & Goodwin, 1966; Rosenberg & Gouaux, 1967; Rosenberg, Gouaux, & Milch, 1966). Production of these lipids is linked to the plastid development and thylakoid membrane amplification and is modulated by light conditions—much higher amounts of these compounds can be found in photosynthetically active cells in comparison to dark-grown or bleached ones (Matson, Fei, & Chang, 1970). Interestingly and in contrast to plants, monogalactosyl and digalactosyl diglycerides are probably synthesized by two different enzymes in *E. gracilis* (Blee & Schantz, 1978). Additionally, the conversion from monogalactosyl to digalactosyl diglycerides is not possible, and the ratio of these galactolipids is significantly skewed towards digalactosyl diglycerides which are synthesized preferentially (Matson et al., 1970). These lipids are essential for the thylakoid assembly and growth, and some have been shown to be miscible with chlorophyll and phytol, major nonprotein components of the thylakoid membranes (Liljenberg & Selstam, 1980).

The major storage compound of euglenids is paramylon, a starch-like polysaccharide, which forms granules of various morphologies in the cytosol and can take up to 90% of the dry cell weight. The paramylon consists of an unbranched, water-insoluble β-1,3-glucan which is rather unusual since most eukaryotes generally synthesize α-glucans (or branched β-1,4- or β-1,6-glucans in case of fungi and some grains) (Barras & Stone, 1968; Barsanti, Vismara, Passarelli, & Gualtieri, 2001; Dwyer & Smillie, 1971; Monfils, Triemer, & Bellairs, 2011; Šantek, Felski, Friehs, Lotz, & Flaschel, 2010). The paramylon biogenesis occurs in all euglenids including nonphotosynthetic species suggesting that it is not evolutionarily related to plastid acquisition. In heterotrophic euglenids, the paramylon is synthesized in mitochondrial prominences and mitochondrion-derived vesicles by the gluconeogenesis pathway (Calvayrac & Briand, 1978). In photosynthetic euglenophytes, the paramylon can be synthesized by both the original mitochondrial pathway and during carbon fixation in pyrenoids of plastids. The close association of plastid and mitochondrion and the eccentrical flow of vesicles derived presumably from both organelles have been observed during the paramylon biogenesis (Calvayrac, Laval-Martin, Briand, & Farineau, 1981). This functional connection of the two organelles is remarkable and can be a potential source of unusual molecular phenomena regarding exchange of metabolites and possibly other compounds. The other major storage compounds of euglenids are wax esters which are not synthesized in plastid (Inui, Miyatake, Nakano, & Kitaoka, 1982; Koritala, 1989; Schneider & Betz, 1985) but whose production is linked to the production of paramylon, since interconversions between these two storage molecules occur regularly and their balance shifts in reaction to light and oxygen availability: in aerobic conditions, the cell preferentially accumulates paramylon; in anaerobic conditions, the wax ester accumulation is more significant (Barras & Stone, 1968; Inui et al., 1982). Both of these compounds are potentially usable in biotechnology, agriculture, nutrition, and even cancer prevention (Karaca et al., 2014; Krajčovič et al., 2015; Kuda, Enomoto, & Yano, 2009; Rodríguez-Zavala, Ortiz-Cruz, Mendoza-Hernández, & Moreno-Sánchez, 2010; Sugiyama et al., 2009; Watanabe et al., 2013).

7. SECONDARY OSMOTROPHY AND PLASTID BLEACHING: PLASTIDS THAT FORGOT HOW TO PLASTID

Secondarily osmotrophic euglenids represent euglenophytes which have lost their photosynthetic pigments and the ability to perform photosynthesis. There are at least five species of euglenophytes that have lost the photosynthesis independently (Marin, 2004; Marin, Palm, Klingberg, & Melkonian, 2003). It is a matter of a debate whether some of these euglenophytes such as *Euglena quartana* (previously *Khawkinea quartana*) or *Phacus ocellatus* (previously *Hyalophacus ocellatus*) have lost the plastid compartments completely or whether they still contain residual plastids with genomes (Marin, 2004). However, the best-studied secondary osmotroph is *E. longa*, a close relative of photosynthetic *E. gracilis* (Mullner, Angeler, Samuel, Linton, & Triemer, 2001), which contains a plastid with a genome (Gockel & Hachtel, 2000).

The circular 73kb plastid genome of *E. longa* is about half the size of a plastid genome of *E. gracilis* and it carries 57 protein-coding genes (Gockel & Hachtel, 2000)—housekeeping genes responsible for transcription and translation (*rrn*, *rpl*, *rps*, *rpo*, and *tuf* genes), tRNA genes, and several ORFs and genes with unknown function (*orf* and *ycf* genes). Transcripts and proteins of various genes, including Rubisco large subunit gene (*rbc*L), have been found suggesting that the *E. longa* plastid possesses functional transcription and translation machinery (Gockel & Hachtel, 2000; Sheveleva & Hallick, 2004; Záhonová et al., 2016). It was recently shown that an intact plastid genome is essential for *E. longa* growth (Hadariová, Vesteg, Birčák, Schwartzbach, & Krajčovič, 2017) but it remains a mystery which gene(s) is/are indispensable for its survival (Gockel & Hachtel, 2000; Hadariová et al., 2017). The photosynthesis-related genes (photosystems I and II, cytochrome *b6f* complex, ATP-synthase) have disappeared from the *E. longa* plastid genome with the exception of *rbc*L (Gockel & Hachtel, 2000). Both *rbc*L and nucleus-encoded *Rbc*S genes are translated but their abundance in cells is very low. Protein sequences of *E. longa rbc*L and *Rbc*S are highly divergent compared to their homologues in the photosynthetic relatives, suggesting that the Rubisco enzyme of *E. longa* probably has an unusual function, if any (Záhonová et al., 2016). Some other nonphotosynthetic organisms retain *rbcL* genes in their plastid genomes and it is hypothesized that these RbcL proteins may either act as oxygenases, be involved in glycine and serine biosynthesis, be required for an alternative lipid biosynthesis pathway (Schwender, Goffman, Ohlrogge, & Shachar-Hill, 2004), or perform another unidentified function (Sanchez-Puerta, Lippmeier, Apt, & Delwiche, 2007; Wolfe & dePamphilis, 1998; Záhonová et al., 2016).

E. longa lacks the eyespot and the paraflagellar swelling, but apart from that, the cell is indistinguishable from bleached mutants of *E. gracilis* under light microscopy, and thus it was formerly viewed as a naturally bleached form *of E. gracilis* (Bodył, 1996). Bleaching of *E. gracilis* is an irreversible process inducible by antibiotics, UV light, high pressure, heat, mutagens, or carcinogens (Krajčovič, Ebringer, & Schwartzbach, 2001). In *E. gracilis*, various morphological and ultrastructural changes can be observed during bleaching. During the switching from autotrophy to heterotrophy, the plastids and plastid DNA are degraded and the colour of the culture changes from green to white (or slightly orange or pink depending on the presence of carotenoids). The bleaching level depends on several factors—pH, content of phosphate in the medium, or the age of the culture. It is possible to reach 100% plastid bleaching in some cases (Cook, Harris, & Nachtwey, 1974; Krajčovič, Ebringer, & Polónyi, 1989). Spontaneous bleaching in *E. gracilis* was observed as early as in 1912 by Ternetz and streptomycin was the first defined agent for controlled bleaching of all cells in a population. In 1961, Ebringer widened a list of bleaching agents by adding erythromycin and some other macrolide antibiotics (Krajčovič et al., 2001).

There are various types of antibacterial drugs with different modes of action. Some of them act as inhibitors of bacterial DNA synthesis (e.g. mitomycin or anthramycin) while the ones with aminohexose molecular structure function as inhibitors of bacterial as well as plastidal protein synthesis (e.g. kanamycin, pactamycin, or neomycin). Their irreversible effect on *E. gracilis* was observed in the past (Ebringer, 1972). Perhaps the most effective eliminators of *E. gracilis* plastids are quinolone antibiotics (inhibitors of bacterial DNA gyrase), especially their new derivatives—fluoroquinolones (Krajčovič et al., 1989). Various ultrastructural and plastid DNA changes occur when different bleaching agents are used (Krajčovič et al., 2001).

For example, the quinolones, but also nitrosoguanidine and furylfuramide, reduce the number of thylakoids and destroy plastids (Polónyi, Ebringer, Krajcovic, & Kapeller, 1990). The plastid DNA degradation has been observed in all bleached *E. gracilis* cultures. Conkling, Thomas, and Ortiz (1993) noticed a gradual loss of cpDNA and the study of Krajčovič et al. (1999) has demonstrated a structural rearrangement of cpDNA in *E. gracilis* cells bleached by elevated temperature. Recently, it was shown that the plastid gene *rpl*16 invariably persists in *E. gracilis* strain *Z* bleached in the long-term by ofloxacin (DNA gyrase inhibitor) or streptomycin (bacterial protein synthesis inhibitor) (Hadariová et al., 2017), which suggested that this gene was retained presumably due to its position on the cpDNA being close to the replication origin (Hadariová et al., 2017). A different gene, 16S rRNA, has been retained in *E. gracilis* FACHB47 strain treated with the same antibiotics (Wang, Shi, & Xu, 2004). It has been reported that plastid transcripts are reduced after treating by various bleaching agents—UV light, temperature, and antibiotics (Geimer et al., 2009).

In the past decades, several stable white mutants derived from *E. gracilis* have been characterized (Polónyi, Ebringer, Dobias, & Krajčovič, 1998; Schiff, Lyman, & Russell, 1971). W_3BUL mutant was induced with UV irradiation of *E. gracilis* var. *bacillaris* strain (Schiff et al., 1971). It contains plastid remnants (Heizmann, Salvador, & Nigon, 1976; Osafune & Schiff, 1980a; Osafune, Schiff, & Hase, 1987; Parthier & Neumann, 1977) and a specific type of sulpholipid in its thylakoid membranes (Saidha & Schiff, 1989). The treatment of *E. gracilis* var. *bacillaris* with streptomycin produced another mutant strain, $W_{10}BSmL$. In contrast to W_3BUL, $W_{10}BSmL$ contains neither the eyespot (Osafune & Schiff, 1980b), carotenoids (Fong & Schiff, 1979), and sulpholipids (Saidha & Schiff, 1989), nor the plastid residues (Osafune & Schiff, 1983). The $W_{gm}ZOflL$ mutant was derived from *E. gracilis* strain *Z* by treatment with an ofloxacin derivative (Polónyi et al., 1998). The common features of $W_{gm}ZOflL$ and W_3BUL mutants include the presence of the eyespot and carotenoids. The giant mitochondria were described in $W_{gm}ZOflL$ in the past (Polónyi et al., 1998) but current microscopic observations suggest that $W_{gm}ZOflL$ mutant does not possess mitochondria of such form anymore (personal experiences). The level of the plastid genome degradation, i.e., plastid genes presence/absence of the nonphotosynthetic mutants of *E. gracilis*, is still unknown but they probably lack most of the, if not all, plastid genes (unpublished data).

The bleaching of *E. gracilis* induced by fluoroquinolone ofloxacin has been shown to be useful in the investigation of potential antimutagens such as flavonoids (Križková, Nagy, Polónyi, & Ebringer, 1998) and antioxidants as ascorbic acid, sodium selenite, and many others (Ebringer et al., 1996; Kogan et al., 2004; Križková, Ďuračková, Šandula, Sasinková, & Krajčovič, 2001; Križková, Mučaji, Nagy, & Krajčovič, 2004; Križková et al., 2006). Ciprofloxacin, besides the effective bleaching of *E. gracilis*, also causes a suppression of the rudimentary plastid (apicoplast) genome replication in a parasite *Toxoplasma gondii* (Roos & Fichera, 1997). Various other quinolones have been demonstrated as useful inhibitors of a malaria-causing apicomplexan *Plasmodium falciparum* (Mahmoudi et al., 2003). Thus, *E. gracilis* plastid bleaching may also have practical implications. The research of the plastid genome degradation in bleached *E. gracilis* may represent a suitable tool for the development of new drugs harmless to humans and successful in fighting apicomplexan parasites that cause millions of fatal disease cases in the third-world countries annually (Krajčovič et al., 2001).

8. CONCLUSIONS

Euglenophytes are a group of phototrophic protists which have been discovered as early as at the end of 17th century. During the last century, their biochemistry and ultrastructure were inspected to a great detail. However, there are still many missing pieces in the puzzle of their molecular biology, genetics, and phylogenetics as some of the methods crucial for these disciplines became widely available only recently and others are still quite limited because of the missing genomic data and insufficient means for genetic transformation. These hurdles are, however, very much worth overcoming because photosynthetic euglenids are potentially very useful models for both applied and basic research. From the applied point of view, they represent easily and environment-friendly cultivated microalgae capable of synthesizing various compounds usable in the production of biofuels and other lipid-based technological materials as well as nutritional and pharmaceutical products. From the basic point of view, they are one of the key groups to understand secondary endosymbioses, establishment and reduction of organelles and evolutionary processes taking part therein. Findings concerning euglenid molecular biology, genetics, and genomics can help to illuminate the evolution of excavates and, by extension, of eukaryotes as whole.

ACKNOWLEDGEMENTS

The authors wish to acknowledge professor Bozena Zakryś who is the author of the drawings used in Fig. 1 and who kindly approved their usage in this publication.

Support for the authors' salaries and stipend came from the project of the Ministry of Education, Youth and Sports of CR within the National Sustainability Program II (Project BIOCEV-FAR) LQ1604 and by the project "BIOCEV" (CZ.1.05/1.1.00/02.0109).

REFERENCES

Adl, S. M., Simpson, A. G. B., Lane, C. E., Lukeš, J., Bass, D., Bowser, S. S., et al. (2012). The revised classification of eukaryotes. *The Journal of Eukaryotic Microbiology*, *59*(5), 429–493. http://dx.doi.org/10.1111/j.1550-7408.2012.00644.x.

Andersson, J. O., & Roger, A. J. (2002). A cyanobacterial gene in nonphotosynthetic protists—An early chloroplast acquisition in eukaryotes? *Current Biology: CB*, *12*(2), 115–119. Retrieved from http://www.ncbi.nlm.nih.gov/pubmed/11818061.

Barras, D. R., & Stone, B. A. (1968). Carbohydrate composition and metabolism in Euglena. In D. E. Buetow (Ed.), *The biology of Euglena: Vol. II. Biochemistry* (149–191). NewYork: Academic Press.

Barsanti, L., Vismara, R., Passarelli, V., & Gualtieri, P. (2001). Paramylon (β-1,3-glucan) content in wild type and WZSL mutant of Euglena gracilis. Effects of growth conditions. *Journal of Applied Phycology, 13*(1), 59–65. http://dx.doi.org/10.1023/A:1008105416065.

Benedetti, A. P., & Checcucci, A. (1975). Paraflagellar body (PFB) pigments studied by fluorescence microscopy in Euglena gracilis. *Plant Science Letters, 4*, 47–51.

Bennet, M., Wiegert, K., & Triemer, R. (2012). Comparative chloroplast genomics between Euglena viridis and Euglena gracilis (Euglenophyta). *Phycologia, 51*(6), 711–718. http://dx.doi.org/10.2216/12-017.1.

Bennett, M. S., & Triemer, R. E. (2015). Chloroplast genome evolution in the Euglenaceae. *Journal of Eukaryotic Microbiology, 62*(6), 773–785. http://dx.doi.org/10.1111/jeu.12235.

Bennett, M. S., Wiegert, K. E., & Triemer, R. E. (2014). Characterization of Euglenaformis gen. nov. and the chloroplast genome of Euglenaformis [Euglena] proxima (Euglenophyta). *Phycologia, 53*(1), 66–73. http://dx.doi.org/10.2216/13-198.1.

Ben-Shaul, Y., Schiff, J. A., & Epstein, H. T. (1966). Studies of chloroplast development in euglena: VII. Fine structure of the developing plastid 1. *Biophysical Journal, 6*(4), 373–383. http://dx.doi.org/10.1038/185825a0.

Bingham, S., & Schiff, J. A. (1979). Events surrounding the early development of Euglena chloroplasts. 15. Origin of plastid thylakoid polypeptides in wild-type and mutant cells. *Biochimica et Biophysica Acta (BBA)—Bioenergetics, 547*(3), 512–530.

Bingham, S., & Shiff, J. A. (1979). Events surrounding the early development of Euglena chloroplasts. 16. Plastid thylakoid polypeptides during greening. *Biochimica et Biophysica Acta (BBA)—Bioenergetics, 547*(3), 531–543.

Blee, E., & Schantz, R. (1978). Biosynthesis of galactolipids in Euglena gracilis: I, incorporation of UDP galactose into galactosyldiglycerides. *Plant Science Letters, 13*, 247–255.

Bodył, A. (1996). Is the origin of Astasia longa an example of the inheritance of acquired characteristics? *Acta Protozoologica, 35*, 87–94, Nencki Institute of Experimental Biology.

Bodył, A., Mackiewicz, P., & Milanowski, R. (2010). Did trypanosomatid parasites contain a eukaryotic alga-derived plastid in their evolutionary past? *Journal of Parasitology, 96*(2), 465–475. http://dx.doi.org/10.1645/Ge-1810.1.

Bolte, K., Bullmann, L., Hempel, F., Bozarth, A., Zauner, S., & Maier, U.-G. (2009). Protein targeting into secondary plastids. *The Journal of Eukaryotic Microbiology, 56*(1), 9–15. http://dx.doi.org/10.1111/j.1550-7408.2008.00370.x.

Bonen, L., & Vogel, J. (2001). The ins and outs of group II introns. *Trends in Genetics, 17*(6), 322–331. http://dx.doi.org/10.1016/S0168-9525(01)02324-1.

Calvayrac, R., & Briand, J. (1978). Paramylon synthesis and the chondriome of Euglena gracilis Z. In G. Ducet & C. Lance (Eds.), *Plant mitochondria* (pp. 435–443). Amsterdam: Elsevier/North Holland.

Calvayrac, R., Laval-Martin, D., Briand, J., & Farineau, J. (1981). Paramylon synthesis by Euglena gracilis photoheterotrophically grown under low O2 pressure. *Planta, 153*(1), 6–13. http://dx.doi.org/10.1007/BF00385311.

Cavalier-Smith, T. (1981). Eukaryote kingdoms: Seven or nine? *Biosystems, 14*(3–4), 461–481. http://dx.doi.org/10.1016/0303-2647(81)90050-2.

Cavalier-Smith, T. (2016). Higher classification and phylogeny of Euglenozoa. *European Journal of Protistology, 56*, 250–276. http://dx.doi.org/10.1016/j.ejop.2016.09.003.

Ciugulea, I., & Triemer, R. E. (2010). *A color atlas of photosynthetic euglenoids*. East Lansing, MI: Michigan State University Press. Retrieved from http://library.wur.nl/WebQuery/clc/1953618.

Conkling, B. A., Thomas, E. J., & Ortiz, W. (1993). Delayed but complete loss of chloroplast DNA in heat-bleaching cultures of Euglena gracilis. *Journal of Plant Physiology, 142*(3), 307–311. http://dx.doi.org/10.1016/S0176-1617(11)80427-X.

Cook, J. R., Harris, P., & Nachtwey, D. S. (1974). Irreversible plastid loss in Euglena gracilis under physiological conditions. *Plant Physiology, 53*(2), 284–290. http://dx.doi.org/10.1104/PP.53.2.284.

Copertino, D. W., Christopher, D. A., & Hallick, R. B. (1991). A mixed group-II/group-III twintron in the Euglena gracilis chloroplast ribosomal protein S3-gene—Evidence for intron insertion during gene evolution. *Nucleic Acids Research*, *19*(23), 6491–6497.
Copertino, D. W., & Hallick, R. B. (1991). Group II twintron: An intron within an intron in a chloroplast cytochrome b-559 gene. *The EMBO Journal*, *10*(2), 433–442. Retrieved from http://www.pubmedcentral.nih.gov/articlerender.fcgi?artid=452664&tool=pmcentrez&rendertype=abstract.
Copertino, D. W., & Hallick, R. B. (1993). Group II and group III introns of twintrons: Potential relationships with nuclear pre-mRNA introns. *Trends in Biochemical Sciences*, *18*(12), 467–471. http://dx.doi.org/10.1016/0968-0004(93)90008-B.
Dabbagh, N., & Preisfeld, A. (2016). The chloroplast genome of Euglena mutabilis-cluster arrangement, intron analysis, and intrageneric trends. *Journal of Eukaryotic Microbiology*, *1993*, 31–44. http://dx.doi.org/10.1111/jeu.12334.
Davies, W. H., Mercer, E. I., & Goodwin, T. W. (1966). Some observations on the biosynthesis of the plant sulpholipid by Euglena gracilis. *The Biochemical Journal*, *98*(2), 369–373. Retrieved from http://www.pubmedcentral.nih.gov/articlerender.fcgi?artid=1264853&tool=pmcentrez&rendertype=abstract.
de M. Bicudo, C. E., & Menezes, M. (2016). Phylogeny and classification of Euglenophyceae: A brief review. *Frontiers in Ecology and Evolution*, *4*(March), 1–15. http://dx.doi.org/10.3389/fevo.2016.00017.
Doetsch, N. A. (2000). *Group III intron structure and evolutionary analysis in euglenoid chloroplast genomes (doctoral dissertation)*. The University of Arizona. http://hdl.handle.net/10150/284139.
Doetsch, N. A., Favreau, M. R., Kuscuoglu, N., Thompson, M. D., & Hallick, R. B. (2001). Chloroplast transformation in Euglena gracilis: Splicing of a group III twintron transcribed from a transgenic psbK operon. *Current Genetics*, *39*(1), 49–60. Retrieved from http://www.ncbi.nlm.nih.gov/pubmed/11318107.
Doolittle, W. F. (1998). You are what you eat: A gene transfer ratchet could account for bacterial genes in eukaryotic nuclear genomes. *Trends in Genetics: TIG*, *14*(8), 307–311. Retrieved from http://www.ncbi.nlm.nih.gov/pubmed/9724962.
Douglas, S. E., & Turner, S. (1991). Molecular evidence for the origin of plastids from a cyanobacterium-like ancestor. *Journal of Molecular Evolution*, *33*(3), 267–273. http://dx.doi.org/10.1007/BF02100678.
Drager, R. G., & Hallick, R. B. (1993). A complex twintron is excised as four individual introns. *Nucleic Acids Research*, *21*(10), 2389–2394. http://dx.doi.org/10.1093/nar/21.10.2389.
Durnford, D. G., & Gray, M. W. (2006). Analysis of Euglena gracilis plastid-targeted proteins reveals different classes of transit sequences. *Eukaryotic Cell*, *5*(12), 2079–2091. http://dx.doi.org/10.1128/EC.00222-06.
Dwyer, M. R., & Smillie, R. M. (1971). β-1,3 glucan: A source of carbon and energy for chloroplast development in Euglena Gracilis. *Australian Journal of Biological Sciences*, *24*(1), 15–22.
Ebringer, L. (1972). Are plastids derived from prokaryotic micro-organisms? Action of antibiotics on chloroplasts of Euglena gracilis. *Journal of General Microbiology*, *71*(1), 35–52. http://dx.doi.org/10.1099/00221287-71-1-35.
Ebringer, L., Dobias, J., Krajčovič, J., Polónyi, J., Križková, L., & Lahitová, N. (1996). Antimutagens reduce ofloxacin-induced bleaching in Euglena gracilis. *Mutation Research/Environmental Mutagenesis and Related Subjects*, *359*(2), 85–93. http://dx.doi.org/10.1016/S0165-1161(96)90255-1.
Egan, J. M., & Carell, E. F. (1972). Studies on chloroplast development and replication in euglena: III. A study of the site of synthesis of alkaline deoxyribonuclease induced during chloroplast development in Euglena gracilis. *Plant Physiology*, *50*(3), 391–395.
Egan, J. M., Dorsky, D., & Schiff, J. A. (1975). Events surrounding the early development of euglena chloroplasts: VI. Action spectra for the formation of chlorophyll, lag elimination

in chlorophyll synthesis, and appearance of TPN-dependent triose phosphate dehydrogenase and alkaline DNase activities. *Plant Physiology*, *56*(2), 318–323.
Ehrenberg, C. G. (1830). *Organisation, Systematik und geographisches Verhältniss der Infusionsthierchen*. Berlin: Druckerei der Königlichen Akademie der Wissenschaften. Retrieved from http://www.biodiversitylibrary.org/item/18341.
Ferroni, L., Baldisserotto, C., Fasulo, M. P., Pagnoni, A., & Pancaldi, S. (2004). Adaptive modifications of the photosynthetic apparatus in Euglena gracilis Klebs exposed to manganese excess. *Protoplasma*, *224*(3–4), 167–177. http://dx.doi.org/10.1007/s00709-004-0072-4.
Fong, F., & Schiff, J. A. (1979). Blue-light-induced absorbance changes associated with carotenoids in Euglena. *Planta*, *146*(2), 119–127. http://dx.doi.org/10.1007/BF00388221.
Fox, L., Erion, J., Tarnowski, J., Spremulli, L., Brot, N., & Weissbach, H. (1980). Communication Eglena gracilis Chloroplast EF-Ts. *The Journal of Biological Chemistry*, *255*(13), 6018–6019. Retrieved from http://www.jbc.org/content/255/13/6018.full.pdf.
Fujita, T., Aoyagi, H., Ogbonna, J. C., & Tanaka, H. (2008). Effect of mixed organic substrate on α-tocopherol production by Euglena gracilis in photoheterotrophic culture. *Applied Microbiology and Biotechnology*, *79*(3), 371–378. http://dx.doi.org/10.1007/s00253-008-1443-0.
Fujita, T., Ogbonna, J. C., Tanaka, H., & Aoyagi, H. (2009). Effects of reactive oxygen species on α-tocopherol production in mitochondria and chloroplasts of Euglena gracilis. *Journal of Applied Phycology*, *21*(2), 185–191. http://dx.doi.org/10.1007/s10811-008-9349-x.
Geimer, S., Belicová, A., Legen, J., Sláviková, S., Herrmann, R. G., & Krajcovic, J. (2009). Transcriptome analysis of the Euglena gracilis plastid chromosome. *Current Genetics*, *55*(4), 425–438. http://dx.doi.org/10.1007/s00294-009-0256-8.
Gibbs, S. P. (1970). The comparative ultrastructure of the algal chloroplast. *Annals of the New York Academy of Sciences*, *175*(1), 454–473. http://dx.doi.org/10.1111/j.1749-6632.1970.tb45167.x.
Gibbs, S. P. (1978). The chloroplasts of Euglena may have evolved from symbiotic green algae. *Canadian Journal of Botany*, *56*(22), 2883–2889. http://dx.doi.org/10.1139/b78-345.
Gibbs, S. P. (1981). The chloroplasts of some algal groups may have evolved from endosymbiotic eukaryotic algae. *Annals of the New York Academy of Sciences*, *361*, 193–208. [1 Origins and E] http://dx.doi.org/10.1111/j.1749-6632.1981.tb54365.x.
Gibor, A., & Granick, S. (1962). The plastid system of normal and bleached Euglena gracilis. *The Journal of Protozoology*, *9*(3), 327–334.
Gockel, G., & Hachtel, W. (2000). Complete gene map of the plastid genome of the nonphotosynthetic Euglenoid flagellate Astasia longa. *Protist*, *151*(4), 347–351. http://dx.doi.org/10.1078/S1434-4610(04)70033-4.
Gomez-Silva, B., Timko, M. P., & Schiff, J. A. (1985). Chlorophyll biosynthesis from glutamate or 5-aminolevulinate in intact Euglena chloroplasts. *Planta*, *165*(1), 12–22. http://dx.doi.org/10.1007/BF00392206.
Gonzalez-Moreno, S., Gomez-Barrera, J., Perales, H., & Moreno-Sanchez, R. (1997). Multiple effects of salinity on photosynthesis of the protist Euglena gracilis. *Physiologia Plantarum*, *101*(4), 777–786. http://dx.doi.org/10.1111/j.1399-3054.1997.tb01063.x.
Griffiths, W. T., Threlfall, D. R., & Goodwin, T. W. (1967). Nature, intracellular distribution and formation of terpenoid quinones in maize and barley shoots. *The Biochemical Journal*, *103*(2), 589–600. Retrieved from http://www.pubmedcentral.nih.gov/articlerender.fcgi?artid=1270445&tool=pmcentrez&rendertype=abstract.
Guiry, M. D., & Guiry, G. M. (2017). *AlgaeBase*. National University of Ireland, Galway: World-ide Electronic Publication.

Hadariová, L., Vesteg, M., Birčák, E., Schwartzbach, S. D., & Krajčovič, J. (2017). An intact plastid genome is essential for the survival of colorless Euglena longa but not Euglena gracilis. *Current Genetics*, *63*(2), 331–334. http://dx.doi.org/10.1007/s00294-016-0641-z.

Hallick, R. B., Hong, L., Drager, R. G., Favreau, M. R., Monfort, A., Orsat, B., et al. (1993). Complete sequence of Euglena gracilis chloroplast DNA. *Nucleic Acids Research*, *21*(15), 3537–3544. http://dx.doi.org/10.1093/nar/21.15.3537.

Hannaert, V., Saavedra, E., Duffieux, F., Szikora, J.-P., Rigden, D. J., Michels, P. A. M., et al. (2003). Plant-like traits associated with metabolism of Trypanosoma parasites. *Proceedings of the National Academy of Sciences of the United States of America*, *100*(3), 1067–1071. http://dx.doi.org/10.1073/pnas.0335769100.

Harris, J. (1695). Some microscopical observations of vast numbers of animalcula seen in water by John Harris, M. A. Rector of Winchelsea in Sussex, and F.R.S. *Philosophical Transactions of the Royal Society of London*, *19*(215–235), 254–259. http://dx.doi.org/10.1098/rstl.1695.0036.

Heizmann, P., Salvador, G. F., & Nigon, V. (1976). Occurrence of plastidial rRNAs and plastidial structures in bleached mutants of Euglena gracilis. *Experimental Cell Research*, *99*(2), 253–260. http://dx.doi.org/10.1016/0014-4827(76)90581-4.

Horrum, M. A., & Schwartzbach, S. D. (1980). Nutritional regulation of organelle biogenesis in euglena: Repression of chlorophyll and NADP-glyceraldehyde-3-phosphate dehydrogenase synthesis. *Plant Physiology*, *65*(2), 382–386. http://dx.doi.org/10.1104/PP.65.2.382.

Howe, C. J., Barbrook, A. C., Nisbet, R. E. R., Lockhart, P. J., & Larkum, A. W. D. (2008). The origin of plastids. *Philosophical Transactions of the Royal Society of London. Series B, Biological Sciences*, *363*(1504), 2675–2685. http://dx.doi.org/10.1098/rstb.2008.0050.

Hrdá, Š., Fousek, J., Szabová, J., Hampl, V., & Vlček, Č. (2012). The plastid genome of eutreptiella provides a window into the process of secondary endosymbiosis of plastid in euglenids. *PloS One*, 7(3), e33746. http://dx.doi.org/10.1371/journal.pone.0033746.

Inagaki, J., Fujita, Y., Hase, T., & Yamamoto, Y. (2000). Protein translocation within chloroplast is similar in Euglena and higher plants. *Biochemical and Biophysical Research Communications*, *277*(2), 436–442. http://dx.doi.org/10.1006/bbrc.2000.3702.

Inui, H., Miyatake, K., Nakano, Y., & Kitaoka, S. (1982). Wax ester fermentation in Euglena gracilis. *FEBS Letters*, *150*(1), 89–93. http://dx.doi.org/10.1016/0014-5793(82)81310-0.

Iseki, M., Matsunaga, S., Murakami, A., Ohno, K., Shiga, K., Yoshida, K., et al. (2002). A blue-light-activated adenylyl cyclase mediates photoavoidance in Euglena gracilis. *Nature*, *415*(6875), 1047–1051.

Jarvis, P., & Robinson, C. (2004). Mechanisms of protein import and routing in chloroplasts. *Current Biology: CB*, *14*(24), R1064–R1077. http://dx.doi.org/10.1016/j.cub.2004.11.049.

Jenkins, K. P., Hong, L., & Hallick, R. B. (1995). Alternative splicing of the Euglena gracilis chloroplast roaA transcript. *RNA (New York, N.Y.)*, *1*(6), 624–633.

Karaca, H., Bozkurt, O., Ozaslan, E., Baldane, S., Berk, V., Inanc, M., et al. (2014). Positive effects of oral β-glucan on mucositis and leukopenia in colorectal cancer patients receiving adjuvant FOLFOX-4 combination chemotherapy. *Asian Pacific Journal of Cancer Prevention*, *15*, 3641–3644. http://dx.doi.org/10.7314/APJCP.2014.15.8.3641.

Karnkowska, A., Bennett, M. S., Watza, D., Kim, J. I., Zakryś, B., & Triemer, R. E. (2015). Phylogenetic relationships and morphological character evolution of photosynthetic Euglenids (Excavata) inferred from taxon-rich analyses of five genes. *Journal of Eukaryotic Microbiology*, *62*(3), 362–373. http://dx.doi.org/10.1111/jeu.12192.

Kasiborski, B. A., Bennett, M. S., Linton, E. W., & Lane, C. (2016). The chloroplast genome of Phacus orbicularis (Euglenophyceae): An initial datum point for the phacaceae. *Journal of Phycology, 52*(3), 404–411. http://dx.doi.org/10.1111/jpy.12403.

Kim, D., Filtz, M. R., & Proteau, P. J. (2004). The methylerythritol phosphate pathway contributes to carotenoid but not phytol biosynthesis in Euglena gracilis. *Journal of Natural Products, 67*(6), 1067–1069. http://dx.doi.org/10.1021/np049892x.

Kim, J. I., Linton, E. W., & Shin, W. (2015). Taxon-rich multigene phylogeny of the photosynthetic euglenoids (Euglenophyceae). *Frontiers in Ecology and Evolution, 3*(August), 1–11. http://dx.doi.org/10.3389/fevo.2015.00098.

Kim, J. I., & Shin, W. (2008). Phylogeny of the euglenales inferred from plastid LSU rDNA sequences. *Journal of Phycology, 44*(4), 994–1000. http://dx.doi.org/10.1111/j.1529-8817.2008.00536.x.

Kirk, J. T. O. (1970). Biochemical aspects of chloroplast development. *Annual Review of Plant Physiology, 21*(1), 11–42. http://dx.doi.org/10.1146/annurev.pp.21.060170.000303.

Kogan, G., Skorik, Y., Zitnanová, I., Krizkova, L., Durackova, Z., Gomes, C., et al. (2004). Antioxidant and antimutagenic activity of N-(2-carboxyethyl)chitosan. *Toxicology and Applied Pharmacology, 201*(3), 303–310. http://dx.doi.org/10.1016/j.taap.2004.05.009.

Koller, B., & Delius, H. (1982). Origin of replication in chloroplast DNA of Euglena gracilis located close to the region of variable size. *The EMBO Journal, 1*(8), 995–998.

Kořený, L., & Oborník, M. (2011). Sequence evidence for the presence of two tetrapyrrole pathways in Euglena gracilis. *Genome Biology and Evolution, 3*(1), 359–364. http://dx.doi.org/10.1093/gbe/evr029.

Koritala, S. (1989). Microbiological synthesis of wax esters by Euglena gracilis. *Journal of the American Oil Chemists Society, 66*(1), 133–134. http://dx.doi.org/10.1007/BF02661801.

Krajčovič, J., Ebringer, L., & Polónyi, J. (1989). Quinolones and coumarins eliminate chloroplasts from Euglena gracilis. *Antimicrobial Agents and Chemotherapy, 33*(11), 1883–1889. http://dx.doi.org/10.1128/AAC.33.11.1883.

Krajčovič, J., Ebringer, L., & Schwartzbach, S. D. (2001). Reversion of endosymbiosis? In *Symbiosis* (pp. 185–206). Dordrecht: Kluwer Academic Publishers. http://dx.doi.org/10.1007/0-306-48173-1_11.

Krajčovič, J., Vacula, R., Steiner, J. M., Löffelhardt, W., Belicová, A., Sláviková, S., et al. (1999). Molecular effects of some stress factors on the chloroplast genetic apparatus of the flagellate Euglena gracilis. In *The chloroplast: From molecular biology to biotechnology* (pp. 121–128). Dordrecht: Springer Netherlands. http://dx.doi.org/10.1007/978-94-011-4788-0_19.

Krajčovič, J., Vesteg, M., & Schwartzbach, S. D. (2015). Euglenoid flagellates: A multifaceted biotechnology platform. *Journal of Biotechnology, 202*, 135–145. http://dx.doi.org/10.1016/j.jbiotec.2014.11.035.

Kraus, B. L., & Spremulli, L. L. (1988). Evidence for the nuclear location of the genes for chloroplast IF-2 and IF-3 in Euglena. *Plant Physiology, 88*(4), 993–995. http://dx.doi.org/10.1104/PP.88.4.993.

Krinsky, N. I., & Goldsmith, T. H. (1960). The carotenoids of the flagellated alga, Euglena gracilis. *Archives of Biochemistry and Biophysics, 91*(12), 271–279. http://dx.doi.org/10.1016/0003-9861(60)90501-4.

Križková, L., Ďuračková, Z., Šandula, J., Sasinková, V., & Krajčovič, J. (2001). Antioxidative and antimutagenic activity of yeast cell wall mannans in vitro. *Mutation Research/Genetic Toxicology and Environmental Mutagenesis, 497*(1–2), 213–222. http://dx.doi.org/10.1016/S1383-5718(01)00257-1.

Križková, L., Mučaji, P., Nagy, M., & Krajčovič, J. (2004). Triterpenoid cynarasaponins from Cynara cardunculus L. reduce chemically induced mutagenesis in vitro. *Phytomedicine, 11*(7–8), 673–678. http://dx.doi.org/10.1016/j.phymed.2003.09.001.

Križková, L., Nagy, M., Polónyi, J., & Ebringer, L. (1998). The effect of flavonoids on ofloxacin-induced mutagenicity in Euglena gracilis. *Mutation Research/Genetic Toxicology and Environmental Mutagenesis*, *416*(1–2), 85–92. http://dx.doi.org/10.1016/S1383-5718(98)00080-1.

Križková, L., Žitňanová, I., Mislovičová, D., Masárová, J., Sasinková, V., Ďuračková, Z., et al. (2006). Antioxidant and antimutagenic activity of mannan neoglycoconjugates: Mannan–human serum albumin and mannan–penicillin G acylase. *Mutation Research/Genetic Toxicology and Environmental Mutagenesis*, *606*(1–2), 72–79. http://dx.doi.org/10.1016/j.mrgentox.2006.03.003.

Kuda, T., Enomoto, T., & Yano, T. (2009). Effects of two storage β-1,3-glucans, laminaran from Eicenia bicyclis and paramylon from Euglena gracili, on cecal environment and plasma lipid levels in rats. *Journal of Functional Foods*, *1*, 399–404. http://dx.doi.org/10.1016/j.jff.2009.08.003.

Lakey, B., & Triemer, R. (2016). The tetrapyrrole synthesis pathway as a model of horizontal gene transfer in euglenoids. *Journal of Phycology*, *53*, 198–217. http://dx.doi.org/10.1111/jpy.12491.

Larkum, A. W. D., Lockhart, P. J., & Howe, C. J. (2007). Shopping for plastids. *Trends in Plant Science*, *12*(5), 189–195. http://dx.doi.org/10.1016/j.tplants.2007.03.011.

Lax, G., & Simpson, A. G. B. (2013). Combining molecular data with classical morphology for uncultured phagotrophic euglenids (Excavata): A single-cell approach. *Journal of Eukaryotic Microbiology*, *60*(6), 615–625. http://dx.doi.org/10.1111/jeu.12068.

Leander, B. S. (2004). Did trypanosomatid parasites have photosynthetic ancestors? *Trends in Microbiology*, *12*(6), 251–258. http://dx.doi.org/10.1016/j.tim.2004.04.001.

Leander, B. S., Esson, H. J., & Breglia, S. A. (2007). Macroevolution of complex cytoskeletal systems in euglenids. *BioEssays: News and Reviews in Molecular, Cellular and Developmental Biology*, *29*(10), 987–1000. http://dx.doi.org/10.1002/bies.20645.

Leander, B. S., Triemer, R. E., & Farmer, M. a. (2001). Character evolution in heterotrophic euglenids. *European Journal of Protistology*, *37*, 337–356.

Leedale, G. F. (1967). *Euglenoid flagellates*. Englewood Cliffs, NJ: Prentice-Hall.

Lefort-Tran, M. (1981). The triple layered organization of the Euglena chloroplast envelope (signification and functions). *Plant Biology*, *94*(1), 463–476.

Liljenberg, C., & Selstam, E. (1980). Interactions of chlorophyll a and terpenoid alcohols with chloroplast acyl lipids in monomolecular films. *Physiologia Plantarum*, *48*(1966), 428–434.

Linton, E. W., Karnkowska-Ishikawa, A., Kim, J. I., Shin, W., Bennett, M. S., Kwiatowski, J., et al. (2010). Reconstructing Euglenoid evolutionary relationships using three genes: Nuclear SSU and LSU, and chloroplast SSU rDNA sequences and the description of Euglenaria gen.nov. (Euglenophyta). *Protist*, *161*(4), 603–619. http://dx.doi.org/10.1016/j.protis.2010.02.002.

Mahmoudi, N., Ciceron, L., Franetich, J.-F., Farhati, K., Silvie, O., Eling, W., et al. (2003). In vitro activities of 25 quinolones and fluoroquinolones against liver and blood stage Plasmodium spp. *Antimicrobial Agents and Chemotherapy*, *47*(8), 2636–2639. http://dx.doi.org/10.1128/AAC.47.8.2636-2639.2003.

Maier, U. G., Zauner, S., & Hempel, F. (2015). Protein import into complex plastids: Cellular organization of higher complexity. *European Journal of Cell Biology*, *94*(7–9), 340–348. http://dx.doi.org/10.1016/j.ejcb.2015.05.008.

Marin, B. (2004). Origin and fate of chloroplasts in the Euglenoida. *Protist*, *155*(March), 13–14. http://dx.doi.org/10.1078/1434461000159.

Marin, B., Nowack, E. C. M., & Melkonian, M. (2005). A plastid in the making: Evidence for a second primary endosymbiosis. *Protist*, *156*(4), 425–432. http://dx.doi.org/10.1016/j.protis.2005.09.001.

Marin, B., Palm, A., Klingberg, M., & Melkonian, M. (2003). Phylogeny and taxonomic revision of plastid-containing euglenophytes based on SSU rDNA sequence comparisons and synapomorphic signatures in the SSU rRNA secondary structure. *Protist*, *154*(1), 99–145. http://dx.doi.org/10.1078/143446103764928521.

Markunas, C. M., & Triemer, R. E. (2016). Evolutionary history of the enzymes involved in the Calvin-Benson cycle in Euglenids. *Journal of Eukaryotic Microbiology, 63*(3), 326–339. http://dx.doi.org/10.1111/jeu.12282.

Martin, W., & Borst, P. (2003). Secondary loss of chloroplasts in trypanosomes. *Proceedings of the National Academy of Sciences of the United States of America, 100*(3), 765–767. http://dx.doi.org/10.1073/pnas.0437776100.

Maruyama, S., Matsuzaki, M., Misawa, K., & Nozaki, H. (2009). Cyanobacterial contribution to the genomes of the plastid-lacking protists. *BMC Evolutionary Biology, 9*(1), 197. http://dx.doi.org/10.1093/bioinformatics/17.8.754.

Maruyama, S., Suzaki, T., Weber, A. P. M., Archibald, J. M., & Nozaki, H. (2011). Eukaryote-to-eukaryote gene transfer gives rise to genome mosaicism in euglenids. *BMC Evolutionary Biology, 11*(1), 105. http://dx.doi.org/10.1186/1471-2148-11-105.

Matson, R. S., Fei, M., & Chang, S. B. (1970). Comparative studies of biosynthesis of galactolipids in Euglena-gracilis strain-Z. *Plant Physiology, 45*(4), 531.

Monfils, A. K., Triemer, R. E., & Bellairs, E. F. (2011). Characterization of paramylon morphological diversity in photosynthetic euglenoids (Euglenales, Euglenophyta). *Phycologia, 50*(2), 156–169. http://dx.doi.org/10.2216/09-112.1.

Monroy, A. F., & Schwartzbach, S. D. (1984). Catabolite repression of chloroplast development in Euglena. *Cell Biology, 81*, 2786–2790. Retrieved from http://www.pnas.org/content/81/9/2786.full.pdf.

Morales, J., Hashimoto, M., Williams, T. A., Hirawake-Mogi, H., Makiuchi, T., Tsubouchi, A., et al. (2016). Differential remodelling of peroxisome function underpins the environmental and metabolic adaptability of diplonemids and kinetoplastids. *Proceedings of the Royal Society of London B: Biological Sciences, 283*(1830). Retrieved from http://rspb.royalsocietypublishing.org/content/283/1830/20160520.long.

Morden, C. W., & Golden, S. S. (1991). Sequence analysis and phylogenetic reconstruction of the genes encoding the large and small subunits of ribulose-1,5-bisphosphate carboxylase/oxygenase from the chlorophyllb-containing prokaryoteProchlorothrix hollandica. *Journal of Molecular Evolution, 32*(5), 379–395. http://dx.doi.org/10.1007/BF02101278.

Mullner, A. N., Angeler, D. G., Samuel, R., Linton, E. W., & Triemer, R. E. (2001). Phylogenetic analysis of phagotrophic, phototrophic and osmotrophic euglenoids by using the nuclear 18S rDNA sequence. *International Journal of Systematic and Evolutionary Microbiology, 51*(3), 783–791. http://dx.doi.org/10.1099/00207713-51-3-783.

Nowack, E. C. M., Price, D. C., Bhattacharya, D., Singer, A., Melkonian, M., & Grossman, A. R. (2016). Gene transfers from diverse bacteria compensate for reductive genome evolution in the chromatophore of Paulinella chromatophora. *Proceedings of the National Academy of Sciences of the United States of America, 113*(43), 12214–12219. http://dx.doi.org/10.1073/pnas.1608016113.

O'Neill, E. C., Trick, M., Henrissat, B., & Field, R. A. (2015). Euglena in time: Evolution, control of central metabolic processes and multi-domain proteins in carbohydrate and natural product biochemistry. *Perspectives in Science, 6*, 84–93. http://dx.doi.org/10.1016/j.pisc.2015.07.002.

Ogawa, T., Tamoi, M., Kimura, A., Mine, A., Sakuyama, H., Yoshida, E., et al. (2015). Enhancement of photosynthetic capacity in Euglena gracilis by expression of cyanobacterial fructose-1,6-/sedoheptulose-1,7-bisphosphatase leads to increases in biomass and wax ester production. *Biotechnology for Biofuels, 8*(1), 80. http://dx.doi.org/10.1186/s13068-015-0264-5.

Osafune, T., & Schiff, J. A. (1980a). Events surrounding the early development of Euglena chloroplasts. *Journal of Ultrastructure Research, 73*(1), 64–76. http://dx.doi.org/10.1016/0022-5320(80)90116-1.

Osafune, T., & Schiff, J. A. (1980b). Stigma and flagellar swelling in relation to light and carotenoids in Euglena gracilis var. bacillaris. *Journal of Ultrasructure Research*, *73*(3), 336–349. http://dx.doi.org/10.1016/S0022-5320(80)90093-3.

Osafune, T., & Schiff, J. A. (1983). W10BSmL, a mutant of Euglena gracilis var. bacillaris lacking plastids. *Experimental Cell Research*, *148*(2), 530–535. http://dx.doi.org/10.1016/0014-4827(83)90176-3.

Osafune, T., Schiff, J. A., & Hase, E. (1987). Light-independent and dependent phases of proplastid development in Euglena gracilis W3BUL. *Cell Structure and Function*, *12*(5), 453–461. Retrieved from http://www.ncbi.nlm.nih.gov/pubmed/3119230.

Parthier, B., & Neumann, D. (1977). Structural and functional analysis of some plastid mutants of Euglena gracilis. *Biochemie und Physiologie der Pflanzen*, *171*(6), 547–562. http://dx.doi.org/10.1016/S0015-3796(17)30349-9.

Polónyi, J., Ebringer, L., Dobias, J., & Krajčovič, J. (1998). Giant mitochondria in chloroplast-deprived Euglena gracilis late after N-succinimidylofloxacin treatment. *Folia Microbiologica*, *43*(6), 661–666. http://dx.doi.org/10.1007/BF02816386.

Polónyi, J., Ebringer, L., Krajcovic, J., & Kapeller, K. (1990). Injured mitochondria in cells of Euglena gracilis after DNA gyrase inhibitors treatment. *Zeitschrift fur Mikroskopisch-Anatomische Forschung*, *104*(1), 61–78. Retrieved from http://www.ncbi.nlm.nih.gov/pubmed/2161591.

Pombert, J. F., James, E. R., Janouškovec, J., & Keeling, P. J. (2012). Evidence for transitional stages in the evolution of Euglenid group II introns and twintrons in the Monomorphina aenigmatica plastid genome. *PloS One*, 7(12). http://dx.doi.org/10.1371/journal.pone.0053433.

Ravel-Chapuis, P., Heizmann, P., & Nigon, V. (1982). Electron microscopic localization of the replication origin of Euglena gracilis chloroplast DNA. *Nature*, *300*(5887), 78–81. http://dx.doi.org/10.1038/300078a0.

Richter, P., Helbling, W., Streb, C., & Häder, D.-P. (2007). PAR and UV effects on vertical migration and photosynthesis in Euglena gracilis? *Photochemistry and Photobiology*, *83*(4), 818–823. http://dx.doi.org/10.1111/j.1751-1097.2007.00134.x.

Rodríguez-Zavala, J. S., Ortiz-Cruz, M. A., Mendoza-Hernández, G., & Moreno-Sánchez, R. (2010). Increased synthesis of α-tocopherol, paramylon and tyrosine by Euglena gracilis under conditions of high biomass production. *Journal of Applied Microbiology*, *109*(6), 2160–2172. http://dx.doi.org/10.1111/j.1365-2672.2010.04848.x.

Roos, D. S., & Fichera, M. E. (1997). A plastid organelle as a drug target in apicomplexan parasites. *Nature*, *390*(6658), 407–409. http://dx.doi.org/10.1038/37132.

Rosenberg, A., & Gouaux, J. (1967). Quantitative and compositional changes in monogalactosyl and digalactosyl diglycerides during light-induced formation of chloroplasts in Euglena gracilis. *Journal of Lipid Research*, *8*, 80–83.

Rosenberg, A., Gouaux, J., & Milch, P. (1966). Monogalactosyl and digalactosyl diglycerides from heterotrophic, hetero-autotrophic, and photobiotic Euglena gracilis, 7, 733–738.

Rybicka, K. K. (1996). Glycosomes—The organelles of glycogen metabolism. *Tissue and Cell*, *28*(3), 253–265. http://dx.doi.org/10.1016/S0040-8166(96)80013-9.

Sagan, L. (1993). On the origin of mitosing cells. 1967. *The Journal of NIH Research: Life Sciences Research and News about the National Institutes of Health and the Alcohol, Drug Abuse, and Mental Health Administration*, *5*(3), 65–72. Retrieved from http://www.ncbi.nlm.nih.gov/pubmed/11541390.

Saidha, T., & Schiff, J. A. (1989). The role of mitochondria in sulfolipid biosynthesis by Euglena chloroplasts. *Biochimica et Biophysica Acta (BBA)—Lipids and Lipid Metabolism*, *1001*(3), 268–273. http://dx.doi.org/10.1016/0005-2760(89)90110-0.

Sanchez-Puerta, M. V., Lippmeier, J. C., Apt, K. E., & Delwiche, C. F. (2007). Plastid genes in a non-photosynthetic dinoflagellate. *Protist*, *158*(1), 105–117. http://dx.doi.org/10.1016/j.protis.2006.09.004.

Šantek, B., Felski, M., Friehs, K., Lotz, M., & Flaschel, E. (2010). Production of paramylon, a β-1,3-glucan, by heterotrophic cultivation of Euglena gracilis on potato liquor. *Engineering in Life Sciences*, *10*(2), 165–170. http://dx.doi.org/10.1002/elsc.200900077.
Schiff, J. A., Lyman, H., & Russell, G. K. (1971). Isolation of mutants from Euglena gracilis. *Methods in Enzymology*, *23*, 143–162. http://dx.doi.org/10.1016/S0076-6879(71)23088-3.
Schneider, T., & Betz, A. (1985). Waxmonoester fermentation in Euglena gracilis T. Factors favouring the synthesis of odd-numbered fatty acids and alcohols. *Planta*, *166*(1), 67–73. http://dx.doi.org/10.1007/BF00397387.
Schwartzbach, S. D. (2017). Photo and nutritional regulation of Euglena organelle development. In S. D. Schwartzbach & S. Shigeoka (Eds.), *Euglena: Biochemistry, cell and molecular biology* (pp. 159–182). Cham, Switzerland: Springer International Publishing. http://dx.doi.org/10.1007/978-3-319-54910-1_9.
Schwartzbach, S. D., & Schiff, J. A. (1974). Chloroplast and cytoplasmic ribosomes of Euglena: Selective binding of dihydrostreptomycin to chloroplast ribosomes. *Journal of Bacteriology*, *120*(1), 334–341.
Schwartzbach, S. D., Schiff, J. A., & Goldstein, N. H. (1975). Events surrounding the early development of euglena chloroplasts: V. Control of paramylum degradation. *Plant Physiology*, *56*(2), 313–317. http://dx.doi.org/10.1111/j.1751-1097.1976.tb06873.x.
Schwender, J., Goffman, F., Ohlrogge, J. B., & Shachar-Hill, Y. (2004). Rubisco without the Calvin cycle improves the carbon efficiency of developing green seeds. *Nature*, *432*(7018), 779–782. http://dx.doi.org/10.1038/nature03145.
Shashidhara, L. S. S., Limsli, S. H., Shackletonii, J. B., Robinsonii, C., & Smith, A. G. (1992). Protein targeting across the three membranes of the Euglena chloroplast envelope. *The Journal of Biological Chemistry*, *267*, 12885–12891.
Sheiner, L., & Striepen, B. (2013). Protein sorting in complex plastids. *Biochimica et Biophysica Acta—Molecular Cell Research*, *1833*(2), 352–359. http://dx.doi.org/10.1016/j.bbamcr.2012.05.030.
Sheveleva, E. V., & Hallick, R. B. (2004). Recent horizontal intron transfer to a chloroplast genome. *Nucleic Acids Research*, *32*(2), 803–810. http://dx.doi.org/10.1093/nar/gkh225.
Shi, L.-X., & Theg, S. M. (2013). The chloroplast protein import system: From algae to trees. *Biochimica et Biophysica Acta*, *1833*(2), 314–331. http://dx.doi.org/10.1016/j.bbamcr.2012.10.002.
Shigeoka, S., Onishi, T., Nakano, Y., & Kitaoka, S. (1986). The contents and subcellular distribution of tocopherols in Euglena gracilis. *Agricultural and Biological Chemistry*, *50*(4), 1063–1065. http://dx.doi.org/10.1271/bbb1961.50.1063.
Slávikova, S., Vacula, R., Fang, Z., Ehara, T., Osafune, T., & Schwartzbach, S. D. (2005). Homologous and heterologous reconstitution of Golgi to chloroplast transport and protein import into the complex chloroplasts of Euglena. *Journal of Cell Science*, *118*(Pt. 8), 1651–1661. http://dx.doi.org/10.1242/jcs.02277.
Stern, A. I., Epstein, H. T., & Schiff, J. A. (1964a). Studies of chloroplast development in Euglena. VI. Light intensity as a controlling factor in development. *Plant Physiology*, *39*(2), 226–231.
Stern, A. I., Schiff, J. A., & Epstein, H. T. (1964b). Studies of chloroplast development in Euglena. V. Pigment biosynthesis, photosynthetic oxygen evolution and carbon dioxide fixation during chloroplast development. *Plant Physiology*, *39*(2), 220–226. http://dx.doi.org/10.1104/pp.39.2.220.
Sugiyama, A., Suzuki, K., Mitra, S., Arashida, R., Yoshida, E., Nakano, R., et al. (2009). Hepatoprotective effects of paramylon, a beta-1, 3-D-glucan isolated from Euglena gracilis Z, on acute liver injury induced by carbon tetrachloride in rats. *The Journal of Veterinary Medical Science the Japanese Society of Veterinary Science*, *71*(7), 885–890. http://dx.doi.org/10.1292/jvms.71.885.

Sulli, C., Fang, Z. W., Muchhal, U., & Schwartzbach, S. D. (1999). Topology of Euglena chloroplast protein precursors within endoplasmic reticulum to Golgi to chloroplast transport vesicles. *Journal of Biological Chemistry, 274*(1), 457–463. http://dx.doi.org/10.1074/jbc.274.1.457.

Sulli, C., & Schwartzbach, S. D. (1995). The polyprotein precursor to the Euglena light-harvesting chlorophyll a/b-binding protein is transported to the Golgi apparatus prior to chloroplast import and polyprotein processing. *Journal of Biological Chemistry, 270,* 13084–13090. http://dx.doi.org/10.1074/jbc.270.22.13084.

Sulli, C., & Schwartzbach, S. D. (1996). A soluble protein is imported into Euglena chloroplasts as a membrane-bound precursor. *The Plant Cell, 8*(1), 43–53.

Teerawanichpan, P., & Qiu, X. (2010). Fatty acyl-CoA reductase and wax synthase from Euglena gracilis in the biosynthesis of medium-chain wax esters. *Lipids, 45*(3), 263–273. http://dx.doi.org/10.1007/s11745-010-3395-2.

Thomas, B. G., & Threlfall, D. R. (1974). Synthesis of polyprenyltoluquinols from homogentisate and polyprenyl pyrophosphates in particulate fractions of Euglena and sugar beet. *Biochemical Journal, 142,* 437–440.

Thompson, M. D., Copertino, D. W., Thompson, E., Favreau, M. R., & Hallick, R. B. (1995). Evidence for the late origin of introns in chloroplast genes from an evolutionary analysis of the genus Euglena. *Nucleic Acids Research, 23*(23), 4745–4752. http://dx.doi.org/10.1093/nar/23.23.4745.

Triemer, R. E., Linton, E., Shin, W., Nudelman, A., Monfils, A., Bennett, M., et al. (2006). Phylogeny of the euglenales based upon combined SSU and LSU rDNA sequence comparisons and description of Discoplastis gen. nov. (Euglenophyta). *Journal of Phycology, 42*(3), 731–740. http://dx.doi.org/10.1111/j.1529-8817.2006.00219.x.

Turmel, M., Gagnon, M.-C., O'Kelly, C. J., Otis, C., & Lemieux, C. (2009). The chloroplast genomes of the green algae Pyramimonas, Monomastix, and Pycnococcus shed new light on the evolutionary history of prasinophytes and the origin of the secondary chloroplasts of euglenids. *Molecular Biology and Evolution, 26*(3), 631–648. http://dx.doi.org/10.1093/molbev/msn285.

Vesteg, M., Vacula, R., Steiner, J. M., Mateásiková, B., Löffelhardt, W., Brejová, B., et al. (2010). A possible role for short introns in the acquisition of stroma-targeting peptides in the flagellate Euglena gracilis. *DNA Research: An International Journal for Rapid Publication of Reports on Genes and Genomes, 17*(4), 223–231. http://dx.doi.org/10.1093/dnares/dsq015.

Walne, P. L., & Arnott, H. J. (1967). The comparative ultrastructure and possible function of eyespots: Euglena granulata and Chlamydomonas eugametos. *Planta,* 77(4), 325–353. http://dx.doi.org/10.1007/BF00389319.

Wang, J., Shi, Z., & Xu, X. (2004). Residual plastids of bleached mutants of Euglena gracilis and their effects on the expression of nucleus-encoded genes. *Progress in Natural Science, 14*(3), 213–217. http://dx.doi.org/10.1080/10020070412331343371.

Watanabe, T., Shimada, R., Matsuyama, A., Yuasa, M., Sawamura, H., Yoshida, E., et al. (2013). Antitumor activity of the β-glucan paramylon from Euglena against preneoplastic colonic aberrant crypt foci in mice. *Food & Function, 4*(11), 1685. http://dx.doi.org/10.1039/c3fo60256g.

Wheeler, G., Ishikawa, T., Pornsaksit, V., & Smirnoff, N. (2015). Evolution of alternative biosynthetic pathways for vitamin C following plastid acquisition in photosynthetic eukaryotes. *eLife, 4,* 1–25. http://dx.doi.org/10.7554/eLife.06369.

Wiegert, K. E., Bennett, M. S., & Triemer, R. E. (2012). Evolution of the chloroplast genome in photosynthetic euglenoids: A comparison of Eutreptia viridis and Euglena gracilis (Euglenophyta). *Protist, 163*(6), 832–843. http://dx.doi.org/10.1016/j.protis.2012.01.002.

Wiegert, K. E., Bennett, M. S., & Triemer, R. E. (2013). Tracing patterns of chloroplast evolution in euglenoids: Contributions from colacium vesiculosum and strombomonas acuminata (euglenophyta). *Journal of Eukaryotic Microbiology*, *60*(2), 214–221. http://dx.doi.org/10.1111/jeu.12025.

Wolfe, A. D., & dePamphilis, C. W. (1998). The effect of relaxed functional constraints on the photosynthetic gene rbcL in photosynthetic and nonphotosynthetic parasitic plants. *Molecular Biology and Evolution*, *15*(10), 1243–1258. http://dx.doi.org/10.1093/oxfordjournals.molbev.a025853.

Wolken, J. J., & Mellon, A. D. (1956). The relationship between chlorophyll and the carotenoids in the algal flagellate, Euglena. *The Journal of General Physiology*, *39*(5), 675–685.

Yamaguchi, A., Yubuki, N., & Leander, B. S. (2012). Morphostasis in a novel eukaryote illuminates the evolutionary transition from phagotrophy to phototrophy: Description of Rapaza viridis n. gen. et sp. (Euglenozoa, Euglenida). *BMC Evolutionary Biology*, *12*(1), 29. http://dx.doi.org/10.1186/1471-2148-12-29.

Yoshida, Y., Tomiyama, T., Maruta, T., Tomita, M., Ishikawa, T., & Arakawa, K. (2016). De novo assembly and comparative transcriptome analysis of Euglena gracilis in response to anaerobic conditions. *BMC Genomics*, *17*(1), 182. http://dx.doi.org/10.1186/s12864-016-2540-6.

Záhonová, K., Füssy, Z., Oborník, M., Eliáš, M., Yurchenko, V., & Stern, D. (2016). RuBisCO in non-photosynthetic alga Euglena Longa: Divergent features, transcriptomic analysis and regulation of complex formation. *PloS One*, *11*(7), e0158790. http://dx.doi.org/10.1371/journal.pone.0158790.

Záhonová, K., Hadariová, L., Vacula, R., Yurchenko, V., Eliáš, M., Krajčovič, J., et al. (2014). A small portion of plastid transcripts is polyadenylated in the flagellate Euglena gracilis. *FEBS Letters*, *588*(5), 783–788. http://dx.doi.org/10.1016/j.febslet.2014.01.034.

Zakryś, B., Milanowski, R., & Karnkowska, A. (2017). Evolutionary origin of Euglena. In S. D. Schwartzbach & S. Shigeoka (Eds.), *Euglena: Biochemistry, cell and molecular biology* (pp. 3–17). Cham, Switzerland: Springer International Publishing. http://dx.doi.org/10.1007/978-3-319-54910-1_1.

CHAPTER TEN

Chlorarachniophytes With Complex Secondary Plastids of Green Algal Origin

Yoshihisa Hirakawa[1]
Faculty of Life and Environmental Sciences, University of Tsukuba, Tsukuba, Japan
[1]Corresponding author: e-mail address: hirakawa.yoshi.fp@u.tsukuba.ac.jp

Contents

Abstract

Diverse algae have evolved 'secondary' plastids through parallel endosymbiotic uptakes of photosynthetic eukaryotes either of green or red algae, and these events are referred to as secondary endosymbioses. Chlorarachniophytes are a group of marine unicellular algae with four-membrane-bound secondary plastids that originated from a green algal endosymbiont. Remarkably, chlorarachniophyte plastids possess a vestigial nucleus termed a 'nucleomorph' in the periplastidal compartment between the second and third membranes that corresponds to the remnant cytoplasm of the endosymbiont. Given that endosymbiont nuclei have disappeared in most secondary plastid-bearing algae, chlorarachniophytes appear to represent an intermediate stage of secondary endosymbiosis and thereby offer an interesting opportunity to study complex plastid evolution. In this chapter, I summarise the current understanding of the evolutionary history of chlorarachniophytes in terms of morphology, phylogeny, and genomics.

Advances in Botanical Research, Volume 84
ISSN 0065-2296
http://dx.doi.org/10.1016/bs.abr.2017.06.006

1. INTRODUCTION

It is widely accepted that two major organelles, plastids (chloroplasts) and mitochondria, are of endosymbiotic origins. Mitochondria have evolved from a single endosymbiosis of proteobacterial ancestry (Embley & Martin, 2006; Gray, 2012), and plastids have arisen from multiple endosymbiotic events of photosynthetic algae (Gould, Waller, & McFadden, 2008; Keeling, 2013). A common ancestor of the Archaeplastida, which includes glaucophytes, rhodophytes (red algae), chlorophytes (green algae), and streptophytes (plants), obtained a primitive plastid through the endosymbiotic uptake of a photosynthetic prokaryote related to the extant cyanobacteria (Ponce-Toledo et al., 2017; Price et al., 2012; Rodríguez-Ezpeleta et al., 2005). Subsequently, certain phagotrophic protists engulfed and integrated an algal endosymbiont related to either the rhodophytes or the chlorophytes, and this led to the spread of 'secondary' plastids in different eukaryotic lineages (Burki et al., 2016; Gould et al., 2008; Keeling, 2013). These eukaryote–eukaryote endosymbioses are referred to as the secondary endosymbioses, in contrast to prokaryote–eukaryote primary endosymbiosis. Multiple secondary endosymbioses have given rise to diverse secondary plastid-bearing organisms, such as haptophytes, heterokonts, cryptophytes, dinoflagellates, apicomplexans, chromerids, euglenophytes, and chlorarachniophytes.

Chlorarachniophytes are marine unicellular algae possessing secondary plastids derived from a green algal endosymbiont. In 1930, the first species of chlorarachniophyte, *Chlorarachnion reptans*, was described by Geitler, and it was originally classified as an amoeboid in the Heterokontophyta. Half a century later, Hibberd and Norris established a new division, Chlorarachniophyta, based on the ultrastructure and pigment composition of *Ch. reptans* (Hibberd & Norris, 1984). To date, 14 species in 8 genera have been described in the phylum Chlorarachniophyta. Chlorarachniophytes are widely distributed in marine habitats (e.g. sandy beaches, coral reefs, coasts, and open oceans) mostly in tropical and temperate regions; no freshwater species have been reported. They show a morphological diversity in cell form that includes flagellate, amoeboid, and coccoid cells (Fig. 1). As a notable feature, chlorarachniophytes possess four-membrane-bound plastids with a vestigial nucleus termed a 'nucleomorph' in the periplastidal compartment between the second and third plastid membranes, a space corresponding to the remnant cytoplasm of the green algal endosymbiont

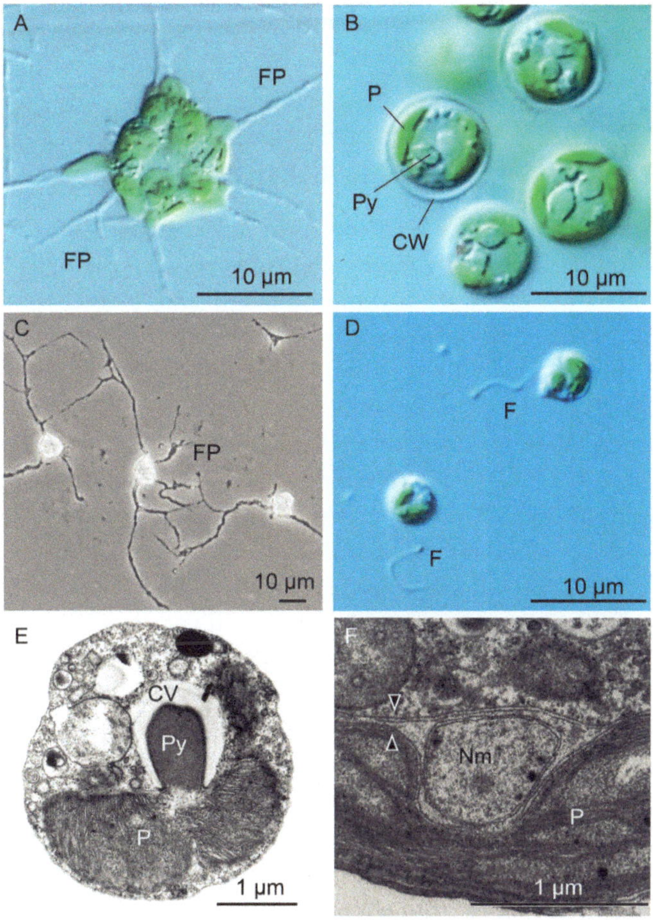

Fig. 1 Microscopic images of chlorarachniophyte cells. (A) Amoeboid cell of *Amorphochlora amoebiformis*. (B) Walled coccoid cells of *Lotharella globosa*. (C) *Chlorarachnion reptans* forming a reticulopodial network. (D) *Bigelowiella natans* flagellate cells. (E, F) Electron microscopic images of *B. natans* showing four-membrane-bound plastid with a projecting pyrenoid (Py) and nucleomorph (Nm). *Arrowheads* represent the inner and outer pairs of plastid membranes. Abbreviations: *CV*, capping vesicle; *CW*, cell wall; *F*, flagellum; *FP*, filose pseudopodia; *P*, plastid.

(Fig. 1). The nucleomorph bears witness to the endosymbiotic theory that one eukaryote is inside another eukaryote (Ludwig & Gibbs, 1989). In secondary endosymbioses, the bulk of the genes carried in the endosymbiont genomes have been either lost or transferred to the host nuclear genomes, and consequently endosymbiont nuclei have disappeared in most secondary endosymbioses. Given that, chlorarachniophyte plastids probably represent a transitional stage from algal endosymbiont to photosynthetic organelle and offer an interesting opportunity to study plastid evolution through secondary endosymbioses.

2. ORIGIN OF CHLORARACHNIOPHYTES

Secondary plastids are believed to have originated from several secondary endosymbioses with either chlorophytes or rhodophytes. Chlorarachniophyte plastids contain chlorophyll *a* and *b* as in chlorophytes (Hibberd & Norris, 1984). Molecular phylogenetic studies have been performed to shed light on the endosymbiont origin of chlorarachniophyte plastids. Based on the phylogenetic tree of plastid 16S ribosomal RNA (rDNA) genes, chlorarachniophyte plastids are inferred to be closely related to extant species of *Chlorella/Nanochlorum* in the Trebouxiophyceae (Van de Peer, Rensing, Maier, & De Wachter, 1996). Nucleomorph 18S rDNA trees showed an affiliation with either the Trebouxiophyceae (Van de Peer et al., 1996) or the Ulvophyceae (Ishida, Green, & Cavalier-Smith, 1999). Trees of nucleus-encoded plastid EF-Tu and PsbO proteins suggest that chlorarachniophyte plastids are closely related to *Bryopsis* of the Ulvophyceae (Ishida, Cao, Hasegawa, Okada, & Hara, 1997) and *Tetraselmis* of the Prasinophyceae (Takahashi et al., 2007), respectively. There is no consensus topology in these single-gene phylogenies. The first complete plastid genome of chlorarachniophytes was reported in *Bigelowiella natans* by Rogers, Gilson, Su, McFadden, and Keeling (2007), and their phylogenetic analysis using 38 plastid-encoded proteins demonstrated that the *B. natans* plastid branched within the clade consisting of the Ulvophyceae, Trebouxiophyceae, and Chlorophyceae (UTC group). Unfortunately, the position of the *B. natans* plastid with regard to specific members of the UTC group was unresolved due to insufficient taxon sampling. Recent progress in plastid genome sequencing in Chlorophyta has solved this issue. A phylogenetic analysis with 55 plastid-encoded protein sequences from 5 chlorarachniophytes and 63 Chlorophyta species has revealed that chlorarachniophyte plastids are closely related to Bryopsidales in the Ulvophyceae (Suzuki, Hirakawa, Kofuji, Sugita, & Ishida, 2016). It is therefore currently thought that chlorarachniophytes acquired their secondary plastids by the uptake of an ancestor of filamentous green algae in the Bryopsidales. Among secondary plastid-bearing organisms, euglenophytes also possess a chlorophyll *a*- and *b*-containing plastid, and Cavalier-Smith hypothesised that chlorarachniophytes and euglenophytes arose from a single common secondary endosymbiosis; this is called the Cabozoa hypothesis (Cavalier-Smith, 1999, 2003). However, molecular phylogenies have not supported the monophyly of chlorarachniophyte and euglenophyte plastids (Rogers et al., 2007; Turmel, Gagnon, O'Kelly, Otis, & Lemieux, 2009). Hence, the Cabozoa hypothesis is now refuted, and it is postulated that two separate acquisitions of green algal endosymbionts generated the chlorarachniophytes and euglenophytes.

As for the host origin of chlorarachniophytes, a phagotrophic ancestry of the Filosa belonging to the Cercozoa has been postulated. Cercozoa is a phylum of the supergroup Rhizaria, and it comprises morphologically diverse protists such as flagellates and filose testate amoebae that are widely distributed in marine, freshwater, and soil environments (Adl et al., 2005; Burki & Keeling, 2014; Cavalier-Smith, 2002). This phylum is divided into two subphyla, Filosa and Endomyxa (Cavalier-Smith & Chao, 2003). Bhattacharya and colleagues initially reported that chlorarachniophytes are closely related to two testate amoebae in the Filosa (*Euglypha* and *Paulinella*) based on a nuclear 18S rDNA tree (Bhattacharya, Helmchen, & Melkonian, 1995). Later phylogenetic studies using 18S rDNA sequences from ~100 cercozoans suggest that chlorarachniophytes occupy a position at the base of the Filosa clade (Bass et al., 2009; Howe et al., 2011; Shiratori & Ishida, 2016). A recent environmental survey has revealed the existence of a tiny bacterivorous protist named *Minorisa minuta* that is the closest relative to the chlorarachniophytes in 18S rDNA phylogeny (del Campo, Not, Forn, Sieracki, & Massana, 2013). This finding has provided new clues on the host origin of chlorarachniophytes. Though *M. minuta* is a colourless protist, it cannot be excluded that the secondary endosymbiosis occurred in a common ancestor of chlorarachniophytes and *M. minuta*, and that the plastid has been lost in *M. minuta*. Future genomic and ultrastructural studies of this interesting organism will address this issue. Overall, it is likely that chlorarachniophytes have evolved by a secondary endosymbiosis in which an ancestral green alga of the Bryopsidales was engulfed and integrated by a phagotrophic ancestor of the Filosa.

3. TAXONOMY OF CHLORARACHNIOPHYTES

In the phylum Chlorarachniophyta, a mere 14 species in eight genera have been described so far. All known species have photosynthetic secondary plastids with a nucleomorph in the periplastidal compartment. In terms of cell morphology, chlorarachniophytes present various cell types such as amoebae with radiating pseudopodia, coccoid cells covered by cell walls, and flagellates with a single laterally inserted flagellum. In several species, two or three cell types are observed over the course of the life cycle. Ishida and colleagues initially proposed a generic classification for chlorarachniophytes based on pyrenoid ultrastructure and nucleomorph location within the plastids (Ishida, Nakayama, & Hara, 1996). At the species level, life cycle patterns and vegetative cell morphology have been emphasised. Here I summarise the morphological characteristics of the known chlorarachniophyte species.

The first species, *Ch. reptans*, was originally described by Geitler (1930). This genus is characterised by a projecting pyrenoid possessing a deep groove created by an invagination of the inner two plastid membranes from the tip, and a nucleomorph located in the groove (Fig. 2A; Hibberd & Norris, 1984; Ishida et al., 1996). The pyrenoid is covered by an additional capping vesicle presumably filled with carbohydrate storage products; similar vesicles are observed in other chlorarachniophyte species (Fig. 1E). *Ch. reptans* was isolated from a sample collected at Puerto Peñasco in Mexico. The vegetative (or asexually reproducing) cells are naked amoebae with filose pseudopodia that are interconnected to form reticulopodial networks (Fig. 1C). Each cell has five to seven bilobed plastids with a central projecting pyrenoid. Cyst-like coccoid cells surrounded by a firm wall are seen especially in old cultures, and flagellate cells bearing a single flagellum have been rarely observed. Hence, *Ch. reptans* presents three different cell types in its life cycle. Interestingly, although chlorarachniophytes are phototrophic algae, *Ch. reptans* is known to be able to engulf small algal prey such as prasinophytes, cryptophytes, dinoflagellates, haptophytes, heterokonts, and cyanobacteria.

Subsequently, *Cryptochlora perforans* was isolated from a sample of the filamentous green alga *Boodleopsis pusilla* collected on the Caribbean coast of Colombia (Calderon-Saenz & Schnetter, 1987, 1989). Unfortunately, pyrenoid ultrastructure and nucleomorph localisation have not been described in this species. The vegetative cells of *Cr. perforans* are naked amoeboid and walled coccoid cells, which reproduce through binary and multiple division. In older cultures, the coccoid cells tend to form thickened and multilayered cell walls that are probably constructed by pectic substances, but not by cellulose. Transformation from the walled to the amoeboid cells, and vice versa, has been confirmed. Flagellate cells are abundantly produced by changing the medium in a crowded old culture, but no cell division has been observed in the flagellate stage. Ingestion of algae and bacteria has been observed. *Cr. perforans* can be distinguished from *Ch. reptans* mainly by the presence

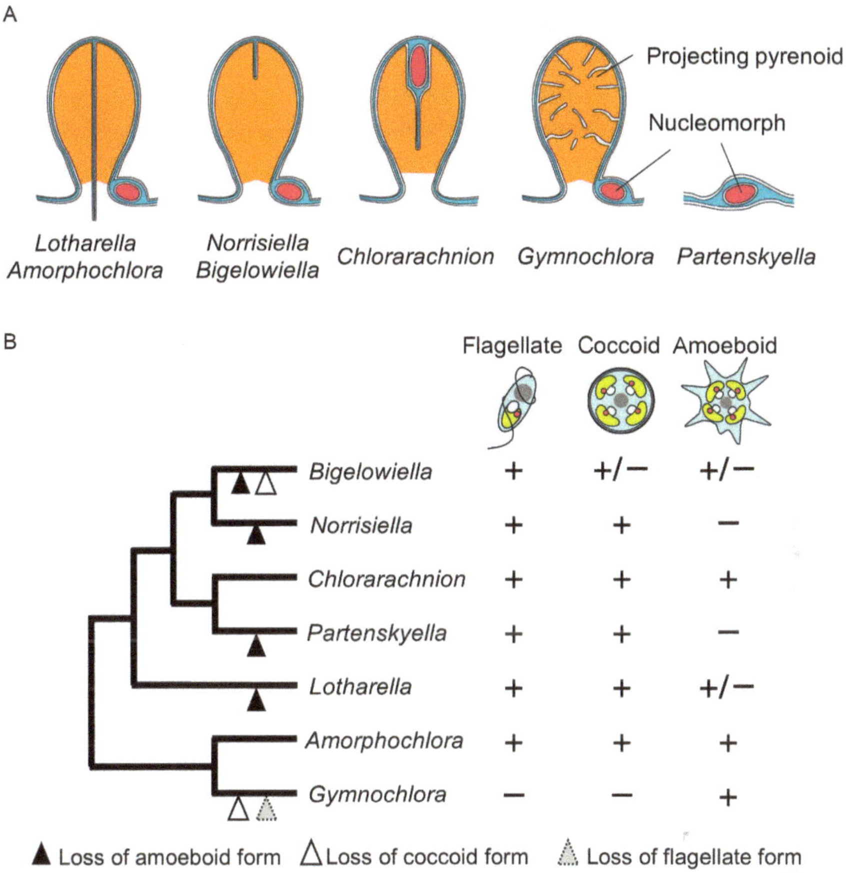

Fig. 2 Diversity of plastid ultrastructure and cell morphology in chlorarachniophytes. (A) Schematic illustration of pyrenoid ultrastructure and nucleomorph localisation in the respective chlorarachniophyte genera. (B) Morphological characters superimposed on a phylogenetic tree of nuclear 18S rDNA sequences. Each column indicate presence (+) or absence (−) of three cell types (amoeboid, coccoid, and flagellate) in life cycles. *Arrowheads* on the tree indicate evolutionary losses of cell forms.

of multiple-layered walls and the absence of reticulopodial networks in *Cr. perforans*.

The genus *Gymnochlora* consists of two species, *G. stellata* (Ishida et al., 1996) and *G. dimorpha* (Ota, Kudo, & Ishida, 2011), and this genus is characterised by a projecting pyrenoid penetrated by many tubular invaginations of the innermost plastid membranes, and a nucleomorph that is located in the pyrenoid base (Fig. 2A). The both species were isolated from samples collected in coral reefs. The vegetative cells of *Gymnochlora* are naked amoebae that do not form reticulopodial networks. Multinucleate giant cells are observed in older cultures, but walled coccoid cells and flagellates have never been found. The two *Gymnochlora* species can be distinguished by their cell sizes (*G. stellata* is 10–20 μm and *G. dimorpha* is 7–14 μm), and by the presence in *G. dimorpha* of two types of amoeboid cells (motile and sessile).

The genus *Bigelowiella* comprises two species, *B. natans* (Moestrup & Sengco, 2001) and *B. longifolia* (Ota, Ueda, & Ishida, 2007b). Both species are planktonic and have been isolated from water samples collected in the open ocean (Sargasso Sea). Their pyrenoids possess a shallow groove created by an invagination of the inner two plastid membranes; the depth of the groove is usually less than half the length of the pyrenoid (Fig. 2A). A nucleomorph is located in the pyrenoid base. This genus is also characterised by flagellate vegetative cells. The cells are 3–8 μm in size and carry one or two bilobate plastids. Although *B. natans* presents only a flagellate form through its life cycle, *B. longifila* has walled coccoid and naked amoeboid stages. Additionally, *Bigelowiella* species possess ejectile organelles (extrusomes) consisting of a short anterior and a long posterior cylinder.

Norrisiella sphaerica was isolated from the coast of Baja California in Mexico (Ota, Ueda, & Ishida, 2007a). The pyrenoid and nucleomorph ultrastructure of this species is similar to that of *Bigelowiella* (Fig. 2A). However, *N. sphaerica* can be distinguished from *Bigelowiella* by its cell division behaviour. The walled coccoid cells of *N. sphaerica* proliferate by producing autospores and zoospores; two to six coccoid or flagellate daughter cells are formed and released from a parental cell via a multinucleate coccoid stage. *N. sphaerica* entirely lacks an amoeboid life stage.

The genus *Lotharella* comprises five species, *L. globosa* (Ishida & Hara, 1994; Ishida et al., 1996), *L. oceanica* (Ota, Silver, Archibald, & Ishida, 2009), *L. polymorpha* (Dietz, Ehlers, Wilhelm, Gil-Rodríguez, & Schnetter, 2003), *L. reticulosa* (Ota & Vaulot, 2012), *L. vacuolata* (Ota, Ueda, & Ishida, 2005), and a variety of *L. globosa* (*L. globosa* var. *fortis*) (Hirakawa, Howe, James, & Keeling, 2011). They have been isolated from various environments including ocean water, coral reef, and sandy substrate. This genus is characterised by a projecting pyrenoid with a deep groove created by an invagination of the inner two plastid membranes; the groove passes through the entire length of the pyrenoid and reaches the plastid stroma (Fig. 2A). A nucleomorph is located in the pyrenoid base. The vegetative cell of *Lotharella* is a naked spherical cell in *L. oceanica* and a walled coccoid cell in the others. Filose pseudopodia have been occasionally observed extending from a pore in the coccoid wall, and interlinked pseudopodia form wide reticulopodial networks. Such cells are referred to as 'walled amoeboid cells'. All species of *Lotharella* possess a flagellate stage that is generally produced by changing the medium.

Amorphochlora amoebiformis was formerly classified in the genus *Lotharella* based on pyrenoid ultrastructure (Ishida, Ishida, & Hara, 2000), but it was subsequently placed into the new genus *Amorphochlora* (Ishida, Yabuki, & Ota, 2011) because the vegetative cell of *A. amoebiformis* is a naked amoeba without a cell wall and is clearly distinguishable from the vegetative cells of *Lotharella*. Flagellate and cyst-like coccoid cells are very rare under general culture conditions in this species.

Partenskyella glossopodia was described as the first picoplanktonic chlorarachniophyte with cells 2–4 μm in diameter (Ota, Vaulot, Le Gall, Yabuki, & Ishida, 2009). It was isolated from a filtered seawater sample collected in the Mediterranean Sea. The vegetative cell is a naked spherical cell possessing a single cup-shaped plastid that unusually lacks a pyrenoid through the life cycle. The naked spherical cells occasionally produce a tongue-shaped pseudopodium (lobopodium) that is different from the filose pseudopodia of other chlorarachniophytes. Flagellate cells and cyst-like coccoid cells are rarely observed. The minute cell size and the absence of a pyrenoid are unique characteristics of this species.

Culture strains of known chlorarachniophyte species except for *Cr. perforans* and *L. polymorpha* are available in the Roscoff Culture Collection (RCC), the National Center for Culture of Marine Phytoplankton (CCMP) at the Bigelow Laboratory for Ocean Sciences, the National Institute for Environmental Studies (NIES), and the Canadian Centre for the Culture of Microorganisms (CCCM) at the University of British Columbia. As outlined earlier, chlorarachniophyte species are difficult to identify at the genus/species level due to their small sizes, complex life cycles, and the fact that electron microscopic analyses are necessary to observe the main diagnostic characters of the plastids. To address this issue, a DNA-barcoding system based on the internal transcribed spacer sequence of nucleomorph ribosomal rDNA has been developed to rapidly assign unidentified cultures of chlorarachniophytes (Gile, Stern, James, & Keeling, 2010). It is expected that this barcoding system will also be useful in environmental surveys to assess the diversity and distribution of chlorarachniophytes.

4. PHYLOGENY AND MORPHOLOGICAL EVOLUTION IN CHLORARACHNIOPHYTES

As mentioned earlier, the chlorarachniophytes are divided into eight genera defined by their pyrenoid ultrastructures and their vegetative cell types. In addition to these morphological characters, molecular phylogenic studies have identified seven of the genera as distinct well-supported clades (the genus *Cryptochlora* is excluded due to the absence of any molecular data) (Gile et al., 2010; Ota & Vaulot, 2012; Ota, Vaulot, et al., 2009). Recent phylogenetic analyses based on nuclear 18S and 28S rDNA sequences have demonstrated a robust affiliation between two naked amoeboid lineages, *Amorphochlora* and *Gymnochlora*, and between *Bigelowiella* and *Norrisiella*, both of which carry a shallow-slit-type pyrenoid. *Chlorarachnion* and *Partenskyella* likely form a sister group to the clade of *Bigelowiella* and *Norrisiella*, while the branching position of the *Lotharella* clade is ambiguous. In order to elucidate the branching order of the respective clades of chlorarachniophytes, multiple-gene phylogenetic analyses will be necessary.

It has been assumed that the ancestor of the chlorarachniophytes had three basic cell types (amoeboid, coccoid, and flagellate) in its life cycle (Hirakawa et al., 2011; Ishida et al., 1999). After the divergence of the genera/species, several lineages presumably lost one or two cell types from their life cycles, resulting in the existing morphological diversity of vegetative cells. Under this scenario, the amoeboid cells of *Gymnochlora* have lost the ability to generate cell walls and flagella, and *B. natans* has discarded the ability to transform into amoeboid and coccoid forms. By contrast, *Chlorarachnion* and *Lotharella* still show all three basic cell types in their life cycles. Given the phylogeny, multiple independent losses of amoeboid and coccoid forms are predicted in chlorarachniophyte lineages, and the flagellate form has been eliminated only in the *Gymnochlora* lineage (Fig. 2B). It has been proposed that the evolution of chlorarachniophyte life cycles is associated with various habitats (Hirakawa et al., 2011; Ishida et al., 1999; Ota et al., 2005). Four planktonic species, *B. natans*, *L. oceanica*, *N. sphaerica*, and *P. glossopodia*, have been collected from pelagic water samples, and none possess an amoeboid form, which makes sense because these species would have no opportunity to attach to substrates (e.g. sand, seaweed, or coral reef). By contrast, three amoeboid species, *G. stellata*, *G. dimorpha*, and *A. amoebiformis*, have been found in coral reefs and sand beaches, and they mostly lack a flagellate form. The coccoid species *L. vacuolata* was isolated from a tidal pool which might be a stressful habitat due to rapid changes in temperature and salinity, and the formation of cell walls might help resist such environmental stresses.

5. NUCLEOMORPH GENOMES OF CHLORARACHNIOPHYTES

Although endosymbiont nuclei are entirely lost in most secondary endosymbioses, vestigial nuclei, referred to as nucleomorphs, are found in two algal lineages, chlorarachniophytes and cryptophytes. Nucleomorphs were first reported in cryptophytes as small double-membrane-bound organelles in the periplastidal compartment between the second and third plastid membranes (Greenwood, 1974; Greenwood, Griffiths, & Santore, 1977). Later on, a nucleomorph-like structure was found in the chlorarachniophyte *Ch. reptans* by electron microscopy (Hibberd & Norris, 1984). Ludwig and Gibbs have demonstrated that the *Ch. reptans* nucleomorph contains DNA by staining with 4′-6-diamidino-2-phenylindole (Ludwig & Gibbs, 1989). Furthermore, their ultrastructural study has revealed that the nucleomorph contains a nucleolus-like region composed of fibrils and granules, the nucleomorph envelope is interrupted by pores, and the nucleomorph divides by an infolding of the double membrane without chromosome condensation or microtubule involvement (Ludwig & Gibbs, 1989). In 1994, McFadden and colleagues have amplified nucleomorph 18S rDNA fragments from chlorarachniophytes, fragments that were distinct from the nuclear ones in size and sequence (McFadden, Gilson, Hofmann, Adcock, & Maier, 1994). This interesting investigation provided the first molecular evidence for the eukaryotic origin of chlorarachniophyte plastids. They have also found that the nucleomorph genomes in the genus *Bigelowiella* consist of three small chromosomes totalling 380 kb by an analysis using pulsed-field gel electrophoresis (PFGE) (Gilson & McFadden, 1995; McFadden et al., 1994). Further investigations with PFGE have revealed that the nucleomorph genomes of different chlorarachniophyte species invariably consist of three small chromosomes totalling from 330 to 1000 kb (Gilson & McFadden, 1999; Ishida, Endo, & Koike, 2011; Silver et al., 2007). It is recognised that chlorarachniophyte nucleomorph genomes are the smallest known nuclear genomes.

So far, complete nucleomorph genomes have been sequenced in four chlorarachniophytes: *B. natans* (Gilson et al., 2006), *L. oceanica* (Tanifuji,

Onodera, Brown, et al., 2014), *L. vacuolata* (Suzuki, Shirato, Hirakawa, & Ishida, 2015), and *A. amoebiformis* (Suzuki et al., 2015). These sequences provide insights into the evolutionary reduction of the endosymbiont genome over the course of secondary endosymbiosis. The nucleomorph genomes are basically conserved in terms of architecture. They consist of three linear chromosomes, each carrying subtelomeric regions that encode an rRNA operon (18S rDNA, 5.8S rDNA, and 28S rDNA) linked to a telomere (Fig. 3). The sequences of six subtelomeric regions are identical in the respective species, presumably as a result of ongoing intra- and interchromosomal recombination between the subtelomeric regions (Gilson & McFadden, 1997). The telomere sequence of *A. amoebiformis* is composed of [TCCTGGG] repeats, whereas the other three possess [TCTAGGG] repeats. These telomere sequences are thought to have been derived from

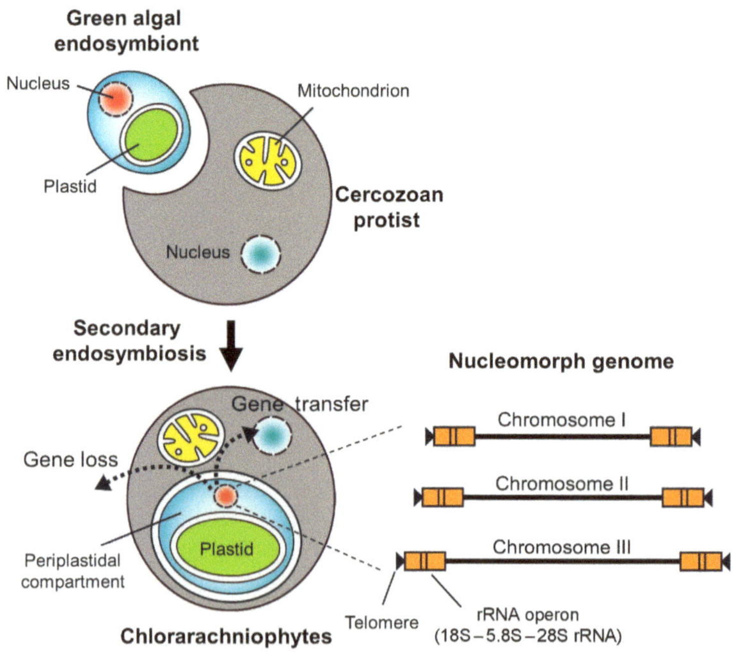

Fig. 3 Evolution of chlorarachniophytes via secondary endosymbiosis. Chlorarachniophyte algae acquired plastids through the endosymbiotic uptake of an ancestral green alga. The bulk of the genes carried in the endosymbiont nucleus have been either lost or transferred to the host nuclear genome, resulting in the vestigial nucleus of the endosymbiont, termed 'nucleomorph'. The nucleomorph genome consists of three short chromosomes, and an rRNA operon exists in the respective subtelomeric regions linked to telomere repeats.

the endosymbiont nuclear genome, because trebouxiophytes possess similar telomeric repeats of [TTTAGGG] (Fulnečková et al., 2012). Compared to other organelle genomes, the nucleomorph genomes have a high A + T content, ranging from 67% to 75%; the A + T content of the nucleomorph genome is 72%, which is higher than that of the plastid genome (70%), the mitochondrial genome (58%), and the nuclear genome (55%) in *B. natans* (Curtis et al., 2012; Gilson et al., 2006; Rogers et al., 2007). Unlike typical eukaryotic genomes, nucleomorph genomes exhibit an extremely gene-dense structure with shrinking noncoding regions. The gene density has been estimated to be 0.8–1.0 genes per kilobase, and the average length of the intergenic spacers is only 86–163 base pairs in the four nucleomorph genomes. Due to this compaction, overlapping transcription of multiple genes exists in nucleomorphs (Williams, Slamovits, Patron, Fast, & Keeling, 2005). Moreover, 800–1000 ultrasmall spliceosomal introns have been found in most nucleomorph genes. These introns range from 18 to 23 nucleotides with a canonical boundary sequence (GT and AG), and they are the smallest known spliceosomal introns in eukaryotes (Gilson & McFadden, 1996; Suzuki et al., 2015). The size reduction of the intergenic spaces and introns would have contributed to the compaction of nucleomorph genomes.

What kinds of genes are retained in nucleomorphs? The nucleomorph genomes of four chlorarachniophytes (*A. amoebiformis*, *B. natans*, *L. vacuolata*, and *L. oceanica*) are predicted to encode approximately 300 proteins as well as 19–22 transfer RNAs (tRNAs), 3–4 small nuclear RNAs (snRNAs), and 6 copies of rRNAs (Gilson et al., 2006; Suzuki et al., 2015; Tanifuji, Onodera, Brown, et al., 2014). The nucleomorph-encoded proteins have been divided into two groups: one is composed of proteins functionally annotated by homology searches against public databases, and the other consists of nucleomorph-specific hypothetical proteins, the so-called ORFans, that have no sequence similarity to any known protein. A total of 198 function-predicted proteins have been annotated from the 4 nucleomorph genomes, and 86% of them (171/198 proteins) are shared among the 4 genomes (Suzuki et al., 2015). These conserved proteins have been predicted to be mainly involved in housekeeping functions such as translation, transcription, DNA/RNA metabolism, and protein folding/degradation. However, these proteins are not sufficient for the maintenance of the nucleomorphs, and nucleus-encoded proteins compensate for the missing nucleomorph genes. Remarkably, the 4 nucleomorph genomes encode the exact same set of 17 plastid-associated proteins that are involved in protein transport (TOC75, TIC20, SecY, and TatC), protein folding and degradation (DnaK, Cpn60, Clp proteases), iron–sulphur cluster formation (SufB), transcription (RpoD), and other functions (MurL and Ycf16). It is thought that nucleomorph-encoded housekeeping proteins exist for the maintenance and expression of these 17 plastid-associated proteins. This means that if these plastid-associated genes were transferred to the host nucleus, the nucleomorphs would eventually disappear. Given the conservation of 200 housekeeping proteins as well as 17 plastid-associated proteins, nucleomorph genome reduction and compaction appear to have occurred early in the evolution of chlorarachniophytes, and the modern-day nucleomorph genomes would have been established before the divergence of the extant chlorarachniophyte species (Suzuki et al., 2015; Tanifuji, Onodera, Brown, et al., 2014).

In addition to function-predicted proteins, the 4 nucleomorph genomes have been estimated to encode 97–154 ORFans, although a mere 18 ORFans have been detected as conserved among the 4 genomes (Suzuki et al., 2015). Most ORFans have been demonstrated to be transcribed into mRNA in chlorarachniophytes (Hirakawa, Suzuki, Archibald, Keeling, & Ishida, 2014; Tanifuji, Onodera, Moore, & Archibald, 2014), but it remains unknown whether the ORFans correspond to real protein-coding genes. One intriguing question is the origins of the nucleomorph ORFans. A comparative analysis of nucleomorph genomes has provided clues to this question. Nucleomorph genomes carry syntenic regions that are blocks of several genes conserved between different species. Several syntenic regions include ORFans that are located at the same positions as the function-predicted genes found in other nucleomorph genomes. For instance, the nucleomorph genome of *L. vacuolata* encodes an ORFan of 776 amino acids between the *mce* and *U5snRNP* genes, whereas *L. oceanica* has the *tbl3* gene (780 amino acids) at the same position (Suzuki et al., 2015). This ORFan is presumed to have been derived from a *tbl3* gene based on its position and size. This suggests that nucleomorph ORFans have evolved from certain functional proteins through drastic changes in their sequences. Nevertheless, it remains an open question whether the syntenic ORFans still have their original functions, and if they have no function, why the extremely reduced nucleomorph genomes still carry a large number of ORFans.

Intriguingly, nucleomorph genes appear to have evolved much faster than the nuclear genes (Patron, Rogers, & Keeling, 2006). Amino acid substitution rates of nucleomorph-encoded ribosomal proteins have been calculated to be approximately three times higher than those of their nuclear counterparts in three chlorarachniophytes (Hirakawa et al., 2014). A previous study suggests that the possible loss of canonical genes involved in nucleomorph DNA replication and repair might be related to the accelerated evolution (Curtis et al., 2012). As another possibility, we have proposed that highly expressed molecular chaperones might have made the high evolutionary rates of nucleomorph genomes possible (Hirakawa et al., 2014). It is known that overexpression of chaperonins masks some of the deleterious effects of mutated proteins, which allows for an increase in the number of accumulating mutations in bacteria (Maisnier-Patin et al., 2005; Tokuriki & Tawfik, 2009). Our transcriptome analysis has indicated that the genes for nucleomorph-encoded heat-shock proteins (NmHsp70 and NmHsp90) are highly expressed compared to other genes in chlorarachniophytes, and the mRNAs derived from these two genes are estimated to comprise approximately 10% of the total nucleomorph-derived mRNAs. It is thus possible to infer that abundant molecular chaperones might compensate for the rapid evolution of nucleomorph genes by buffering mutations (Hirakawa et al., 2014).

Although chlorarachniophyte nucleomorph genomes share conserved features among different species, they exhibit remarkable differences in genome size with estimates ranging from 330 to 1000 kb (Gilson & McFadden, 1999; Ishida, Endo, et al., 2011; Silver et al., 2007). What has caused this size variation? The known complete nucleomorph genome sequences are 372, 374, 432, and 611 kb in *B. natans*, *A. amoebiformis*, *L. vacuolata*, and *L. oceanica*, respectively (Gilson et al., 2006; Suzuki et al., 2015; Tanifuji, Onodera, Brown, et al., 2014). Comparative analyses of these genomes have revealed that the primary factor leading to the size variation is the existence of multiple duplicated genes spreading in the nucleomorph genomes of *Lotharella*. The *B. natans* nucleomorph genome has no duplicated protein-coding genes, and only the *dnaK* gene is found to be duplicated in the *A. amoebiformis* nucleomorph genome. In contrast, 18 and 61 protein-coding genes are duplicated in the *L. vacuolata* and *L. oceanica* nucleomorphs, respectively; some of these are shared between these two species. These observations suggest that expansions of the nucleomorph genomes have occurred through gene duplication in the *Lotharella* lineages before and after the species divergence. However, there are no clues as to the functional significance of those duplicated genes, because most of them are ORFans of unknown function.

One of the most fundamental questions is whether reductive nucleomorph evolution is ongoing or has approached an endpoint. The obvious conservation of function-predicted nucleomorph genes, including the exact same set of plastid-associated genes, suggests that massive genome reduction occurred in the common ancestor of chlorarachniophytes prior to the divergence of the extant species. A limited number of lineage-specific gene losses have been observed in the respective species. These missing nucleomorph genes have been surveyed in the nuclear genomes, but no evidence has been found for recent gene transfers from the nucleomorph to nuclear genomes (Suzuki et al., 2015; Tanifuji, Onodera, Brown, et al., 2014). One possible explanation is that the difference in A+T content between the nucleomorph and nuclear genomes may be a barrier to successful gene transfer (Suzuki et al., 2015). Overall, these observations suggest that chlorarachniophyte nucleomorph genomes have almost reached an endpoint in reductive evolution, and it is unlikely that nucleomorphs will eventually disappear.

6. CONVERGENT EVOLUTION OF NUCLEOMORPH GENOMES IN CHLORARACHNIOPHYTES AND CRYPTOPHYTES

As noted earlier, vestigial endosymbiont nuclei, termed nucleomorphs, have been found in chlorarachniophytes as well as in cryptophytes. In cryptophytes, complete nucleomorph genomes have been sequenced in four species: *Guillardia theta* (Douglas et al., 2001), *Hemiselmis andersenii* (Lane et al., 2007), *Cryptomonas paramecium* (Tanifuji et al., 2011), and *Chroomonas mesostigmatica* (Moore, Curtis, Mills, Tanifuji, & Archibald, 2012). Chlorarachniophytes and cryptophytes clearly arose by separate secondary endosymbioses of a green and a red alga, respectively (Douglas, Murphy, Spencer, & Gray, 1991; Douglas & Penny, 1999; Rogers et al., 2007). Despite their independent origins, the nucleomorph genomes of these two lineages are strikingly similar in architecture (reviewed by Archibald & Lane, 2009; Gilson, 2001; Moore & Archibald, 2009). One of the most obvious similarities is the presence of three short chromosomes of several hundred kilobases. The nucleomorph genomes have reductively converged on the same number of chromosomes from different algal endosymbionts that presumably harboured different number of chromosomes. It remains unknown whether this similarity is a strange coincidence or a necessary consequence. Douglas and colleagues have proposed that nucleomorph chromosomes may have a size limitation for proper segregation, and this may be related to the chromosome number (Douglas et al., 2001). Most eukaryotic chromosomes exhibit a high degree of condensation during mitosis, and condensed chromosomes of that kind have never been observed in nucleomorphs by electron microscopy (Hansmann, Falk, Scheer, & Sitte, 1986; Ludwig & Gibbs, 1989; Meyer, 1987), which is supported by the finding that nucleomorph histone proteins lack the multiple phosphorylation sites that are involved in mitotic chromosome condensation (Marinov & Lynch, 2016). Therefore, three short chromosomes, instead of an aggregated large one, are likely to be appropriate for segregation within a small nucleomorph of less than 1 μm (Douglas et al., 2001). However, the largest nucleomorph chromosome known, approximately 450 kb in size, has recently been found in the chlorarachniophyte *P. glossopodia*, a species that does possess a typical small nucleomorph; this chromosome is more than four times as large as the smallest known one (98 kb) in *B. natans* (Ishida, Endo, et al., 2011). This implies that the size limitation is unlikely to be the principal reason for three short chromosomes in nucleomorphs.

As another remarkable similarity, nucleomorph chromosomes commonly possess conserved subtelomeric regions including an rRNA operon (18S rDNA, 5.8S rDNA, and 28S rDNA) and a telomere. In cryptophyte nucleomorph genomes, a 5S rDNA has also been found next to telomeres consisting of either [GA] or [$(AG)_7AAG_6A$] repeats. The cryptophyte subtelomeric regions can be distinguished from the chlorarachniophyte ones strictly by their telomere sequences and the presence/absence of 5S rDNA. Nevertheless, the basic architecture of the subtelomeric regions is similar between chlorarachniophytes and cryptophytes, and again these have evolved independently from distinct endosymbiont genomes. It is interesting to note that an rRNA operon linked to every telomere has also been found in the reduced nuclear genomes of parasitic protists such as the microsporidia *Encephalitozoon* (Brugere, Cornillot, Metenier, & Vivares, 2000; Corradi, Pombert, Farinelli, Didier, & Keeling, 2010; Katinka et al., 2001) and the diplomonad *Giardia* (Adam, Nash, & Wellems, 1991; Upcroft, Abedinia, & Upcroft, 2005). Moreover, tandemly arrayed rRNA operons associated with telomere repeats are frequently found in some eukaryotic genomes of plants and animals (Dvořáčková, Fojtová, &

Fajkus, 2015; Ocalewicz, 2013), and it has been suggested that such telomeric rDNA sequences may contribute to the stabilisation of chromosome termini in plants (Pich, Fuchs, & Schubert, 1996). Taken together, the organisation of subtelomeric rRNA operons might have some advantages that have led to convergent evolution in these reduced nuclear genomes.

The nucleomorph genomes of chlorarachniophytes and cryptophytes show a reasonable similarity in gene composition. When the conserved genes in 4 chlorarachniophytes are compared with those in 4 cryptophytes, 93 protein genes (~50%) are found to be shared, and half of these are predicted to be involved in translation (Suzuki et al., 2015). With respect to plastid-associated nucleomorph genes, 31 genes are found in three cryptophytes (excluding the nonphotosynthetic *Cr. paramecium*), and 17 genes exist in the 4 chlorarachniophytes (Suzuki et al., 2015; Tanifuji, Onodera, Brown, et al., 2014). Among chlorarachniophytes and cryptophytes, four plastid-associated nucleomorph genes (*clpP1*, *clpP2*, *cpn60*, and *rpoD*) are shared. Although it is difficult to say whether the number of conserved genes count as 'a lot' or 'a little', nucleomorph genomes have independently retained a particular set of genes in both chlorarachniophytes and cryptophytes.

We recently investigated the ploidy of nucleomorph genomes, and our PCR-based quantification demonstrated that nucleomorphs contain multiple sets of three chromosomes (Hirakawa & Ishida, 2014). The *B. natans* nucleomorph genome is diploid ($2n=6$ chromosomes) and the cryptophyte *G. theta* carries a tetraploid nucleomorph genome ($4n=12$ chromosomes). Two endosymbiotically derived organelles, plastids and mitochondria, usually contain higher genome copy numbers compared to their bacterial relatives. Given that, it seems likely that genomic polyploidisation is a general feature in endosymbiotically derived organelles including nucleomorphs (Hirakawa & Ishida, 2014). It is possible that the polyploidy of nucleomorph genomes affects the amount of gene product, because the mRNA transcript levels of nucleomorph genes are significantly higher than their counterparts in haploid nuclear genomes (Hirakawa et al., 2014; Tanifuji, Onodera, Moore, et al., 2014).

Overall, the nucleomorph genomes of chlorarachniophytes and cryptophytes have obvious similarities in architecture, gene content, and ploidy. Given that nucleomorph genomes have convergently evolved via different routes from different starting points, there must have been similar reductive pressures on nucleomorph genomes in both algal groups.

7. PLASTID GENOMES OF CHLORARACHNIOPHYTES

A decade ago, Rogers and colleagues reported the first complete plastid genome in the chlorarachniophyte *B. natans*, and this genome at 69.2 kb was the smallest known photosynthetic plastid genome at that time (Rogers et al., 2007). To date, four more plastid genomes have been sequenced in *L. oceanica* (Tanifuji, Onodera, Brown, et al., 2014), *L. vacuolata* (Suzuki, Hirakawa, et al., 2016), *G. stellata* (Suzuki, Hirakawa, et al., 2016), and *P. glossopodia* (Suzuki, Hirakawa, et al., 2016). These five plastid genomes display a high degree of conservation in size, gene content, and gene order, in contrast to nucleomorph genomes. The chlorarachniophyte plastid genomes are circular and they range from 67.4 to 72.6 kb, which is smaller than the plastid genomes of core chlorophytes, which range from 94 to 521 kb (Brouard, Otis, Lemieux, & Turmel, 2010; Turmel, Otis, & Lemieux, 2015). The chlorarachniophyte plastid genomes are all predicted to encode almost the same set of 59–61 proteins, 6 rRNAs, and 29–31 tRNAs, thus representing an unusually high gene density (gene-coding regions occupy 85%–87% of the genomes). In comparison with chlorophytes, chlorarachniophytes lack 13 plastid genes that are generally conserved in plastid genomes of core chlorophytes: *petL*, *psaM*, *psbZ*, *rpl12*, *rpl32*, *rps9*, *infA*, *ccsA*, *cemA*, *chlB*, *chlL*, *chlN*, and *ftsH* (*ycf2*). Five homologues of these missing genes (*rpl12*, *rpl32*, *rps9*, *infA*, and *ftsH*) are found in the nuclear genome of *B. natans*, suggesting that some of the plastid genes have been either lost or transferred to the nuclear genome via secondary endosymbiosis (Curtis et al., 2012; Suzuki, Hirakawa, et al., 2016). Given the conservation of plastid gene content among species, this gene loss and transfer would likely have occurred in the common ancestor of chlorarachniophytes. As an aside, in contrast to the plastid genomes, recent migration of mitochondrial genes encoding two ribosomal proteins (*rps1* and *rpl16*) has been observed in the nuclear genome of *L. oceanica* but not *B. natans* (Tanifuji, Archibald, & Hashimoto, 2016). Additionally, seven DNA fragments (less than 300 nucleotides) of the *B. natans* mitochondrial genome have been found in its nuclear genome, suggesting recent endosymbiotic gene transfer (EGT) from the mitochondria to the nucleus (Curtis et al., 2012). However, no recent gene transfers from the plastid and nucleomorph genomes have been detected (Curtis et al., 2012). These findings support the idea that DNA transfer from mitochondria to nucleus is more frequent than from plastids or nucleomorphs.

The gene order is mostly identical among the five plastid genomes of chlorarachniophytes, which exhibit a canonical quadripartite structure consisting of two inverted repeats encoding a small and large subunit (SSU and LSU) rRNA, a 5S rRNA, four tRNAs, and a couple of protein genes. Several rearrangement events have been observed among the five genomes, e.g., an inversion of *petB* and *petD* genes and a strand switch of *trnS* genes between *P. glossopodia* and *L. vacuolata*. However, the rearrangement frequency in the chlorarachniophyte plastid genomes is estimated to be clearly lower than in chlorophyte plastid genomes (Suzuki, Hirakawa, et al., 2016). This may be explained by the high gene density of chlorarachniophyte plastid genomes, which apparently increases the risk of gene disruption via rearrangement.

In terms of introns, chlorarachniophyte plastid genomes carry self-splicing group I and group II introns (Suzuki, Hirakawa, et al., 2016). A group I intron has been found in a tRNA gene (*trnL*) of the plastid genomes of *L. vaculolata*, *L. oceanica*, and *P. glossopodia*. This intron appears to have arisen from the endosymbiont, because diverse chlorophytes also have a group I intron in their *trnL* genes at the same position. Presumably, *B. natans* and *G. stellata* have independently lost their group I intron after the divergence of the species. Furthermore, group II introns are found to be present in two genes (*ycf3* and *psbM*) of three chlorarachniophytes, *G. stellata* carries only the *psbM* intron, and both introns are absent in *B. natans*. The intron positions in the respective genes are identical among the different chlorarachniophytes, but differ from their positions in chlorophytes. Given that, chlorarachniophytes are likely to have obtained their group II introns in a common ancestor, and these introns have been secondarily lost in some of the species. Overall, chlorarachniophyte plastid genomes seem to have been reduced and compacted by discarding several protein genes, noncoding regions, and introns through the secondary endosymbiosis.

8. PROTEIN TARGETING INTO CHLORARACHNIOPHYTE PLASTIDS

In secondary plastid-bearing algae, many nuclear genes encoding plastid proteins underwent two rounds of EGT. During the primary endosymbiosis, massive numbers of genes disappeared from the integrated endosymbiont genome, and the bulk of the essential genes for plastid maintenance were relocated to the host nuclear genomes (Martin et al., 1998; Timmis, Ayliffe, Huang, & Martin, 2004). Such gene movements from endosymbiont to host have also occurred during subsequent secondary endosymbioses (Archibald, 2015). Therefore, the protein products of EGT-derived genes are now targeted back to the plastids, and the development of protein transport systems is considered a fundamental requirement in the transition from an endosymbiont to a host-dependent organelle (Bhattacharya, Archibald, Weber, & Reyes-Prieto, 2007; Cavalier-Smith & Lee, 1985). Because plastids are surrounded by multiple envelope membranes, nucleus-encoded proteins synthesised in the cytoplasm must be able to overcome these membrane barriers.

The primary plastids of plants and some algae are enclosed in double membranes, and plastid protein import is achieved by translocon complexes named TOC and TIC that reside in the outer and inner envelope membranes, respectively. The TOC complex is postulated to consist of three core proteins, TOC34, TOC75, and TOC159 (the numbers correspond to their calculated molecular weights) (Andrès, Agne, & Kessler, 2010; Schwenkert, Soll, & Bölter, 2011; Shi & Theg, 2013). The components of the inner membrane translocon are still under debate (Bölter & Soll, 2016; Nakai, 2015). Two proteins, TIC20 and TIC110, have been proposed to be alternative channel-forming core components: TIC110 is associated with TIC40 and the molecular chaperone Hsp93 (Kovács-Bogdán, Soll, & Bölter, 2010), while TIC20 forms a 1 MDa complex consisting of TIC56, TIC100, and TIC214 (Kikuchi et al., 2013). Three additional proteins, TIC32, TIC55, and TIC62, are supposed to play a role in redox-dependent import regulation (Hörmann et al., 2004; Küchler, Decker, Hörmann, Soll, & Heins, 2002). For proper targeting, the most nucleus-encoded plastid proteins are synthesised as precursor proteins (preproteins) carrying N-terminal transit peptides; they vary in length and have no consensus motifs, and are generally characterised by very few acidic residues and abundant hydroxylated serine residues (Bruce, 2000; Li & Teng, 2013). Transit peptides are recognised by the receptor proteins TOC34 and TOC159 and are cleaved upon import into the stroma.

In contrast to plants, chlorarachniophytes have more complex plastids that are surrounded by four membranes. It is postulated that the inner two membranes originated from the primary plastid membranes, and the first and second outermost membranes correspond to the former phagosomal membrane of the host and the former plasma membrane of the endosymbiont, respectively (Cavalier-Smith, 2000; McFadden et al., 1994). A nucleomorph resides in the periplastidal compartment between the outer and inner pairs of plastid membranes, which corresponds to the remnant cytoplasm of the endosymbiont. Such structural complexity raises the question of how nucleus-encoded proteins are imported into the plastids. The first examples of nucleus-encoded plastid proteins in chlorarachniophytes, light-harvesting complex (LHC) proteins, were reported in *B. natans* (Deane et al., 2000). The LHC proteins carry an N-terminal extension, referred to as the bipartite presequence, that encodes a hydrophobic leader sequence, followed by a transit peptide-like sequence with very few acidic residues and relatively abundant serine residues (Deane et al., 2000). Following this discovery, Archibald and colleagues identified 78 putative plastid-targeted proteins, most of which contain bipartite presequences in their N-termini (Archibald, Rogers, Toop, Ishida, & Keeling, 2003). The hydrophobic readers of these proteins are chemically similar to the signal peptides of endoplasmic reticulum (ER)-targeted proteins and are predicted to carry possible cleavage sites (Rogers et al., 2004). The transit peptide-like sequences have no consensus sequences and are enriched in serine and arginine but not in acidic residues, and so are reminiscent of plant transit peptides (Rogers et al., 2004). A genetic transformation system using a green fluorescent protein (GFP) gene was developed for *A. amoebiformis*, and this opened up a door to investigate protein transport in chlorarachniophytes (Hirakawa, Kofuji, & Ishida, 2008). In a previous study, we confirmed the plastid-targeting capability of N-terminal bipartite presequences by using GFP fusions (Hirakawa, Nagamune, & Ishida, 2009). The hydrophobic leader sequences are involved in targeting preproteins to the ER, and transit peptide-like sequences are necessary for import into the plastid stroma. Furthermore, site-directed mutagenesis has revealed that a pronounced net positive charge generated by several basic amino acids in the transit peptide-like sequences is essential for passage across the inner two of the four plastid membranes (Hirakawa et al., 2009). Contemporaneously, Gile and Keeling reported that nucleus-encoded EFL proteins targeted to the periplastidal compartment (PPC) are also equipped with an N-terminal bipartite presequence like the plastid-targeted preproteins (Gile & Keeling, 2008). Transit peptide-like sequences of PPC-targeted EFL and plastid-targeted preproteins exhibit similar amino acid compositions, with the exception of their glutamic acid residues. It has since been demonstrated that several acidic amino acids counteracting a net positive charge in transit peptide-like sequences are essential for stopping protein import by the PPC without passing through the inner two membranes (Hirakawa, Gile, Ota, Keeling, & Ishida, 2010). In addition, the EFL preproteins are found to carry a curious C-terminal extension rich in lysine and aspartic acid residues that functions as a PPC-retention signal (Gile & Keeling, 2008; Hirakawa et al., 2010). Taken together, these investigations suggest that PPC- and plastid-targeted preproteins are transported via a similar ER-mediated pathway, and the net charges of transit peptide-like sequences play an important role in the accurate sorting of these two types of preproteins in the PPC (Fig. 4). This model

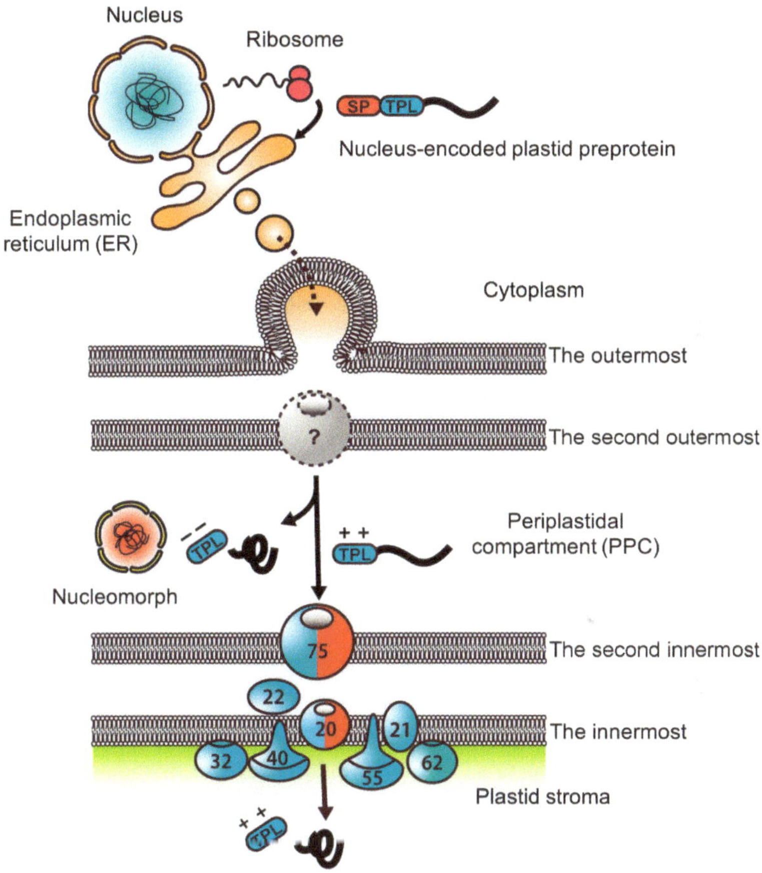

Fig. 4 A model of plastid targeting of nucleus-encoded proteins in chlorarachniophytes. Nucleus-encoded preproteins carrying an N-terminal bipartite presequence, a signal peptide (SP), and a transit peptide-like sequence (TPL) are first cotranslationally imported into the ER concomitant with the removal of the SP. Subsequently, these preproteins are possibly transported by vesicles to the second outermost membrane of the plastid. Once in the PPC, plastid-targeted proteins are sorted from PPC-targeted proteins according to the charge of their TPL. Several plant-like TOC and TIC proteins may be involved in transport across the inner two membranes. Nucleus- and nucleomorph-encoded translocons are highlighted with *blue* and *red*, respectively.

is also supported by the fact that basic residues, especially lysine, are abundant in the putative N-terminal presequences of nucleomorph-encoded plastid-targeted proteins (Gile & Keeling, 2008).

Interestingly, N-terminal bipartite presequences and ER-mediated transport pathways of nucleus-encoded plastid proteins have been broadly reported in secondary plastid-bearing organisms (Bolte et al., 2009; Patron & Waller, 2007; Sheiner & Striepen, 2013). Similar protein-targeting systems have independently evolved by distinct secondary endosymbioses, which may be a consequence of the similar integration processes of the endosymbionts. Engulfed endosymbionts are assumed to be initially enclosed in a phagocytic vacuole, and the outermost membrane of secondary plastids is expected to correspond to the phagosomal membrane, suggesting that the secondary plastids still lie topologically outside the host cells (Cavalier-Smith, 2000; van Dooren, Schwartzbach, Osafune, & McFadden, 2001). Therefore, it makes sense that nucleus-encoded proteins are transported through the ER-mediated secretory pathway en route to the secondary plastids. It is worth noting that the outermost plastid membrane in cryptophytes, haptophytes, and heterokonts is continuous with the host ER decorated with 80S ribosomes. To explain this observation, Cavalier-Smith has hypothesised that the outermost phagosomal membrane has fused with the host ER (Cavalier-Smith, 2000), while Gould and colleagues have speculated instead that the outer two membranes surrounding an endosymbiont have been replaced by double-layered membranes arising from elongated host ER (Gould, Maier, & Martin, 2015). In terms of transit peptide-like sequences, there is a remarkable difference in functional domains between chlorarachniophytes and other lineages possessing red algal-derived secondary plastids, the so-called red secondary plastids. It has been found that transit peptide-like sequences carry a conserved aromatic amino acid, typically phenylalanine, at the first position in heterokonts (Kilian & Kroth, 2005), cryptophytes (Gould et al., 2006), haptophytes (Patron, Waller, & Keeling, 2006), apicomplexans (Ralph, Foth, Hall, & McFadden, 2004), dinoflagellates (Patron, Waller, Archibald, & Keeling, 2005), and chromerids (Woehle, Dagan, Martin, & Gould, 2011), which has been experimentally demonstrated to be required for passage across the inner two plastid membranes (Gould et al., 2006; Gruber et al., 2007; Kilian & Kroth, 2005). A phenylalanine motif is commonly observed in transit peptides of rhodophytes and glaucophytes within a few residues of their N-termini, but it is absent from chlorophyte transit peptides (Patron & Waller, 2007). These observations suggest that the essential aromatic signal in transit peptide-like sequences is preserved in the 'nongreen' lineages, and chlorarachniophytes make use of an alternate signal consisting of basic amino acids interspersed with the transit peptide-like sequences, which would have been derived from the green algal endosymbiont.

In recent years, the nuclear genome sequencing of *B. natans* has been completed, thereby providing valuable information (Curtis et al., 2012). According to in silico predictions based on N-terminal presequences, approximately 700 and over 1000 nucleus-encoded proteins are estimated to be targeted to the plastid and the PPC, respectively (Curtis et al., 2012; Suzuki, Ishida, & Hirakawa, 2016). A proteomic analysis of isolated *B. natans* plastids has demonstrated that over 100 putative plastid-targeted proteins with N-terminal bipartite presequences exist in the plastids (Hopkins et al., 2012). To import the bulk of the proteins into chlorarachniophyte plastids, a translocation apparatus is presumed to exist in the respective membranes (Fig. 4). As the first step, nucleus-encoded plastid preproteins would be cotranslationally transported into the ER in the same way as other secretory proteins, and possibly across the outermost plastid membrane via ER-derived vesicle transport (Hirakawa et al., 2009). The translocon at the second outermost membrane remains largely unknown in chlorarachniophytes. Symbiont-derived ER-associated protein degradation (ERAD) machinery, termed SELMA, has been identified as the translocon for crossing the second outermost membrane of red secondary plastids (Hempel, Bullmann, Lau, Zauner, & Maier, 2009; Stork et al., 2012). However, core components of SELMA, such as derlin proteins, have not been found in the *B. natans* genome, suggesting that chlorarachniophytes have evolved an alternative translocon for the second outermost membrane (Hirakawa, Burki, & Keeling, 2012b). It is important to elucidate this translocon to gain a better understanding of the endosymbiotic process in chlorarachniophytes. As for putative translocons of the inner two membranes, plant-like TOC and TIC proteins have been identified in *B. natans* (Hirakawa et al., 2012b). Homologous genes for channel-forming TOC75 and TIC20 are found in both the nuclear and nucleomorph genomes, and other components (e.g. TIC21, TIC22, TIC32, TIC40, TIC55, and TIC62) are encoded by the nuclear genome (Fig. 4). These endosymbiotically derived TOC and TIC homologues may be involved in protein transport across the inner two plastid membranes. Although the outer membrane of plant plastids is equipped with transit peptide receptors (TOC34 and TOC159) to distinguish plastid preproteins from the large pool of cytoplasmic proteins, such receptors have not been detected by homology searches in the chlorarachniophytes. Given that the PPC is massively reduced compared to its original cytoplasm in terms of protein composition, chlorarachniophytes may not require a strict sorting mechanism of plastid preproteins at the second innermost plastid membrane (Hirakawa et al., 2012b). Nevertheless, it is certainly true that chlorarachniophytes have an unknown mechanism for recognising transit peptide-like sequences in the PPC.

Lastly, I describe two noncanonical plastid-targeting signals found in chlorarachniophytes. The first, the RubisCO small subunit protein of *A. amoebiformis* has an N-terminal bipartite presequence as well as an internal targeting signal in its mature protein that is required for transport from the ER to the plastids (Hirakawa & Ishida, 2010). This internal signal may be specialised for targeting the highly abundant RubisCO proteins. The second is a dual targeting signal for plastids and mitochondria found in two aminoacyl-tRNA synthetases of *B. natans* (Hirakawa, Burki, & Keeling, 2012a). Genes for a histidyl-tRNA and glycyl-tRNA synthetases alternatively generate two products translated by either the first or the second ATG start codons, and these are equipped with a bipartite presequence and a mitochondrial targeting peptide, respectively. It is likely that dual protein targeting to two different organelles is a common feature in primary and secondary plastid-bearing organisms, despite their evolutionary divergence.

9. EVOLUTIONARY MOSAICISM OF CHLORARACHNIOPHYTE NUCLEAR GENOMES

As described earlier, chlorarachniophyte plastids strongly depend on nucleus-encoded proteins, many of which are assumed to be derived from the green algal endosymbiont via EGT. Nevertheless, a significant proportion of plastid-associated proteins exhibit phylogenetic affinity to lineages other than chlorophytes. It has been reported that 11 of 62 plastid-associated proteins in *B. natans* are closely related to those of streptophytes, rhodophytes, or red secondary plastid-bearing algae (Archibald et al., 2003), and phylogenetic analyses of several cyanobacterial-derived proteins involved in the Calvin cycle (Yang, Maruyama, Sekimoto, Sakayama, & Nozaki, 2011), the haem pathway (Cihlář, Füssy, Horák, & Oborník, 2016), and plastid division (Hopkins et al., 2012; Yang, Matsuzaki, Takahashi, Qu, & Nozaki, 2014) suggest an affiliation between chlorarachniophytes and red lineages. Furthermore, the nuclear genome of *B. natans* has been found to carry 45 genes classified as being derived from red algae, and these are called 'red' genes (Curtis et al., 2012). This mosaicism raises the question of how chlorarachniophytes have acquired such 'nongreen' genes. Two major hypotheses address this question. First, it has been proposed that multiple lateral gene transfer (LGT) events have occurred, mainly from red lineages and possibly by means of an ancestral mixotrophic chlorarachniophyte feeding on algal prey (Archibald et al., 2003). Indeed, two extant chlorarachniophyte species, *Ch. reptans* and *Cr. perforans*, are able to engulf small algae (Calderon-Saenz & Schnetter, 1989; Hibberd & Norris, 1984). Alternatively, a cryptic endosymbiosis of chlorarachniophytes has been proposed. Under this hypothesis, the common ancestor of chlorarachniophytes might have carried a red plastid prior to the secondary endosymbiosis of a green alga, and the 'red' genes in modern-day chlorarachniophyte nuclear genomes might be footprints of EGT from a preexisting red algal endosymbiont (Yang et al., 2014). A similar situation has been described in red secondary plastid-bearing algae; a large number of 'green' genes have been predicted in diatom genomes, possibly representing a cryptic endosymbiosis involving a green alga before the establishment of red secondary plastids (Moustafa et al., 2009). However, this hypothesis has been found to be unconvincing according to a reevaluation by (Deschamps & Moreira, 2012).

It is worth mentioning that two types of nucleus-encoded proteins essential for maintaining the plastids and nucleomorphs have been found to be 'nongreen' genes. The first is the plastid division-related FtsZ proteins, which are known to assemble into a ring on the stromal surface of the inner plastid membrane and generate contractile force for membrane constriction in primary plastids (TerBush, Yoshida, & Osteryoung, 2013). It has been confirmed that FtsZ proteins also form a ring-like structure that could be involved in partial constriction of the inner pair of plastid membranes in chlorarachniophytes (Hirakawa & Ishida, 2015). Intriguingly, the FtsZ proteins of chlorarachniophytes show a phylogenetic affiliation with red secondary plastid-bearing algae, such as haptophytes and heterokonts (Hirakawa & Ishida, 2015; Hopkins et al., 2012; Yang et al., 2014). The other nucleus-encoded protein is the core component for nucleomorph DNA replication. Recently, a homologue of DNA polymerase delta in *B. natans* has been demonstrated to be targeted to the nucleomorph (Suzuki, Ishida, et al., 2016). It is quite surprising to find that the nucleomorph DNA polymerase appears to have a close relationship with viral sequences rather than eukaryotic genes (Blanc, Gallot-Lavallée, & Maumus, 2015; Suzuki, Ishida, et al., 2016). Taken together, these observations suggest that some key genes for the maintenance of chlorarachniophyte plastids have been replaced by laterally transferred genes. Although the reasons for these gene replacements remain unknown, LGT likely had an impact on the evolutionary process of chlorarachniophytes.

REFERENCES

Adam, R. D., Nash, T. E., & Wellems, T. E. (1991). Telomeric location of *Giardia* rDNA genes. *Molecular and Cellular Biology*, *11*, 3326–3330.

Adl, S. M., Simpson, A. G. B., Farmer, M. A., Andersen, R. A., Anderson, O. R., Barta, J. R., et al. (2005). The new higher level classification of eukaryotes with emphasis on the taxonomy of protists. *Journal of Eukaryotic Microbiology*, *52*, 399–451.

Andrès, C., Agne, B., & Kessler, F. (2010). The TOC complex: Preprotein gateway to the chloroplast. *Biochimica et Biophysica Acta*, *1803*, 715–723.

Archibald, J. M. (2015). Genomic perspectives on the birth and spread of plastids. *Proceedings of the National Academy of Sciences of the United States of America*, *112*, 10147–10153.

Archibald, J. M., & Lane, C. E. (2009). Going, going, not quite gone: Nucleomorphs as a case study in nuclear genome reduction. *Journal of Heredity*, *100*, 582–590.

Archibald, J. M., Rogers, M. B., Toop, M., Ishida, K., & Keeling, P. J. (2003). Lateral gene transfer and the evolution of plastid-targeted proteins in the secondary plastid-containing alga *Bigelowiella natans*. *Proceedings of the National Academy of Sciences of the United States of America*, *100*, 7678–7683.

Bass, D., Chao, E. E., Nikolaev, S., Yabuki, A., Ishida, K., Berney, C., et al. (2009). Phylogeny of novel naked filose and reticulose Cercozoa: Granofilosea cl. n. and Proteomyxidea revised. *Protist*, *160*, 75–109.

Bhattacharya, D., Archibald, J. M., Weber, A. P. M., & Reyes-Prieto, A. (2007). How do endosymbionts become organelles? Understanding early events in plastid evolution. *BioEssays*, *29*, 1239–1246.

Bhattacharya, D., Helmchen, T., & Melkonian, M. (1995). Molecular evolutionary analyses of nuclear-encoded small subunit ribosomal RNA identify an independent rhizopod lineage containing the Euglyphina and the Chlorarachniophyta. *The Journal of Eukaryotic Microbiology*, *42*, 65–69.

Blanc, G., Gallot-Lavallée, L., & Maumus, F. (2015). Provirophages in the *Bigelowiella* genome bear testimony to past encounters with giant viruses. *Proceedings of the National Academy of Sciences of the United States of America*, *112*, E5318–E5326.

Bolte, K., Bullmann, L., Hempel, F., Bozarth, A., Zauner, S., & Maier, U. G. (2009). Protein targeting into secondary plastids. *The Journal of Eukaryotic Microbiology*, *56*, 9–15.

Bölter, B., & Soll, J. (2016). Once upon a time—Chloroplast protein import research from infancy to future challenges. *Molecular Plant*, *9*, 798–812.

Brouard, J. S., Otis, C., Lemieux, C., & Turmel, M. (2010). The exceptionally large chloroplast genome of the green alga *Floydiella terrestris* illuminates the evolutionary history of the Chlorophyceae. *Genome Biology and Evolution*, *2*, 240–256.

Bruce, B. D. (2000). Chloroplast transit peptides: Structure, function and evolution. *Trends in Cell Biology*, *10*, 440–447.

Brugere, J. F., Cornillot, E., Metenier, G., & Vivares, C. P. (2000). Occurrence of subtelomeric rearrangements in the genome of the microsporidian parasite *Encephalitozoon cuniculi*, as revealed by a new fingerprinting procedure based on two-dimensional pulsed field gel electrophoresis. *Electrophoresis*, *21*, 2576–2581.

Burki, F., Kaplan, M., Tikhonenkov, D. V., Zlatogursky, V., Minh, B. Q., Radaykina, L. V., et al. (2016). Untangling the early diversification of eukaryotes: A phylogenomic study of the evolutionary origins of Centrohelida, Haptophyta and Cryptista. *Proceedings of the Royal Society B: Biological Sciences*, *283*, 20152802.

Burki, F., & Keeling, P. J. (2014). Rhizaria. *Current Biology*, *24*, R103–R107.

Calderon-Saenz, E., & Schnetter, R. (1987). *Cryptochlora perforans*, a new genus and species of algae (Chlorarachniophyta), capable of penetrating dead algal filaments. *Plant Systematics and Evolution*, *158*, 69–71.

Calderon-Saenz, E., & Schnetter, R. (1989). Morphology, biology, and systematics of *Cryptochlora perforans* (Chlorarachniophyta), a phagotrophic marine alga. *Plant Systematics and Evolution*, *163*, 165–176.

Cavalier-Smith, T. (1999). Principles of protein and lipid targeting in secondary symbiogenesis: Euglenoid, dinoflagellate, and sporozoan plastid origins and the eukaryote family tree. *The Journal of Eukaryotic Microbiology*, *46*, 347–366.

Cavalier-Smith, T. (2000). Membrane heredity and early chloroplast evolution. *Trends in Plant Science*, *5*, 174–182.

Cavalier-Smith, T. (2002). The phagotrophic origin of eukaryotes and phylogenetic classification on protozoa. *International Journal of Systematic and Evolutionary Microbiology*, *52*, 297–354.

Cavalier-Smith, T. (2003). Genomic reduction and evolution of novel genetic membranes and protein-targeting machinery in eukaryote-eukaryote chimaeras (meta-algae). *Philosophical Transactions of the Royal Society B: Biological Sciences*, *358*, 109–134.

Cavalier-Smith, T., & Chao, E. E. (2003). Phylogeny and classification of phylum Cercozoa (protozoa). *Protist*, *154*, 341–358.

Cavalier-Smith, T., & Lee, J. J. (1985). Protozoa as hosts for endosymbioses and the conversion of symbionts into organelles. *The Journal of Protozoology*, *32*, 376–379.

Cihlář, J., Füssy, Z., Horák, A., & Oborník, M. (2016). Evolution of the tetrapyrrole biosynthetic pathway in secondary algae: Conservation, redundancy and replacement. *PLoS One*, *11*, e0166338.

Corradi, N., Pombert, J.-F., Farinelli, L., Didier, E. S., & Keeling, P. J. (2010). The complete sequence of the smallest known nuclear genome from the microsporidian *Encephalitozoon intestinalis*. *Nature Communications*, *1*, 77.

Curtis, B. A., Tanifuji, G., Burki, F., Gruber, A., Irimia, M., Maruyama, S., et al. (2012). Algal genomes reveal evolutionary mosaicism and the fate of nucleomorphs. *Nature*, *492*, 59–65.

Deane, J. A., Fraunholz, M., Su, V., Maier, U. G., Martin, W., Durnford, D. G., et al. (2000). Evidence for nucleomorph to host nucleus gene transfer: Light-harvesting complex proteins from cryptomonads and chlorarachniophytes. *Protist*, *151*, 239–252.

del Campo, J., Not, F., Forn, I., Sieracki, M. E., & Massana, R. (2013). Taming the smallest predators of the oceans. *The ISME Journal*, *7*, 351–358.

Deschamps, P., & Moreira, D. (2012). Reevaluating the green contribution to diatom genomes. *Genome Biology and Evolution*, *4*, 683–688.

Dietz, C., Ehlers, K., Wilhelm, C., Gil-Rodríguez, M. C., & Schnetter, R. (2003). *Lotharella polymorpha* sp. nov. (Chlorarachniophyta) from the coast of Portugal. *Phycologia*, *42*, 582–593.

Douglas, S. E., Murphy, C. A., Spencer, D. F., & Gray, M. W. (1991). Cryptomonad algae are evolutionary chimaeras of two phylogenetically distinct unicellular eukaryotes. *Nature*, *350*, 148–151.

Douglas, S. E., & Penny, S. L. (1999). The plastid genome of the cryptophyte alga, *Guillardia theta*: Complete sequence and conserved synteny groups confirm its common ancestry with red algae. *Journal of Molecular Evolution*, *48*, 236–244.

Douglas, S., Zauner, S., Fraunholz, M., Beaton, M., Penny, S., Deng, L. T., et al. (2001). The highly reduced genome of an enslaved algal nucleus. *Nature*, *410*, 1091–1096.

Dvořáčková, M., Fojtová, M., & Fajkus, J. (2015). Chromatin dynamics of plant telomeres and ribosomal genes. *The Plant Journal*, *83*, 18–37.

Embley, T. M., & Martin, W. (2006). Eukaryotic evolution, changes and challenges. *Nature*, *440*, 623–630.

Fulnečková, J., Hasíková, T., Fajkus, J., Lukešová, A., Eliáš, M., & Sýkorová, E. (2012). Dynamic evolution of telomeric sequences in the green algal order Chlamydomonadales. *Genome Biology and Evolution, 4*, 248–264.
Geitler, L. (1930). Ein grünes Filarplasmodium und andere neue Protisten. *Archiv für Protistenkunde, 69*, 615–636.
Gile, G. H., & Keeling, P. J. (2008). Nucleus-encoded periplastid-targeted EFL in chlorarachniophytes. *Molecular Biology and Evolution, 25*, 1967–1977.
Gile, G. H., Stern, R. F., James, E. R., & Keeling, P. J. (2010). DNA barcoding of chlorarachniophytes using nucleomorph its sequences. *Journal of Phycology, 46*, 743–750.
Gilson, P. R. (2001). Nucleomorph genomes: Much ado about practically nothing. *Genome Biology, 2*, R1022.
Gilson, P., & McFadden, G. I. (1995). The chlorarachniophyte: A cell with two different nuclei and two different telomeres. *Chromosoma, 103*, 635–641.
Gilson, P. R., & McFadden, G. I. (1996). The miniaturized nuclear genome of eukaryotic endosymbiont contains genes that overlap, genes that are cotranscribed, and the smallest known spliceosomal introns. *Proceedings of the National Academy of Sciences of the United States of America, 93*, 7737–7742.
Gilson, P. R., & McFadden, G. I. (1997). Good things in small packages: The tiny genomes of chlorarachniophyte endosymbionts. *BioEssays, 19*, 167–173.
Gilson, P. R., & McFadden, G. I. (1999). Molecular, morphological and phylogenetic characterization of six chlorarachniophyte strains. *Phycological Research, 47*, 7–19.
Gilson, P. R., Su, V., Slamovits, C. H., Reith, M. E., Keeling, P. J., & McFadden, G. I. (2006). Complete nucleotide sequence of the chlorarachniophyte nucleomorph: Nature's smallest nucleus. *Proceedings of the National Academy of Sciences of the United States of America, 103*, 9566–9571.
Gould, S. B., Maier, U. G., & Martin, W. F. (2015). Protein import and the origin of red complex plastids. *Current Biology, 25*, R515–R521.
Gould, S. B., Sommer, M. S., Kroth, P. G., Gile, G. H., Keeling, P. J., & Maier, U. G. (2006). Nucleus-to-nucleus gene transfer and protein retargeting into a remnant cytoplasm of cryptophytes and diatoms. *Molecular Biology and Evolution, 23*, 2413–2422.
Gould, S. B., Waller, R. F., & McFadden, G. I. (2008). Plastid evolution. *Annual Review of Plant Biology, 59*, 491–517.
Gray, M. W. (2012). Mitochondrial evolution. *Cold Spring Harbor Perspectives in Biology, 4*, a011403.
Greenwood, A. D. (1974). The Cryptophyta in relation to phylogeny and photosynthesis. In J. V. Sanders & D. J. Goodchild (Eds.), *Electron microscopy* (pp. 566–567). Canberra: Australian Academy of Sciences.
Greenwood, A. D., Griffiths, H. B., & Santore, U. J. (1977). Chloroplasts and cell compartments in Cryptophyceae. *British Phycological Journal, 12*, 119.
Gruber, A., Vugrinec, S., Hempel, F., Gould, S. B., Maier, U. G., & Kroth, P. G. (2007). Protein targeting into complex diatom plastids: Functional characterisation of a specific targeting motif. *Plant Molecular Biology, 64*, 519–530.
Hansmann, P., Falk, H., Scheer, U., & Sitte, P. (1986). Ultrastructural localization of DNA in two Cryptomonas species by use of a monoclonal DNA antibody. *European Journal of Cell Biology, 42*, 152–160.
Hempel, F., Bullmann, L., Lau, J., Zauner, S., & Maier, U. G. (2009). ERAD-derived preprotein transport across the second outermost plastid membrane of diatoms. *Molecular Biology and Evolution, 26*, 1781–1790.
Hibberd, D. J., & Norris, R. E. (1984). Cytology and ultrastructure of *Chlorarachnion reptans* (Chlorarachniophyta division nova, Chlorarachniophyceae classis nova). *Journal of Phycology, 20*, 310–330.
Hirakawa, Y., Burki, F., & Keeling, P. J. (2012a). Dual targeting of aminoacyl-tRNA synthetases to the mitochondrion and complex plastid in chlorarachniophytes. *Journal of Cell Science, 125*, 6176–6184.

Hirakawa, Y., Burki, F., & Keeling, P. J. (2012b). Genome-based reconstruction of the protein import machinery in the secondary plastid of a chlorarachniophyte alga. *Eukaryotic Cell*, *11*, 324–333.

Hirakawa, Y., Gile, G. H., Ota, S., Keeling, P. J., & Ishida, K. (2010). Characterization of periplastidal compartment-targeting signals in chlorarachniophytes. *Molecular Biology and Evolution*, *27*, 1538–1545.

Hirakawa, Y., Howe, A., James, E. R., & Keeling, P. J. (2011). Morphological diversity between culture strains of a chlorarachniophyte, *Lotharella globosa*. *PLoS One*, *6*, e23193.

Hirakawa, Y., & Ishida, K. (2010). Internal plastid-targeting signal found in a RubisCO small subunit protein of a chlorarachniophyte alga. *The Plant Journal*, *64*, 402–410.

Hirakawa, Y., & Ishida, K. (2014). Polyploidy of endosymbiotically derived genomes in complex algae. *Genome Biology and Evolution*, *6*, 974–980.

Hirakawa, Y., & Ishida, K. (2015). Prospective function of FtsZ proteins in the secondary plastid of chlorarachniophyte algae. *BMC Plant Biology*, *15*, 276.

Hirakawa, Y., Kofuji, R., & Ishida, K. (2008). Transient transformation of a chlorarachniophyte alga, *Lotharella amoebiformis* (chlorarachniophyceae), with *uid*A and *egfp* reporter genes. *Journal of Phycology*, *44*, 814–820.

Hirakawa, Y., Nagamune, K., & Ishida, K. (2009). Protein targeting into secondary plastids of chlorarachniophytes. *Proceedings of the National Academy of Sciences of the United States of America*, *106*, 12820–12825.

Hirakawa, Y., Suzuki, S., Archibald, J. M., Keeling, P. J., & Ishida, K. (2014). Overexpression of molecular chaperone genes in nucleomorph genomes. *Molecular Biology and Evolution*, *31*, 1437–1443.

Hopkins, J. F., Spencer, D. F., Laboissiere, S., Neilson, J. a. D., Eveleigh, R. J. M., Durnford, D. G., et al. (2012). Proteomics reveals plastid- and periplastid-targeted pro teins in the chlorarachniophyte alga *Bigelowiella natans*. *Genome Biology and Evolution*, *4*, 1391–1406.

Hörmann, F., Küchler, M., Sveshnikov, D., Oppermann, U., Li, Y., & Soll, J. (2004). Tic32, an essential component in chloroplast biogenesis. *The Journal of Biological Chemistry*, *279*, 34756–34762.

Howe, A. T., Bass, D., Scoble, J. M., Lewis, R., Vickerman, K., Arndt, H., et al. (2011). Novel cultured protists identify deep-branching environmental DNA clades of Cercozoa: New genera *Tremula*, *Micrometopion*, *Minimassisteria*, *Nudifila*, *Peregrinia*. *Protist*, *162*, 332–372.

Ishida, K., Cao, Y., Hasegawa, M., Okada, N., & Hara, Y. (1997). The origin of chlorarachniophyte plastids, as inferred from phylogenetic comparisons of amino acid sequences of EF-Tu. *Journal of Molecular Evolution*, *45*, 682–687.

Ishida, K., Endo, H., & Koike, S. (2011). *Partenskyella glossopodia* (Chlorarachniophyceae) possesses a nucleomorph genome of approximately 1 Mbp. *Phycological Research*, *59*, 120–122.

Ishida, K., Green, B., & Cavalier-Smith, T. (1999). Diversification of a chimaeric algal group, the chlorarachniophytes: Phylogeny of nuclear and nucleomorph small-subunit rRNA genes. *Molecular Biology and Evolution*, *16*, 321–331.

Ishida, K., & Hara, Y. (1994). Taxonomic studies on the Chlorarachniophyta. I. Chlorarachnion globosum sp. nov. *Phycologia*, *33*, 351–358.

Ishida, K., Ishida, N., & Hara, Y. (2000). *Lotharella amoeboformis* sp. nov.: A new species of chlorarachniophytes from Japan. *Phycological Research*, *48*, 221–229.

Ishida, K., Nakayama, T., & Hara, Y. (1996). Taxonomic studies on the Chlorarachniophyta. II. Generic delimitation of the chlorarachniophytes and description of *Gymnochlora stellata* gen. et sp. nov. and *Lotharella* gen. nov. *Phycological Research*, *44*, 37–45.

Ishida, K., Yabuki, A., & Ota, S. (2011). *Amorphochlora amoebiformis* gen. et comb. nov. (Chlorarachniophyceae). *Phycological Research*, *59*, 52–53.

Katinka, M. D., Duprat, S., Cornillot, E., Méténier, G., Thomarat, F., Prensier, G., et al. (2001). Genome sequence and gene compaction of the eukaryote parasite *Encephalitozoon cuniculi*. *Nature*, *414*, 450–453.

Keeling, P. J. (2013). The number, speed, and impact of plastid endosymbioses in eukaryotic evolution. *Annual Review of Plant Biology, 64*, 583–607.

Kikuchi, S., Bédard, J., Hirano, M., Hirabayashi, Y., Oishi, M., Imai, M., et al. (2013). Uncovering the protein translocon at the chloroplast inner envelope membrane. *Science, 339*, 571–574.

Kilian, O., & Kroth, P. G. (2005). Identification and characterization of a new conserved motif within the presequence of proteins targeted into complex diatom plastids. *The Plant Journal, 41*, 175–183.

Kovács-Bogdán, E., Soll, J., & Bölter, B. (2010). Protein import into chloroplasts: The Tic complex and its regulation. *Biochimica et Biophysica Acta, 1803*, 740–747.

Küchler, M., Decker, S., Hörmann, F., Soll, J., & Heins, L. (2002). Protein import into chloroplasts involves redox-regulated proteins. *The EMBO Journal, 21*, 6136–6145.

Lane, C. E., van den Heuvel, K., Kozera, C., Curtis, B. A., Parsons, B. J., Bowman, S., et al. (2007). Nucleomorph genome of Hemiselmis andersenii reveals complete intron loss and compaction as a driver of protein structure and function. *Proceedings of the National Academy of Sciences of the United States of America, 104*, 19908–19913.

Li, H. m., & Teng, Y. S. (2013). Transit peptide design and plastid import regulation. *Trends in Plant Science, 18*, 360–366.

Ludwig, M., & Gibbs, S. P. (1989). Evidence that the nucleomorphs of *Chlorarachnion reptans* (Chlorarachniophyceae) are vestigial nuclei: Morphology, division and DNA-DAPI fluorescence. *Journal of Phycology, 25*, 385–394.

Maisnier-Patin, S., Roth, J. R., Fredriksson, A., Nyström, T., Berg, O. G., & Andersson, D. I. (2005). Genomic buffering mitigates the effects of deleterious mutations in bacteria. *Nature Genetics, 37*, 1376–1379.

Marinov, G. K., & Lynch, M. (2016). Conservation and divergence of the histone code in nucleomorphs. *Biology Direct, 11*, 18.

Martin, W., Stoebe, B., Goremykin, V., Hapsmann, S., Hasegawa, M., & Kowallik, K. V. (1998). Gene transfer to the nucleus and the evolution of chloroplasts. *Nature, 393*, 162–165.

McFadden, G. I., Gilson, P. R., Hofmann, C. J., Adcock, G. J., & Maier, U. G. (1994). Evidence that an amoeba acquired a chloroplast by retaining part of an engulfed eukaryotic alga. *Proceedings of the National Academy of Sciences of the United States of America, 91*, 3690–3694.

Meyer, S. R. (1987). The taxonomic implications of the ultrastructure and cell division of a stigma-containing *Chroomonas* sp. (Cryptophyceae) from Rocky Bay, Natal, South Africa. *South African Journal of Botany, 53*, 129–139.

Moestrup, Ø., & Sengco, M. (2001). Ultrastructural studies on *Bigelowiella natans*, gen. et sp. nov., a chlorarachniophyte flagellate. *Journal of Phycology, 37*, 624–646.

Moore, C. E., & Archibald, J. M. (2009). Nucleomorph genomes. *Annual Review of Genetics, 43*, 251–264.

Moore, C. E., Curtis, B., Mills, T., Tanifuji, G., & Archibald, J. M. (2012). Nucleomorph genome sequence of the cryptophyte alga *Chroomonas mesostigmatica* CCMP1168 reveals lineage-specific gene loss and genome complexity. *Genome Biology and Evolution, 4*, 1162–1175.

Moustafa, A., Beszteri, B., Maier, U. G., Bowler, C., Valentin, K., & Bhattacharya, D. (2009). Genomic footprints of a cryptic plastid endosymbiosis in diatoms. *Science, 324*, 1724–1726.

Nakai, M. (2015). The TIC complex uncovered: The alternative view on the molecular mechanism of protein translocation across the inner envelope membrane of chloroplasts. *Biochimica et Biophysica Acta, 1847*, 957–967.

Ocalewicz, K. (2013). Telomeres in fishes. *Cytogenetic and Genome Research, 141*, 114–125.

Ota, S., Kudo, A., & Ishida, K. (2011). *Gymnochlora dimorpha* sp. nov., a chlorarachniophyte with unique daughter cell behaviour. *Phycologia, 50*, 317–326.

Ota, S., Silver, T. D., Archibald, J. M., & Ishida, K. (2009). *Lotharella oceanica* sp. nov.—A new planktonic chlorarachniophyte studied by light and electron microscopy. *Phycologia, 48*, 315–323.

Ota, S., Ueda, K., & Ishida, K. (2005). *Lotharella vacuolata* sp. nov., a new species of chlorarachniophyte algae, and time-lapse video observations on its unique post-cell division behavior. *Phycological Research, 53*, 275–286.

Ota, S., Ueda, K., & Ishida, K. (2007a). *Norrisiella sphaerica* gen. et sp. nov., a new coccoid chlorarachniophyte from Baja California, Mexico. *Journal of Plant Research, 120*, 661–670.

Ota, S., Ueda, K., & Ishida, K. (2007b). Taxonomic study of *Bigelowiella longifila* sp. nov. (Chlorarachniophyta) and a time-lapse video observation of the unique migration of amoeboid cells. *Journal of Phycology, 43*, 333–343.

Ota, S., & Vaulot, D. (2012). *Lotharella reticulosa* sp. nov.: A highly reticulated network forming chlorarachniophyte from the Mediterranean Sea. *Protist, 163*, 91–104.

Ota, S., Vaulot, D., Le Gall, F., Yabuki, A., & Ishida, K. (2009). *Partenskyella glossopodia* gen. et sp. nov., the first report of a Chlorarachniophyte that lacks a pyrenoid. *Protist, 160*, 137–150.

Patron, N. J., Rogers, M. B., & Keeling, P. J. (2006). Comparative rates of evolution in endosymbiotic nuclear genomes. *BMC Evolutionary Biology, 6*, 46.

Patron, N. J., & Waller, R. F. (2007). Transit peptide diversity and divergence: A global analysis of plastid targeting signals. *BioEssays, 29*, 1048–1058.

Patron, N. J., Waller, R. F., Archibald, J. M., & Keeling, P. J. (2005). Complex protein targeting to dinoflagellate plastids. *Journal of Molecular Biology, 348*, 1015–1024.

Patron, N. J., Waller, R. F., & Keeling, P. J. (2006). A tertiary plastid uses genes from two endosymbionts. *Journal of Molecular Biology, 357*, 1373–1382.

Pich, U., Fuchs, J., & Schubert, I. (1996). How do Alliaceae stabilize their chromosome ends in the absence of TTTAGGG sequences? *Chromosome Research, 4*, 207–213.

Ponce-Toledo, R. I., Deschamps, P., López-García, P., Zivanovic, Y., Benzerara, K., & Moreira, D. (2017). An early-branching freshwater cyanobacterium at the origin of plastids. *Current Biology, 27*, 386–391.

Price, D. C., Chan, C. X., Yoon, H. S., Yang, E. C., Qiu, H., Weber, A. P. M., et al. (2012). *Cyanophora paradoxa* genome elucidates origin of photosynthesis in algae and plants. *Science, 335*, 843–847.

Ralph, S. A., Foth, B. J., Hall, N., & McFadden, G. I. (2004). Evolutionary pressures on apicoplast transit peptides. *Molecular Biology and Evolution, 21*, 2183–2194.

Rodríguez-Ezpeleta, N., Brinkmann, H., Burey, S. C., Roure, B., Burger, G., Löffelhardt, W., et al. (2005). Monophyly of primary photosynthetic eukaryotes: Green plants, red algae, and glaucophytes. *Current Biology, 15*, 1325–1330.

Rogers, M. B., Archibald, J. M., Field, M. A., Li, C., Striepen, B., & Keeling, P. J. (2004). Plastid-targeting peptides from the chlorarachniophyte *Bigelowiella natans*. *The Journal of Eukaryotic Microbiology, 51*, 529–535.

Rogers, M. B., Gilson, P. R., Su, V., McFadden, G. I., & Keeling, P. J. (2007). The complete chloroplast genome of the chlorarachniophyte *Bigelowiella natans*: Evidence for independent origins of chlorarachniophyte and euglenid secondary endosymbionts. *Molecular Biology and Evolution, 24*, 54–62.

Schwenkert, S., Soll, J., & Bölter, B. (2011). Protein import into chloroplasts—How chaperones feature into the game. *Biochimica et Biophysica Acta, 1808*, 901–911.

Sheiner, L., & Striepen, B. (2013). Protein sorting in complex plastids. *Biochimica et Biophysica Acta, 1833*, 352–359.

Shi, L.-X., & Theg, S. M. (2013). The chloroplast protein import system: From algae to trees. *Biochimica et Biophysica Acta, 1833*, 314–331.

Shiratori, T., & Ishida, K. (2016). *Trachyrhizium urniformis* n. g., n. sp., a novel marine filose thecate amoeba related to a cercozoan environmental clade (novel clade 4). *Journal of Eukaryotic Microbiology, 63*, 722–731.

Silver, T. D., Koike, S., Yabuki, A., Kofuji, R., Archibald, J. M., & Ishida, K. (2007). Phylogeny and nucleomorph karyotype diversity of chlorarachniophyte algae. *The Journal of Eukaryotic Microbiology*, *54*, 403–410.

Stork, S., Moog, D., Przyborski, J. M., Wilhelmi, I., Zauner, S., & Maier, U. G. (2012). Distribution of the SELMA translocon in secondary plastids of red algal origin and predicted uncoupling of ubiquitin-dependent translocation from degradation. *Eukaryotic Cell*, *11*, 1472–1481.

Suzuki, S., Hirakawa, Y., Kofuji, R., Sugita, M., & Ishida, K. (2016). Plastid genome sequences of *Gymnochlora stellata*, *Lotharella vacuolata*, and *Partenskyella glossopodia* reveal remarkable structural conservation among chlorarachniophyte species. *Journal of Plant Research*, *129*, 581–590.

Suzuki, S., Ishida, K., & Hirakawa, Y. (2016). Diurnal transcriptional regulation of endosymbiotically derived genes in the chlorarachniophyte *Bigelowiella natans*. *Genome Biology and Evolution*, *8*, 2672–2682.

Suzuki, S., Shirato, S., Hirakawa, Y., & Ishida, K. (2015). Nucleomorph genome sequences of two chlorarachniophytes, *Amorphochlora amoebiformis* and *Lotharella vacuolata*. *Genome Biology and Evolution*, 7, 1533–1545.

Takahashi, F., Okabe, Y., Nakada, T., Sekimoto, H., Ito, M., Kataoka, H., et al. (2007). Origins of the secondary plastids of Euglenophyta and Chlorarachniophyta as revealed by an analysis of the plastid-targeting, nuclear-encoded gene *psbO*. *Journal of Phycology*, *43*, 1302–1309.

Tanifuji, G., Archibald, J. M., & Hashimoto, T. (2016). Comparative genomics of mitochondria in chlorarachniophyte algae: Endosymbiotic gene transfer and organellar genome dynamics. *Scientific Reports*, *6*, 21016.

Tanifuji, G., Onodera, N. T., Brown, M. W., Curtis, B. A., Roger, A. J., Ka-Shu Wong, G., et al. (2014). Nucleomorph and plastid genome sequences of the chlorarachniophyte *Lotharella oceanica*: Convergent reductive evolution and frequent recombination in nucleomorph-bearing algae. *BMC Genomics*, *15*, 374.

Tanifuji, G., Onodera, N. T., Moore, C. E., & Archibald, J. M. (2014). Reduced nuclear genomes maintain high gene transcription levels. *Molecular Biology and Evolution*, *31*, 625–635.

Tanifuji, G., Onodera, N. T., Wheeler, T. J., Dlutek, M., Donaher, N., & Archibald, J. M. (2011). Complete nucleomorph genome sequence of the nonphotosynthetic alga *Cryptomonas paramecium* reveals a core nucleomorph gene set. *Genome Biology and Evolution*, *3*, 44–54.

TerBush, A. D., Yoshida, Y., & Osteryoung, K. W. (2013). FtsZ in chloroplast division: Structure, function and evolution. *Current Opinion in Cell Biology*, *25*, 461–470.

Timmis, J. N., Ayliffe, M. A., Huang, C. Y., & Martin, W. (2004). Endosymbiotic gene transfer: Organelle genomes forge eukaryotic chromosomes. *Nature Reviews Genetics*, *5*, 123–135.

Tokuriki, N., & Tawfik, D. S. (2009). Chaperonin overexpression promotes genetic variation and enzyme evolution. *Nature*, *459*, 668–673.

Turmel, M., Gagnon, M. C., O'Kelly, C. J., Otis, C., & Lemieux, C. (2009). The chloroplast genomes of the green algae *Pyramimonas*, *Monomastix*, and *Pycnococcus* shed new light on the evolutionary history of prasinophytes and the origin of the secondary chloroplasts of euglenids. *Molecular Biology and Evolution*, *26*, 631–648.

Turmel, M., Otis, C., & Lemieux, C. (2015). Dynamic evolution of the chloroplast genome in the green algal classes Pedinophyceae and Trebouxiophyceae. *Genome Biology and Evolution*, 7, 2062–2082.

Upcroft, J. A., Abedinia, M., & Upcroft, P. (2005). Rearranged subtelomeric rRNA genes in *Giardia duodenalis*. *Eukaryotic Cell*, *4*, 484–486.

Van de Peer, Y., Rensing, S. A., Maier, U. G., & De Wachter, R. (1996). Substitution rate calibration of small subunit ribosomal RNA identifies chlorarachniophyte endosymbionts as remnants of green algae. *Proceedings of the National Academy of Sciences of the United States of America, 93*, 7732–7736.

van Dooren, G. G., Schwartzbach, S. D., Osafune, T., & McFadden, G. I. (2001). Translocation of proteins across the multiple membranes of complex plastids. *Biochimica et Biophysica Acta, 1541*, 34–53.

Williams, B. A. P., Slamovits, C. H., Patron, N. J., Fast, N. M., & Keeling, P. J. (2005). A high frequency of overlapping gene expression in compacted eukaryotic genomes. *Proceedings of the National Academy of Sciences of the United States of America, 102*, 10936–10941.

Woehle, C., Dagan, T., Martin, W. F., & Gould, S. B. (2011). Red and problematic green phylogenetic signals among thousands of nuclear genes from the photosynthetic and apicomplexa-related *Chromera velia*. *Genome Biology and Evolution, 3*, 1220–1230.

Yang, Y., Maruyama, S., Sekimoto, H., Sakayama, H., & Nozaki, H. (2011). An extended phylogenetic analysis reveals ancient origin of "non-green" phosphoribulokinase genes from two lineages of "green" secondary photosynthetic eukaryotes: Euglenophyta and Chlorarachniophyta. *BMC Research Notes, 4*, 330.

Yang, Y., Matsuzaki, M., Takahashi, F., Qu, L., & Nozaki, H. (2014). Phylogenomic analysis of "red" genes from two divergent species of the "green" secondary phototrophs, the chlorarachniophytes, suggests multiple horizontal gene transfers from the red lineage before the divergence of extant chlorarachniophytes. *PLoS One, 9*, e101158.

SUBJECT INDEX

Note: Page numbers followed by "*f*" indicate figures, and "*t*" indicate tables.

D

E

N

O

P

R

S

T

U

V

W

X

Z

AUTHOR INDEX

Note: Page numbers followed by "*f*" indicate figures, and "*t*" indicate tables.

B

C

D

E

H

L

M

N

O

P

Q

R

S

T

U

V

W

X

Y

Z

FSC
www.fsc.org
MIX
Papier | Fördert
gute Waldnutzung
FSC® C083411

Kolibri 360 GmbH
Ferdinand-Jühlke-Straße 7
99095 Erfurt, Deutschland
produktsicherheit@kolibri360.de